MW01485854

Commercial
Aviation Safety

Commercial Aviation Safety

Anthony Lawrenson
Clarence C. Rodrigues
Shem Malmquist
Matthew Greaves
Graham Braithwaite
Stephen K. Cusick

Seventh Edition

New York   Chicago   San Francisco
Athens   London   Madrid
Mexico City   Milan   New Delhi
Singapore   Sydney   Toronto

Commercial Aviation Safety, Seventh Edition

2 3 4 5 6 7 8 9   LCR   28 27 26 25 24

Library of Congress Control Number: 2023903686

ISBN 978-1-264-27870-1
MHID 1-264-27870-5

Sponsoring Editor Ania Levinson	**Indexer** MPS Limited
Editorial Supervisor Janet Walden	**Production Supervisor** Lynn M. Messina
Project Manager Poonam Bisht, MPS Limited	**Composition** MPS Limited
Acquisitions Coordinator Olivia Higgins	**Illustration** MPS Limited
Copy Editor MPS Limited	**Art Director, Cover** Jeff Weeks
Proofreader MPS Limited	

Contents at a Glance

Contents

Part III   Human Safety

About the Authors

Anthony Lawrenson, PhD, is the lead author of the seventh edition of *Commercial Aviation Safety*. He is an ex-Royal Air Force pilot and A2 Qualified Flying Instructor. Over his 35-year career in aviation, he has flown 11 military and civil aircraft types, most recently as a captain on Boeing 777s for a major international airline. His PhD and published research focus on the relationship between safety culture and corporate liability in commercial aviation. For the last two decades, Anthony has lectured and trained in human factors, aviation law, and safety risk management as a visiting lecturer at Cranfield University. His consulting and training company, Salus Intelligence, has delivered training to senior executives in various safety critical roles. Over a 5-year project in elite sport, he has trained head coaches and performance directors of majority of the UK's Olympic and national sports teams. He is a director of a deep-tech research and development company, Volant Autonomy which develops flight guidance software for next-generation aircraft.

Clarence Rodrigues, PhD, is a Health and Safety Consultant in the College of Public Health, and an Adjunct Professor in the Department of Industrial and Management Systems Engineering at University of South Florida (USF), Tampa, Florida. He is also an independent international EHS consultant. Prior to the above, he has held faculty appointments at Khalifa University in Abu Dhabi, United Arab Emirates; Embry-riddle Aeronautical University in Daytona Beach, Florida; the Indiana University of Pennsylvania, and the University of Pennsylvania in Pennsylvania. He also had an industry appointment as worldwide engineering manager for ergonomics and system safety at Campbells, USA. Dr. Rodrigues has authored or co-authored numerous publications, including the fourth through the sixth editions of *Commercial Aviation Safety*. He is a US professional engineer (PE), a certified safety professional (CSP), and a certified professional ergonomist (CPE), who has conducted professional work in the United States, UAE, Kuwait, Canada, England, India, Malaysia, Mexico, and Scotland.

Shem Malmquist is an instructor at the Florida Institute of Technology who teaches aviation safety and accident investigation and is involved in researching topics involving system safety engineering. In addition, he is a Boeing 777 Captain flying worldwide. Captain Malmquist has published numerous technical and academic articles stemming from his work on flight safety and accident investigation. His most recent work has involved approaches to risk analysis and accident prevention. His past work includes several committees of the US Commercial Aviation Safety team. He also has either led or been deeply involved in several major aircraft accident investigations, performing operations, human factors, systems and aircraft performance analysis. An elected Fellow of the Royal Aeronautical Society, as well as full member of ISASI, the Resilience Engineering Association, AIAA, the Human Factors and Ergonomics Society, IEEE, the Flight Safety Foundation, and SAE where he is an active member of the Flight Deck and Handling Quality Standards for Transport Aircraft and several other committees involving aircraft certification standards.

Matthew Greaves, PhD, is an independent consultant in the field of aviation safety and accident investigation. His most recent work includes operational safety data analysis in helicopters. He is a Visiting Fellow at Cranfield University where he worked for 10 years, finishing as the Head of the Safety and Accident Investigation Centre. His work there included teaching on the Masters level Accident Investigation and FDM courses and leading industrial research projects including the novel use of FDM data and internal sensors for helicopter gearboxes. Prior to joining Cranfield, he worked at QinetiQ, the science and technology research organization. Matthew holds a degree and PhD in Engineering and an MSc by Research in Accident Investigation. He has authored numerous technical and academic papers and is a Chartered Engineer, a Fellow of the Royal Aeronautical Society, and Member of the International Society of Air Safety Investigators.

Graham Braithwaite is a professor and Director of the Transport Systems Theme, and Head of the Safety and Accident Investigation Centre, at Cranfield University, UK. He holds a PhD in aviation safety management from Loughborough University and his work focuses on safety management, accident investigation, human factors, and the influence of culture on safety. Graham has supported airlines, regulators, and safety investigation agencies around the world and has led two successful bids for the Queen's Anniversary Prize—the highest honor for a UK University. In addition to his Cranfield role, Graham is the Independent Safety Adviser to the British Airways Board and for TUI Northern Region Airlines. He is a Fellow of the Royal Aeronautical Society and the International Society of Air Safety Investigators.

Stephen K. Cusick, J.D., recently retired as aviation safety professor from the Florida Institute of Technology, College of Aeronautics in Melbourne, Florida. He is a retired US Navy Captain, Naval Aviator, and Surface Warfare Officer. An experienced search and rescue pilot with Commercial, Multiengine, Instrument, and Helicopter flight ratings from the FAA, he has been engaged in aviation safety research in a variety of areas including Safety Management Systems, Aircraft Loss of Control in Flight, Fatigue Risk Management, and Helicopter Operations. During his legal career he served as an attorney with the US Navy General Counsel's Office and as corporate counsel serving high technology aerospace companies. He is a recipient of the General Dynamics Award for Aviation Excellence.

Foreword

I was originally appointed to be a Board Member of the National Transportation Safety Board (NTSB) in 2006 by US President George W. Bush. One week into the job, I was called early in the morning by the NTSB Response Operations Center. There had been an airline accident within the past hour. The airplane was still on fire and multiple fatalities were expected.[1] As a newly appointed Board Member, I was to accompany the NTSB Go Team to the crash location in Lexington, Kentucky. Several hours later, I found myself staring at the burned-up hull of a Bombardier CRJ where 49 people died just hours earlier.

So, there it began—the start of what became 15 years as a Board Member of one of the world's premier accident investigation agencies, including the final 4 years as chairman. Over those years, I launched with the Go Team on three dozen accidents in all modes of transportation and was involved with deliberating and determining probable cause of close to 250 accidents.

My universe of aviation safety, as was that of previous editions of *Commercial Aviation Safety*, has largely been US focused. This seventh edition adds authors who bring an important international perspective to the text. After all, commercial aviation is a global enterprise. When a large airliner crashes, victims are often citizens of several nationalities. In March 2019, for example, an Ethiopian Airlines Boeing 737 Max crashed minutes after departure from Addis Ababa, claiming the lives of all 157 onboard. There were 35 nationalities, including 32 Kenyans, 18 Canadians, 9 Ethiopians, 8 Chinese, Americans, and Italians who lost their lives.[2]

There is a real face to safety. The hardest, most difficult, and painful conversations I had were with family members who were grieving the loss of a loved one following a transportation accident. Oftentimes I met with them within hours of the mishap, their emotions raw. The only slither of inspiration I could offer at that point was my commitment to find out what happened so we could prevent it from happening again.

Sometimes change is brought about by findings and recommendations from the official investigation. Other times, the changes result from political bodies passing legislation to force change. In recent past, I've seen two cases where victims' family groups have banded together to lobby US Congress for changes. Some of the most widespread aviation regulatory changes in decades were the results of such lobbying by family groups.

---

[1] NTSB. (2007b). *Aircraft accident report: Attempted takeoff from wrong runway, Comair flight 5191, Bombardier CL-600-2B19, N431CA Lexington, Kentucky, August 27, 2006.* (NTSB Report No. NTSB/AAR/07-05). Washington, DC: Author.

[2] https://www.cnn.com/2019/03/10/africa/ethiopian-airlines-crash-victims/index.html

In February 2009, Colgan Air flight 3407 crashed on approach to Buffalo, New York.[3] Fifty lives were lost. The families were so outraged and vocal about the circumstance of the crash, they lobbied Congress for radial overhauls to the ways airline pilots are selected, trained, and tested. With the misguided belief that fatigue was a causal factor, there was a total revamp of pilot flight and duty times regulations which were aimed at mitigating fatigue.[4] Whereas previously the minimum flight hours to be hired as an airline pilot was less than 300 flight hours, Congress mandated it be changed to a minimum of 1500 hours.[5] In the wake of the March 2019 Boeing 737 Max crash involving Ethiopian flight 302, a similar effort was undertaken by families of victims from that crash. Their efforts led to Congress passing the Aircraft Certification Reform and Accountability Act in 2020. That law mandated sweeping changes to commercial aircraft certification standards and practices, as well as Federal Aviation Administration oversight of those practices.[6]

As I was writing this foreword, I recalled the first aviation human factors safety symposium I attended. On the way back to the airport, I told the taxi driver why I had been in Washington. I then proceeded to point out how safe commercial aviation was. "There probably won't be a crash today, and there probably won't be one tomorrow," I boldly stated. Unfortunately, I was wrong: The next day my airline suffered a fatal mishap. With that in mind, it is with great foolishness that I mention that the United States and other regions of the world have had an extraordinary safety record in the recent past. That record didn't always exist, however. During the 15 years I served on the NTSB (2006–2021), for example, there were just over 100 fatalities on US operated scheduled passenger air carriers. However, in the 15 years that preceded my time at NTSB, there were nearly 1100 fatalities on US scheduled air carriers. The authors of this book do an excellent job of discussing safety initiatives that, no doubt, have contributed to this remarkable decline in commercial aviation accidents.

Despite how good the commercial aviation safety record has been, each dawn brings a new day. Each day, the counter is reset. The history is just that—it's history— it's something in the past. What matters the most is what happens today. As I'm writing this foreword, just a few days ago, two jets nearly collided on a fog shrouded runway in Texas. A few weeks before this, there was a potentially serious runway incursion that could have resulted in a runway collision at New York's John F. Kennedy International Airport. Once you feel you're safe, the tendency is to get complacent and that's when bad things happen.

---

[3] NTSB. (2010). *Aircraft accident report: Loss of control on approach, Colgan Air, Inc., Operating as Continental Connection Flight 3407, Bombardier DHC 8 400, N200WQ, Clarence Center, New York, February 12, 2009.* (NTSB Report No. NTSB/AAR/10-01). Washington, DC: Author.

[4] Interestingly, although there was the widespread belief that fatigue was a factor, the NTSB could not establish that fatigue was a causal factor in the crash. According to NTSB, "The pilots' performance was likely impaired because of fatigue, but the extent of their impairment and the degree to which it contributed to the performance deficiencies that occurred during the flight cannot be conclusively determined."

[5] A US House of Representatives bill, H.R. 5900, was passed by Congress and signed into law by President Barack Obama on August 1, 2010. It is now known as Public Law 111-216, Airline Safety and Federal Aviation Administration Extension Act of 2010. See https://www.govinfo.gov/app/details/PLAW-111publ216#:~:text=An%20act%20to%20amend%20the,safety%2C%20and%20for%20other%20purposes.

[6] H.R. 8408 Aircraft Certification Reform and Accountability Act. See https://www.congress.gov/bill/116th-congress/house-bill/8408/text

For this book's authors and for me, aviation safety is personal. Before I was appointed to the NTSB, I was a 24-year airline pilot for a major US-based airline. It was a wonderful career, but one that had safety, human factors, and accident investigation interwoven throughout. Following a merger with another airline in 1989, the airline suffered five fatal crashes over the next 5 years.

The first of these occurred 6 weeks after the merger. A Boeing 737-400 overran the runway at New York's LaGuardia and ended up being impaled on the approach lights and partially in the water. A mother and her adult daughter died in that crash. The investigation determined that while sitting at the gate between flights, somehow a person or object inadvertently moved the rudder trim knob to the full left trim position. Because the rudder, rudder pedal, and nose wheel are all interconnected, this movement of the trim knob caused the nose to be deflected four degrees left. During takeoff roll, because of the canted nose wheel, the first officer, who was initiating his first takeoff since completing training 39 days earlier, had difficulty steering the aircraft down the runway. In a clumsy exchange of controls and half-hearted rejected takeoff, the airline sailed off the runway at 34 knots and onto the approach lighting stanchions that towered above the dark waters below.[7]

The NTSB determined the probable cause of this accident was "the captain's failure to exercise his command authority in a timely manner to reject the takeoff or take sufficient control to continue the takeoff, which was initiated with a mis-trimmed rudder." The Safety Board also noted that the captain should have detected the mis-trimmed rudder before takeoff.

Sixteen months later, an air traffic controller at Los Angeles International Airport (LAX) cleared a SkyWest Airlines flight 5569, a Fairchild Metroliner commuter airline, to taxi into position and await takeoff clearance on runway 24L at the intersection of taxiway 45 and runway 24L. That intersection is about 3000 feet (approximately 915 meters) from the runway threshold.[8] The controller had another aircraft crossing runway 24L downfield and she needed to allow that aircraft to cross before issuing takeoff clearance to the SkyWest 5569 crew. Day had just turned into night at the airport, and the controller strained in the glare of terminal building lights to see where SkyWest 5569 was holding on the runway.

While SkyWest sat idling on the runway, the crew of USAir 1493, flying a Boeing 737-300, checked onto the tower frequency and were cleared to land by the controller on that same runway. Meanwhile, the controller had become preoccupied with another aircraft and forgot about SkyWest 5569 sitting on the runway.

The 737's first officer later recounted to investigators that as he lowered the 737 nose onto the runway, he saw the Metroliner was immediately in front of him. The landing lights of his airplane were reflecting off the propellers of the Metroliner. There was simply no time to take evasive action. The 737 collided with the standing Metroliner at 112 knots. Thirty-four lives were lost, including all aboard the Metroliner.[9]

In his testimony to NTSB, the first officer referred to the Metroliner, saying, "It wasn't there. It was invisible." NTSB learned that the Metroliner's wing mounted

[7] NTSB. (1990). *Aircraft accident report: USAir, Inc., Boeing 737-400. LaGuardia Airport, Flushing, New York. September 20, 1989.* (NTSB Report No. NTSB/AAR/90-03). Washington, DC: Author.
[8] To expedite departures, airports often use intersection departures for smaller aircraft.
[9] NTSB. (1991). *Aircraft accident report: Runway collision of USAir flight 1493 and Skywest flight 5569 Fairchild Metroliner. Los Angeles International Airport, Los Angeles, California. February 1, 1991.* (NTSB Report No. NTSB/AAR/91-08). Washington, DC: Author.

position lights blended in perfectly with the in-runway lighting, making it indistinguishable from the runway lights when viewed from the cockpit of the landing 737.[10]

It's true that the air traffic controller's error was the proximate cause, or the "active failure," as described by Dr. James Reason, that ultimately led to the accident. However, NTSB's investigation found there were deeper systemic factors that, according to NTSB, "created an environment in the Los Angeles Air Traffic Control tower that ultimately led to the failure of the [controller] to maintain an awareness of the traffic situation, culminating in the two inappropriate clearances and the subsequent collision of the USAir and SkyWest aircraft." Those factors, as noted by NTSB, were "failures of Los Angeles Air Traffic Control Facility Management to implement procedures that provided redundancy to prevent such errors from leading to accidents, and ATC management at that facility to provide sufficient oversight and effective quality assurance of the ATC system." Although not part of the probable cause, NTSB found that glare from terminal lighting made it difficult for air traffic controllers to discern aircraft at the location where the Metroliner sat awaiting takeoff clearance.

Thirteen months later, during a later winter snowstorm, a Fokker F-28 crashed at New York's LaGuardia Airport. It had been a slow taxi-out due to weather conditions, during the extended taxi, considerable snow had accumulated on the wings. Finally, 35 minutes after deicing, the crew was cleared for takeoff. With snow-contaminated wings, the airplane stalled immediately after liftoff and the Fokker crashed into the water adjacent to runway. I had flown with the captain years earlier, and one of the flight attendants. Both were killed, along with 25 passengers.

The year 1994 brought two additional crashes for our airline. As the DC-9 approached the runway for landing on a mid-summer afternoon, the crew encountered heavy rain on short final approach and elected to go around. They immediately encountered a heavy downdraft windshear. The crew mishandled the go-around maneuver and the plane plunged into trees. Thirty-seven of the fifty-seven occupants perished.[11] One of the surviving passengers was my brother-in-law.

Sixty-nine days later, a Boeing 737-300 plunged into a Pennsylvania hillside. One-hundred, thirty-two lives were lost. After a prolonged but thorough investigation, NTSB determined that a mechanical issue caused the rudder to move opposite to the direction commanded by the crew.[12] I served as a representative of the Air Line Pilots Association, and worked as part of the NTSB's investigation. Assigned to the NTSB's human performance group, our job was to try to understand the crew's behavior and performance. We painstakingly spent excruciating days and weeks going through the cockpit voice recording, listening and trying to analyze what was going on in the moments leading up to the crash. Although I didn't know that crew, I felt like I knew them because we delved into personal details of their lives.

---

[10] NTSB. (1993). *Aircraft accident report: Takeoff stall in icing conditions, USAir flight 405, Fokker F-28, N485US. LaGuardia airport, Flushing, NY, March 22, 1992.* (NTSB Report No. NTSB/AAR/93-02). Washington, DC: Author.

[11] NTSB. (1995). *Aircraft accident report: Flight into terrain during missed approach USAir 1016, DC-9-31, N954V, Charlotte/Douglas International Airport Charlotte, North Carolina July 2, 1994.* (NTSB Report No. NTSB/AAR/95-03). Washington, DC: Author.

[12] NTSB. (1999). *Aircraft accident report: Uncontrolled descent and collision with terrain, USAir flight 427. Boeing 737-300, N513AU. Near Aliquippa, Pennsylvania, September 8, 1994.* (NTSB Report No. NTSB/AAR/99-01). Washington, DC: Author.

As discussed in Chapter 4, the purpose of an accident or incident investigation isn't to assign fault or blame; it's to learn what happened so measures can be put in place to prevent future accidents or incidents. Safety investigations can be hampered when fault, blame, or threat of criminalization are entered into the equation. When people know they may be punished for their actions, they may not be forthcoming with truthful information that could help explain the accident.

While it is true that a pilot, air traffic controller, or a mechanic or engineer may have committed an error that led to an accident or serious incident, Chapter 9 discusses terms such as "pilot error" or "human error" oversimplify the factors that may have led to the crash. Accidents are usually the result of several factors that combine to create the accident sequence.

When I was at NTSB, framed in my office were the powerful words of my friend, Captain Daniel Maurino: "The discovery of human error should be considered as the starting point of the investigation, and not the ending point that has punctuated so many investigations of the past." Once the human error is identified, the prevailing question should become "why was the error committed?" Only after you have answered these questions can the human error become understood and the underlying conditions corrected.

Human error does not occur in a vacuum. It must therefore be examined in the context in which the error occurred. In other words, if an error occurs in the workplace, the workplace must be examined to look for conditions that could provoke error. What were the physical conditions at the workplace? Was lighting adequate to perform the task? Were the procedures and training adequate? Did the organizational norms and expectations prioritize safety over competing goals? Was the operational layout of the workplace conducive for error?

I was involved in the final deliberation of an accident involving SpaceShipTwo, a commercial space vehicle that suffered an inflight breakup during a test flight in 2014. From the onboard video recorder it was evident that the copilot prematurely moved a lever which led to an uncommanded movement of the vehicle's tail feather—a device similar to a conventional aircraft's horizontal and vertical stabilizer. The feather is actuated by a cockpit lever to pivot the tail feather upward 60 degrees relative to the longitudinal axis of the aircraft; its purpose is to stabilize the aircraft during reentry phase of flight. However, if the feather is deployed at the wrong time, as in this case, the resulting aerodynamic loads on the aircraft will lead to catastrophic inflight breakup. The obvious "cause" of the accident was that the copilot committed an error of unlocking the feather at the wrong time which led to uncommanded actuation of the feather. However, this finding alone would serve no useful purpose for preventing similar errors in the future. After all, the copilot was killed, so surely he would not commit this error again.[13]

By digging deeper, the investigation found that influencing the copilot's error was the high workload he was experiencing during this phase of flight, along with time pressure to complete critical tasks from memory—all while experiencing vibration and g-loads that he had not experienced recently. On the broader perspective, the spaceship designer/manufacturer, Scaled Composites, did not consider that this single error could lead to an unintended feather activation.

---

[13] NTSB. (2015). *Aircraft accident report: In-Flight Breakup During Test Flight Scaled Composites SpaceShipTwo, N339SS; Near Koehn Dry Lake, California; October 31, 2014* (NTSB Report No. NTSB/AAR/15-02). Washington, DC: Author.

Although the copilot had practiced for this flight several times in the simulator, this premature movement of the feather unlock lever occurred on the fourth powered flight of the SpaceShipTwo, indicating to me that the likelihood of this error was high. "By not considering human error as a potential cause of uncommanded feather extension of the SpaceShipTwo, Scaled Composites missed opportunities to identify the design and/or operational requirements that could have mitigated the consequences of human error during a high workload phase of flight." Because of the underlying design implications, the NTSB issued a safety recommendation to the US Federal Aviation Administration to ensure that commercial space flight entities identify and address "single flight crew tasks that, if performed incorrectly or at the wrong time, could result in a catastrophic hazard." Additionally, the manufacturer added a safety interlock to ensure that this lever could not be activated during this critical flight regime.

I'll conclude by briefly mentioning safety culture. Chapter 12 does a nice job of describing different types of cultures, including safety culture. When speaking to groups about safety leadership, I oftentimes ask a simple question: Do you have a good safety culture? Although a simple question, the answer can be complex. I encourage participants to carefully consider their answer and to keep their reply to themselves. I suspect every audience has a few who feel smug—overly confident they are doing all the right things. I pity them. To quickly shake them from a potentially dangerous misconception, I show the participants a quote by Professor James (Jim) Reason: "It is worth pointing out that if you are convinced that your organization has a good safety culture, you are almost certainly mistaken…. A good safety culture is something that is striven for but rarely attained…. [and] the process is more important than the product."[14]

Dr. Reason's statement is profound in many ways. It provides a cautionary tale: once we feel we are "there," we tend to get complacent. Once complacency sets in, we lose focus on operating safely.

In his writings,[15] Reason mentions a "chronic unease" that, although perhaps unsettling, keeps us alert to potential safety problems. Complacency is truly an enemy in any high-risk business, but especially aviation. Reason's quote also makes the point that there is no "there" with safety culture. Like safety itself, there is no destination; it's the constant journey, the processes, and the struggles associated with it that keeps us on our toes.

So, a word of caution: Be wary of anyone who touts their safety culture as being strong. I vividly remember one crash that we investigated. The business aviation operator appeared to have all of the right things in place. "Safety culture within the department is shared among all team members," wrote the auditor who performed an audit of their operations. "Open reporting of hazards is consistently encouraged by management," was another audit comment, followed by, "Solid safety program, maturing nicely." Despite glowing comments such as this, our investigation revealed that this aviation department wasn't even doing basic things like completing flight deck checklists or flight control checks before takeoff—both of which were causal to the accident.[16]

I'm a big fan of Jim Reason. I still fondly remember a lovely dinner in 2008 with Jim and his daughter when I visited London. However, I believe there's one critically

---

[14] Reason, J. (1997). *Managing the risks of organizational accidents.* Burlington: Ashgate

[15] Ibid.

[16] NTSB. (2015). *Aircraft accident report: Runway overrun during rejected takeoff, Gulfstream Aerospace Corporation G-IV, N121JM; Bedford, MA; May 31, 2014* (NTSB Report No. NTSB/AAR/15-03). Washington, DC: Author.

important element that is absent from his model of safety culture, as described in Chapter 12. What is that element, you ask? It's top-level leadership support and commitment. It's a commitment that in the face of competing priorities, we will live our safety culture values. Without top-level leadership support and commitment, a healthy safety culture cannot be sustained. I've seen leaders who say all the right things, yet faced with challenges, those words are nothing more than lip service. Talk is cheap. Leadership actions and attitudes are what counts.

As I wrap up and turn the controls over to the book's authors, I'd like to note that the authors are experts in their fields. There is a wealth of knowledge between them, and they have done a marvelous job of assembling their considerable expertise and knowledge to create the seventh edition of *Commercial Aviation Safety*. With that, I leave it to you—the next generation of aviation safety professionals. We're counting on you!

Robert L. Sumwalt
Distinguished Fellow in Aviation Safety
Executive Director
Center for Aviation and Aerospace Safety
Embry-Riddle Aeronautical University
Daytona Beach, Florida

February 10, 2023

Preface

As the seventh edition of *Commercial Aviation Safety* is published, it is appropriate to reflect on the history and contributors to the 33 years of this enduring educational textbook. The first edition was published by McGraw Hill in 1991 and it has been updated regularly since that time. The original author, Alexander T. Wells, of Embry-Riddle Aeronautical University was a prolific textbook author or co-author of 14 aviation educational textbooks whose vision and contributions to the field of aviation safety will remain a profound legacy for many generations to come. He was the sole author of the first three editions.

The fourth edition introduced co-author Clarence C. Rodrigues, who also was a professor in the College of Aviation at Embry-Riddle Aeronautical University. This edition provided a major rewrite of the previous three editions, which set a new tone and direction for the text. In addition to updating and making current safety regulations and security information contained in previous editions, it established new changes in format, content, and order of the chapters. This edition introduced essential regulatory information on the Occupational Safety and Health Administration (US-OSHA), the Environmental Protection Agency (US-EPA), and the Transportation Security Administration (US-TSA).

The fifth edition introduced co-author Stephen K. Cusick, who was a member of Florida Institute of Technology's College of Aeronautics faculty in Melbourne, Florida. This edition updated aviation safety information contained in previous editions and broadened the field of study to include regulatory information on ICAO and Safety Management Systems (SMS) procedures that are essential subjects for today's practicing aviation safety professional.

The sixth edition introduced co-author Antonio I. Cortés, who was well known in this area as Associate Dean of the College of Aviation at Embry-Riddle Aeronautical University. His vast experience as an ex-military and airline pilot, aviation safety program manager, and human factors researcher added greatly to the depth and scope provided in this edition.

The seventh edition introduces four new authors and colleagues whose experience and credits are indeed impeccable. Anthony Lawrenson, Shem Malmquist, Mathew Greaves, and Graham Braithwaite bring a wealth of academic and operational experience to this completely updated edition, to keep pace with the latest in aviation safety concepts and practices. Our colleagues from Cranfield University, a British postgraduate public research university specializing in science, engineering, design, technology, human factors, and management, have provided a truly international perspective to the book.

We hope it will be helpful to students and educators training future professionals in this critical field.

Features of the Seventh Edition

The seventh edition of *Commercial Aviation Safety* was written during a period of intense turmoil not only for the aerospace industry but for humanity. The COVID-19 pandemic grounded the majority of the world's commercial aviation fleet and stands as the greatest impact event of the industry's history. It is with this background that the seventh edition begins to look at commercial aviation safety within its wider social context. The incredible achievements of commercial aviation in terms of innovation, technical ambition, and reliability could not have happened without external influences and broader social change. The constantly evolving risk landscape is reflected in the changing nature of threats that commercial aviation now must face. Those threats are evolving, and logically the way in which we manage them must also evolve. No longer can we look for faults in the machinery or broken components to explain why major aviation catastrophes occur. Commercial aviation is a system, and unless we understand how and why the system works, we cannot begin to understand how and why the system fails.

This book breaks the subject down into four parts, reflecting the author's combined perspectives of commercial aviation safety. The intention here is to allow ease of access for those wishing to use the book as a study reference. The order of the four parts is designed to walk any newcomer through the essential elements before delving into the more esoteric areas of each subject. We look at commercial aviation safety from a theoretical, technical, and human factors viewpoint but this culminates in the increasingly important systems perspective. Where we felt further explanation was beneficial, we have added footnotes at the end of each page. We have provided referencing at the end of each chapter rather than risk losing the thread in one singular reference source at the end of the book.

In the final chapter, we look forward toward the future of commercial aviation safety. We have written this book to be the one we would have liked to have been given when we all first entered this fascinating area as a student or as a practitioner. While the authors have all spent decades studying and working within aerospace, we are left with far more questions than answers.

Writing this book as a team has stimulated much debate and discussion from which we have all benefitted; we hope it stimulates even more debate and discussion from our readers. The questions we have posed at the end of each chapter are as much a reflection on important issues as they are a simple check of understanding. We will continue to discuss many of the issues we have raised through a series of short videos that will accompany this book through the McGraw-Hill Professional website at https://www.mhprofessional.com/CAS7-9781264278701. Maintaining these crucial conversations is the primary objective of the seventh edition of *Commercial Aviation Safety*.

Acknowledgments

The authors of the seventh edition are proud and humbled to be part of this ongoing legacy in the field of commercial aviation safety. In this edition of the book, we have tried to condense the work of some of the most influential thinkers across the many specialist subjects that are combined in this professional and academic field. In doing so we have attempted to include new and emerging research ideas, particularly from our colleagues at Embry-Riddle Aeronautical University, in the United States, and from the staff at British Airways, TUI, and the Safety and Accident Investigation Centre at Cranfield University in the United Kingdom. The full list of academics and professionals that have contributed either directly or indirectly in the writing of this book is too numerous to mention, but it is a credit to the safety mindfulness within our industry that help or advice was never more than a phone call or email away.

As with any project that takes away precious family time we can only express our sincere gratitude to those we love and who have supported us on those days when it was most needed. Thank you, Vicky, Albert, Iris, Abigail, Joe, Nichola, Suzanne, Caitlin, Finlay, Lucy, and Jean.

Introduction

Safety is sometimes described as an emergent property of a system, that is, it's not something you necessarily control, but a product of what you put in place. Commercial aviation safety is not something left to chance, it is the combined product of science, engineering, management, psychology, and many other skills and disciplines. This complex interdependence of so many specialist subjects can present a perplexing combination to those new to the industry. Even those who have spent their lives immersed in aviation regularly find that there is always more to learn and plenty to debate.

On the 18th of November 2022, the runway collision between a Latam Airbus A320 NEO and an airport fire truck, at Jorge Chávez International Airport, Lima, reminded the aerospace community of the ongoing trend to criminalize aircraft accidents. The Latam pilots were arrested and within hours of the accident, the Peruvian prosecutor's office announced an investigation into potential manslaughter of two fire fighters. Meanwhile, in France, Airbus and Air France are engaged in court proceedings which commenced on the 10th of October 2022, for the manslaughter of the 228 passengers crew of Air France 447, killed in June 2009 after crashing into the Atlantic Ocean, off the north coast of Brazil. Finally, it has recently emerged in the United States that the Boeing Corporation will likely face a total cost in excess of $65 billion following their conduct in the design, production, and selling of the Boeing 737 MAX aircraft. The high-profile failure of Boeing's safety culture which resulted in the death of 346 people in two accidents now looks likely to become the most expensive corporate failure in history.

These examples demonstrate that despite the phenomenal success of commercial aviation safety, there is still considerable work to be done in understanding how these complex systems interact and occasionally produce undesirable behaviors. There has to be more focus on capturing and understanding the sociopolitical, legal, and environmental influences that drive some of these system behaviors if we are going to effectively manage the emerging threats of the near future. Equally, if not more importantly we need to understand why this phenomenally complex system ordinarily works so well and achieves incredible levels of safety performance.

Concepts of Safety

When settling into our seats onboard a modern commercial airliner, few of us ponder on the complexities of making the typically three million components of these machines work harmoniously together. The precise navigation functions that guide across the changing climates of countries and continents are taken for granted. Before watching the in-flight entertainment, most of us do not consider the energies that have to be expended and controlled in elevating the hundreds of tons of aircraft, people, freight, and fuel into the earth's stratosphere. The global armada of people and machines transports the equivalent of the world's population every two years and does this safely; in fact, so safely that it is described as an ultrasafe system exceeding the safety performance of many of the more mundane of human activities. This book is a narrative of how scientific advances in commercial aviation have made the extraordinary into the ordinary.

Despite the technological and safety performance achievements of commercial aviation system, it remains vulnerable to failure. But are these failures inevitable? The strengths and weaknesses of the system of commercial aviation are reflected in the interaction between people and advanced technology sometimes referred to as a sociotechnical system. This interaction is of particular focus within this seventh edition of *Commercial Aviation Safety*. The interrelationship is not unique to aviation but is becoming an area of increasing academic and commercial interest as society follows the process of increasing digitization. Today we are presented with an increasing array of technological capabilities from artificial intelligence through to commercial spaceflight, but it is important to recall the basics of how safety is achieved. We follow this path by initially considering some of the philosophical underpinnings of what safety and risk mean. Rather than dwelling on what causes *accidents*, we attempt to balance this discussion by asking what causes *safety*?

If there are any misgivings about the motivations that drive this sector of the aerospace industry, the clue is in the title: commercial. It would be easy to subscribe to the oft proclaimed principle from airlines, manufacturers, operations, and maintenance companies that safety comes first but that is quite clearly not always the case. If safety always came first, the commercial viability of the industry would be in question. This raises an obvious question of where safety actually comes. The two intuitively opposed objectives of safety and commercial must be balanced. How this balance is achieved in commercial aviation is as much a challenge as many of the technical challenges faced by the operators and manufacturers. We consider how safety and success might ideally coexist by the alignment of strategic objectives. This is not simply through business efficiencies but the absorption of safety into mainstream commercial practices.

One of the paradoxes of commercial aviation safety is that no matter how impressive the safety performance statistics appear, interest in them is overshadowed by the absolute fascination in system failure. This is particularly the case when system failure results in fatal accidents. If we are going to understand the mechanics of commercial aviation, it is crucial that we establish how our preconceptions of the system are framed. There is something particularly visceral about the crash of a commercial aircraft which has been shown to have a widespread impact on, among other things, popular media, mental health, and even global trading markets. This book attempts to bring some rationality to what often becomes an emotive and disturbing subject of discussion. We aim to challenge the credibility of some of the assumptions and occasional myths that have perpetuated within the safety management of commercial aviation as well as the validity and relevance of many of the metrics that are used.

Technical Safety

However, before we can delve too deeply into the human and organizational side of safety, we need to ground ourselves in the technical side, to give a solid base from which to build.

The global aviation system has built its enviable reputation by trying, sometimes failing, and then understanding and fixing the problem for the future. This is particularly true in the technical domain.

Aviation is first and foremost a technical pursuit, and modern aircraft are incredibly reliable—but that has not always been the case. In the early years of aviation, there were a large number of catastrophic accidents, many of which had a technical origin. By studying the failures, increasing our understanding, and modifying the rules and regulations surrounding the design, maintenance, and operation of aircraft, we have built systems that can reasonably be expected to fail less than once per billion flight hours (equivalent to more than 100,000 years of continuous operation).

The aviation system is a highly reliable and a highly regulated system—some would argue *because* it is a highly regulated system. The many thousands of pages of regulations that go into the operation and licensing of design, manufacturing, operating, maintenance, and navigation organizations are the tip of the iceberg. And while no one person can be expected to be intimately familiar with every aspect of every regulation, it is crucial that people working in the safety industry understand the context in which they work. We will look at what it takes to design, certify, maintain, and operate a modern aircraft.

The improvement in aircraft technical performance has been accompanied by an increase in complexity. Improved automation to support pilots, improved onboard processing, new communication equipment, and increased attention on emissions, fuel efficiency, and weight-saving have all increased the technical challenges beyond merely improving the failure rates of a 1960s aircraft. This again leads us to balance safety against the other competing priorities but this time in the technical domain.

Another transformation that has challenged safety is the increase in traffic. The number of aircraft in the air at one time has increased hugely since the advent of aviation and 2018 saw more than 40 million flights, more than 100,000 per day. The number of commercial flight hours flown per year has doubled between 2002 and 2018. We will look at the airport and air traffic system to see how processes work today and what improvements and challenges are around the corner.

Another change that cannot be ignored is the proliferation of drones and the proposed introduction of Urban Air Mobility vehicles. With disruptive new players entering the market, the system faces a challenge to retain its hard-won safety lessons with a new set of stakeholders who have been fortunate enough to see fewer failures to date. How does regulation and the airspace system accept change without loss of performance?

This also highlights a further difficulty. When examining how to solve a current problem in aviation, it is tempting to conclude "I wouldn't start here!" The technical history, evolution, and legacy of aviation are embedded within most of the systems in use today. The reality of aviation is that the system cannot simply change overnight and while it is easy to envisage a future from a clean sheet of paper, the truth is that we also have to engineer the journey to that future. That means bringing along legacy aircraft and transitioning legacy systems, without costing too much, requiring too much change to the status quo, or, most importantly, compromising safety. The technical challenge today is no smaller than it ever was—it's just a different challenge.

Human Safety

Ask any aviation enthusiast what their favorite film is, and *Top Gun* would likely be high on their list. The incredible flying skills of the lead character, Captain Pete Mitchell, are only matched by his charisma and good looks. But imagine boarding your next commercial flight and being greeted on board by the aircraft commander, resplendent in their neatly pressed uniform, and spotting that their name badge read "Maverick"! The two images of what makes for an "ace pilot" are very different—one is always ready to take risks and push the envelope whereas the other is trained to follow Standard Operating Procedures, use checklists, and work as a team. Commercial aviation has achieved a high level of safety by selecting and training professionals to be more like the latter while the former is generally left on the silver screen.

This has not always been the case. Flying aircraft in the early days carried an enormous amount of personal risk for pilots and their passengers. Pilots needed to be bold and fearless! The limitations of the aircraft in terms of maneuverability, structural integrity, ability to cope with weather, and so on often meant that pilots lost control or were unable to make a safe landing. As the industry learned from its mistakes, it also increased complexity. There was less than 50 years between the first powered flight at Kitty Hawk in 1903 and the first commercial jet aircraft entering service in 1951. During a period of rapid transition, the limitations of the human being became all too apparent.

The terms "human error" and "pilot error" became synonymous with the cause of aircraft accidents. By the 1970s it was clear that something was amiss and organizations including the Flight Safety Foundation and the International Air Transport Association convened conferences to see what could be done. The world's worst aircraft accident at Tenerife in 1977 claimed the lives of 583 and highlighted that some of the best people can make the worst mistakes. Since then, crew training and aircraft design have changed radically, but human performance limitations remain.

In understanding the role of the human in the air transportation system, we need to consider the roles of physiology and psychology—the human machine and its operating system! These range from the frailties of cognition or operating under stress to the ingenuity and adaptability of humans. It is a story of extremes and of an industry that has evolved to manage certain threats only to introduce new ones. For the commercial aviation manager, humans arguably represent both their greatest asset and most insidious threat. The challenge is to try and train- or engineer-out as many of the potential human errors or violations while preserving the benefits that human operators can bring.

This is a story not just of pilots, but of cabin crew, air traffic controllers, design engineers, maintainers, managers, and beyond. The complex sociotechnical system that is aviation draws upon a myriad of skillsets and personality types, spread across multiple cultures around the world. It is a highly dynamic industry where new technologies create both new opportunities such as ultra-long-haul flights and corresponding threats such as crew fatigue. By gaining a deeper understanding of the things that influence human behavior and what, if anything, can be done to enhance human performance, aviation safety professionals can offer a range of solutions.

We explore a range of accidents and near misses to illustrate where humans have been both the challenge and the solution. There is a great deal that can be learned from the detailed forensic analysis of human behavior that technology such as the flight data recorder has afforded us. Our failures and successes remind us of the critical role that humans play at the heart of the aviation system.

Organizational and System Safety

What if our basic concepts of how systems work were fundamentally flawed through oversimplification? We create a challenge to ourselves when we try to understand how complex accidents develop using overly simplistic models. Most of the safety analysis in the past 20 years has been based on simplistic linear models. These were often called "event chains"; they are based on the premise that incidents and accidents can be traced through a series of events, each one based on the previous one. This logic suggests that we can replicate this process and use it to predict futures system failures. To make a system safe, it follows that all we have to do is to ensure that each link in the chain performs in a reliable way. Although the concept is intuitive, today we are finding that accidents will occur despite all the components working exactly as expected and designed. How can this be?

Fortunately, other models have been developed that can reflect how complex systems actually work. Using a theoretic model based on system theory, the behavior of complex systems can be viewed as a hierarchical control structure, where controllers in the system, either human or software, are used to constrain the behavior of the component below them. Through this approach it becomes possible to accurately model how complex systems actually behave in the real world. This approach has deep applications in understanding the human role in modern systems and explains why sometimes automation does not go as expected. It can also provide insights into complex sociotechnical processes such as safety management.

Safety management systems (SMS) are now widely accepted in commercial aviation. In Chapter 14 we review the current approaches to safety management, describing the different facets that go into a safety management system. An effective way to prevent an accident is by identifying unsafe trends in an organization in advance and then implementing changes. SMS was created to identify such trends and enforce safety in an organization. We will review the history of SMS before taking a deeper look at the component pillars of SMS: Safety Policy, Safety Assurance, Safety Promotion, and Safety Risk Management.

An important element of SMS are safety data and statistics. We will review sources that identify trends in the industry. After that, we explore what possible risks may be involved in the use of statistics and what traps we should be wary of. Statistics can be accurate but how we present them can tell a misleading story. It is important for all safety professionals to be able to identify some applications when the statistics might lead us astray.

We will also consider proactive safety. Although the SMS has proven to be effective in some types of risks, current methods are limited by the linear modelling that underlie many of them. A large part of this challenge is that modern commercial aviation is extremely safe, and, although that is a good thing, it also means that we are left scraping for any meaningful data that might indicate a safety trend. Small sample sizes can mean that any outlying bit of data has the potential to dramatically skew results. Several of the more forward-thinking approaches to get around this problem avoid the use of statistics altogether.

We will take an in-depth look at the concept of *High Reliability Organizations (HRO)*. These are organizations that appear to defy the odds, by operating in high-risk environments yet having much lower-than-expected accident or incident rates. Researchers have considered that it is possible to learn from what these organizations are doing, how they are doing it, and then take that information and apply it to other organizations. A complementary idea is that of adaptability, which means that an organization, or even the crew and systems on a single aircraft, can be designed such that they are adaptable to unexpected changes. This leads us into a look at the concept of *resilience engineering*. Here we take an organization, or a design, and build into it resilience, that is, the ability to absorb changes and disruptions without an accident occurring. It is worth emphasizing that the role of humans in all of these system designs is considered an asset, rather than a problem.

In 2018, David Woods, a research engineer at Ohio University developed a theoretical construct called *graceful extensibility.* This model is founded on the concept that there are only finite resources available to any component of a system, whether it be human, software, or hardware. These are the operational limits, but to be able to absorb very unexpected circumstances, the component and system as a whole need to be able to extend their capacity in a way that is smooth and graceful, not abrupt, as the latter can lead to more problems.

If there is insufficient data to study accidents, an alternative might be to study data from when there are no accidents or incidents? This is the premise put forward by the Danish academic Erik Hollnagel, in a model he calls *Safety II.* The idea is that we study and learn from what goes right, when no accidents or incidents occur. There are many of these types of flights, in fact the vast majority of flights go as planned. The question then becomes, "why does it go right?"

A related area of research involves approaches to improve the performance of humans in a system. Called *cognitive systems engineering,* researchers have focused on improving aspects such as automation design, procedures, and policies to enable humans to perform better. To understand how to better design our systems, as a whole and at a more granular level, we need data. Where can data be collected? Here we review the types of data that can be collected, both automatically and through reporting systems. We will consider some approaches to safety management that are designed to measure what does not seem to exist. Some have argued that the information is there, we just need sufficient computing power to identify it. Here we will look at the problems in this concept, and then look at how we might identify actual leading indicators of an accident using system theory.

Commercial aviation is safe not just due to the proximate factors that we have already discussed, but also due to the broad network of organizations that provide standards, oversight, and guidance. These include the government organizations and specialist regulators, the legal system and the many industry trade associations, labor unions, and the liability and insurance industries. For the most part, this combination of varied stakeholders combines to provide a robust and resilient system. But not always.

A global trend of the increasing use of the criminal justice system has forced a debate between the primacy of system safety and social justice.

Future Safety

Having been introduced to the fascinating world of commercial aviation safety we conclude this seventh edition with a single chapter that focusses on the developing and emerging themes of the future. One might assume the technology of commercial aviation naturally equips the industry to face the many challenges of a rapidly digitizing society. However, there are many aspects of the industry that are very conservative and slow to evolve.

Commercial aviation has always seen itself as a high technology and innovative industry, but the current trends suggest this is not a sustainable position. The standard equipment pack of an average business traveler will often exceed the processing capability of the aircraft on which they are flying. The combination of exceptionally high standards required of any avionic or control hardware and software, the decades of lead time from design to deployment, and the longevity of a modern commercial aircraft's lifespan means that other forms of technology will increasingly run ahead. While healthy conservatism is one of the many reasons for the ultrasafety and reliability levels we rely on as air travelers, we will consider how commercial aviation can keep abreast of the accelerating process of digitization.

Commercial aviation safety has significant challenges to face if it is to maintain or even improve on its incredible performance record. The principal challenges are complexity and resource limits. The anticipated level of accelerated industrial growth is in the face of global pressure to preserve our environment. This situation limits many of the traditional approaches of technology to solve these problems. This book intends to open that discussion and encourage debate about the priorities of our industry. We hope to enjoy reading and debating these subjects and we look forward to receiving your comments.

The CAS-7 Team
December, 2022

Concepts of Safety

CHAPTER 1

An Introduction to Commercial Aviation Safety

Introduction

Despite significant improvements to commercial aviation safety over the last 25 years, major accidents still occur. This chapter introduces the sociotechnical field of safety and how it has evolved through four eras of focus over the decades. The chapter starts by showcasing an accident that forced the industry to question some basic beliefs in the safety system that has been created. Modern commercial aviation safety is anything but common sense. It is instead complex, multidisciplinary, and science-based.

1.1 Air France 447

1.1.1 A Story of an Accident

At approximately 02:10 UTC on June 1, 2009, an Airbus A330-200, call sign Air France (AF) 447, began emitting automatic system health signals to its home engineering base near Paris. Four minutes later, without any radio transmissions from its crew, the aircraft had crashed into the Atlantic Ocean off the North-East coast of Brazil killing all 228 passengers and crew. Figure 1-1 shows recovery crews working with a piece of wreckage from AF447 far out in the Atlantic Ocean.

This Airbus A330-200, registration F-GZCP, was 4 years old and the youngest aircraft in the Air France A330 fleet. This aircraft had taken off from Rio de Janeiro at 22:29 UTC and after navigating up the east Atlantic coast of Brazil entered the infamous Intertropical Convergence Zone (ITCZ). The ITCZ can produce intense thunderstorms which provide an array of significant threats to aircraft safety; all pilots undergo meteorological training, and it is ingrained in them from the beginning to avoid thunderstorms whenever possible. No modern commercial airliner can outclimb many of the larger thunderstorms, so pilots use weather radars which measure the size of raindrops to determine the worst areas of turbulence and navigate around them. As any typical airline crew would do in such circumstances, the pilots flying AF447 were focused on the task of keeping away from the worst parts of the storms within the ITCZ. As professional pilots, this task was at the forefront of their minds, so it is worth considering how violent these thunderstorms can be.

FIGURE 1-1 AF447 wreckage found in the Atlantic Ocean, after a 22-month search over 17,000 square kilometers, and depths of up to 4000 meters. (Source: Brazilian Air Force)

In aviation, turbulence is divided into three levels: light, moderate, and severe. To understand the feeling during these three levels of turbulence, imagine you are driving a car, perhaps a big 4×4 vehicle, which takes an exit off the freeway and joins a bumpy backroad—this is "light" turbulence. It's annoying and can spill your coffee but it won't seriously hurt you. Now you take another turn, and leave the backroad to drive across a plowed field. It's now very bumpy but in a typically understated aviation language it is called "moderate" turbulence. You're just able to drive in the right direction, but spill your coffee on your shirt and feel more apprehensive than annoyed. You are glad to be seated and wearing your seatbelt which stops you from hitting your head or arms or legs on something hard. Now as you drive across the field, you attempt to navigate a steep downhill embankment. Your car is destabilized and it rolls over and over. You become so disorientated that you can't really discern up from down. You are unable to control your coffee-stained limbs and loose objects in the car keep hitting and hurting you. This is "severe" turbulence, which can injure and occasionally kill the occupants of an aircraft. Severe turbulence can cause significant damage and, in extreme cases, it has led to the in-flight break up of aircraft. The crew of AF447 were well aware of the potential threat from the ITCZ. Although apprehension about severe turbulence may have influenced their actions, it wasn't turbulence but rather ice that became the primary threat to the aircraft's safety.

Ice is perhaps not a threat many people associate with flying in tropical air, but a modern aircraft cruises between 35,000 and 40,000 feet (10,668–12,192 meters) where the ambient outside air temperature is approximately −50°C (−58°F). Given the billions of gallons of water suspended in a typical thunderstorm, a gigantic icy grip can overwhelm an aircraft flying close to, or within it, in seconds. Ice can clog up engines, restrict flight controls, and blind an aircraft's instruments. Another threat comes from the fact that the air can be laced with supercooled water, that is, water that is well below freezing temperature but still maintains its liquid form. The moment this water touches a solid object, such as an aircraft, it solidifies. Tons of ice can accumulate within seconds, making the aircraft dangerously overweight. In addition, the slightest accumulation of ice on the smooth wings of an aircraft reduces its ability to produce lift. According to a paper written by leading Boeing researchers,[1] a covering of ice that turns the surface from smooth to coarse grained like a sandpaper can reduce the lift capability of a wing by up to 25%. Inches of ice can instantaneously attach to the aircraft's windscreen obscuring the pilot's view—greatly increasing dependency on the simultaneously failing instruments. Icing conditions can really spoil your day's flying.

Another more insidious risk from icing, particularly at higher altitudes, is ice crystal icing (ICI). In 2009, ICI was not particularly well understood by the aviation industry. It forms in clouds of microscopic ice crystals which, although tend to bounce off the airframe in flight, can stick to the rotating blades of jet engines or accumulate inside speed sensors. To mitigate the risk from icing, modern aircraft have an array of anti-icing systems on the wings, in the engines, and around the aircraft's crucial speed and altitude sensors. What makes ICI such a deadly threat is that unlike the more conventional airframe icing mentioned earlier, ICI can be difficult for the aircraft to detect and, therefore, for the flight crew to diagnose the problem. The threat posed to AF447 from ICI illustrates: the more sophisticated the aircraft technology becomes, the more vulnerable it becomes under a certain combination of events. This observation is a very brief outline of Normal Accident Theory which, with a number of other accident theories, we shall discuss in more depth in Chapter 3. The theory attributed to the American academic and author Professor Charles Perrow suggests that complex systems can produce exotic failure combinations that are practically impossible to predict. As he succinctly stated, "Unfortunately, most warning systems do not warn us that they can no longer warn us."[2] This quotation from Perrow suggests that technological advances if left unchecked will increasingly produce more surprises, and occasionally catastrophic outcomes.

The AF447 flight crew consisted of the captain and two copilots. The extra pilot meant each of them could take a period of rest to mitigate fatigue while still ensuring that two pilots were always alert and manning the primary crew stations. Traditionally, for a flight of this length, one of the copilots is nominated by the airline's crew control as "the heavy" or relief pilot. On the night of the 1st of June, this pilot was David Robert, an experienced copilot, who had taken the first rest period. Robert then took over from Captain Marc Dubois who planned to take the second, middle rest period. Following Air France standard procedures, Robert returned to the flight deck and sat in the left-hand seat as pilot-not-flying (PNF) to take over the decision-making role that had been covered by the captain. In the right-hand seat and at the controls of the aircraft, the pilot flying (or PF) was a less experienced and younger copilot, Pierre-Cedric Bonin.

According to the official Bureau of Enquiry and Analysis (BEA)[i] report, before leaving the flight deck to take rest, Captain Dubois listened to the handover brief delivered by Bonin to Robert but did not specifically state who had overall command in his absence, nor did he discuss what strategies should be employed by the two copilots in order to navigate across the ITCZ safely. Bonin had expressed some discomfort about the characteristics of the ITCZ and its potential for hazardous weather, especially the large cumulonimbus buildups, commonly known as "CBs" or thunderstorms. The BEA

---

[i] The French aviation accident investigation organization: Bureau of Enquiry and Analysis for Civil Aviation Safety or to give its correct French title: Bureau d'Enquêtes et d'Analyses pour la sécurité de l'aviation civile.

report suggested the lack of clarity from the captain's briefing may have influenced the limited crew coordination that occurred when, later in the flight, the situation began to deteriorate.

After Captain Dubois had left the flight deck, Bonin, perhaps expressing his own justifiable caution about flying through the ITCZ, contacted the cabin crew, warning them to "take care" and take their seats until the anticipated turbulence had subsided. From the left-hand seat Robert adjusted the weather radar and detected a thunderstorm on the current course ahead; he suggested a slight change to the aircraft's course to the PF, Bonin. The aircraft adjusted its track to the north but not enough to avoid the turbulent and ice-filled air. Figure 1-2 displays a combination of infrared and other weather imaging to illustrate the band of thunderstorms and turbulent air produced by the ITCZ that Robert and Benin were navigating through.

An aircraft's airspeed is critical to its ability to fly safely and is sensed by its pitot tubes. A pitot is a metal tube like a drinking straw that points forward into the airstream. The sensed dynamic pressure gives an indication of how fast the aircraft is moving through the air. The aircraft's altitude is determined by static pressure sensors that are positioned around the forward fuselage of an aircraft which measure the change in ambient pressure as the aircraft climbs or descends. This combination of dynamic and static pressure is crucial to any aircraft, especially at night when flying at high altitude these sensor systems feed information to the autopilot and due to minimal visual clues, the pilots are often totally dependent on their flight instruments.

The most common anti-icing methods used to protect the various aircraft systems are high-powered electrical heaters which, in the case of pitot tubes, are continuously heated or, as in other systems, switched on automatically when ice is detected. Ensuring these heater systems are fully functional and sufficiently reliable is an essential part of the aircraft's certification program. Despite these modern automatic and highly efficient

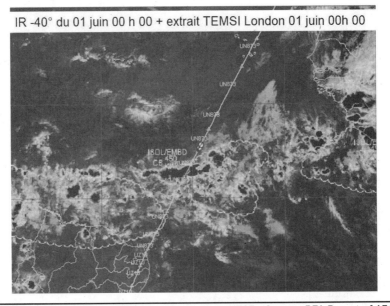

IR -40° du 01 juin 00 h 00 + extrait TEMSI London 01 juin 00h 00

Figure 1-2   Weather depiction of AF447 path across the ITCZ. (Source: BEA Report of AF447)

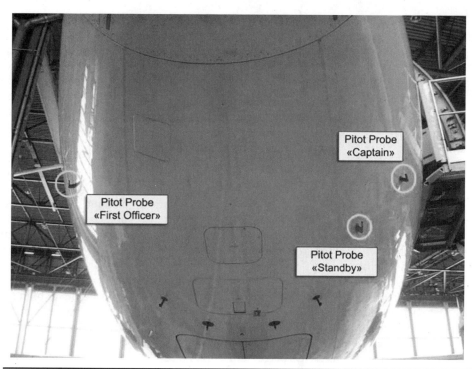

FIGURE 1-3 Location of Airbus A330 pitot probes as seen from the nose of the aircraft looking aft. (Source: BEA Report of AF447)

anti-ice systems, they can sometimes be overwhelmed by the sheer volume and type of ice accumulation. They can also fail. As AF447 entered the area of unstable air detected by Robert, the pitot tubes filled with microscopic ice crystals, partially blocking them.

Using its sensor suite, similar to the one depicted in Figure 1-3, the Airbus A330 detected multiple faults and automatically relayed this information electronically to its home engineering base. The automatic transmission of data about system health is a routine activity for modern aircraft. It allows maintenance personnel to prepare for the aircraft's arrival, source replacement parts, and allocate personnel in advance. The system is designed so that faults can be rectified in the short turnarounds when the aircraft is being prepared for its next journey. Some commercial aircraft operators can keep their long-haul aircraft airborne for over 15 hours in any given day; since aircraft on the ground don't make money, efficient turnarounds are key to a profitable operation.

The electronic signals from AF447 suggested that there was a discrepancy between the aircraft's measured airspeeds which tell the crew, and crucially the aircraft, how fast the aircraft is traveling through the air. If the aircraft is traveling too fast, it could sustain structural damage, and if it is too slow, there could be insufficient lift generated from the wings for the aircraft to maintain altitude and to sustain controlled flight. Balancing the aircraft's speed in between these parameters is sometimes referred to by pilots as the "coffin corner" since exiting the range of acceptable speeds can result in catastrophic loss of the aircraft. If there was a discrepancy between the multiple airspeed sensors, it meant that neither the crew nor the aircraft could determine what speed to fly in order to maintain safe and controlled flight. The aircraft's airspeed detectors were temporarily blocked with ice crystals and started to give erroneous information to the aircraft's flight computers and in turn to the pilot's instruments.

Several significant events occurred in close succession: the aircraft's autopilot disconnected, triggering a loud warning noise and alert caption; the autothrottle disconnected with subsequent airspeed reduction; the right-hand altitude indication dropped by 330 feet; the aircraft's control laws changed from "normal" to "alternate"[ii]; and most significantly, a nervous and shaken Bonin, the PF took control to manually fly the aircraft.

The BEA accident report suggests these multiple warnings came as a shock to both pilots. Ordinarily, the cruise phase of a long-haul flight is a time of low workload. The crew dynamics are deliberately slowed down to conserve mental energies for the higher workload demands of approach and landing. With the loss of airspeed and autopilot control, Bonin and Robert were suddenly faced with multiple and confusing inputs which did not make sense to them; they were startled.[iii] Neither pilot verbalized the significant changes to their environment and the aircraft's operating mode. As the BEA report notes, in order to retrospectively understand an individual's behavior during such events, it is worth remembering that humans react to their perception of their environment rather than how the world actually is. For example, just talk to two groups of sports fans right after a game. They both watched the same game, and both are telling "a truth" but those truths can significantly vary. The bounded rationality of individual perception can be very different to the actual reality of any given situation. This natural subjectivity is particularly enhanced when humans experience fear.

Perhaps preloaded with his discomforts about flying through the ITCZ, Bonin reacted intuitively to the displayed altitude loss and sharply pulled back on the sidestick controller, like the type depicted in Figure 1-4, pitching the aircraft nose up by 10° above the horizon to regain the aircraft's cruise altitude. Possibly relying on his inaccurate beliefs about the inability to stall an A330,[iv] built up during his training, Bonin continually pulled back on the sidestick control throughout the episode, making potential diagnosis by the other pilots even more complicated. About 30 seconds after the autopilot disengagement and becoming increasingly concerned about the situation, Robert called the captain back from his rest.

Robert became distracted by prioritizing the recall of the captain to the flight deck. His colleague, Bonin, continued to manually climb the aircraft to the limit of its operational ceiling, apparently unaware that they were losing critical airspeed. At 37,500 feet (11,430 meters), the ambient air is too thin for the Airbus A330 to create sufficient lift or

---

[ii] A significant and occasionally controversial characteristic of Airbus aircraft design is that their aircraft are engineered to guard against the variability of human behavior. In "normal law," the aircraft will override any human pilot inputs that take the aircraft outside of a safe flight envelope, for example, trying to fly too fast or slow or flying the aircraft at a dangerous angle of attack. Some malfunctions, such as that encountered by AF447, induce "alternate law" where no flight envelope protection is provided.

[iii] Startle is a particular condition which will be discussed in more depth later in the book. At the time of the AF447 accident, it was an under-researched phenomenon but was succinctly defined in a later BEA report concerning a separate event (F-GLZU, AF471, Airbus A340-313, July 22, 2011, page 12), "The 'startle effect' generated by a sudden event such as an aural warning may produce a 'reflex'-type reaction with some pilots. Sometimes this effect sparks primal instinctive reaction, instant and inadequate motor responses. These basic reflexes may prove to be incorrect and difficult to correct under time pressure and may affect the pilot's decision-making ability."

[iv] In "normal law" this assumption would be correct as the aircraft's flight envelope protection prevents departure from controlled flight. However, the pilots were either unaware or had not noted that their A330 aircraft was then in "alternate law" with no engaged flight envelope protection.

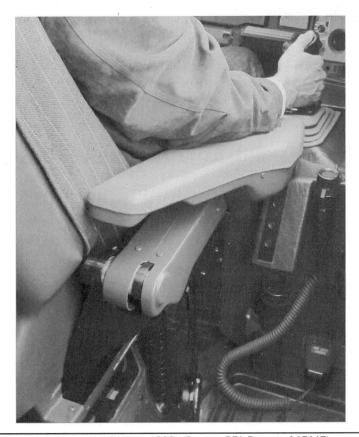

Figure 1-4   Right-seat sidestick of Airbus A330. (Source: BEA Report of AF447)

the engines to produce sufficient thrust at the nose-up attitude demanded by Bonin's manual input. The nose of the aircraft was then at approximately 12° above the horizon but neither pilot noticed that the aircraft was flying dangerously slow. After 46 seconds from losing reliable airspeed information, the aircraft airspeed reduces to a point where it cannot maintain sufficient lift. The aircraft has left its safe flight envelope and the airflow over the wings is now significantly disrupted—it has stalled. The stall warning activates, and the aircraft begins to descend rapidly toward the sea.

Descending over the dark Atlantic Ocean with little or no visual clues, Bonin and Robert are totally dependent on their flight instruments, but what the instruments are telling them is making no sense to them. The aircraft's attitude is now at 15° nose up with full power on both engines, but the aircraft is descending rapidly at around 10,000 ft/min. Descending with the nose-up attitude means that the aircraft's angle of attack, or alpha,[v] the angle between the wings and the airflow, is at 40°, and it is in a continuous stall. This condition produces vibration through the airframe, known as

---

[v] The angle by which the nose of the aircraft is pointing is the attitude. That is what is displayed to the pilots but is not the same as the angle of attack relative to the airflow: that is "alpha." At 40° alpha, the aircraft is in a stall and would have to reduce to well below 0° alpha to recover. From this point in the flight, AF447's alpha never goes below 35°.

buffet, and noise from the disrupted air. From the cockpit voice recorder (CVR) transcript, it appears Bonin and Robert initially interpret this "airflow" noise as an indication the aircraft is overspeeding—in fact, the exact opposite is occurring. Despite having full power and a nose-up attitude, the lack of forward airspeed meant the aircraft is too slow and therefore descending uncontrollably.

A minute and a half after the loss of airspeed and autopilot disconnect, Captain Dubois returns to the flight deck. He is faced with two confused and frightened copilots as well as an array of sounds and warnings, including an intermittent stall warning. Robert tells him they have lost control of the aircraft. The aircraft is in a stall, but the crew fail to recognize any of the classic stall characteristics, despite the audible and visual warning. The crew try to diagnose what is happening, but no one mentions even the possibility of a stall—they cannot see any of the crucial signs. Dubois cannot understand why the aircraft is behaving in this manner, but he lacks one piece of crucial information—Bonin is still pulling back on the sidestick and maintaining the aircraft in a continuous stall. By the time the crew begin to realize what is going on, it is too late to initiate a recovery. At 02:14:28 UTC, 4 minutes and 23 seconds after the initial upset and autopilot disengagement, AF447, an Airbus A330-200 with 228 passengers and crew onboard, crashes into the Atlantic Ocean.

1.1.2   The AF447 Accident Report

In its comments on the causes of the crash of AF447, the BEA report notes that the philosophy of aircraft certification is predicated on the assumption that when non-normal situations occur, the operating flight crew will take appropriate action. In its summary of the causes of the accident, the report describes the output of the aircraft's design, initial crew training, and recurrent training that "... does not generate the expected behavior in any acceptable reliable way."[3] The report infers that the crew's unexpected, prolonged reaction to the initial event undermines the whole certification process of commercial aviation. In effect, it puts the whole commercial aviation safety model into "common failure mode."

The comment could be interpreted to suggest that if we are to determine a cause for this tragic event and prevent its recurrence, then a detailed analysis of the role and the performance of the humans involved is where we can find the answer and the key to prevent future accidents. However, the broader recommendations of the report and subsequent related investigations suggest that the role of the crew only provides one aspect and one perspective on this tragic accident.

The three pilots—Dubois, Robert, and Bonin—were aware of the mortal danger they faced and yet were unable to diagnose what some commentators subsequently judged to be obvious. They faced a frightening and rapidly evolving situation which, although exaggerated by inappropriate, subconscious, and intuitive responses, was typically and predictably human. As Gary Klein[vi] has highlighted through his extensive research into decision-making that when experts are placed under conditions of stress, it is not their ability to make rapid and accurate decisions that fails, but their ability to make sense of their environment. Imagine, for example, being awakened at night by a loud bang. Your first thoughts may race to the conclusion that there is someone breaking into your

---

[vi] Gary Klein is one of the most influential researchers in decision-making following his work within the military, firefighting, and aviation. See: Klein, G. (1996). The effect of acute stressors on decision making. In J. E. Driskell & E. Salas (Eds.), *Stress and human performance* (pp. 49–88). Hillsdale, NJ: Lawrence Erlbaum Associates, Inc.

house; so you race to get dressed or call the police—or in panic try to do both! It takes a little time for you to control your breathing and recall that there was a forecast of strong winds and the yard gate often unlatches in strong wind conditions. Situations can arise where even experts (or sometimes especially experts) cannot build sufficient situational awareness under time pressure to enable rational choice.[vii]

This accident produced a high level of shock around the world—that a modern and capable aircraft, operated by a reputable airline with a highly trained crew, could simply crash into the sea. With such global interest, the high-profile AF447 accident stimulated considerable research and commanded extensive resources. These resources didn't stop at the estimated $160 million cost of the 2-year recovery program of the airframe from the bottom of the Atlantic Ocean but extended to the extensive number of simulated research and safety review programs across the global aviation industry. The analysis of the broader implications of the accident is still ongoing, many years after the immediate media demands for explanations.

1.1.3   Other Stories of AF447

Humans like to be told stories. Stories can allow people to understand their complex lives and give meaning to life's events. Stories enable people to move on after events like the crash of AF447 which shake their beliefs in establishments and systems like commercial aviation. The official accident report, written by the BEA, is a story: a version of the truth, albeit a version authored by respected subject matter experts and based on established scientific methodologies and evidence.

Many other stories appeared in mainstream and industry media sources that relied far more on speculation and supposition. Some of these stories placed blame on the competence or even the lifestyle of the AF447 crew. One editorial comment in a reputable international aviation magazine suggested that since Robert's seat was found by the crash investigators motored back, this suggested a lack of professional rigor. If we read the accident report and take the time to contextualize Robert's behavior, it is more likely that as his captain, Dubois, re-entered the panicked flight deck, Robert, simply acting on instinct, motored the seat backward thinking Dubois would immediately retake his place in the left-hand seat. If we aim to learn authentic lessons from such tragic accidents, we should rather than selecting facts that suit our retrospective view of a situation accept that the evidence will not always produce a neat, logical, and sequential story for us to take away. We have to be directed by the evidence which may be uncomfortable, messy, and complex, but following the evidence means that we are able to improve the system and ultimately reduce the probability of future disasters.

If we were to conclude that the flight crew of AAF447 were in some way ill-equipped to deal with a transient technical fault, then it raises some profound questions about the regulatory status of all professional pilots. Should we perhaps consider why and how the loss of situational awareness transpired? What training had they undergone and of what quality? How did they compare with others in their organization and across the broader global profession? The BEA report cites related research into airline crew's response to similar scenarios and concludes this incident was not simply a one-off but more a representation of a total system failure.

[vii] We have far less control of our actions under stress than we might think. For an excellent introduction to behavior under stress, try Peters, S. (2013). *The chimp paradox: The mind management program to help you achieve success, confidence, and happiness.* TarcherPerigee.

1.1.4   A Deeper Understanding

To develop a better understanding of the loss of AF447 within the context of commercial safety, we may wish to investigate the background of the system in which this accident occurred rather than purely on the immediate events prior to the accident. As we will discuss in Part IV of this book, to learn how to reduce the probability of complex accidents like this happening again, we must look not only at the proximate events and the proximate actors but to the emergent properties of the total system in which they operate. Given the long lead times between design, manufacture, and years of operation, these properties can have incubation periods lasting for decades.

There are substantial lessons to be learned from the human factors at play during events like the loss of AF447, but equally important are the lessons from the character of the operating organization, the cultural influences,[viii] the regulatory structure, and the design characteristics of the aircraft and those of its manufacturer and maintainer. After the publication of the BEA report, the French prosecutor received evidence that some Airbus A330s had experienced recurrent faults in the pitot heating system for over a year before the AF447 fatal accident. Unless we begin to understand the dynamics and reasoning behind the decision makers that influence the serviceability of modern commercial aircraft, we are not really addressing the underlying causes within the system in which they operate. Air France and Airbus are currently being tried in the French courts for corporate manslaughter. The legal process may now provide an alternative perspective and explanation of the accident that was not covered in the BEA report.[ix]

To quote the International Civil Aviation Organization (ICAO) Safety Management Manual, "Aviation safety is a dynamic concept, since new safety hazards and risks are continuously emerging and need to be mitigated. Safety systems to date have focused largely on individual safety performance and local control, with minimal regard for the wider context of the total aviation system. This has led to growing recognition of the complexity of the aviation system and the different organizations that all play a part in aviation safety. There are numerous examples of accidents and incidents showing that the interfaces between organizations have contributed to negative outcomes."[4]

Another characteristic of accident analysis is the difference in timescales between the desire for explanation and the scientific validation of evidence. The appetite for information in the immediate aftermath of an air accident or major incident runs at an entirely different pace to the necessary scientific investigation and analysis; the media satisfy the former while industry professionals and safety researchers must wait for the outcome of the latter.

The void between the two timelines often creates different explanations—different stories. After accidents, frequent references are made to the cause or sometimes causes

---

[viii] Air France commissioned an independent review board to assess and report on its own safety culture. The Independent Safety Review Team (ISRT) was established in December 2009. It consisted of a number of global (non-French) aviation safety experts and produced 35 recommendations for Air France. Despite an initial commitment by Air France, the documents containing the recommendations have never been released to the public. Source: http://news.aviation-safety.net/2011/02/09/air-france-acts-on-independent-safety-review-team-recommendations.

[ix] In 2014, multiple manslaughter charges were brought by French prosecutors against Air France and Airbus for their alleged role in the loss of 228 lives. Those charges were dropped by French prosecutors in 2019; however, on the 12th of May 2021, the crash victim's association, headed by Daniele Lamy, won an appeal resulting in fresh charges. As of November 2022, Air France and Airbus are facing the prospect of a criminal trial for manslaughter.

of events, but the sheer complexity of interactions confounds what the public really wants, that is a simple explanation. This aspect of the "accident story" is not merely trivializing the rigors of safety investigation. The story is part of the process that is needed to allow human intuitive tendencies to establish a rapid, simple, and satisfactory understanding, with which society can move on.[5] It is a basic human need, but one we must be mindful of this need.

The mechanism of commercial aviation safety is not a simple story—it is complex, is nuanced, and is rapidly evolving with new technologies and new system properties. The AF447 accident serves as an illustration of how gaining insight into why accidents happened requires a multidisciplinary approach. Effective safety analysis requires a deep understanding of the technology, the interaction of that technology with humans, how humans interact with each other, how safety performance is influenced by the various organizations and stakeholders, and ultimately how this complex system functions as a whole entity. This entity was not designed; it has evolved, and it is still evolving.

1.2   Safety Evolution

1.2.1   Four Eras of Commercial Aviation Safety

This multidisciplinary approach is reflected in the established source of internationally recognized standards regarding safety management in commercial aviation, the ICAO Safety Management Manual.[6] Now into its fourth edition, the document describes four eras of safety management based on contemporary aviation safety knowledge. As described in Figure 1-5, each era represents where the primary focus of safety analysis has emerged and how this focus has developed over the decades.

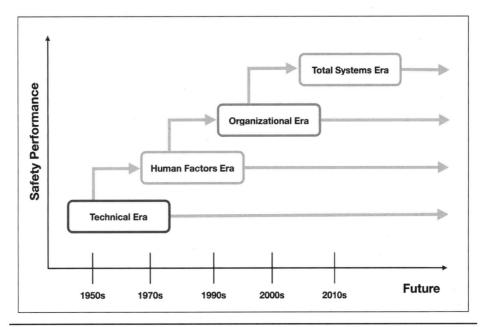

FIGURE 1-5   The evolution of commercial aviation safety analysis. (Source: Authors)

From the early 1900s to the late 1960s, the technical era saw mechanical failure as the primary reason for aircraft accidents. From the early 1970s to the mid-1990s, the human factors era emerged after a significant number of perfectly serviceable aircraft were involved in fatal accidents and investigators started analyzing the limitations of the humans who operated them. From the mid-1990s to the start of the 21st century, the organizational era saw increasing recognition of the role played by the mechanisms and cultures of the operating organizations. Subsequently from the start of the 21st century to the present day, the total system era started considering a more holistic perspective—seeing the complex interdependence of commercial aviation as a total entity.

Improvements in aviation safety performance attributed to the first two eras, technical and human factors, have seen an associated development of regulatory regimes and safety philosophy, which can specify performance standards. The latter two eras present considerably more challenges to the traditional regulatory regimes. These challenges will be discussed throughout the book as they indicate how commercial aviation is likely to evolve in the future. After addressing some of the more fundamental concepts of safety in Part I, this book will predominantly focus on the technical era in Part II. It will focus on the human factors era in Part III and consider the evolution and challenges presented by organizational and total system issues in Part IV. The following section presents a brief overview of each era of commercial aviation safety.

1.2.1.1 The Technical Era

The evolution of safety standards witnessed over these past few decades has its roots stemming back to the very beginning of commercial aviation in the early twentieth century. On the 17th of September 1908, the propeller of Orville Wright's "Model A" struck a control wire during a demonstration flight; the aircraft crashed, killing Wright's single passenger, First Lieutenant Thomas Selfridge. The US Army appointed Frank Lahm as the accident investigator, thus starting over one hundred years of lineage in investigating aircraft accidents.

Air travel was seen in its early days as a risky but glamorous activity for the rich and adventurous. A handful of early adopters of this nascent technology of flying began scheduled commercial flights as early as 1914 between St. Petersburg and Tampa, Florida. The First World War saw a period of rapid expansion in not only the number of aircraft but in the scale of innovation. In 1921 the US Army was the first to keep statistics on aviation accidents. They recorded those 361 major accidents that occurred in 77,000 hours of flight during one single year. The accident rate was approximately one for every 215 hours flown. Aviation was indeed a dangerous occupation by any modern standards.

As demand for air travel grew, the need to improve reliability became apparent. In 1924 there was a fatal accident for every 13,500 miles flown by commercial flyers—if one were to compare that to today's volume of flying, that would equate to over 8000 fatal accidents per year. After regulations including regular inspections and engine overhauls were introduced to the Air Mail Service of the United States, the fatal accident rate dropped to one fatal accident per 789,000 miles.[7] Significant resources from governments around the world flowed into aircraft development as the potential of aviation as a mode of transportation became apparent.

Not only was commercial aviation seen as an industry with potential for massive growth but also as a representation of national technological prowess and power projection. By the Second World War, aviation was recognized not only as a crucial asset for

FIGURE 1-6   A TWA 1958 Lockheed Constellation at New York's JFK Airport. This aircraft was once the pinnacle of aviation technology. (Source: Zhukovsky, 2019, New York, Dreamstime)

warfighting but also as an essential logistical network. The costs of early transatlantic flights such as the Lockheed Constellation, shown in Figure 1-6, were approximately tenfold more expensive than the costs today, but more and more government investment saw the development of faster, cheaper, and more reliable aircraft. By the 1950s, the jet engine was emerging as a more powerful and safer replacement to its piston predecessor. Aircraft could now achieve reliable and safe long-distance intercontinental flights cutting travel times down to hours rather than days. Government-backed innovation continued and accelerated through the enhanced national competition encountered during the Cold War. Not only did aircraft capability benefit from purely military applications but government-backed research and development evolved into high-altitude and supersonic flight.

Part II of this book will take a close look at today's legacy from the technical era, portraying how aircraft safety systems, airport design, and air traffic systems all help create a safety net in which commercial aviation operates.

By the late 1960s and into the 1970s, another political stimulant motivated better and safer aerospace technology—the national airline. Huge prestige was gained by the richer and more technologically capable nations for being able to produce and operate the latest commercial aircraft. These aircraft had to be not only technically capable but also reliable and above all safe. Whenever aircraft accidents occurred, particularly involving multiple fatalities, governments went to considerable lengths to determine the cause of crashes in order to prevent recurrence. This strong association with government agencies, particularly the military, throughout the evolution of commercial aviation meant the adoption of rigid and hierarchical operators, governed by stringent regulation.

Senior politicians wanted to ensure their reputations were not sullied by cavalier business practices where political currency was so high. Low flight hours and poor-quality training were generally seen as major causes of losses in military aviation; thus, government-backed safety programs and regulations were initiated. These new and well-resourced courses lasted for years and produced professional qualifications in flying, engineering, navigation, and air traffic control. The vast military establishments of the mid- to late 20th century produced experienced and technically competent personnel who were able to supplement the output from civilian training organizations.

1.2.1.2   The Human Factors Era

In the 1970s the commercial aviation industry was well established with multiple operators vying for the growing demand for international air travel. They operated the latest aircraft technology with aircraft now capable of carrying hundreds of people for thousands of miles. International jet transportation was almost becoming commonplace—that is, until it went drastically and spectacularly wrong. The size of the modern aircraft and the associated number of passengers meant aircraft accidents were often major disasters. Despite having the benefit of highly reliable aircraft and well-trained operators, accidents were still happening. Not only were they prevalent but the profile of the commercial aviation accident was growing.

This increased public awareness of safety was not necessarily borne of failures within the industry. As the mass fatalities associated with the two World Wars faded from the public's mindset, societal tolerance of risk was changing. A commercial aircraft accident had always been politically damaging and bad for business but increasingly the stakes were higher. New technology had always been the hallmark of aircraft development and now that spirit of innovation focused on a new era of accident investigation. Investigators started to notice something different in many recent accidents: up until the crash, there was nothing technically wrong with the aircraft.

Aircraft accident investigators were generally chosen for their technical expertise rather than their behaviorist knowledge.[x] They developed increasingly sophisticated techniques which described in great detail what happened to doomed aircraft and their unfortunate occupants. What was often the more complex question was why an aircraft's crew behaved in such a way that resulted in a crash—a particularly difficult question if the crew did not survive. Although safety regulators began to mandate the carriage of CVRs, the deeper motivations were often the most obscure but offered critical insight into an accident. Accident reports were technologically comprehensive but not necessarily insightful. They competently described what happened but gave few clues as to why and therefore what lessons could be learned. New techniques in risk management across many business sectors were already starting to recognize the significance of human behavior and its influence on commercial performance.

The human element of aircraft control engineering and design began to appear during the Second World War but rapid developments in psychology and behavioral science throughout the latter half of the 20th century were able to enhance and manipulate the vagaries of human behavior research. This is an important concept which was adopted under the umbrella term of Cockpit Resource Management (CRM), which was intended to promote best practices for NASA test pilots and space mission crew.

---

[x] The highly respected Air Accidents Investigation Branch of the United Kingdom recruited their first internal human factors expert in 2017.

FIGURE 1-7   Enhanced crew coordination and communication training greatly improved safety. (Source: Pixabay Jonny Belvedere)

Soon, these CRM programs were filtering into airline training courses. These courses formally recognized the requirement for a high degree of crew coordination and communication as described in scenarios such as the engine start seen in Figure 1-7. By the end of the twentieth century, CRM (now referred to as *crew* resource management) was becoming mandated for flight and cabin crew air traffic controllers and engineers. Early training programs gave candidates insight into what accident investigators were focusing on in terms of accident causation: the human factor.

Psychological profiling tests started to challenge some of the more traditional assumptions on what was effective or even acceptable behavior for aviation professionals. As many individuals and many behavioral traditions had evolved from the military, the rigidity of command hierarchy present on commercial flight decks was challenged. An early "target" of these training initiatives was the traditionalist, authoritative, and dominating captain. The formality and deference associated with some traditional command styles were seen as major inhibitors to effective teamwork on the flight deck, in the cabin, in the air traffic center, and in the engineering development base. Command gradient error (also known as trans-cockpit authority gradient) was cited as a significant contributory factor in numerous accidents through inhibition of free communication between team members. Although many of these early attempts to challenge embedded behavior and cultures did not go without pushback, the tide had turned away from the hierarchical and technocratic emphasis of commercial aviation toward a more inclusive environment in the cockpit. Part III of this book will explore the topic of human safety in detail, discussing whether humans are the problem in safety or potentially the solution for accident prevention.

1.2.1.3 *The Organizational Era*

Tying the evidential outputs from accident investigation into training objectives for aviation professionals might seem to be an obvious strategy to improve safety performance; however, this process is still under development. Further and more contemporary research into human factors started to recognize that an individual's behavioral patterns do not evolve in a bubble. Humans are sociable creatures and develop much of their behavioral traits from observation and engagement with others.

Safety researchers started focusing attention on the role of organizations themselves in accident prevention in the late 1980s and early 1990s as a function of several high-profile events. The nuclear accident at Chernobyl in 1986 called attention to how a poor safety culture had contributed to such a nuclear disaster. Similar events, such as the King's Cross Underground fire in London in 1987, the Piper Alpha oil platform explosion in the North Sea in 1988, and the crash of Continental Express Flight 2574 in Texas in 1991, all highlighted the critical role of management and organizational dynamics in accident causation.

Lisa Feldman-Barret, a researcher from Northeastern University, is among the top 1% of the world's most cited scientists. Her work on the brain and affective sciences has led her to suggest certain areas of our neural topography are designated as redundant. She suggests that this is likely to allow us to adopt and absorb cultural norms into our thought processes, allowing us to benefit from common knowledge and wisdom. Rather than invent everything ourselves in every new environment, we can draw on the experience of others. The significance of the environment in human decision-making has spawned interest and research into the influence of organizations in aviation safety. Organizational safety culture has now been established as a major causal influence in numerous accidents, but the industry (and many others) is still coming to terms with how best to effectively invoke this insight into regulation or industry's best aviation safety practice.

1.2.1.4 *The Total Systems Era*

The adoption of new practices to constantly enhance safety in commercial aviation is one of the many features of the high-performing system that we see today. The constant evolution of commercial aviation safety has seen it adopt and adapt performance-enhancing techniques and procedures throughout its history. The early utilization of military structures and procedures ensured compliance. Stringent regulation meant adherence to very exacting standards of engineering and operation. Government resources and political influences ensured an almost constant supply of new technology and advanced techniques in accident investigation, which promoted constant learning. The self-reflection of the industry has meant not only changes in emphasis but also changes in direction as scientific research has provided opportunities to make incremental gains and occasionally great strides forward in safety performance. Over the last 15 years this self-reflection has engendered a focus on safety as the property of a system versus an individual component. The total system era looks at accident prevention as the cumulative effect of interaction of components throughout their life cycles, asking whether otherwise highly reliable components can actually be at odds with safety. It recognizes that accident causes are complex and involve numerous socio-technical factors. One theme that has remained constant throughout this evolution in safety management is how best to promote the importance of individual professionalism and attention to detail, as illustrated in Figure 1-8; it is this collective mindset which underpins the whole system.

Figure 1-8   A complex system of protections needs constant attention. (Source: Pixabay ScottSLM)

In an effort to understand how this incredibly successful and intricate system really works, it is unsurprising that subject material and research often overfocus on examples of spectacular failures; after all, that's what normally piques the attention of society and the world's media. However, the irony is that despite the focus on disaster, the total system is a spectacular success. Although there are considerable lessons and legitimate questions raised by accidents like AF447, to grasp how this complex system really functions we should, perhaps, start with an entirely different question: Why does it almost always go so right?

1.3   The Right Stuff

1.3.1   Commercial Aviation Today

Since those very early days of the Wright Brothers at Kitty Hawk and of the Brazilian Alberto Santos-Dumont flights in Paris, aviation has come a long way. Today, the aviation industry supports a lot of people and results in substantial revenue. The International Civil Aviation Organization (ICAO) is the agency of the United Nations which monitors and regulates today's international commercial aviation system. One of its roles is to maintain accurate statistical records of the global civil aviation system's performance.[8]

Prior to the COVID-19 outbreak in early 2020, commercial aviation had generally seen decades of sustained growth and was preparing for decades more of continuous expansion. ICAO had forecast a doubling of commercial air passengers between 2020 and 2040 to reach an annual total of approximately 10 billion passengers per year. That forecast has been invalidated by the recent COVID global pandemic which has obviously had a massive impact on the industry. The peak of growth in commercial aviation was therefore reached in 2019. In this year, the ICAO records show that there were approximately 4.5 billion commercial air passengers: that's the equivalent of flying more people than the population of the earth, every 2 years. These passengers flew on 38 million flights; that's the equivalent of more than one flight taking off every second throughout the year. Total global air freight amounted to some 57.6 million metric ton: that's the equivalent mass of 150 Empire State Buildings. Commercial aviation has indeed been a massive and growing industry.

1.3.2   Safety Statistics

What is even more remarkable are the statistics that focus on the commercial aviation system's overall safety performance. During 2019 there were only a total of 115 aircraft accidents of scheduled commercial air transport operations.[xi,8] This resulted in a global commercial aviation accident rate of three accidents per million departures. Throughout 2019 there were a total of six fatal accidents resulting in 239 deaths. In contrast, according to another agency of the United Nations, the World Health Organization, every year there are 1.35 million people killed on the roadways of the world.

With such low sampling data from fatal accidents in commercial aviation, annual statistics can be misleading without looking at broader trends to normalize the data. For example, if in the early years of aviation, we have a year where the world experiences 100 fatal aviation accidents, followed by a year where 103 fatal aviation accidents take place, it suggests a marginal increase in risk over the 2 years. However, if we have one year where there is one fatal aviation accident, followed by the next year when there are three fatal accidents, it suggests a massive 300% increase in risk over the 2 years. That might make a great headline story but doesn't necessarily represent safety performance or the longer-term trend. For a true comparison you have to normalize the data by flight hours, distance flown, or numbers of passengers. Even with a consistent overall improvement in safety performance, year on year safety metrics can vary significantly. As an example, in 2017, of the 4 billion air passengers on scheduled flights, only eight were killed; a year later in 2018 that figure had risen to 474.[9] Professor Arnold Barnett of the Massachusetts Institute of Technology is a global expert on aviation safety statistics. He recommends that a better perspective of safety performance trends is achieved from the review over decades rather than annual figures. He has calculated that the worldwide risk of being killed in a commercial aviation accident has dropped by a factor of 2 every decade, but in the last decade that reduction has become closer to 3. When comparing the fatality risks of flying, Barnett also prefers to look at the number of deaths per numbers of people boarding an aircraft rather than deaths per mile or per flight. In his own words, this is because "… it literally reflects the fraction who perished during air journeys."[9]

What Barnett's analysis shows us is that statistically commercial aviation safety has continually and consistently improved to levels of system performance that were previously unimaginable. Since the late 1980s the system has seen a doubling in safety performance every decade. From an actuarial science perspective, commercial aviation has been described as an ultrasafe system, defined as a system with accident rates that

---

[xi] These figures apply to commercial air transport aircraft with a certified takeoff mass of over 7500 kilograms.

have fallen to between one accident per 10^5 events to one accident per 10^6 events.[xii] Since Amalberti's 2001 paper[10], in some parts of the world safety performance has improved by an order of magnitude, with accident rates on scheduled commercial aircraft approaching one accident for every 10^7 flights.

Between 1988 and 1997, global accident statistics[11] showed that there was one passenger death for every 1.3 million people that boarded a commercial aircraft. During the next decade that figure had improved so that between 1998 and 2007 there was one death for every 2.7 million passengers. Between 2008 and 2017, airline passenger deaths fell to one in 7.9 million passengers boarded. Professor Arnold Barnett calculates that in the highest performing regions, such as North America or Europe, areas that he describes as "traditional first world," a passenger who boards a random commercial flight every day would take approximately 79,000 years before being killed in a commercial aircraft accident.

1.4 Why It Goes So Right

1.4.1 Suggestions as to Why

How commercial aviation safety has developed to such advanced levels of safety performance is not an easy question to answer. In a little over a century, it has evolved from a nascent field of transport research into an industry that managed to produce $841 billion per year[12] in revenues and is ultrasafe. Some commentators have suggested the culture in commercial aviation rates highly when compared with some of the attitudes and practices in other industries. Unfortunately, willful blindness or insufficient levels of psychological safety have been highlighted as underlying causes to numerous catastrophes in various sectors.

While these explanations of system performance have considerable merit, the unique journey and nature of commercial aviation suggests a more contextual explanation of the system's success. Other industries have looked at commercial aviation and attempted to mimic some of the characteristics which are thought to have contributed to its current enviable safety status. But like most complex issues, the reasons for the system's performance are multiple and interrelated. The whole system has changed over time allowing adaptation and lessons to be learned. In the following sections, we suggest some plausible reasons why the safety performance of commercial aviation safety has continued to improve. The section is then followed by "Safety Evolution," a chronology of how the industry has evolved over the years, not only as new technologies have emerged, but as new concepts and philosophies have been developed and adapted to improve system performance. Throughout this book we will continue to challenge and discuss whether commercial aviation safety is sufficiently receptive and adaptive to continue and even improve on its very impressive performance as the rate of technological change accelerates.

1.4.2 Fly by Rules

One of the most commonly cited reasons for such high standards of safety performance is the level of stringent regulation that is applied around the world at very consistent levels. Thought to be the most consistently regulated of industries, the laws and

---

[xii] Amalberti refers to "events" being contextual to the activity in question, e.g., the number of flights, rail journeys, or sailings.

standards governing aviation at the international level are prescribed by ICAO. We shall look at ICAO in more detail later in the book but this is a good place to start when trying to understand the regulatory mechanisms of commercial aviation. The ICAO system allows national governments to adapt their varying levels of resources to comply in certain key areas, namely safety standards, while giving more flexibility in less critical areas. The regulatory structure emphasizes the international nature of the industry. Without a great deal of international cooperation in a global infrastructure, safety standards would become inconsistent and, in some cases, meaningless.

Even by the standards of other safety-critical industries such as medicine, nuclear power, or petrochemical production, commercial aviation is exceptionally highly regulated. However, while conventional regulation has proved to be a successful mechanism to measure and manage technical and human performance standards, questions have been raised about whether it can be effectively applied to organizational and system safety. Le Coze and Wiig[13] suggest that "overproceduralization" can be counterproductive in certain types of complex systems. Significant doubts exist that some of the increasingly important characteristics of complex systems can be effectively measured and therefore managed by prescriptive regulation. Therefore, future regulations should gravitate to a performance-based approach as compliant systems do fail. Compliance is necessary but not sufficient. In their published paper, on the relationship between safety culture and criminalization, Lawrenson and Braithwaite[14] suggest regulators should consider a more contextual approach to the management of safety culture in commercial aviation.

1.4.3   Political Currency

Commercial aviation holds significant political currency in national and international fields. There is a considerable amount of political will which not only drives the cross-border regulations but also ensures the industry is comparatively well resourced and therefore acceptably safe.[xiii] Airlines are seen as the lifeblood of modern economies, moving affluent and influential businesspeople around the international economic community and carrying large volumes of time-sensitive cargo throughout the globe.

In the latter part of the 20th century, the national airline emerged as a symbol of state pride and technological prowess. With the notable exception of the United States, the national airline became the *de facto* ambassador of the nation state. The national airline became the must-have ticket to the international community, so a consistently safe operation was an essential element of national pride. Many national airlines are still government owned and all have at some stage relied on their political masters for financial or policy support. Governments have also been selectively beneficial to their aerospace manufacturers, a policy which has occasionally initiated political controversy over international state aid regulation. The most notable and high-profile example in the long-running competition between the two global leaders in commercial aviation manufacturers is the case of Boeing and Airbus.[xiv] While competition plays an essential role in promoting innovation, the entrenchment of differing design philosophies cannot always be the optimal way forward to enhance safety performance. However, this

---

[xiii] The relationship between resources and safety performance is well illustrated in Arnold Barnett's (2020) comparative analysis of global safety statistics.
[xiv] For an engaging and comprehensive discussion on this long-running industrial feud, see *Boeing versus Airbus* by John Newhouse (2008).

does not suggest commercial interests and safety enhancement cannot be complementary. Innovation is a key requirement for both resilient operations and business efficacy. Throughout this book, we will consider how commercial considerations can be effectively deployed to enhance safety performance.

1.4.4   Constant Innovation

Throughout this book we will provide examples of how aviation has benefited from the almost constant development and refinement of new technologies. Because the average research and development cycle in major aviation production is approximately 20 years, government funding is usually essential to keep national aviation industries from succumbing to international competition.

In a European Commission report[15] on the European vision for aviation in 2050 it was estimated that over €250 billion of public and private money was needed for European research and development in the sector. Innovation has been a hallmark of the aviation industry which is often an early adopter of new and emerging technology. As an example, consider car braking systems such as Maxaret, which rapidly apply then reapply brake pressure preventing the wheels of an aircraft from "locking-up." Such early systems reduced aircraft stopping distances by approximately 30% not only allowing greater takeoff weights but increasing safety margins. Heavier aircraft could now use shorter runways and the emergency stopping distance of aircraft was considerably reduced. The braking system, developed by Dunlop, was adopted by aviation manufacturers in the early 1950s but did not appear on commercial car production until the 1970s. Research and development programs run by national governments have meant that commercial aviation has benefited from spin-off technologies initially intended for the military, advanced aerospace, and space flight operations.

1.4.5   Fear of Flying

Whatever the statistical description of commercial aviation safety might look like, there is no doubt that for many people, even the remotest possibility of being involved in an aircraft accident is a very real and visceral fear. The commercial aircraft accident has evolved within our cultural landscape as a rare but nonetheless terrifying event. There is no end to dramatic movie scenes where our otherwise safe and controlled modern lives are placed in jeopardy by a catastrophic safety event on commercial aircraft. No number of statistical memes about heart attacks, cancer, or crocodiles will mitigate the fear that commercial flying invokes in significant sections of society.

The media have no small part to play in accentuating this fear as they know there is a willing and captivated audience. Behind the coverage of the February 2022 Russian invasion of Ukraine and the January 6, 2021, attack on the US Capitol, CNN's most widely followed story in their broadcasting history was the loss of Malaysian Airlines MH370 in 2014. From an objective and statistical perspective, this fear is irrational but is prevalent at least partially due to a very common mental shortcut, or heuristic known as availability.[16] Availability means that humans can often distort the risk from some activities because they are too easy to recall. The high cultural profile of the commercial aircraft crash makes it very easy for humans to wildly overestimate our chances of being injured or even killed, despite the overwhelming evidence to the contrary.

Accidents are bad for any business, but airlines are particularly vulnerable. Fear of flying is one of the most common phobias and is a sensitive issue in the marketing of airlines. When an engine exploded on Qantas Flight (QF) 32 shortly after takeoff from

Singapore, the company's share price immediately dropped 5% (AU$200 million) based on the mere rumor that the aircraft had crashed while the aircraft's crew were still successfully dealing with the emergency above the skies of Singapore.[17] One study[18] looking at the impact of commercial aviation accidents on stock market prices found stock market prices significantly fell after a major aircraft accident. Those losses could exceed 60 times the actual economic loss, suggesting there is a deeper psychological impact on society. The researchers Clark and Rock[19] reported that studies have recorded flying-related anxiety affecting up to 40% of the population, with approximately 2.5% of people having a clinically diagnosed condition.

Given the potential impact of fear on the commercial viability of commercial aviation, it is undoubtedly an area in which airlines need to tread carefully. Not only do commercial aviation operators need to be safe, but they also need to be seen to be safe. But fear is not necessarily a negative influence on the industry, it can also be a great motivator toward improving and maintaining safety performance. The American neuroscientist Joseph LeDoux has commented that fear is the most powerful motivator in the human psyche. Fear experienced by those individuals who are responsible for the safe operation of aircraft has been highlighted by safety researchers.[20,21] Some have noted the desirable state of chronic unease among safety managers and accountable executives as an effective and positive influence on safety performance. The chronic unease of those individuals who have a direct influence on commercial aviation safety is considered by some as an effective way by which safety policy and regulation are rigorously enforced, safety innovation is supported, and constant vigilance is maintained.

1.5   The Impact of COVID-19

1.5.1   The Face Value

Given the enormity of the COVID-19 global pandemic, it is entirely appropriate that we assess the impact of this outbreak on commercial aviation. Given the nature of commercial aviation, it has been one of the most affected sectors in terms of its ability to operate under the circumstances. The International Air Transport Association (IATA) represents some 290 of the world's airlines comprising 82% of the global air traffic. In a July 2021 press release, the IATA Director General, Willie Walsh, summarized IATA's view of the situation by saying, "We are seeing movement in the right direction, particularly in some key domestic markets. But the situation for international travel is nowhere near where we need to be. June (2021) should be the start of peak season, but airlines were carrying just 20% of 2019 levels. That's not a recovery, it's a continuing crisis caused by government inaction." Despite similar criticisms from across the industry of the regulations that restrict international travel, governments have generally been consistent in their conservative approach to COVID-19 border controls. Whatever the relative merits of government policy and operators' frustration, the pandemic numbers paint a bleak picture.

1.5.2   The Cash Value

The COVID-19 pandemic resulted in an unprecedented decline in the world's commercial aviation activities, which despite extensive vaccination programs declined by over 40% for 2021. As suggested by the image of London's Heathrow airport (Figure 1-9) during the height of the pandemic, the world's airline flying schedules were extensively reduced.

FIGURE 1-9   The pandemic's impact on aviation has been unprecedented. (London Heathrow, March 2021, Source: Authors)

The impact on the industry has dwarfed the combined effects of the 1973 oil crisis, the 1982 Iran-Iraq war, the 1991 Gulf crisis, the 1997 Asian crisis, the 9/11 terror attacks, the 2003 SARS outbreak, and the 2008 financial crisis. The total loss of income from the reduction in passenger numbers since the beginning of the pandemic is estimated by ICAO to be in the region of $742 billion. Even the more optimistic economic scenarios put forward by ICAO predict that after the second year of the COVID-19 outbreak, commercial aviation will still be operating approximately 20% below 2019 performance figures.

1.5.3   The Impact on Technical Safety

As we have stated, the COVID-19 pandemic is an unprecedented and long-term event. The international system of commercial aviation has simply never experienced anything like this level of disruption to its operation. The world's airlines have reduced commercial passenger capacity by over 50% for sustained periods, creating a whole new variety of threats to aviation safety.

Bloomberg reported in April 2020 that 16,000 of the world's 26,000 commercial aircraft had been grounded. The grounding of two-thirds of the world's commercial fleet for long periods of time brings a whole new array of technical problems. It is not simply the logistical nightmare of trying to find parking slots for so many big aircraft. These complex and intricate machines were never designed to spend substantial time on the ground. Their normal working cycle involves being filled with tons of aviation fuel, engines started, hulls pressurized, and flown when the whole aircraft literally stretches itself out. They fly to altitudes where the humidity is practically zero and the ambient temperature sits at approximately −50°C (−58°F); there can be fewer more efficient ways to sterilize a 300-ton machine than taking it to 40,000 feet (12,192 meters) for a day out.

Commercial aircraft need lots of work and attention while in storage, and their hydraulic and flight control systems need to be regularly serviced. The humidity erodes parts and electrical components at much higher rates than original maintenance programs were designed for. The aircraft also make attractive homes for wildlife. The large structure makes ideal nesting spots, and the various nooks and crannies are great for harboring families of insects as well as bacterial growth. Although this situation creates a lot of fun and novelty for this new wave of wildlife inhabitants, it is less than ideal for aircraft serviceability and airworthiness.

1.5.4   The Impact on Human Safety

As well as grounding their aircraft, commercial aviation operators have been forced to significantly reduce their workforce. Professionals engaged in commercial aviation often take years to train and can be very expensive. The skill-sets involved in flying, maintaining, and controlling aircraft are highly tuned and require regular assessment and practice. Skills which might be second nature in a normal environment, such as preparing a crew briefing, delivering an Air Traffic Control (ATC) clearance, or carrying out a routine inspection, can suddenly become more onerous, time consuming, and inevitably vulnerable to error.

A phenomenon known as "skill fade" has been regularly discussed within operator training programs and on social media forums. One safety performance metric used by safety analysts is the "unstable approach," where pilots have been unable to get the aircraft to the correct speed and height to ensure a safe landing. IATA reported that the rate of these incidents had sharply risen during 2020. Typically, the rate of unstable approaches sits at about 13–14 per 1000 flights but in May this figure rose to 35 per 1000 flights. It is apparent that with disuse, people just get out of practice. Not only did the pandemic affect the skill level of aviation personnel but also the environment of those who were able to continue to operate. With so many less aircraft being flown, the whole dynamic of commercial aviation changed. Everything ran at a different pace. COVID-19 protection procedures have to be learned and adhered to. In complex environments, humans are highly reliant on routines and the pandemic broke that routine.

1.5.5   The Impact on Organizational and System Safety

The psychological impact of COVID-19 across all aviation occupations has been massive. Organizations even remotely connected to the industry have had to embark on employee layoffs, furlough programs, and the cancellation of major contracts. It is not difficult to imagine where many aviation professionals' concentration has been focused when livelihoods are on the line. Enduring large-scale change saps precious organizational resources and redirects energy away from operational matters.

The bonds of trust between the employers, unions, and workforces of commercial aviation are often stretched in relatively normal times, but when organizations are facing collapse and financial ruin, mutual empathy can run very thin. The stress and distraction for fear of losing a job, with the associated fatigue from insomnia, is only one aspect of the equation.

One of the bedrocks of commercial aviation safety is the intelligence gleaned from operational staff through safety reporting. We shall explore this subject in more depth in Part III when we focus on human safety but suffice it to say that individuals are less likely to draw attention to themselves over safety concerns if they feel their jobs are at risk. The whole system of commercial aviation relies on safety intelligence from highly trained professionals who ordinarily pick up on lead indicators of system performance. However, over the last 2 years, the nature of the system has changed.

Previous disruptive events that have affected commercial aviation have meant that some aspects or sections have been temporarily suspended or have significantly reduced capacity. The COVID-19 pandemic has meant that the whole system is out of practice. The longer-term effects from this phase of aviation history are yet to be fully assessed—the ability of the established systems of safety risk management will be placed under pressure as will the effectiveness of safety cultures across the system rather than just within organizations. In short, how we manage and emerge from the pandemic will be a revealing test of the resilience of the total system of commercial aviation safety.

Key Terms

CB

CRM

ICAO

Ice Crystal Icing

ITCZ

Non-Normal Situations

Safety Evolution

Skill Fade

Startle Effect

Weather Radar

Topics for Discussion

1. List the most significant contributory factors regarding the loss of Air France 447.
2. Describe the four phases of the evolution of aviation safety management as described by ICAO.
3. How are stock markets affected by significant fatal commercial aviation accidents?
4. When did the human element of aircraft control and design first start to appear?
5. What is the "startle effect"?
6. Name some of the accidents and incidents which led to an increased levels of research interest into organizational influences on accidents.
7. What characteristics of the commercial aviation system have made it as safe as it is today?
8. How was commercial aviation safety affected by the COVID-19 pandemic?

References

1. Lynch, F. T., & Khodadoust, A. (2001). Effects of ice accretions on aircraft aerodynamics. *Progress in Aerospace Sciences*, *37*(8), 669–767.
2. Perrow, C. (2011). *Normal Accidents*. Princeton, NJ: Princeton University Press, p. 138.
3. BEA (Bureau d'Enquêtes et d'Analyses Pour la sécurité de l'aviation civile). (2012). On the accident on 1st June 2009 to the Airbus A330-203 registered F-GZCP operated by Air France flight AF 447 Rio de Janeiro—Paris. https://www.bea.aero/en/investigation-reports/notified-events/detail/event/accident-of-an-airbus-a330-203-registered-f-gzcp-and-operated-by-air-france-crashed-into-the-atlantic-ocean/
4. https://www.icao.int/safety/safetymanagement/Pages/default.aspx, p. 2.
5. Dekker, S. W., & Nyce, J. M. (2012). Cognitive engineering and the moral theology and witchcraft of cause. *Cognition, Technology & Work*, *14*(3), 207–212.
6. https://www.icao.int/safety/safetymanagement/Pages/default.aspx
7. Commons, N. (1978). Bonfires to beacons: Federal civil aviation policy under the Air Commerce Act 1926-1938. US Dept of Transport, Federal Aviation Administration, p. 25.
8. *2019 ICAO statistics*. https://www.icao.int/annual-report-2019/Pages/the-world-of-air-transport-in-2019.aspx
9. Barnett, A. (2020). Aviation safety: a whole new world? *Transportation Science*, *54*(1), 84–96.
10. Amalberti, R. (2001). The paradoxes of almost totally safe transportation systems. *Safety Science*, *37*(2-3), 109–126.
11. Boeing Company. (2018). Statistical summary of commercial jet airplane accidents worldwide operations, 1959-2017. https://www.boeing.com/resources/boeingdotcom/company/about_bca/pdf/statsum.pdf. Accessed September 20, 2019.
12. Boeing Company. (2018). Statistical summary of commercial jet airplane accidents worldwide operations, 1959-2017 (n 18). https://www.boeing.com/resources/boeingdotcom/company/about_bca/pdf/statsum.pdf. Accessed September 20, 2019.
13. Le Coze, J. C., & Wiig, S. (2013). Beyond procedures: Can safety culture be regulated? In: *Trapping Safety into Rules: How Desirable and Avoidable Is Proceduralization of Safety?* Farnham: Ashgate, Chapter 12.
14. Lawrenson, A. J., & Braithwaite, G. R. (2018). Regulation or criminalisation: What determines legal standards of safety culture in commercial aviation? *Safety Science*, *102*, 251–262.
15. https://ec.europa.eu/transport/sites/default/files/modes/air/doc/flightpath2050.pdf
16. Tversky, A., & Kahneman, D. (1973). Availability: A heuristic for judging frequency and probability. *Cognitive Psychology*, *5*, 207–232. doi:10.1016/0010-0285(73)90033-9.
17. *The Sydney Morning Herald*. https://www.smh.com.au/business/qantas-shares-dive-then-recover-after-crash-scare-20101104-17f7l.html
18. Kaplanski, G., & Levy, H. (2010). Sentiment and stock prices: The case of aviation disasters. *Journal of Financial Economics*, *95*(2), 174-201.
19. Clark, G. I., & Rock, A. J. (2016). Processes contributing to the maintenance of flying phobia: A narrative review. *Frontiers in Psychology*, *7*, 754.
20. Fruhen, L., Flin, R., & McLeod, R. (2013). Chronic unease for safety in managers: A conceptualisation. *Journal of Risk Research*, *17*, 969–979. doi:10.1080/13669877.2013.822924.
21. Reason, J. T. (1997). *Managing the Risks of Organizational Accidents*. Vol. 6. Aldershot: Ashgate.

CHAPTER **2**

What Is Safety?

Introduction

It can be a little tedious to start any chapter with definitions and clarifying concepts, but as we begin our fascinating journey into the world of commercial aviation, we think you will find this effort worth it. Throughout this book we will be referring to and referencing various terms, theories, and models associated with safety; so by spending a little time now on explaining some of the basic ideas, we can begin to grasp many of the underlying principles behind real-world examples. This chapter intends to clarify the term *safety* and the many related terms that are used within the industry and associated with safety management. Safety is not an absolute concept—it means very different things to different people. It is a highly contextual concept that is worth thinking through before we start analyzing the case studies throughout this book.

2.1 Safety Philosophy

2.1.1 A Tale of Two Accidents

Accidents in commercial aviation are rare. When they do happen, they generally attract considerable media attention and all too often unqualified speculation, sometimes disguised as expert analysis. There is an immediate public demand for an explanation and reaffirmation that the commercial aviation system is safe. In the words of a former inspector general of the US Department of Transport (DOT), "When a plane goes down in flames and dozens or hundreds of lives are lost, what the public wants is reassurance - reassurance that the accident was a fluke, that flying is statistically the safest way to travel and that someone is watching over aviation to guarantee it is safe."[1] We will discuss later in the book why such a commentary is so important to society's need to understand the cause of accidents, but we will gain some insight by comparing two examples of aircraft accidents and their relative media responses.

The first of these events took place on the morning of July 6, 2013, at San Francisco International Airport[i]. In good weather, a fully serviceable Boeing 777, Flight AAR 214, operated by Asiana Airlines, crashed into the undershoot of runway 28L. Following a mishandled visual approach and the operating flight crew's confusion about the aircraft's automatic modes, the aircraft's tail plane hit the seawall of the airport sending the bulk of the aircraft and significant amounts of debris onto the runway. Three people

---

[i] The NTSB produced an informative animation which provides a detailed explanation of the Flight AAR214 investigation here: https://www.youtube.com/watch?v=QVaQYhd_Qy0

Figure 2-1   AAR214 wreckage at San Francisco International Airport. (Source: NTSB Report)

were killed including two who were ejected from the aircraft during the collision. A further 187 people were injured, 49 of them seriously. Although the human loss of this accident is tragic, when we look at the extent of the damage to the airframe, seen in Figure 2-1, it is remarkable that these numbers were not significantly higher.

The accident was particularly notable as after 17 years of commercial service this was the first fatal accident involving a Boeing 777. The story dominated the world's media with harrowing footage of the heavily damaged aircraft cartwheeling across the airfield. The accident received further interest over the following few days after early reports suggested that one of the surviving passengers was accidently killed by an airport rescue vehicle during the recovery operation. The Asiana Airlines crash remains one of the most high-profile commercial aviation accidents of the early 21st century.

To engage with the high level of media interest in the crash, the then-chairman of the NTSB,[ii] Deborah Hersman, gave an extensive press briefing 2 days after the accident, on July 8, 2013, explaining how the crash would be investigated.[2] The subsequent NTSB report on the accident recommended Boeing, the aircraft's manufacturer, to review the Boeing 777 autothrottle design and Asiana Airlines to review the effectiveness of its pilot training program. We shall review this accident from a human factors perspective in Chapter 9.

On the following day, another commercial aviation accident occurred on US territory with over three times the fatalities of the Asiana crash. This accident received little

[ii] The National Transportation Safety Board (NTSB) is an independent US federal government agency charged with determining the probable cause of transportation accidents and promoting transportation safety and assisting victims of transportation accidents and their families.

FIGURE 2-2   Rediske Air crash wreckage at Soldotna Airport. (Source: NTSB, July 2013)

US national media attention and practically no international coverage. A de Havilland DHC3 Otter, operated by Rediske Air, crashed at Soldotna Airport, approximately 80 miles south-west of Anchorage in Alaska. The subsequent investigation, seen in Figure 2-2, found that the aircraft was significantly overweight. The aircraft stalled almost immediately after takeoff due to excessive trim forces caused by the inappropriate cargo weight and loading. All 10 people onboard, including the single operating pilot, were killed. The comparison of the different levels of media scrutiny applied to the two fatal accidents, barely separated by 24 hours, suggests that the relationship between how society perceives and subsequently regulates acceptable levels of safety for commercial aviation is far from simple.

2.1.2   Numbers and Distance

Statistics about the number of deaths are not necessarily related to the scale of emotional impact on the public. Paul Slovic, a psychologist at the University of Oregon, has been studying a psychological trait known as "psychic numbing" for decades. He stated in an interview with the BBC,[3] "If we're talking about lives, one life is tremendously important and valuable and we'll do anything to protect that life, save that life, rescue that person. But as the numbers increase, our feelings don't commensurately increase as well." The comparative reaction to the Asiana and the Rediske Air accidents suggests that the number of fatalities in any given accident is not the only factor at play in generating the level of public interest and prompting regulatory scrutiny.

Another potential explanation for the contrast in the levels of media and public interest is a psychological phenomenon known as "difference through distancing." Linked to social identity theory developed by Henri Tajfel, research has shown that we feel more empathy to those in society with which we identify. In contrast, by differentiating between our own life experience and those of victims of an accident, we create a barrier of comfort. The barrier allows us to create the illusion that we do not have to worry about a similar set of circumstances threatening our well-being. For example, when we hear of an automobile fatality, we often search for factors that make us feel safe, such as perhaps the accident happened at night and the casualty was not wearing a seat belt. Knowing such facts, we may be tempted to say, "Well I never drive at night and always wear a seatbelt, so such an event would never happen to me."

Perhaps in the regulator's and in the public's mind-set, flying in a single-engine air taxi aircraft in Alaska is a higher but tolerable[iii] level of risk. It is therefore logical and appropriate to set a lower standard of regulated safety for the air taxi than that of being onboard a multiengine jet airliner flying into a large international airport. Subjecting the generally larger multiengine aircraft of international travel to the highest standards of regulation makes practical sense; the potential for harm is significantly greater. Similarly, flying a smaller regional air taxi service and operating what are often smaller single-engine aircraft would ensure relatively less risk. However, the examples do provoke questions about what motivates the regulator and how the public assesses relative risk. At what point do either of these operations become safe enough? Where does the line of acceptable safety and risk sit and who decides where that is? What factors are at play in making that decision? These questions need to be answered and the subject of safety and risk requires further analysis.

2.1.3   Different Rules for Different Flights

At the time of the accident, both operators, Asiana and Rediske Air, were licensed common carriers[iv] and subject to the national and international regulatory regimes of commercial aviation. But there were differences in the regulatory requirements of each operator—differences the NTSB found to be significant. The Rediske Air accident report stated that the probable cause of the crash was related to a lack of regulatory requirement for the aircraft's weight and balance to be documented for each flight. As the operation was carried out using a single-engine aircraft, the flight was exempt.

The Federal Aviation Administration (FAA)[v] mandates documentation of weight and balance for every single flight involving multiengine aircraft, but this does not apply to single-engine aircraft operations. The NTSB had recommended this regulation be extended to all passenger carriers, including single-engine aircraft over 10 years prior to the Rediske Air crash following a succession of related safety reports, incidents, and accidents. As is common for all national aviation authorities, the FAA states that its

---

[iii] Tolerable risk and acceptable risk have subtle but significantly different meanings. "Tolerability" does not mean "acceptability." It refers to a willingness to live with a risk so as to secure certain benefits and in the confidence that it is being properly controlled. For a risk to be "acceptable," on the other hand, means that for purposes of life or work, we are prepared to take it pretty well as it is: The tolerability of risk from nuclear power stations, UK HSE, 1992.

[iv] Defined in an FAA circular, the four elements of common carriage include "(1) a holding out of a willingness to (2) transport persons or property (3) from place to place (4) for compensation." Fed. Aviation Admin., Advisory Circular No. 120-12A: Private Carriage versus Common Carriage of Persons or Property (Apr. 24, 1986).

[v] The Federal Aviation Administration (FAA) is the agency of the United States Department of Transportation responsible for the regulation and oversight of civil aviation within the United States, as well as operation and development of the National Airspace System. Its primary mission is to ensure safety of civil aviation.

primary mission is to ensure safety of civil aviation. While this mission statement seems ethically commendable, it is perhaps an overly simplistic goal. The comparison of these two accidents, occurring in the same country within one day of each other, suggests that the relationship between social perception of risk, economic efficacy, and regulatory standards in commercial aviation is a complex issue. When asked by *USA Today* why the FAA didn't update more of its older regulations, the then-chairwoman of the NTSB, Deborah Hersman, stated, "We actually are waiting for more people to be killed before we can do something that makes sense. We don't kill enough people in aviation to merit regulatory changes."[4]

2.2   Safety Thinking

2.2.1   Why Is Safety Important?

This may seem self-evident but before diving into semantics it is worth considering what safety means to us and considering its psychological importance. Like many of the most important things in life it is something that we can all easily take for granted. We may not regularly give much deliberate thought to the subject, but we certainly talk about it. Of the approximate 170,000 words recorded in the *Oxford English Dictionary*, *safe* or *safety* are among the most commonly used in normal communication. Safety is the key theme of this book and no doubt if you have a professional or academic interest in commercial aviation it is a concept that you will have given some thought, discussed, and perhaps written about. The General Service List[5] provides a list of the most important 2000 words needed by new students of the English language. "Safety" and "safe" both appear in that list suggesting they are not only commonly used but seen as essential to basic communication. We use these words daily and assume their meaning is conveyed without much need for explanation or clarification. But when we think about what the word means we are often met with considerable complexity and sometimes contradictory ideas.

Safety is an essential part of our lives. One thing that became apparent during the recent COVID-19 pandemic is that when lives are threatened, everything else in life becomes of secondary importance. The recent focus on health and well-being spurred by the pandemic has illustrated how significant it is for us all to feel safe. We need to feel safe in order to realize our objectives—it is a practical requirement. The quest for certainty in our lives and in our environment is a basic human need.[6] Throughout history, this need has been satisfied by social order and structure. To greater and lesser extents across our various social backgrounds, we have contracted away some elements of our social liberty in order to enjoy the psychological freedom of feeling safe from harm. These psychological underpinnings are crucial and permit commercial aviation to function effectively and safely.

When seated in a commercial aircraft, people are generally very compliant with the procedures imposed on them by the operator. They sit down, put on their seat belts, be quiet and listen when asked, switch off their phones, hand over personal possessions, give out personal data, and generally comply with requests made by the crew. It's difficult to think of another social situation where people are so obedient. Certainly, there are stringent legal requirements onboard an aircraft, but law enforcement interventions are so infrequently needed that they generally make the news if they do! People comply with the regulations onboard aircraft because they want to—they want to feel safe.

2.2.2   What Is Safety?

When lecturing to students about safety or opening a discussion with professionals, sometimes with years of experience in this field, it remains one of the most challenging tasks to define the term. The traditional view of safety is that it reflects the absence of something rather than being something tangible. This presents a significant problem for any professional involved in the management of safety. It was Karl Weick who famously described system reliability (later paraphrased to safety) as a dynamic nonevent.[7] As insightful as the phrase maybe from a theoretical perspective, try persuading a manager with a tight budget to approve funding for something that won't happen and if successful becomes invisible to us. This issue remains a constant source of turmoil for proponents of safety in all industries; how do we actively promote a nonevent?

The *Merriam-Webster* dictionary defines safety as "the condition of being safe from undergoing or causing hurt injury or loss." The *Collins Dictionary* states that safety "… is the state of being safe from harm or danger." Both definitions suggest that safety is achieved in the absence of various negative states—hurt, injury, loss, harm, or danger. So, from this perspective, rather than stating what safety is, there is a tendency to mold its definition around what it isn't. This tendency extends beyond day-to-day use and into the professional and academic arena. As one of the world's most influential safety thinkers, Professor Nancy Leveson defines "…safety as the absence of accidents, where an accident is defined as an event involving an unplanned and unacceptable loss."[8] The predominant activity in the management of commercial aviation safety has been to measure, classify, and analyze accidents. This traditional approach to safety measurement is probably best summed up by another influential safety thinker, Professor James Reason,[9] who stated that "safety is defined and measured more by its absence than its presence."

2.2.3   What Is Risk?

In the words of Charles Linbergh, the first person to fly solo across the Atlantic, "If one took no chances, one would not fly at all. Safety lies in the judgment of the chances one takes."[10] As Linbergh's words suggest, whichever definition of safety we prefer, it is invariably linked to the concept of risk. If safety is achieved by removing or reducing risk to some acceptable level, then it invites an inevitable question—what is risk? The *Merriam-Webster* dictionary provides a very succinct definition of risk as the "possibility of loss or injury."[11] This link with possibility often overlaps with the term *hazard*. Anything that has the potential of causing "loss or injury" is referred to as a hazard, but the terms *risk* and *hazard* are so often used interchangeably it can cause confusion. If a manhole was left uncovered in the middle of a busy sidewalk, it presents a hazard. If we put up warning signs, the "hazard" remains but the "risk" is reduced.

We humans have an intuitive sense of risk.[12] While some academic research has shown we are not always best equipped to assess the precise mathematical probabilities of the modern world, our instincts have generally served us well throughout our evolutionary journey.[vi] We must have had to be able to naturally evaluate various day-to-day risks otherwise saber-toothed tigers would have been fatter, and we would be a much scarcer species. The importance of considering our risk instincts as well as statistical

---

[vi] Any explanation of the fascinating study of human heuristics and biases should always include any published work by Nobel Prize winners Daniel Kahneman and Amos Tversky.

data is reflected by the United Kingdom's Health and Safety Executive (HSE) which collates safety data and carries out safety research on behalf of the British government. The UK HSE's position is that statistics should complement intuitive risk assessment as humans have developed their natural processes, "... mechanisms that reflect our personal preferences and the values of the society in which we live."[13] The statement is an affirmation that effective risk management is not only about the numbers but is a combination of numerous disciplines, including considering how safe we feel.

Over the last 100 years, our understanding of mathematical probability has advanced considerably. Prior to this advancement we largely relied on luck, fate, or belief.[14] But statistics are only one of the many increasingly sophisticated techniques we have developed to measure and manage risk. The range of modern risk management techniques is vast, but formal risk management techniques roughly fall into two main approaches—qualitative or quantitative. Qualitative risk assessment often utilizes a subject matter expert to provide an overview. Although it tends to involve a greater degree of subjectivity than quantitative methodologies, it can also provide a greater insight into complex or novel risks. It can also be used to tap into expert intuition when trying to identify emerging risks. This approach is not a guessing game and will often involve a formal process of risk classification. However, from a functional perspective, aviation risk management has tended to favor a quantitative approach.

Perhaps because quantitative approaches are perceived as more objective and therefore "scientific" they refer to systematic processes which evaluate risks. Because established methodologies of risk management have predominantly evolved from the financial and legal sectors, they have been predominantly statistical and rule based. This is certainly the case that within the aviation industry much of the literature on safety performance has emerged from engineering and technology disciplines.[15]

As Figure 2-3 illustrates, from a safety management perspective, risk is frequently described as a combination of severity and probability or, sometimes, impact and likelihood. This approach allows us to visualize and compare the scale or quantum of risk in order to make rational risk reduction measures and sensible safety policies. This approach is often displayed as a two-scale diagram sometimes referred to as a heat map or RAG or Red, Amber, and Green diagram (similar to the black and white version) depicted in Figure 2-3. These risk matrices are useful as they provide a qualitative (where descriptors of the quantum of risks are used) and semiquantitative (where numeric values are allocated) representation of the size, or quantum, of risk.

		PROBABILITY				
		Extremely Improbable	Extremely Remote	Remote	Reasonably Probable	Frequent
SEVERITY	Catastrophic	Review	Unacceptable	Unacceptable	Unacceptable	Unacceptable
	Hazardous	Review	Review	Unacceptable	Unacceptable	Unacceptable
	Major	Acceptable	Review	Review	Review	Review
	Minor	Acceptable	Acceptable	Acceptable	Acceptable	Review

FIGURE 2-3   An example of a "heat map" used to visualize risk. (Source: Authors)

For example, I could leave my house and walk to my local store and be hit on the head by a meteorite with a force that was sufficient to kill me. This is an extremely low probability event, but a catastrophic event, should it occur. What is considerably more likely is that I am hit by a car with a force sufficient (assumed) to kill me. In either case described above I would be killed, but the probability of occurrence of the second event is greater than that of the first event hence it would be ranked as a higher risk event.

Risk matrices are an excellent medium for describing and comparing simple or single-system risks. They allow comparison and ranking of hazards by levels of risks in order to allocate suitable resources for risk mitigation. A common misconception occurs when trying to assess the tolerability of risks, when arguments are made for each separate or individual risks rather than assessing the collective or an individual's aggregate risk from all causes.[16] Used incorrectly the process can drift away from risk management into risk spreading. The implications of this misuse of risk and safety management principles will be considered throughout this book and will feature in several case studies.

2.2.4   Target Zero

Rather than engage in the grey areas of risk management and be driven by the rather wooly concepts of acceptability and tolerability, some aviation safety managers have taken up the challenge of entirely eradicating harm by adopting a "Target Zero" philosophy. By taking the more idealist absence of accidents view, the approach does not tolerate any compromise to safety—it is absolute in promoting the intrinsic value of human life. While ethically commendable, this approach has been widely criticized within safety science literature for its impracticality and potential for counterproductive influence.

One notable critic of this approach has been Dutch academic, Professor Sidney Dekker.[17] Dekker sees the "Target Zero" philosophy as fundamentally flawed for two main reasons. The first is that Target Zero policies have little to no scientific underpinning. We shall address some of the more prevalent myths of safety management later in the book, but suffice it to say, Dekker has a strong point here. The second reason for his opposition to such policies is that they encourage a culture of target chasing rather than addressing serious risks to an operation. Safety, or rather chasing the absence of adverse events becomes a management target rather than a shared organizational value. Such cultures can inadvertently suppress open reporting because employees are inhibited from candid disclosure of the emergent threats produced in safety-critical systems like commercial aviation. If crude safety metrics are introduced that promote a reduction in the number of safety reports, a "Target Zero" perspective becomes an anathema to an open reporting culture.

Imagine sitting in a room full of investment bankers when the manager stands up to proclaim that their department will not accept any losses over the next quarter. Such a manager would surely lack any credibility in misunderstanding the nature of their business; investment banking, just like commercial aviation, requires participants to manage risk not ignore it. In any environment where risk has to be managed, it first has to be acknowledged. Sitting back and assuming risks are managed because proclamations have been made about zero tolerance produces a far greater potential for disaster. In the words of one of aviation's pioneers, pictured in Figure 2-4, Wilbur Wright, "In flying I have learned that carelessness and overconfidence are usually far more dangerous than deliberately accepted risks."[18]

FIGURE **2-4**   Flying always encompasses some level of risk. The Wright Brother's first flight, Kitty Hawk Field, December 1903. (Source: Pixabay)

The myth of absolute safety leaves us with little scope for maneuver in terms of how we manage safety in a day-to-day environment where risks are ever present and constantly evolving. As a junior military pilot, one of this book's authors recalls a graffitied flight safety poster on the wall of his squadron. The picture showed an accusing finger pointing at the reader and asking the question, "What are you doing for flight safety today?". One student had written a pithy reply on the poster, "Going home!". Although written in jest, the graffitied poster makes a succinct point about aviation—it always involves some level of risk.

2.3   Measuring Safety

2.3.1   Regulatory Metrics

As we have discussed, maintaining public confidence in the system is an essential element of safety management and cannot or should not be ignored. How risk metrics are framed is highly influential to public attitudes on commercial aviation safety. The need to demonstrate high levels of safety performance is evidenced right through the hierarchy of aviation regulation. At the top of the international regulatory hierarchy is ICAO. The organization generally defines an accident as either the loss or significant damage of an aircraft or an event which results in the death or injury of those onboard. The counting and classification of accidents is described as "...a key safety indicator for commercial aviation operations worldwide ..."[19] Although safety indicator, representing the prevalence of bad outcomes, accidents are not the sole indicator of safety, nor are they always the most accurate. To maintain an acceptable level of safety in an increasingly complex risk environment, ICAO recognizes that further and more diverse measures are required.

ICAO defines safety as "the state in which harm to persons or of property damage is reduced to, and maintained at or below, an acceptable level through a continuing process of hazard identification and risk management."[20] This definition highlights two significant features of the regulator's perspective of safety management. The first is that safety can be measured, and by implication, managed through the development of predictive models. The second feature is that the focus of aviation safety management is to maintain a level of loss to an acceptable level. Two questions arise. First, *how* are we achieving this process of hazard identification in order to manage risk, and second, what is the acceptable level to which we are aiming? We will discuss levels of acceptable risk in the following section of this chapter but first let's consider how we are identifying and classifying/ranking these hazards to manage risk.

2.3.2   Measuring Ultrasafe Systems

We have discussed the Target Zero approach and concluded an element of risk will always exist in any form of aviation. We have also observed that accident rates have consistently reduced to the point that commercial aviation is described as an ultrasafe system. However, as commercial aviation accidents have become such increasingly rare events, they provide little to no broad sampling data. Retrospectively analyzing accidents can result in considerable uncertainty over emergent patterns or leading indicators. As described in the previous chapter, this issue creates a paradox in the traditional process of aviation safety management.[21] This paradox, described by Amalberti, is that the current performance of the system denies the application of the traditional indicators of safety performance: accidents. The safer commercial aviation has become, the less accident data is produced. The less accident data is produced, the less reliably we can predict future accidents.

Overreliance on the measurement of accident numbers as an indicator of system safety can therefore provide the illusion of degree of system control that does not yet exist. Amalberti highlighted that the ongoing process of collating larger and larger accident databases does not improve safety prediction. He claims that the processes have started to move away from their original intended purpose of simply monitoring accident rates rather than to continually linking accidents to incidents and then minor incidents.[21] Overdependence on quantified risk assessment has been observed by researchers and institutions across the entire aviation industry: "The quantification of risk is unfortunately sometimes seen as a 'numbers game,' relying on questionable data, crude modeling of scenarios and subsequent simplistic mathematical treatment."[22]

While the commercial aviation industry can be justifiably proud of the level of safety performance it has achieved to date, it now must adopt methodologies that are relevant to today's system in order to enable continuous improvement. The dissonance between promoted safety statistics and that of actual safety performance is no conspiracy or deception. The promotion of safety statistics is an absolutely necessity for the purposes of calming social anxieties and maintaining public confidence. However, retrospection does not necessarily produce the perceived level of predictive accuracy[23] that safety institutions might infer.[24] Although the illusion of control over safety performance is an essential element of commercial aviation's ongoing viability, providing better predictive accuracy through the engagement of safety science should be the long-term goal of the industry's safety management strategy. We have now reached a position where many of the traditional metrics have run their useful course and now need to be supplemented, and in some cases replaced by newer and more relevant leading indicator metrics.

2.3.3   Reactive, Proactive, and Predictive Safety

ICAO's definition of safety is often favored by those involved in the management of safety as it infers that safety is something we can measure and therefore control. However, if hazard identification and risk management processes were merely reliant on accident statistics, we would be constantly looking back at previous incidents rather than looking forward at future potential threats. That approach has in the past served us well but must now be reviewed and complemented by other safety performance metrics. Learning from accidents was once the only way by which safety was promoted; we now have to find others.

Whether driven by morbid curiosity or a desire for genuine professional development, safety management was often supported by a vast library of cautionary tales of bad luck or bad airmanship. Conversations that once promoted best practices, knowledge, and experiences in aviation were often shared verbally. When people made mistakes, the story soon spread within the team, the organization, and in the case of high-profile incidents, across the industry. People learned lessons from others', sometimes unfortunate, experiences. When accident reports were released, people would read, reflect, and relate to their own experience. Many unfortunate individuals paid the ultimate price which enabled the rest of the aviation community to learn and attempt to prevent reoccurrence. It was a predominantly a reactive system.

There is a great deal of logic in *not* waiting for the next crash to progress to the next level of positive safety measures. This collective realization has prompted the evolution in safety philosophy which is reflected in the more contemporary versions of ICAO's Safety Management Manual (SMM). Today's SMM describes three broad safety management approaches, classified as reactive, proactive, and predictive. While we can learn a considerable amount from the reactive study of particular accidents and incidents, they have always to be interpreted within the context in which the event occurred. They do not provide us with a panacea—a general solution for future safety, just clues, and pointers that might avert further accidents. Being proactive means enacting what we do know and have learned and engaging safety stakeholders, promoting safety, and spreading the word. Predictive measures are the ultimate goal of future safety management. If we could accurately predict safety performance, then we could accurately prevent accidents. In this book, we will explore the progress and ambition of these emerging safety programs and consider the considerable potential presented by new technologies, not least that of artificial intelligence.

This latter idea is extremely appealing to the world of safety risk management. Imagine a world where we could practically eradicate the risk of aviation accidents. Enhanced confidence in the development of new ambitious systems could spawn the development of newer and more efficient aircraft with innovative design and radical performance. But things aren't that simple. New technologies bring a wealth of possibilities, but they also bring with them a level of complexity and system interdependence that we have never before encountered. The goal of predictive safety is not just an idealist's wish list but an essential unpinning of future aircraft development. As technologies such as AI, quantum computing, robotics, Internet of Things, 5G, and virtual and augmented reality are introduced, it is increasingly important that we understand how the commercial aviation system learns, adapts, and evolves.

2.3.4   Safety Reporting Initiatives

There has been a long-standing recognition in aviation that engaging a variety of safety initiatives would provide the best way to promote system learning. A basic principle of system safety is that predictive management requires timely and accurate safety intelligence rather than a reactive system that looks back; you need to be able to recognize any degradation in system performance rather than ponder over an accident that has already happened. As commercial aviation becomes increasingly complex, it becomes increasingly difficult to centrally monitor system health. As system activities become increasingly specialized and contextual, any health monitoring system needs expert and firsthand intelligence from those humans who are most familiar with the system. The best-placed and most sensitive individuals to potential threats are those most directly involved in their day-to-day operation. These individuals are the most tuned into what normal looks like and therefore able to detect early threats before they get out of control (this concept of system controls will be discussed in depth when we discuss systems theory later in the book).

Moving away from the analysis of past accidents has required a considerable change in mind-sets across organizational management in the traditionally hierarchical structures of commercial aviation. No longer are the central system managers necessarily the subject matter experts on system safety performance, but those individuals who are exposed to the nuances of these complex systems need to use their experience to embrace the responsibility of proactive reporting. It also requires considerable promotion from management to create the requisite environment of trust in order for effective reporting systems to function. Reporting systems are fragile as they are almost totally reliant on the authentic trust and confidence of the workforce. They are vulnerable to failure on three fronts: fear, futility, and functionality (the three "fs"). Fear of repercussions (whether real or imagined) will stop any reporting for obvious reasons. The perceived lack of response from the management system can make reporting seem a futile act for the reporter, humans need recognition for their actions otherwise they will find psychological reward elsewhere. Finally, the system must function efficiently for the reporter. The design emphasis of effective reporting systems must prioritize ease and intuition for the user, over that of the safety department's analysts. Invariably the reports are made contemporaneously, that is at, or close to the time a reportable incident occurs so all relevant information can be gathered in a timely and accurate manner. As any police investigator or journalist will tell you, there is a finite amount of time that people can retain an accurate recollection of events.

We shall discuss the significant impact of positive reporting cultures in Part III. The first stage of development toward predictive safety management is the gathering of safety intelligence commonly known as safety reporting. There are several national safety reporting initiatives across the world. The following list represents members of the International Confidential Aviation Safety System (ICASS) group. The group collectively recognizes the safety value of confidential reporting and the sharing of that information as an effective means of promoting aviation safety:

- UNITED STATES—Aviation Safety Reporting System (ASRS)
- UNITED KINGDOM—Confidential Human Incident Reporting Program (CHIRP)

- CANADA—Confidential Aviation Safety Reporting Program (SECURITAS)
- AUSTRALIA—REPCON Confidential Reporting Scheme
- BRAZIL—Flight Safety Report System (RCSV)
- JAPAN—Aviation Safety Information Network (ATEC).
- FRANCE—Confidential Environment for Reporting (REC)
- TAIWAN—Taiwan Aviation Confidential Safety Reporting System (TACARE)
- KOREA—Korean confidential Aviation Incident Reporting System (KAIRS)
- CHINA—Sino Confidential Aviation Safety System (SCASS)
- SINGAPORE—Singapore Confidential Aviation Incident Reporting (SINCAIR)
- SPAIN—Safety Occurrence Reporting System (Sistema de notificación de sucesos)

In addition, there are global and regional initiatives which further the process of merging safety intelligence. Two examples are ECCAIRS focused on Europe and GSIP mainly focused on the Pan-American and Asia-Pacific regions:

- EUROPE—European Coordination Centre for Accident & Incident Reporting Systems (ECCAIRS)
- GLOBAL (Predominantly Pan-American/Asia-Pacific)—Flight Safety Foundation—Global Safety Information Project (GSIP)

These initiatives concentrate on the gathering of reporting data. To turn that data into useable information requires further processes of analysis and meaningful representation. As the data can come from various backgrounds, it has to be understood in context. In addition, the flow of independent data sources will never be the same from such a variety of cultural sources—it is practically impossible to ascertain the authenticity of all reporting sources. As a general principle of any form of research, it is always a good idea to use independent methodologies to cross-check the data sources. This has led to a broader range of methodologies engaging in safety performance monitoring.

2.3.5   Safety Monitoring Initiatives

Two examples of alternative system monitoring to traditional safety reporting processes are the Line Operations Safety Audit (LOSA) program and Flight Operations Quality Assurance (FOQA). LOSA is a unique flight safety initiative developed at the University of Texas Human Factors Research Project, in Austin, Texas, under the leadership of Professor Robert Helmreich. A former PhD student of Professor Helmreich, Dr. James Klinect, built on the principles of Crew Resource Management (CRM) developed by his mentor and used them to focus both on the systemic and front-line factors that enable safety and efficiency in commercial aviation operations. LOSA is a nonpunitive, peer-to-peer flight deck observation program that collects safety-related flight data during normal operations. The program is designed to detect early indicators of potential safety problems and recommend improvement measures where appropriate. The basic process involves the use of highly experienced pilots, who sit on the

FIGURE 2-5 A modern pilot's role is to continually look for threats and errors. (Source: Unsplash, Blake Guidry)

flight deck jump seat during normal operations to record the management of threats by crews during day-to-day line operations. The skill-set the observers are most focused on is generically referred to as Threat and Error Management (TEM). It includes among other skills, communications, workload management, decision-making, and problem solving. In effect it provides a real-world snapshot of front-line operations, such as the scenario seen in Figure 2-5, where pilots are constantly looking for potential threats and errors which can compromise a safe operation. This highly respected program is today coordinated by the LOSA Collaborative[25] which today is an independent organization.

There are many different versions of FOQA in use today. Often referred to as FDM (Flight Data Monitoring)[vii] the process uses the analysis of routine flight data to detect, measure, and mitigate aviation hazards. It is based on the same principles of open reporting systems insofar as it is (intended to be) nonpunitive and allows learning about system performance without waiting for an accident. FOQA entails the gathering of data from a flight data recorder on multiple flights and using sophisticated software to detect emerging trends. These emergent properties might indicate potential threats to a safe operation and although the analysis of individual flights is generally unnecessary, further investigation can shed light on potential hazards.

[vii] There are numerous versions/titles of FOQA programs specializing on different areas of flight operations—examples include operation flight data monitoring (OFDM), flight data analysis (FDA), maintenance operational quality assurance (MOQA), military flight operational quality assurance (MFOQA), simulator operational quality assurance (SOQA), corporate flight operational quality assurance (CFOQA), helicopter flight data monitoring (HFDM), and helicopter operational monitoring program (HOMP). It's fair to say safety managers like acronyms.

Under ICAO Annex 6, regional legislation requires all commercial airlines to implement a FOQA program.[viii] Despite significant research suggesting the positive effect of FOQA programs (some of which has been sponsored by the FAA itself), the FAA has yet to implement this recommendation as a mandatory requirement for US aviation operators. The Administration's approach is contained in a circular,[26] which suggests various approaches to implementing FOQA.

Operators can administer stand-alone FOQA programs or choose to participate in national and some international safety initiatives. The FAA's program, the Aviation Safety Information Analysis and Sharing platform (ASIAS),[27] or IATA's Flight Data Exchange (FDX),[28] means that rather than being restricted to the data provision of a single operator, or even of those of a single nation, millions of data contributors can combine their redacted findings within a single platform. Merging FOQA data is intended to provide far greater sampling sources and in turn, greater capacity to identify emergent patterns and threats within commercial aviation.

2.4 Limits of Safety

2.4.1 Are We Safe Enough Yet?

In this chapter we have considered the merits of using accident data to reflect safety performance in commercial aviation. These statistics are not the only metric available and accordingly we will consider what alternatives might supplement or even replace this approach to give safety management some form of lead indicators of system health. We have also seen how reporting and safety monitoring programs can produce immense amounts of system data to identify trends and system anomalies before they get out of control. The process of constant learning and adaptation is one of the reasons that safety performance in commercial aviation has developed such an enviable reputation among other safety-critical industries. We might even consider that having read the probabilities of injury or harm in the first chapter of this book one might be forgiven for thinking that this system might now be safe enough. However, before we jump ahead it is worth reflecting on two assumptions that we may be making at this juncture. The first is that we fully understand the mechanisms that facilitate safety performance at its current levels and second that the current levels of safety performance will continue to be adequate for society's needs and expectations in the future.

New technology can often bring many advantages and can improve safety performance when applied intelligently. However, it can also create social anxiety as new levels of complexity make it harder for us to understand the mechanisms of risk. In Part II we observe how some of the emerging technologies in commercial aviation means that many of those even intricately involved in its operation do not understand how it works. How and why technology affects us in this way was extensively considered by Ulrich Beck in his famous book *Risk Society*.[29] Beck observed that while we generally welcome the benefits of new technologies into society, the increasingly distant relationship

---

[viii] It would be inaccurate to refer to the mandating of an ICAO annex due to a quirk of legal procedure. While the main body of the ICAO Charter is mandated under the normal principles of international law, the Annexes remain recommendations of best practice for the commercial aviation industry. As an example, Annex 5 of the charter recommends SI units (metric) to be used throughout aviation. In practice this is almost universally ignored with each state citing their preferred units of measurement.

between ourselves and the institutions that control these new associated risks create considerable anxiety. It is important that safety management professionals frequently reflect on the diverse range of internal and external stakeholders their activities serve.

2.4.2   Changing Standards

The changing level of acceptable risk is an issue that has been comprehensively considered and discussed by the American historian, economist, and author Peter Bernstein.[30] Bernstein describes emergent themes in how our society thinks about potential harm and what levels of risk are acceptable to us. He states that it is the fear of uncertainty which has evolved from the recognition that we don't always understand our environment. This misunderstanding of the modern world drives many of societies' concerns about new technology and uncertainty about the future. The emergence of technology such as jet-powered international travel accentuates societies' collective concerns and these concerns are amplified given the high-profile nature of commercial aviation's accidents.

Bernstein gives a thorough explanation of deeper historical foundations dating back to Greek and Roman philosophies when risk was translated through signs from nature and stories about the antics of the gods. These stories or parables were metaphors of reality which allowed ordinary people to understand life's changes and complexities in a way that science and statistics seems incapable of. It is an oft-heard cliché of management consulting that statistics and numbers never change behaviors as effectively as well-told stories.

However, while science hasn't totally dominated the way humans behave, it has had considerable influence over the way we have started to think, particularly over the last century. Humans have increasingly turned to science to explain life's complexities and particularly when we face adversity. But scientific evidence alone is inadequate to maintain society's continued confidence in adequate levels of safety for commercial aviation. The recent pandemic has served to illustrate that scientific, evidence-based arguments do not totally satisfy societal anxiety about our perception of threats and what is acceptable.

2.4.3   Acceptable Level of Safety

In trying to define what is an acceptable level of safety for commercial aviation we almost immediately hit a barrage of social, economic, ethical, regulatory, and legal challenges. If we consider social acceptability, there seems to be no benchmark that we can anchor as to what is acceptable to the public. If we consider the image in Figure 2-6 of holidaymakers sunbathing, while a large commercial aircraft flies a few meters above them, would we expect to see or accept these peoples' right to conduct the same activity near the runways of Schiphol Airport in Amsterdam or Chicago's O'Hare Airport? Risk tolerance seems to a be highly contextual and occasionally irrational subject. As Bernstein has identified, it changes over time and overall it seems to have trended toward increasing levels of risk aversion. But that trend seems to be highly contextual to what activity we engage in. If we consider other forms of transport, such as driving automobiles, as a society we seem to be content with remarkably high death rates. According to the National Highway Safety Administration, in 2020 there were over 38,000 fatalities from automobile accidents in the United States.[31] According to the Association for Safe International Road Travel, US road crashes are the leading cause of death for 1- to 54-year-olds, leading to societal and economic costs of some $871 billion. It would be difficult to imagine a sustainable commercial aviation system that produced that level of damage and loss.

FIGURE 2-6   People can have very different ideas about what is acceptable risk. An Air France Boeing 747 landing at St. Maarten. (Source: Unsplash, Ramon Kagie)

Calculating the economic costs from injury and loss of life may be a disturbing subject for some but if we are going to assess what is acceptable then it would seem to make sense. To provide and maintain any level of safety assurance, some level of economic capability needs to be factored into the equation. Rather than adopt an absolutist approach to human safety as "Target Zero" promotes, a pragmatic acceptance of the limits of economic capability could be considered.[32] However, as compelling as this pragmatic approach may seem, it has been rejected in some areas of safety risk management. The tolerability of risk based on principles of cost/benefit analysis have on occasion proved to be particularly unpalatable by society. In his analysis of the safety performance of the Australian commercial aviation sector, Professor Graham Braithwaite identified one possible influence on the high levels of performance as the rejection by key figures in the Civil Aviation Authority on formulaic approaches to the risk tolerability of loss of human life.[33]

2.4.4   Tolerable, Acceptable, and ALARP (As Low As Reasonably Practicable)

We are now beginning to see some of the philosophical dilemmas presented to those involved in safety management. Given that on one side of the spectrum, absolute safety is generally considered an unachievable goal, while on the other, methods of calculating the cost of accidents are often unpalatable, some middle ground must be reached. Achieving this balance is a constant challenge to regulators when trying to establish parameters for safety performance. Several "terms of art"[ix] have evolved which try to represent this balance. We shall address this terminology throughout the book but begin with three of the more fundamental and widely used, tolerable, acceptable, and ALARP.

[ix] A "term of art" is one that has a particular meaning within the scope of a subject matter, in this case the safety risk management of commercial aviation.

As part of any risk management process, once a risk is identified and assessed, it can be classified into one of three groups of tolerability: acceptable, undesirable (or tolerable), or unacceptable.

An acceptable level of safety (sometimes referred to as ALOS) represents that safety is a dynamic process. The more traditional regulatory approach to safety risk management has been for regulators to set safety goals and targets then periodically audit the organization for compliance. ALOS expresses the objectives of a safety regulator or oversight authority while recognizing the organization's need to fulfill its core business function. This objective-led form of regulation was born out of an increasing recognition that the technologies and operational complexities involved in activities like commercial aviation were becoming impractical to constantly monitor and supervise. In commercial aviation, safety performance targets are defined individually to recognize the highly contextual nature of differing types of operations. "Each agreed established level of safety should be commensurate with the complexity of individual operator/service providers' operational contexts, and the level to which safety deficiencies can be tolerated and realistically addressed."[34]

To manage an undesirable risk, certain mitigations must be put into place. These mitigations may make the risk acceptable or if the activity is considered socially desirable because of the benefits brought to society, while it might never be an acceptable activity, it may be considered tolerable. Certain forms of military flight operations may fall under this categorization of risk tolerability. For example, given the need to train crews to fly competently at low level, a risk assessment would never consider this acceptable in a civilian environment, but given the consensus that military flight crews need exposure to this intense training environment, it would be considered a tolerable risk, given its social desirability, and providing suitable mitigation is employed. If there is little or no social requirement and/or sufficient mitigation cannot be practically applied, a risk is simply undesirable and therefore unacceptable.

One of the most important acronyms in safety risk management is that of ALARP.[x] A company can discharge its liability for damage if it can demonstrate that the cost of removing a risk factor was grossly disproportionate to the quantum of risk and that it has reduced that quantum of risk to be *as low as is reasonably practicable*. The issue of what is reasonable cost should consider issues of resource, time available, and finance. Commercial aviation is regarded as the most stringently regulated of industries, but this does not absolve operators, MROs (Maintenance, Repair, and Overhaul Organizations), or traffic control organizations from a continual self-assessment of the appropriateness of their risk management practices.

2.5 Commercial Safety

2.5.1 Safety or Success?

As the title of this book suggests, safety and commercial considerations are intrinsically related; after all, organizations are involved in commercial aviation to make profit. One element cannot exist without the other as success in commercial aviation is impossible without acceptable levels of risk. In an interview with the *Observer* newspaper in

---

[x] Also known as ISFARP—In so far as is reasonably practicable. The ALARP principle was based on an English law case: Edwards v National Coal Board [1949] 1 All ER 743 CA

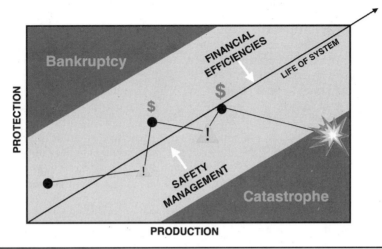

FIGURE 2-7   Originally conceived by Professor James Reason, the safety space represents the balance between potential bankruptcy and potential disaster (Source: ICAO SMM (2017:325)[20])

August 2000, Stelios Haji-Ioannou, the founder of EasyJet, was asked about how an accident in his former role as a shipping line owner affected his approach to managing safety in his new airline: "I tell them if you think safety is expensive, try an accident. You can be worth nothing just because of one mistake."[35] There is a fine balance which must be navigated by airline managers between investment in safety processes and the constant drive to contain costs. Maintaining this balance is referred to by Professor James Reason as remaining in the safety space, as this illustration, in Figure 2-7, adapted from his excellent book, *Managing the Risks of Organizational Accidents* shows. The diagram now features in ICAO's SMM,[20] and depicts the life cycle of an organization during which the emphasis on costs competes with focus on safety. If the balance is lost, then either bankruptcy or, as in the case shown in the diagram, catastrophe is the result.

If one were to corner an unsuspecting airline chief executive officer (CEO) at a social occasion and ask him or her what their primary obligation was as leader of their organization, would their answer be safety or success? It would be surprising if any CEO didn't reply with the third choice, that is, both; after all, achieving that balance is their job. But for CEOs it is ultimately profit and loss which defines their success, and it is that element of their role which attracts incentive payments.[36] The rarity of major accidents means underinvestment in adequate safety measures can often continue for years without significantly adverse consequences. But if you're not so lucky as CEO, board member, or senior manager, the repercussions can come back and threaten your job, your liberty, and potentially your company's very existence. Under many jurisdictions of the world, one singular individual must be identified from within an organization as the accountable manager. The purpose of this regulatory requirement is to avoid any lack of clarity as who carries the primary responsibility for safety. There may be multiple post holders with all manner of grand titles but only one identifiable individual "carries the can" for safety.[xi]

[xi] The phrase "carrying the can" is an old French Army term. The individual charged with carrying the unit's gunpowder to keep it safe and dry. Hence someone who *carries the can* held an important position of responsibility.

No equivalent incentive exists to maximize safety performance—unless of course we include the deterrent effect from one of those very rare events—a fatal accident, but by then it's all too late. In September 2021 a Delaware judge ruled that board members of the Boeing Corporation must face a lawsuit, not directly for breaching safety standards, in designing, producing, and selling aircraft with significant safety flaws, but for the financial losses incurred by Boeing shareholders. The ruling follows the US Department of Justice (DoJ) investigation into the manner in which Boeing had withheld crucial information from the FAA in connection with the fatal crashes of two of its Boeing 737-MAX aircraft.[37] Boeing's lawyers were able to avoid corporate manslaughter charges by the DoJ for the deaths of the 346 passengers who died in the two Boeing 737 MAX crashes of Lion Air Flight 610 and Ethiopian Airlines Flight 302. They agreed a plea bargain settlement for approximately $2.5 billion in January 2021 by admitting to fraud and conspiracy charges.

In the words of Acting Assistant Attorney General David P. Burns, "Boeing's employees chose the path of profit over candor by concealing material information from the FAA concerning the operation of its 737 MAX airplane and engaging in an effort to cover up their deception."[38] The $2.5 billion is a lot of money, but to put it into perspective, following the BP Deepwater Horizon explosion, which resulted in the death of 11 workers on the oil rig and the biggest environmental disaster in US history, in 2012 BP settled for a $4 billion fine with two of its senior managers facing manslaughter charges.[39] We shall discuss how Boeing's safety culture allowed for such activities to materialize in Chapter 12 and consider the legal consequences of that deterioration of safety culture in Chapter 16.

2.5.2 Commercial Pressure

Allocating sufficient resources to safety training, promotion, and equipment sits predominantly within the remit of senior management within commercial aviation. Management mind-sets, particularly at senior level tend to focus on the financial viability of their organization. Operational staff also have a big influence on maintaining this balance; however, their primary focus tends to remain within the operational realm. While neither side are oblivious to the sometimes-conflicting requirements of operational safety and financial viability, the differing emphasis between these working groups can create tensions.

The last two decades have seen a pattern of deregulation across the commercial aviation sector. The emergence of the low-cost carrier is one example where market forces have started to alter the traditional field of competition in the industry. The overall effect has been to stretch resources to survive the increasingly fierce competition.[40] While competition can be very healthy in any sector, including transport, there can also be a level of unhealthy competition. If, for example, many businesses within the sector are operating at a loss, fundamental questions need to be asked about the sustainability of the overall model and the potential impact on safety resources. All commercial activities operate within limited resources, so financial viability is an important area of focus for national aviation authorities.

One area where commercial pressure can influence the integrity of flight operations is in fuel carriage. Behind labor costs at 32.3%, fuel burn is the second highest area of airline expenditure at 17.7% so, it is an obvious area for diligent management to minimize extraneous fuel burn.[41] Pilots use company-produced fuel plans to guide their decision-making around how much fuel they should carry. Their decision process will have to consider weather at departure, during the route, and at the destination airfield. They may also have an acceptable technical defect on the aircraft which results in a greater fuel burn or require the use of the auxiliary power unit (APU)[xii] throughout the flight. Pilots may also take into consideration any likely delays at their intended destination airport, based on their experience and judgment. Complex operating environments may also require more flexibility, such as a change in the intended landing runway which requires more fuel or simply in the overall judgment of the captain, a combination of all these factors justify extra fuel to allow subsequent inflight "thinking time" to manage a safe operation.

Being the final arbiter on the fuel carriage is an important responsibility for flight crews. They must consider a plethora of operational considerations but also the cost of carrying extra fuel. There are real financial penalties to pay as the carriage of fuel means extra weight and extra weight means greater fuel burn. In short, it costs fuel to carry fuel and too much fuel carriage is regarded by airline management as a waste. From the pilot's perspective, carrying insufficient fuel (on an off day when everything else gets difficult) is the sort of decision you only get wrong once. One common pilot saying is that "the only time you are carrying too much fuel, is if you're on fire!" The way that airlines tend to balance competing demands around fuel carriage is by providing flight crews with timely and relevant information with which they can form a balanced and sensible decision. Most of this information covers operational factors but also includes the actual cost of the extra fuel burn so the penalty can be weighed against the advantage of extra fuel. However, as we have discussed early in this chapter, when we manage risk, overly focusing on one single aspect of risk-balanced decisions is poor practice. As a general principle, the better the quality of information, the better the decision.

2.5.3 Safety and Success

Perhaps because safety has been elevated by some in the industry to the primary goal of commercial aviation it has been treated differently than other competing interests within the aviation business. But taking the alternate view that safety is just one of many process elements that are required for a successful operation suggests there is no reason for safety and profit to not complement each other, as the following quote from an earlier edition of ICAO's SMM suggests, "A misperception has been pervasive in aviation regarding where safety fits, in terms of priority, within the spectrum of objectives that aviation organizations pursue, regardless of the nature of the services that aviation organizations might deliver. This misperception has evolved into a universally accepted stereotype: in aviation, safety is the first priority. While socially ethically and morally impeccable because of its inherent recognition of the supreme value of human life, the stereotype, and the perspective that it conveys do not hold ground when considered from the perspective that the management of safety is an organizational process."[42]

The earlier example of improving information to improve the quality of decision-making about fuel burn is one example where safety and commercial success can

---

[xii] An auxiliary power unit (APU) is a mini jet engine, often located in the aircraft's tail section. The APU can provide air conditioning, hydraulic and electrical power when the main engines are not running on the ground. The APU can also operate during flight to provide an alternative source of pneumatic, electrical or hydraulic power.

complement each other. Better awareness of the financial and operational value of fuel tends to improve efficiencies across the whole operation. Optimizing the required flight level and minimizing APU fuel use are small but important changes in the way a crew can operate but make big differences when these behaviors are altered across a whole airline. Simply by making crews more aware of the resources at hand and leaving them to allocate resources to their operational needs can result in a win-win scenario.

Increasing productivity and improving efficiency are not necessarily negative influences on safety. The requisite investment can mean better equipment, improved working practices, and more accurate information for the operation. Aviation attracts certain types of employees who are able to adapt and learn throughout their career. Being able to accommodate change and make decisions under conditions of uncertainty are the hallmarks of aviation professionals. Not only is this the norm for many people employed in commercial aviation but being able to maintain a high standard of safe operation is often the source of professional pride. The promotion of a culture of risk ownership is deeply embedded into aviation professionals when it comes to safety but there is no reason why that culture cannot extend to ownership of efficiency.

Just as the control of fuel carriage can be addressed by good management practice, the FAA has identified areas of safety that can comfortably coexist with profitable policies.[43] It's not just the aversion of the potential cost of a major accident, but the measures brought in to manage safety can be equally leveraged to improve business intelligence and therefore decision-making. For example, promoting the reporting of safety issues is a good way that organizations can monitor any related commercial issues that might otherwise have been overlooked. Another example is through the maintenance of standard operating procedures (SOPs) for safety purposes which can provide a good foundation for other commercial processes to be standardized, reducing error, and saving time. Good (safe) practices around engine handling can reduce wear and tear as well as maintenance costs. Insurance premiums are sometimes reduced where operators can demonstrate good practices and reduced claims. In short, good safety practice is good business practice.

2.5.4   Marketing Safety

When we think of safety programs and promotions, our thoughts can focus on those rather irritating and meaningless posters with slogans like "safety first" or "lookout for trouble." Although the intent is often well meaning, the tone and content can tend toward the condescending. It suggests that professionals need some kind of external motivation to do a good job within the constraints of the system they work in before they embark on a day's work. But attempting to market unrealistic and unachievable targets can become counterproductive if not damaging to an organization's safety performance. Safety promotion programs are littered with meaningless slogans which exhort safety principles but do little in the way of directing safety practices. In the words of W. Edwards Demming, "Eliminate slogans, exhortations, and targets for the workforce asking for zero defects and new levels of productivity. Such exhortations only create adversarial relationships, as the bulk of the causes of low quality and low productivity belong to the system and thus lie beyond the power of the work force."[44]

Poorly managed or crudely articulated safety campaigns can accentuate the differences between what management personnel are trying to promote as principle and what operational staff experiences in terms of adequate resources. This dissonance can prompt a cynical backlash from personnel rather than rejuvenate safety mind-sets.

If safety is to be promoted both internally and externally by an organization, it's crucial to have the right skill sets and adequate resources involved from the beginning. The appointment as head of safety or promoter of safety programs can be seen as something of a thankless task. This has at least partially to do with the fact that it's very difficult to show tangible benefits from successful safety management, particularly in complex ultrasafe systems. The other issue is that safety can become synonymous with increasing cost rather than making profit. For these two reasons, safety is perhaps not seen as a route to success within many organizations.

There are however exceptions. During Jerry Lederer's remarkable career in aviation safety, he organized and headed the Flight Safety Foundation and later went on to head safety for all of NASA's operations. Lederer had a successful career by understanding that maintaining the right balance of communication and even marketing safety was in many ways more important than trying to scare people into safety rule conformity. Reflecting on his career in aviation safety management, in 2004, he said, "When word got around that I was starting up, some people said that I should not get into this stuff, that I would be sitting on a keg of dynamite, that it would ruin my career and that safety was not a saleable object—shows you how safety was a hard sell in those days. You mentioned safety and you scared people away. That is the big thing that I had to overcome—by diplomacy, mostly, and by not putting out things that would scare people."

Lederer was aware of the crucial importance of safety to successful program operations in aviation, but he was equally cognizant of the need for people to understand the how and the why of safety management. While safety is certainly an area that carries considerable potential to create anxiety, not just for passengers and crew but for those involved in managing commercial aviation operations. Safety must be marketed through effective communication throughout the system and not just imposed as the burden of front-line workers.

As we have previously stated, success in commercial aviation is not just about being safe, it's about being seen to be safe. As we have discussed earlier in this chapter, feeling safe is an essential requirement to humans. Without it, we can no longer function effectively as we are constantly distracted by fears for our own safety and those we care about. Safety has been identified by several empirical studies as the key determining factor when customers are choosing an airline.[45,46] Using safety performance as a marketing tool is generally avoided by commercial aviation business; however, there is a complex relationship between safety and commercial viability that extends beyond fear of personal harm. As safety is such a powerful and emotive influence, it should come as no surprise that there is strong evidence linking overall customer satisfaction with perceived levels of safety. Rather than treating safety as purely a cost, it should be considered an asset with which airlines and airports can enhance the link between safety and customer satisfaction.

2.5.5 The Cost of Accidents

A major aviation accident with multi-fatalities represents the ultimate waste of resources, reduction in social cohesion, and failure of public and private governance. Aviation operators are significantly affected by these traumatic events and some organizations never recover from the economic or public relations impact. Following two successive fatal crashes in July 2014 and February 2015, TransAsia airlines ceased operations in November 2016. In July 2014, a TransAsia ATR-GIE, Flight GE222, was on approach to

FIGURE 2-8 TransAir Flight GE235 caught on a car's dashcam as it impacts a bridge and crashes into the Keelung River. (Source: Air Safety Council Report)

Magong Airport when it impacted terrain within 1 kilometer of the airport. Of the 54 passengers and 4 crew, only 10 people survived.[47]

Eight months later in February 2015, a TransAsia ATR72, Flight GE235, lost control following an engine failure after takeoff. Seen in a still from the dashcam footage of a passing car, in Figure 2-8, the aircraft's wingtip collided with a taxi on an overpass before crashing into the Keelung River.[48] All 43 occupants were fatally injured including all 3 flight crew, 1 cabin crew, and 39 passengers.[49] It is not just a case of simple cash flow which brings these organizations down but the loss in confidence from investors, customers, and sometimes their own staff which makes their continued operation unviable.

Although rare, and depending on the severity of the crash, the cost of a single commercial aviation accident can reach over half a billion US dollars. According to the research company Cirium, the commercial aviation insurance sector paid out $919,000 million in "all-risks" hull and liability claims during 2020, despite the COVID-19 pandemic reducing the operating global fleet activity by approximately 60%. For a more accurate reflection of typical losses in the sector, the average cost per year across the world is $1.51 billion over the previous 10-year period. But again, the toll of accidents, particularly fatal accidents, extends far beyond financial cost.

Following the fatal crash of two of its recent versions of the Boeing 737, the 737-MAX, the Boeing Corporation has faced considerable financial losses, estimated to be in the region of $65 billion, which represents a devaluation of approximately one-quarter of its market capitalization.[50] The total cost to Boeing looks likely to exceed that of BP following the explosion of the Deepwater Horizon which stands at $68 billion. This would constitute the largest corporate failure in history. Even a huge company like Boeing cannot guarantee its ongoing survival following this level of attrition. The financial media have repeatedly commented on the irrationality of shareholders abandoning such a traditionally stoic company and have pointed to its long and generally successful history as a pioneer and leader of the US aircraft manufacturing industry. These comments of atonement have merit, but they must also be placed into the context of today's level of risk tolerance and societal tolerance of corporate conduct. In Chapter 16 we shall explore the phenomena of changing attitudes to the level of liability of those corporate entities in commercial aviation that do not achieve acceptable standards of public safety. This issue will become increasingly important to the future success of commercial aviation through the 21st century.

Key Terms

Acceptable Risk

ALOS

Heat Map

Risk

Safety

Safety Management Manual

Target Zero

Tolerable Risk

Topics for Discussion

1. Why were the Asiana crash at San Francisco on July 6, 2013, and the Rediske Air crash at Soldotna Airport on July 7, 2013, under different standards of aviation regulation?
2. What makes safety so important to humans?
3. Provide a simple definition of risk.
4. How does the international regulatory body ICAO define safety?
5. Give some examples of how we measure safety and risk in commercial aviation.
6. What does ICAO Annex 6 require regional legislation to have implemented?
7. What are the two most significant costs to airlines?
8. What would you say is the highest priority of an airline's senior management? Please give evidence.
9. How can airlines successfully balance commercial success and good safety practice? What measures would you implement?
10. What is ALOS and where is it defined? Is it the same standard for each type of commercial aviation operator?
11. What is the relationship between Hazard and Risk?

References

1. Schiavo, M. (1997). *Flying Blind, Flying Safe*. New York: William Morrow.
2. NTSB press briefing. https://www.youtube.com/watch?v=d9MTLlzf8Co
3. The BBC (British Broadcasting Corporation). https://www.bbc.com/future/article/20200630-what-makes-people-stop-caring
4. *USA Today*. June 20, 2014. https://eu.usatoday.com/story/news/nation/2014/06/14/unfit-for-flight-part-3/10533813/. Accessed March 2, 2022.
5. https://www.cambridge.org/elt/blog/2018/05/29/general-service-list/. Accessed May 25, 2022.
6. Kruglanski, A. W., & Orehek, E. (2012). The need for certainty as a psychological nexus for individuals and society. In: M. A. Hogg & D. L. Blaylock (eds), *Extremism and the Psychology of Uncertainty*. Oxford: Wiley-Blackwell, pp. 3–18.
7. Weick, K. E. (1987). Organizational culture as a source of high reliability. *California Management Review, 29*(2), 112–127.
8. Leveson, N. G. (2004). A new accident model for engineering safer systems. *Safety Science, 42*(4), 237–270.
9. Reason, J. (2000). Safety paradoxes and safety culture. *Injury Control and Safety Promotion, 7*(1), 3–14.
10. Charles Lindbergh, journal entry August 26, 1938, published in *The Wartime Journals*, 1970.
11. https://www.merriam-webster.com/dictionary/risk
12. https://www.mdpi.com/2071-1050/11/21/6024/htm
13. HSE (Health & Safety Executive). (2001). Reducing risks and protecting people. http://www.hse.gov.uk/risk/theory/r2p2.pdf. Accessed August 12, 2017.
14. Bernstein, P. L. (1996). *Against the Gods: The Remarkable Story of Risk*. New York: Wiley.
15. Stolzer, A. J., Halford, C. D., & Goglia, J. J. (eds). (2011). *Implementing Safety Management Systems in Aviation*. Farnham, Surrey: Ashgate Publishing, Ltd.
16. Wilkinson, G., & David, R. (2009). Back to basics: Risk matrices and ALARP. In: C. Dale & T. Anderson (eds), *Safety-Critical Systems: Problems, Process and Practice*. London: Springer, pp. 179–182.
17. https://safetydifferently.com/oil-and-gas-safety-in-a-post-truth-world/
18. Wilbur Wright in a letter to his father, September 1900.
19. ICAO (International Civil Aviation Organisation). (2016). Safety report: A co-ordinated risk-based approach to improving global aviation safety. https://www.icao.int/safety/Documents/ICAO_Safety_Report_2015_Web.pdf
20. ICAO (International Civil Aviation Organisation). (2017). *Safety Management Manual (SMM)*. 4th ed. Doc 9859 AN/474. https://www.icao.int/safety/SafetyManagement/Documents/SMM%204th%20edition%20highlights.pdf
21. Amalberti, R. (2001). The paradoxes of almost totally safe transportation systems. *Safety Science, 37*(2), 109–126.
22. FAA/EUROCONTROL. (2007). ATM safety techniques and toolbox safety action plan 15. https://www.eurocontrol.int/eec/gallery/content/public/document/eec/report/2007/023_Safety_techniques_and_toolbox.pdf
23. Lofquist, E. A. (2010). The art of measuring nothing: The paradox of measuring safety in a changing civil aviation industry using traditional safety metrics. *Safety Science, 48*(10), 1520–1529.
24. Hollnagel, E. (2014). *Safety-1 and Safety-2: The Past & Future of Safety Management*. Farnham, Surrey: Ashgate.

25. https://www.losacollaborative.com/about/
26. https://www.faa.gov/regulations_policies/advisory_circulars/index.cfm/go/document.information/documentid/23227
27. https://www.asias.faa.gov/apex/f?p=100:1
28. https://www.iata.org/en/services/statistics/gadm/fdx/
29. Beck, U. (1992). *Risk Society: Towards a New Modernity*. London: Sage.
30. Bernstein, P. L. (1996). *Against the Gods: The Remarkable Story of Risk*. New York: Wiley.
31. https://www.reuters.com/world/us/us-traffic-deaths-soar-38680-2020-highest-yearly-total-since-2007-2021-06-03/
32. Calabresi, G. (2008). *The Cost of Accidents*. New Haven, CT: Yale University Press. Also see Reducing Risks and Protecting People—R2P2. https://www.hse.gov.uk/managing/theory/r2p2.htm
33. Braithwaite, G. (2001). *Attitude or Latitude? Australian Aviation Safety*. Farnham, Surrey: Ashgate.
34. ICAO Safety Management Manual Doc 9859, Annex 11, Attachment E.
35. https://www.theguardian.com/business/2000/aug/27/transportintheuk.theobserver
36. Hopkins, A., & Maslen, S. (2019). *Risky Rewards: How Company Bonuses Affect Safety*. Boca Raton, FL: CRC Press.
37. https://www.reuters.com/business/aerospace-defense/shareholders-may-pursue-some-737-max-claims-against-boeing-board-2021-09-07/
38. https://www.justice.gov/opa/pr/boeing-charged-737-max-fraud-conspiracy-and-agrees-pay-over-2.5-billion
39. https://www.justice.gov/opa/pr/bp-exploration-and-production-inc-agrees-plead-guilty-felony-manslaughter-environmental
40. International Air Transport Association. (2009). Improved profitability but margins still pathetic: Europe continues to lag. Press Release no. 57, Montreal, December 14. https://www.iata.org/en/pressroom/pressroom-archive/2010-press-releases/2010-12-14-01/
41. https://transportgeography.org/contents/chapter5/air-transport/airline-operating-costs/
42. ICAO (International Civil Aviation Organisation). (2009). *Safety Management Manual (SMM)*.
43. https://rgl.faa.gov/Regulatory_and_Guidance_Library/rgAdvisoryCircular.nsf/0/6485143d5ec81aae8625719b0055c9e5/$FILE/AC%20120-92.pdf
44. Demming, W. E. (1986). *Out of the Crisis*. Cambridge, MA: Massachusetts Institute of Technology.
45. Atalik, Ö., & Özel, E. (2007). Passenger expectations and factors affecting their choice of low-cost carriers—Pegasus Airlines. Paper presented at the Northeast Business and Economics Association Conference, Central Connecticut State University, New Britain, November 7–9.
46. Gilbert, D., & Wong, R. K. C. (2003). Passenger expectations and airline services—A Hong Kong based study. *Tourism Management*, 24(5), 519–532.
47. https://www.ttsb.gov.tw/media/3194/asc-aor-16-01-002-en.pdf
48. https://www.youtube.com/watch?v=I8l1_ZDzxcM. Accessed November 2022
49. https://www.ttsb.gov.tw/media/3209/asc-aor-16-06-001-en.pdf. Accessed November 2022
50. https://edition.cnn.com/2020/11/17/business/boeing-737-max-grounding-cost/index.html. Accessed November 2022.

Why Do Accidents Happen?

Introduction

Throughout the first two chapters we have considered the subject of commercial aviation safety from its foundations. One of the most common questions asked in the aftermath of an accident is "why did it happen"? As we have discussed, there are often a multitude of experts who are willing to step into the media limelight and explain complex events in nothing more than a few well-rehearsed headlines. Although the human need to understand these often-tragic events is compelling, to take authentic lessons from them and to continue to improve commercial aviation safety performance, we must have a solid grasp of the basic principles of what accidents are and what causes them. From this position, we can then begin to understand some of the more widely used models and theories of safety, used to explain and understand the principles of commercial aviation safety.

3.1 What Is an Accident?

3.1.1 Defining Accidents

The word *accident* is commonly used to describe something unintended or unexpected and generally harmful or unwelcome. The word first appeared in 14th-century French literature and would appear to be a derivative of the Latin word "accidēns" meaning "falling down" or descending, a particularly apt definition within the context of aviation safety. It is also referred to as an event, a happening, or as something taking place or occurring. The definition of accident featured in the Oxford English dictionary introduces the associated negative effects of an accident: an unpleasant event, especially in a vehicle, that happens unexpectedly and causes injury or damage.

The word has strong emotional connotations; it makes us feel uncomfortable, as by definition it is something that can cause damage or harm, but it is something we have little or no control over. In the promotion of safety management, a well-known cliché often appears in promotional material or lectures—"safety is no accident." This simple and apparently intuitive phrase may seem self-evident, but as we are beginning to discover in commercial aviation safety, very little of the subject is either simple or self-evident. While we acknowledge the purpose of safety management is to reduce the number of accidents to a minimum, managing the safety performance of modern commercial aviation systems requires more sophisticated concepts and management techniques than simply recounting adverse events.

Annex 13 of the ICAO Convention[1] defines an accident as an occurrence associated with the operation of an aircraft which takes place between the time any person boards the aircraft with the intention of flight until such time as all such persons have disembarked:

- in which a person is fatally or seriously injured[i];
- in which an aircraft sustains damage or structural failure requiring repair;
- after which the aircraft in question is classified as being missing.[ii]

Annex 13 also provides us with a definition of an incident as opposed to an accident. An incident is an occurrence other than an accident, associated with the operation of an aircraft, which affects or could affect the safety of operation.[1] The Annex goes on to define a "serious incident" as an incident involving circumstances that an accident nearly occurred.[iii] The first accompanying note to the definition of a serious incident clarifies the point, that the difference between an accident and a serious incident lies only in the severity of outcome. The second note gives reference to examples of what might be considered as serious, suggesting there is some level of subjective judgment as to what constitutes a serious incident.

Effective aviation safety management must constantly reevaluate system performance to maintain and improve safety standards. The French academic René Amalberti has suggested that aviation safety management has placed too much emphasis on trying to learn from specific accidents rather than developing other metrics of system health.[2] If we were to rely too heavily on the simple metrics of accident numbers, particularly in an ultrasafe system,[iv] there is a danger we are not getting an accurate picture of the underlying safety system performance. If accidents are anomalous events, he suggests that we are only looking at the overall system from a narrow perspective.

According to the definitions in Annex 13, both accident and (serious) incident data are crucial to effective system monitoring, but while an aviation accident is highly unlikely to escape detection, the report of an incident, even a serious incident, is reliant on the effectiveness of system monitoring and the organization's safety culture. While many serious incidents will be automatically picked up by system monitoring devices, a significant number will rely on the effectiveness and authenticity of the reporting culture of the relevant operating organization. We shall look at the impact of culture on effective safety management in some depth in Chapter 12 and explore why it is so important in the increasingly complex system of commercial aviation.

While the *output* of a safe system may well be no accidents, overreliance on this reactive metric to *manage* system safety will invariably inhibit our ability to develop more predictive safety management processes. Throughout this book, we will explore the way that effective safety management has continued to evolve by adopting an

---

[i] For statistical uniformity only, an injury resulting in death within thirty days of the date of the accident is classified as a fatal injury by ICAO.

[ii] An aircraft is considered to be missing when the official search has been terminated and the wreckage has not been located such as the loss of Malaysian Airlines MH370 on March 8, 2014.

[iii] Examples of "serious incidents" can be found in Attachment C of Annex 13 and in the Accident/Incident Reporting Manual (Doc 9156).

[iv] Amalberti describes an ultrasafe system as one where the risk of a disastrous accident is below one accident per 10 million events (1×10^{-7}).

increasingly broad array of trend-monitoring techniques and system management processes that continue to improve safety standards.

3.1.2 The Perception of Accidents

How we perceive the nature of accidents has a significant effect on how we try to interpret why and how they happen. Danish safety researchers Jonas Lundberg, Carl Rollenhagen, and Erik Hollnagel refer to the "What-you-look-for-is-what-you-find" principle.[3] They assert that the way we make assumptions about the nature of an accident is highly influential in what we "discover" during investigations. In this chapter, we shall explore how the framing of an accident by media and even official reports can dramatically alter our interpretation of a series of events. Even the word *accident* itself brings with it several preconceived ideas which influence our thinking and our explanations.

The use of the term "accident" was actively promoted by car manufacturers in the 1930s during the early days of widespread motor vehicle use. As cars began to invade spaces traditionally occupied by people and horses, serious injuries resulted from these increasingly frequent collisions. This new transport technology was treated with significant suspicion by the public, and court cases tended to find car drivers, especially those of larger cars, to be at fault following injury to pedestrians. This judicial bias was perceived as a threat to the developing industry, so car manufacturers began to provide "training" and sometimes incentives to the journalists that reported these events as *car accidents* rather than *car crashes*. The logic behind these campaigns was that describing collisions between people and cars as accidents, public opinion shifted away from the presumption that these events were avoidable and therefore blameworthy.[4]

In recent years, some safety promoters have campaigned and applied pressure on the media to ban the word *accident* as they see that the use of the word implies an event driven by fate or luck rather than an event that could have been prevented. This viewpoint has been challenged by academics as there is little to no empirical evidence to suggest that by changing the description of an event by removing the word "accident" from media reporting improves safety performance. A nationwide survey carried out in the United States found that 83% of people continue to naturally associate the word *accident* with preventability rather than associating it with fate or luck.[5]

What can be established from safety research is that commonly held assumptions about accidents can be framed to influence our understanding of how these events unfold.[6] The influence of the media on our thinking about safety is particularly relevant to aviation because of the significant impact aviation has on our social or collective psyche. Although automobile accidents present significantly more risk to people than modern commercial aviation, an aircraft accident attracts significantly more media coverage.[7] As we saw in the previous chapter, media reports of aviation accidents (and public interest levels) can significantly vary in the amount of profile they are afforded, and as we shall discover the way in which they are framed.

3.1.3 Framing the Accident

In basic terms, there are two general approaches taken by the media in describing accidents: episodic framing and thematic framing. The differences between the two are important to understand when we read or hear about aviation accidents. The most common approach is that of episodic framing which tends to take a narrower and more personalized perspective on individual case studies and discrete events. This approach can encourage the blaming or victimization of those involved in an accident as it focuses on individual details rather than the broader situation. For example, the media might report on the crash of a commercial aircraft and focus on those directly affected by the tragedy of one incident. Episodic framing would not include or discuss broader issues of flight safety across the industry or risk technology changes which might give the story more context. The lesser used approach of thematic framing takes the much wider, systems perspective by focusing on the relevant issues and identifying trends and themes over time. When we are analyzing aviation accident reports from investigators or from the media it is important that we are conscious of how the story is being framed and to consider what actual evidence is being presented.

It is important to remember when reviewing aviation accident reports either from governmental or media sources, the story will be framed in a particular way which is appropriate to the context and sometimes intent of the report. Given the nature of public interest in aviation accidents, which is often intense but often short-lived, a great deal of resource is used to provide a forensic examination and objective explanation, but this perspective invariably focuses on the immediate and proximate events prior to an accident. There is a tendency to report and interpret accidents through the lens of episodic framing, focusing in detail on the immediate links to the catastrophe.

There are practical reasons why the broader social and political influences on commercial aviation safety are often not explored in great depth. Like any public body, investigative institutions are resource limited, and it would be perhaps naïve to assume they are immune from some level of political or economic influence. In the accident summary at the end of this chapter, it is worth noting that accidents involving Western-built aircraft attract more resources for investigation than those occurring in developing nations and involving non-Western-built aircraft. Safety researchers have noted that broader political, organizational, or system considerations are generally only analyzed in depth after high-profile accidents (generally involving Western-built aircraft) when resources are made available due to intense public demands, Western media interest, and political acquiescence.[8]

Major accidents can give rise to multimillion dollar inquiries and produce hours of taped and thousands of pages of transcript evidence underlining sociopolitical connection to regions of global affluence.[9] As those individuals that are conducting enquiries are often given significantly more resources and free reign as to what area they are investigating, broader influences such as organizational culture, social policy, or regulatory competence are brought into the explanation of why an accident occurred.[v] Those left with more limited investigation resources must focus on the more proximate explanation of what happened rather than exploring the broader context of why commercial aviation accidents occur.

3.1.4 The Mount Erebus Disaster

One of the most famous examples of how even major inquiries can alter the way in which an accident is interpreted is in the investigation report and subsequent Royal Commission of Inquiry into the crash of Air New Zealand Flight TE-901. The aircraft, a McDonnell Douglas DC-10, flew into Mount Erebus in Antarctica during a sightseeing

---

[v] An inexhaustive list of such high-profile aviation accidents could include the Tenerife PanAm-KLM ground collision, Air France 447, Air Ontario Flight 1363 (the Dryden Inquiry), Air New Zealand Flight 901 (discussed in this chapter).

Chapter 3: Why Do Accidents Happen? **61**

flight on November 28, 1978. All 237 passengers and 20 crew were killed in the impact and it remains New Zealand's deadliest peacetime disaster. Although the remains of the victims were painstakingly recovered from the scene for forensic identification, the crash site was so remote and inaccessible that much of the wreckage and debris from the crash remain at the site even today.

The aircraft's route had been reprogrammed by the airline's operational management team to fly toward the 12,500-foot (3810 meters) mountain rather than the original and much lower route following the McMurdo Sound. The reroute had not been discussed with the Flight TE-901 crew. Although the forecast for blowing snow and the potential for white-out conditions had been mentioned, it was inadequately briefed. The lack of effective communication within the Air New Zealand operations staff managing Flight TE-901 resulted in the flight crew unknowingly operating below a minimum safe altitude in conditions of poor meteorological visibility.

The initial accident inquiry was conducted by New Zealand's chief inspector of air accidents, Ron Chippindale. Chippindale's report cited pilot error as the main cause of the accident—in particular, Captain Jim Collin's decision to descend below the customary minimum safe altitude level of 6000 feet (1830 meters) when the crew were uncertain of their position. However, a set of circumstances prior to, and during, the flight had led Captain Collins to believe the aircraft's position was over the sea of McMurdo Sound, as depicted in Figure 3-1. Several pilots disagreed with Chippendale's report findings, most notably another Air New Zealand Captain, Gordon Vette, who, having become increasingly suspicious of the motive and direction of the official report, conducted his own investigation into the crash. The release of the official accident report on June 12, 1980, prompted a public outcry and the New Zealand government immediately initiated a Royal Commission led by Justice Peter Mahon.

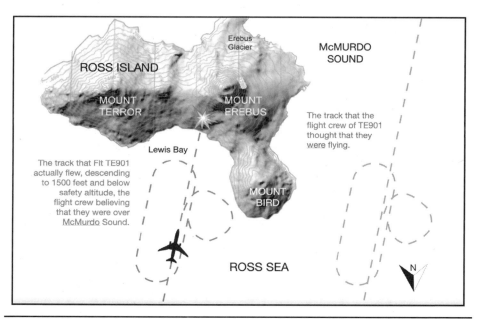

FIGURE 3-1 The track that Flight TE901 flew and an approximation of the track the crew thought they were following. (Source: Authors)

The Mahon Inquiry controversially concluded that Air New Zealand executives and senior pilots had engaged in a conspiracy of lies and coverups from accident investigators. Mahon's report cleared the crew of blame and instead focused on the airline and broader regulatory framework of the airline industry. Although he observed 10 specific factors that were all required in order for the accident to have occurred,[10] Mahon made the following overall statement in his concluding remarks:

> In my opinion therefore, the single dominant and effective cause of the disaster was the mistake made by those airline officials who programmed the aircraft to fly directly at Mount Erebus and omitted to tell the aircrew. That mistake is directly attributable, not so much to the persons who made it, but to the incompetent administrative airline procedures which made the mistake possible.[11]

The initial accident report conducted by Ron Chippindale is an example where the episodic framing of an accident report is narrowed by the restricted area of inquiry. Although it was certainly not helped by the considerable level of obstruction put in place through the withholding of crucial information, and at times, outright deceit of some of Air New Zealand's executives, it serves to illustrate how a narrower view of investigation can obscure the context in which an accident takes place. The broader remit, or thematic framing, of Justice Peter Mahon's major inquiry gave an entirely different explanation of the Mount Erebus disaster. These differing perspectives of the initial accident report and the public inquiry lead us to two entirely different causal explanations of why that accident happened. It is now appropriate that we start to unravel what we mean by accident causation.

3.2   Accident Causation

3.2.1   Designing Our Own Accident

To start to develop our knowledge of causation, let's design our own (fictitious and minor) accident and explore some of the basic issues of causation that we experience in day-to-day life. Imagine you have been offered an interview for a new job. It's a job you have been wanting for a long time, so you want to give the interview your best shot. The interview is face-to-face and a few hours' drive from your home. Wanting to be your best on the day, you think ahead and contact a relative who lives a few miles from the interview site. The plan is to travel by train the day before and stay the night with your relative ensuring you are not too tired by a long car drive. You arrive in the evening and have dinner. It's been a long time since you have seen your relative and the conversation goes on for a couple of hours, well after dinner. After a while, you realize you don't have enough time to give a final review of the interview notes that you had prepared and bring the conversation with your talkative relative to a close. Going to bed and getting a good night's sleep is now the priority. However, your thoughts keep going back to the notes you hadn't had time to read over, and struggle to get to sleep. The bed feels very different to your own and there is a lot of wind and rain noise due to a storm in the night.

In the morning you feel like you had a bad night's sleep and are feeling a little anxious. Your relative tells you there is a bus strike today, but you have plenty of time to get a taxi. You are a little abrupt with your relative and decline breakfast but grab a strong coffee to try to get you going for the important day. In your haste, you spill the coffee down your white shirt. Thankfully, your relative has another similar (albeit slightly smaller) white shirt but you find it difficult to hide the fact that you are getting quite frustrated. The taxi company is overbooked due to the bus strike and wet weather so despite the heavy rain you decide to walk the two miles to the interview. The 2-mile walk is an estimated distance from the map on your phone but doesn't show that the entrance to the company site is on the opposite side from where you are walking; it adds another half mile to your walk. You are now late, soaking wet, and anxious about the interview which does not go well. The interviewers found you to seem a little distracted, underprepared, and nervous. You receive a rejection letter within a week.

So, what *caused* this to happen? Was it the poor night's sleep? Your talkative relative? The bus strike? The weather? The coffee spill? Your emotions? Misinterpreting navigation data on your phone? From your perspective, you had done your best. You started the process well prepared (notes and how to get to the interview) and gave thought to how to overcome some perceived difficulties (timings and fatigue). You used all available resources (stayed at a relative's house) and coped with minor setbacks (coffee spill and shirt change), adapted as the situation changed (taxi instead of bus), and had a backup plan (walk instead of ride). Was there one event that we could pick out of this short and simple story and elevate its importance to the cause of failure? There seem to be several reasons why the interview didn't go well, some of them interrelated and others just contributed to the outcome, but nothing seems to stand out as a singular cause that explains why things worked out the way they did.

Other perspectives on the reason for your poor performance might provide a completely contrasting causal explanation. We could investigate by asking those closest to the story what their thoughts might be. Your relative might describe you as a little abrupt and perhaps a little rude (overly task-focused) during your visit; these qualities would not have helped in an interview (poor interpersonal skills). They could cite as evidence the way you abruptly closed down the evening's conversation when they were about to explain about the bus strike (lack of effective communication). You might even be blamed for the coffee spill in your haste to leave their house (distracted) or not having your own spare shirt (lack of resources), for not checking the transport situation yourself in advance (poor planning). The interviewer's perspective was not challenged by you explaining why you were late (lack of assertiveness) saw you as nervous (overly emotional or stressed) and underprepared despite your best efforts to avoid fatigue and a last-minute rush (lack of preparation).

The problem we face by taking this arbitrary approach in picking out any or all of these plausible explanations is that the selection of which antecedents or preceding events should be considered is a highly subjective process. If we held any kind of prior agenda and wanted to establish blame for an event, we could pick out from the list of explanations any combination which suited our preconceived or precontrived explanation of any event. What it does not give us is any kind of scientific explanation of why an event or an accident has occurred—we can simply decide on whatever plausible explanation best suits our needs without really learning why things happen. This is not to say this approach is always invalid, but it is certainly not without its vulnerabilities. Perhaps the first step to improving this methodology is simply to accept that an element of subjectivity is always present when deciding what is relevant. Some level of bias could be mitigated by using an evidence-based argument, professionalism, and good judgment but we humans are naturally creatures of biases and emotion rather than process and logic.

3.2.2   Accidents and Human Behavior

Some studies that have focused on the precursors of aviation accidents have attempted to identify behavioral patterns and even the personality traits of the individuals associated with the event. In fact, one of the first scientific theories which attempted to explain the reason that accidents occurred is known as "accident proneness theory."[12] The theory posited that certain individuals had a special susceptibility to accidents. However, accident proneness theory has been largely discredited; in addition to lacking empirical evidence, the theory failed to consider the disparities that exist between different individual subject's exposure to risk which can have considerable variation. For example, if a flight crew were going to regularly operate in good weather environments with good training facilities, high-quality ATC support, and on routes they were familiar with, they would not be exposed to the same level of risk as crews that had to operate in poor weather, minimal training, no ATC support, and continually changing destinations.

One approach that has attempted to link human behavior characteristics and error in aircraft maintenance is the "Dirty Dozen,"[13] a concept developed by Gordon Dupont, then an employee of Transport Canada. Through successive case studies Dupont identified 12 of the most common characteristics of accidents which have been subsequently used as the basis of many human factors training courses[vi] in aircraft maintenance around the world. ICAO has developed a considerably more comprehensive list of conditions containing over 300 human error precursors.[14] We shall consider the relationship between human factors and accident theory in depth in Part III, but for now we are still left with the problem of trying to identify what elements of the preexisting circumstances can be considered a cause rather than simply considered a preexisting condition. Causation theory is a vast and complex subject. While this book is intended to keep a practical perspective about commercial aviation safety, it would do you, the reader, a great disservice were we not to consider some of the theoretical basics of accident causation.

3.2.3   Defining Cause

In the definition section of Annex 13 of the Chicago Convention, cause is described as "[A]ctions, omissions, events, conditions, or a combination thereof, which led to the accident or incident."[15] The definition, like others which attempt to encapsulate a definition of causation, is wide and leaves the gate open to our personal interpretation. Humans have a deep psychological need to understand how the world around them works. This is particularly true when things happen that we didn't anticipate or don't really understand—we need to find a causal explanation. However, the principles of causation are often counterintuitive, and humans are highly reluctant (and sometimes unable) to expend the cognitive effort to deal with this level of complexity. This is largely to do with the way we are designed. The way a human brain operates is a masterclass in the efficient use of limited resources. Neuroscientists estimate that the brain uses around 20 watts or the same amount of electrical power it takes to run a dim electric bulb. The brain is lazy by design. This minimal use of energy has significant consequences for the way in which we perceive the world: we all prefer the quick and easy solutions. When we consider the complexities of causal theory and the amount of

---

[vi] One example is found in the United Kingdom's Civil Aviation Authority training guide for human factors in aircraft maintenance: https://publicapps.caa.co.uk/docs/33/CAP715.PDF

thought and effort it requires, it perhaps comes as no surprise that we grab hold of the first plausible explanation of events and stubbornly stick to it.

The intensity of this phenomenon is driven by a fear of the unknown. According to the celebrated 19th-century German thinker Friedrich Nietzche, "First principle: any explanation is better than none. Because it is fundamentally just our desire to be rid of an unpleasant uncertainty, we are not very particular about how we get rid of it: the first interpretation that explains the unknown in familiar terms feels so good that one 'accepts it as true.' We use the feeling of pleasure ('of strength') as our criterion for truth. A causal explanation is thus contingent on (and aroused by) a feeling of fear."[16]

However, if we are to develop our understanding of safety, then we must give some thought to the concepts of causation, not only to understand why things go wrong, but perhaps more importantly, why things go right. We have seen that whether we take a narrow or broad perspective of an accident, the framing of events can lead us to very different conclusions. We have also determined that accidents are often formally defined by their outcome or their aftermath rather than the related events that preceded them. It would be very easy at this point to overlook this complex and thorny subject but without the basic knowledge, causation keeps coming back to inhibit our understanding of aviation and safety science.

3.2.4   Causal Theory

In his essay "On the Notion of Cause,"[17] the famous British philosopher Bertrand Russell noted that progress in science consists of recognizing "a continually wider circle of antecedents" as necessary for the precise calculation of any event. In other words, in order to predict and manage future events (such as preventing accidents), we must consider an increasing number and more broadly-related set of preexisting conditions. It should be emphasized that there is no academic consensus on what defines cause. Some philosophers, notably the empiricists David Hume, John Locke, and George Berkeley, have argued there is no such thing as causation as they believe all human knowledge has its base in experience, so if we cannot sense or experience causation, then it doesn't exist. This empirical approach requires that we see or experience a phenomenon firsthand for it to be real—as we cannot see causation, it is not real.

This perspective is challenged by the rationalists who argue that human knowledge can be derived through pure reasoning, and we can therefore make some assumptions based on our knowledge of the world. It is fair to say that most practitioners engaged in accident investigations would fall into the rationalist camp. If we arrived at an accident site at the side of a mountain, where the aircraft had clearly impacted at high speed and has disintegrated into small pieces, it is probably safe to assume what was the immediate cause of the aircraft's destruction. We wouldn't recrash another similar aircraft to prove that hitting a mountain at high speed will cause catastrophic damage. Just because we didn't directly sense the event doesn't mean we cannot *rationally* infer some elements of cause. But to understand the wider context of why and how the destroyed aircraft ended up where it did, we must, as Bertrand Russell suggests, cast a much wider net.

Trying to develop an objective way to link cause and effect has haunted some of our greatest minds so we should not attempt to entirely resolve the issue here. What seems to be consistent is that our understanding of any event is predetermined by our perspective of the world. Our subjectivity isn't just a human flaw—it's a human design characteristic. Our short-cuts to understanding, or heuristics, are an efficient way of making sense of the complexities of our world. However, it does mean that we are not naturally equipped to delve into the intricacies of causal philosophy—it's just too much effort.

This issue was extensively considered by the 19th-century English philosopher John Stuart Mill.[18] He described causation as a cluster of related concepts, preceded by multiple related antecedents (preceding events). Mill states that trying to decide between what is a cause and what is a condition is a waste of time as humans will simply make the pattern of events fit their own perspective. In other words, when we choose what is a cause and what is simply a condition, we do so to suit our own purposes. Whatever merits, Mill's argument might have, it illustrates the complex and sometimes circular arguments that discussing cause can invoke. But in order to understand the various accident models that we shall review in this chapter, a basic understanding of the issues surrounding causation is extremely useful.

3.2.5   Necessary and Sufficient Conditions

When we try to explain an accident, we can define the preconditions as either a "necessary" or "sufficient" condition. Let us imagine that after reading this fascinating book we were to examine you on its contents, and you needed to pass the examination to complete your commercial aviation safety qualification. It would be a *necessary* condition for you to have read the book and be familiar with the contents, but it would not be a *sufficient* condition for you to pass the examination as you may not understand the contents of the book, the exam questions, or you may not even turn up for the exam.

Another way of understanding the difference is to imagine two soda-dispensing machines. Let us imagine that the dispensing of the can of soda is the "event" or "accident" we are trying to explain and the insertion of a coin as the cause. Each machine has a different fault so that one machine requires a coin to dispense a can of soda, but this does not always work. The inserting of a coin is *necessary* but not *sufficient* condition with which to obtain the soda. The other machine will dispense a soda every time a coin is inserted but it also randomly and periodically dispenses a soda of its own accord. In this machine the insertion of the coin is a *sufficient*, but not always *necessary* condition of obtaining your drink. Some safety theorists argue that both necessary and sufficient conditions need to be present before an event can be described as a cause.

Let us try to apply these concepts into a simple aviation example and then decide whether this is an appropriate approach to accident investigation. If we considered an aircraft that had a design fault in its primary controls which *could* cause a loss of control. The fault is significant but has not yet manifested itself in the loss of control of the aircraft. The fault may cause control failure at any time in normal operation, but it is more likely to appear if the aircraft is flown aggressively at or near its intended design limits. On one day, a flight is undertaken with a pilot who, unaware of the design flaw, flies the aircraft in an unnecessarily aggressive fashion and the design fault triggers a loss of control and subsequent crash of the aircraft. What was the cause of the accident?

First, let us consider the pilot's actions. Flying an aircraft in an unnecessarily aggressive fashion is highly unprofessional, particularly in commercial aviation where the manner of one's professional conduct can directly threaten the safety and the confidence of the traveling public. In this example, some might naturally want to blame the pilot for the loss of life and damage, but the important question is, did the pilot actually *cause* the accident? To establish liability or blame, we might carry out research and point to similar accidents where aggressive flying had resulted in accidents or serious incidents. But in order to establish liability, we would have to link the aggressive flying of the aircraft to this event and this actual crash; we would have to establish a causal link. To establish that link, we might consider whether the pilot's actions were *sufficient* and *necessary* elements of this accident.

As the design fault may have manifested itself without any aggressive flying, we must conclude that the pilot's actions were not a *necessary* element to cause the crash. The pilot in question flew (albeit aggressively and unprofessionally) within the design limits of the aircraft—the pilot's actions were not solely sufficient to initiate the fault. In contrast, when we consider the design flaw of the aircraft, it may well have manifest itself without any unusual or aggressive flying; the flaw is therefore solely *sufficient* to cause the accident. The design fault was also a *necessary* element of the accident as the aggressive flying of the pilot may well have caused discomfort for those onboard but would not in itself have caused the crash.

The pilot's manner of flying may well have increased the probability of the accident, but it is the interrelationship with the design fault which would seem to provide the most logical explanation for the crash. Even in this fictitiously simple example, the complexities of establishing why accidents happen begin to surface. Real-world examples are rarely straightforward and serve to emphasize that the practical application of causal theory is a difficult process by any standards. What many individuals involved in law, investigation, and even within academia have come to accept is that causation is a multifaceted and complex subject. It is therefore worthy of very careful consideration before any rapid conclusions are drawn about how and why an accident occurs. You will note from studying some of the accident reports linked at the end of this chapter, contemporary accident investigators refer to nothing more definite than "probable cause." We will revisit some contemporary approaches to the subject of complexity and complex systems throughout this book.

To further illustrate these points, let us consider the role of human error in an accident. Human error is a widely cited cause of accidents, but one which is often misconstrued by some casual commentators of safety science. An investigator might look for human error associated with an accident. Given the propensity for humans to regularly make errors, such a search is rarely unfruitful; there is always someone to blame. But what then? Where does that lead us? Does the investigation continue to establish why human errors were made? Is human error a cause of the accident or a symptom of broader system-based flaws? To address some of these questions, we need to understand what are the underlying assumptions that are driving the perspective of an investigation; these assumptions fundamentally shape our collective view or model of our world.[19] Because we are driven to understand our environment, we need these models to help us deal with the sheer complexity of accidents. The principle doesn't just cover accident theory but extends across the whole landscape of human cognition. Perhaps best described by the American political and organizational theorist Herbert Simon and his concept of the "bounded rationality" of the individual[20]: "The world you perceive is a drastically simplified model of the real world."[21]

3.3 Developing Perspectives

3.3.1 Evolving Accident Theories

Safety as a scientific subject has gradually developed over the last one hundred years but interest accelerated during the latter decades of the 20th century. Writers such as Ulrich Beck and Anthony Giddens articulated increasing public concerns about the

FIGURE 3-2   Beech Aircraft in production circa 1942. The prevailing scientific view of the early 20th century saw the removal of human intervention as an effective means of improving efficiency and safety performance. (Source: Alden Jewell, Creative Commons)

consequences of increasing dependency on new technology and, in particular, what happens when it fails.

Over the last century, the way in which we perceive the cause and nature of accidents has largely been determined by the predominant social and scientific theories of our time. With the emergence of theories of modern psychology at the beginning of the 20th century, academics, such as the American John Broadus Watson, focused attention on the limitations of the human being. The early accident theories were based on the principles of behaviorism and emergent ideas surrounding human cognition.[vii] Increasing faith in the efficacy of process, machines, and automation inviting interest in theories such as Taylorism.[22] The basic models and techniques of the 1920s were originally intended to deal with lost time accidents. Based on that logic, safety was thought to improve by reducing the vagaries and unpredictable influence of humans in industrial systems, replacing human intervention with technical solutions. In fact, throughout most of the 20th century the predominant perspective was that removing the human element was an effective way to increase productivity and safety performance (Figure 3-2).

By the 1970s safety science developed an increasing focus on the control and direction of energy as the fundamentals of accident causation in models such as Bowtie (discussed later in this chapter) and Tripod Beta. But that period also witnessed a renewed focus on the role of the human within system failures. Although commercial aviation had seen significant improvements in the reliability of aircraft, the effectiveness of the operating crews within these highly automated environments was increasingly

---

[vii] The 1960s saw the further development of Hale and Surry's models which were expanded theories based on the basics of human cognition.

the subject of scrutiny. Against this background, James Reason's Swiss Cheese Model (SCM) promoted the idea of human error as a system threat and spurred a whole series of barrier-based theories and led the way toward organizational and system theories. The latter group of models will be discussed in more detail in Part IV of this book as this theoretical genre represents the predominant focus for today's researchers of safety science.

3.3.2   New Accidents

A phrase has occasionally emerged from some safety commentators that "there are no new accidents." This rather pessimistic summary takes the viewpoint that as many accidents share common characteristics, we cannot be learning from accident investigations; there is almost an inevitability to reoccurrence. The phrase "There are no new accidents" suggests that many patterns or common themes keep reoccurring and are discovered during accident investigation. However, Professor Nancy Leveson has repeatedly suggested[23] that the very nature of accidents in modern complex systems is changing. According to Leveson, it is not just the accelerating developments in technological advancement but a whole array of interrelated social and technological changes that necessitate a change in our mindset about the etiology, or cause, of modern accidents. Leveson has proposed seven developing sociotechnical themes of the 21st century that collectively mean we need to fundamentally shift our previous assumptions about accidents and adopt newer, system-based strategies. Leveson suggests the following reasons why a change of perspective is necessary:

- Fast pace of technological change.
- The changing nature of accidents.
- The emergence of new types of hazards.
- Decreasing social tolerance for single accidents.
- Increasing system complexity and coupling.
- More complex relationships between humans and automation.
- Changing regulatory and public views of safety.

Each of these areas are discussed throughout this book, particularly in Part IV, which will focus on organizational and system safety.

3.3.3   New Investigations

In addition to systems-based approaches to safety and risk management, new perspectives on accident investigation have emerged. Developing theories about the nature of accidents have been recognized and discussed from different perspectives by safety researchers trying to further our understanding of the nature of accidents and how we interpret emerging patterns. Because the scope of investigations tended to sit within the three categories of human, technical, and organizational, many investigations tended to take a "bottom up" perspective in their explanations. However, to focus on only these parts of a complex event limits the investigation and prevents the necessary "top down" or system perspective. Many contemporary safety researchers have concluded that only by adopting a more holistic approach to accident investigation can we begin to manage new and more complex systems.

The problem with this approach is that, although it is scientifically derived, it runs against the natural grain of human cognition. While many accident theorists such as Hollnagel and Leveson have repeatedly called for a more systems-based perspective, there has been limited progress on how that perspective can be related to the real world. Over the past few decades, we have seen trends in theory gradually altering the emphasis and perspective of how we see accidents, but the more these models attempt to reflect system complexity, the more they become an abstract construct to the point that humans cannot link them to the world around them. The closer the models become to the rapidly increasing complexity of the real world, the less practical they become in telling the story in a way that humans can understand and relate to. As we have discussed earlier in this and previous chapters, the purpose of accident investigation is as much about telling the story (so we may learn) as it is about finding some ultimate or absolute truth.

Professor James Reason, whose theories we shall discuss later in this chapter, noted that "... the pendulum may have swung too far in our present attempts to track down possible errors and accident contributions that are widely separated in both time and place from the events themselves."[24] Reason stresses the importance of "keeping it real" when we use these models to explain accidents. To use a well-used (if not over-quoted) idiom, accident models can make a great servant but a poor master. They are tools with which we can understand and explain the world, but they are not the complete story.

3.3.4   Emerging Patterns

When researching data from multiple accident scenarios, it is perhaps unsurprising that some safety analysts will begin to recognize patterns and trends in these complex events. The recognition of these patterns and trends is essential when conducted with appropriate scientific rigor. This is an accepted methodology of safety science and one which stimulates new pathways for future research. However, there is a balance to be struck here between recognizing emergent patterns or behaviors and imposing new theories onto the facts as some form of enlightenment about accident causation.

In the same way humans explain their world through stories or myths, and we also create models or simplifications of our surroundings. We rely on these models as a way of explaining and shaping our understanding of how events happen. When we are dealing with complex areas like accident causation, we use models to understand and explain the how and the why of the world. No one single model can explain the world that it is trying to represent, and each model will have its strengths and weaknesses. In this book, we are not attempting to represent all of the many models that have been used in commercial aviation but only a selection of what we consider to be the most useful and to provide a synopsis of their strengths and weaknesses. Just as we discussed the different ways of "framing" when we report an event, the model we choose to explain an accident can influence how we see and interpret the event and its antecedents.

In the following sections of this chapter, we shall discuss several established accident models. Although there are many approaches to classifying accident models, we have taken a pragmatic approach and simply divided them into two groups. In the first group, there are linear causation models, those that tend to have a recognizable and intuitive sequence of events approach. The second group of models are those models which have attempted to encapsulate complexity through a more holistic or systems-based approach. While acknowledging the need to recognize the changing societal and technological environment, as described by Professor Leveson and others, we also recognize the importance of relatability. How we balance these two critical requirements is a challenge for the future safety science and as a student of safety science, a debate to which we hope you will make your own contribution.

3.4 Linear Causation Models

3.4.1 Accident Chains and Dominoes

One of the first theorists to use models to explain how accidents happened was the American safety engineer Herbert William Heinrich who worked for the Travelers Insurance Company. Heinrich collated data from industrial accidents of the early 20th century. His Domino Theory suggested that accidents occur following a succession of related events sometimes referred to as a chain of events. Having identified what each of these events are through the study of previous or related events, safety management focuses on the removal or suppression of one of these dominoes to stop the sequence of events from continuing. The model is intuitive, relying on a linear causal trajectory and the associated conditions (or dominoes) are identified retrospectively. It makes several assumptions which are at odds with many of the points raised during our discussion of causation. For example, looking back at an event presents us with the reassuring clarity of hindsight rather than the disturbing reality of real-world complexity. We attribute the relevance of preceding events (or antecedents) with the certain knowledge of what will come to pass. Causation is only linear when we look back because we make it so; we do that because as Friedrich Nietzche has explained, it creates anxiety and discomfort for us all when we cannot explain why things happen. Linear causation is a construct of our own minds and of our imagination.

A major objection to the domino explanation of accidents is that it often implies that the individuals most proximate were the cause; if you remove the "faulty" individuals, then you remove the problem. The search for someone to blame after an accident is often short and rarely unfruitful. As we have mentioned, behavioral studies into accidents were carried out between the 1920s and the 1960s assessing the validity of the theory of "accident proneness" in individuals.[25] These studies were never validated by the scientific community and safety researchers have continually pointed to the extent of harm to safety management systems is caused by focusing attention of human error. As Nancy Leveson stated, "Blame is the enemy of safety."[26] The domino explanation of causation limits our ability to look beyond what we preconceive to be the immediate and obvious explanation of an accident. It frames accidents in an episodic way that points a simplistic finger of blame.

3.4.2 The Accident Triangle

Heinrich's most famous theory is the accident triangle, which was first published in a 1931 book *Industrial Accident Prevention*. The theory is based on his analysis of approximately 50,000 accidents where he posited that one fatal accident was associated with 29 minor injury incidents and 300 no-injury incidents. Heinrich's book lasted four editions, and the fourth edition showed a stylized graphical rendering of a "path" to injury. In this edition his triangle began to look something like a pyramid, seen here in Figure 3-3, earning another title for his theory, the "Accident Pyramid".

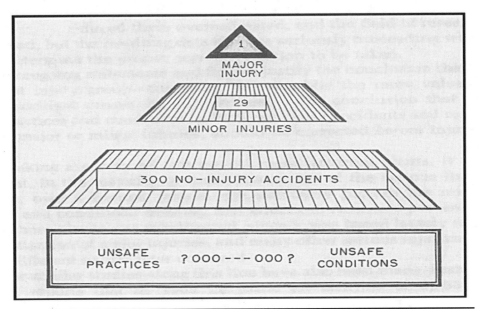

Often mistakenly attributed to Heinrich is a 1969 study carried out by Frank E. Bird, an employee of the Insurance Company of North America. Bird's far more comprehensive study analyzed approximately 1.8 million accidents within 297 cooperating companies engaged in 21 different industrial activities. Bird set out four groups of outcomes: a fatal accident, serious accident, an accident, and an incident. Bird determined the ratio between the different groups to be 1:10:30:600. By categorizing the significance of events by the nature of the outcome both Heinrich and Bird contradict one of the basic laws of causation and risk management.[27] Simply put, sometimes the biggest mistakes will often result in the most inconsequential outcomes and contrastingly, sometimes the tiniest of errors can have catastrophic results.

In attempting to translate the complex subject of causation into a simple model, Heinrich produced what is arguably the most controversial of all accident models. The intuitive sense of the relationship between regular but low-consequence events (trips or minor falls) and rare but high-impact events (fatal accidents) has brought Heinrich many followers and although retracted from the ICAO Safety Management Manual, the model is still used in some circles of safety training and management even today. Scientific papers have repeatedly questioned the validity and predictive capability of the approach[28,29] and some have even suggested that following several attempts to replicate relationship identified by Heinrich that the data is fabricated.[30] In the field of aviation an extensive 10-year study by Arnold Barnett and Alexander Wang, two highly respected MIT professors, showed a complete lack of correlation when attempting to use such data as a predictive tool. In fact, Barnett and Wang noted that in relation to major or fatal accidents the relationship was closer to inverse rather than any positive correlation. Despite its simplicity and many detractors, the model has served as a medium to explain elements of safety within organizations, but it does not stand up to scientific scrutiny as having any basis for a predictive tool. The problem arises, as with many models, that it is sometimes used as if it were science and used in safety management by individuals who do not recognize the model's inherent limitations.

Perhaps the biggest criticism of the belief in Heinrich's model is that it promotes focus onto individual events rather than broader system safety. Let's imagine an airline noted an increasing number of low-speed events which occurred when pilots flew without the aircraft's autothrottle engaged.[viii] These events mean the aircraft get very close to a stall which is a condition when insufficient lift is generated by the wings. Left to deteriorate, these events could result in a loss of control of the aircraft so are clearly a significant threat. Let us assume the airline's safety management staff refer to the relationship suggested by the accident triangle between these low-impact events and the increased potential for a high-impact accident. The pilots were disconnecting their autopilots to maintain their handling skills in preparation for flights when the autothrottle is unavailable because of unserviceability.

The airline produces an internal notice that the autothrottle must never be deliberately disconnected for "handing practice." Over the next few months, the number of low-speed events reduces. Is the airline any safer? Well, if one believes that Heinrich's Triangle is the basis of your policy then yes, you will believe you are safer—there are fewer reported low-speed events. However, what is the broader and longer-term effect on the operation or system of banning autopilot practice? There are less low-speed events but there is likely to be a general reduction in throttle/speed handling skills across your pilot workforce as they are no longer keeping up regular and nonassessed practice of flying without autothrottle during real-world operations. Pilots are most likely to need those skills when other systems have failed. Arguably the operation will be less resilient when the pilot's basic handling skills are needed most.

3.4.3   Root Cause Analysis

Another approach to rationalizing some of the complexities of causation in accident investigation is through Root Cause Analysis (RCA). The basic idea of RCA is that we can differentiate between the more proximate, obvious, or intuitive causes of an accident and those which are distant from the scene of an accident, either physically or temporally. Rather than focus on the obvious, RCA encourages investigators to look back in time and establish what other factors and influences came into play well before an accident occurred. A professional investigator is therefore not blinkered by what nonprofessionals may assume is obvious and instead casts a wider and deeper net in the search for an explanation. Modern accident investigators are aware of the multicausal nature of accidents and investigations have often extended way beyond any proximate explanation of events. However, as we have discussed, the remit of any investigation is often defined by the availability of resources rather than the extent and complexity of related evidence.

The RCA approach serves as a useful reminder to those analyzing accidents to avoid the oversimplification of causation. The search for deeper causal influences is entirely appropriate given what we know about human tendencies to create simple and intuitive causal relationships, but like any methodology, its effectiveness is governed by

---

viii A basic function of an autothrottle is to maintain the target speed selected by the pilot by automatically adjusting engine(s) thrust.

how objectively RCA is used. The more "obvious" explanations such as human error or technical failure can obscure more insidious and complex preconditions such system or management failures. As an approach, RCA can give us the illusion of control over our world but several questions can be levied at the assumed objectivity of RCA. For example, who decides which preconditions are relevant and which should be excluded from an investigation? What criteria are used in making this decision? When do we stop looking at deeper influences? As we discussed earlier in this chapter, the philosopher John Stuart Mill reminds us that we are not particularly good at maintaining objectivity when it comes to deciding what is a cause and what is a condition. In suggesting a rather misleading form of objectivity, RCA provides little in the way of protection from a common tendency to frame and explain an accident in a way that promotes our individual or organizational agenda.

3.4.4   Reason's Swiss Cheese Model

Arguably the most famous and influential of all accident models is the Swiis Cheese Model (SCM). Developed in the 1980s by the British safety researcher Professor James Reason, the model aims to explain how even complex and well-defended organizations can fail. Reason's academic career started in medicine but migrated toward psychology and was heavily influenced by the cognitive models of Danish researcher Jens Rasmussen.[ix] Reason's SCM incorporates three approaches to safety analysis, from the perspectives of engineering, the organization, and the person. The emergence of Reason's concept took place with the background of successive high-profile organizational disasters that occurred in the 1980s.[x]

The model promotes multiple layers of defense, or barriers, within an organization intended to protect it from failure. This system engineering perspective is sometimes described as "defenses in depth" as these barriers exist throughout an organization's hierarchy from the level of senior management right through to the front-line operators. In the SCM barriers not only protect the organization at multiple layers but the latent flaws within them facilitate the mechanisms of organizational failure. These latent flaws typically refer to the system's designers, policy, and decision makers and managers. The model therefore illustrates disasters such as aircraft crashes as the product of multiple factors interrelating at various levels within organizations (Figure 3-4).

Undoubtedly influenced by Reason's early medical studies and his research work into the psychology of human error within the medical profession, he labeled these latent flaws within the model's defenses as pathogens. To build on this chapter's earlier explanation of accident causation (at 3.2), these pathogens, or latent errors, are necessary but not sufficient causes of organizational failure. To overcome the organizational, technical, and human defenses within a system, it is the combination of latent and active errors within a system that result in organizational failure. As human error is a relatively constant feature of sociotechnical systems, Reason deduced that focus on latent errors was the most effective way to develop organizational resilience.

[ix] Rasmussen's SRK model envisaged three types of cognitive activity based on Skill, Rules, and Knowledge.

[x] These disasters included the massive chemical explosion at Bhopal in India (1984), the nuclear reactor explosion at Chernobyl, in Ukraine (1986), the loss of the Challenger Space Shuttle, Florida, United States (1986), the fire at the Kings Cross rail station, London (1987), the sinking and oil spillage from the Exxon Valdez tanker (1987), the sinking of the Herald of Free Enterprise car ferry in the English Channel (1988), and the explosion of the Piper Alpha oil rig in the North Sea off the Scottish coast (1988).

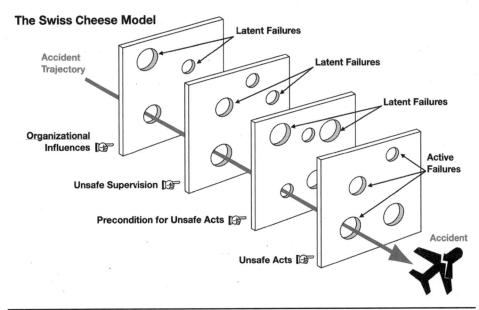

The Swiss Cheese Model

FIGURE 3-4 Arguably the most famous, influential, and perhaps most commonly misinterpreted accident model. Reason's SCM changed many of the early assumptions of risk and safety management. (Source: Authors)

The SCM stands as the most widely used and popular of accident models. In fact, the compelling and intuitive nature of the graphic has in some ways undermined what Reason was trying to propose. While the image suggests a linear causal route to accidents, Reason's written explanation is significantly more contextual. He recognized the model was not supposed to be a scientific model but one designed to articulate a complex concept. It was derived and deployed during the 1990s when Reason was not only an academic but also a professional risk consultant working with companies like British Petroleum (latterly BP) and British Airways to articulate the multifaceted influences on organizational safety. Several hypotheses were derived from Reason's SCM which spurred further research into the role of the organization in accident causation:

- Simpler systems with weaker defenses are more vulnerable to pathogens than complex and better defended systems.
- The more resident pathogens in any given system, the more likely it is that an accident will occur.
- The more complex the system, the more pathogens it can accommodate.
- The higher and more powerful an individual is within a system, the greater his or her potential is to generate pathogens.
- The pathogens present in a system can be detected before an accident, unlike active errors that are difficult to predict and are often found "after-the-fact."

Despite the acclaim and popularity of Reason's SCM, it has been subject to significant levels of criticism. The main thrust of this criticism has been on the model's oversimplicity[31] and the SCM's suggestion of linear causation.[32] But when viewed from the perspective of the model's influence on safety research and how we view accident causation we must consider the significant influence SCM has had on practical safety management. It has not only underpinned safety programs like the Human Factors Analysis and Classification System (HFACS, discussed next in this chapter) but has been a central tenet to ICAO's Safety Management Manual (SMM) and therefore adopted into hundreds of operator's safety management systems. The intuitive nature of SCM has developed a heuristic—an ability to short-cut conversation to the complexities of organizational influence on flight safety. The SCM is incredibly influential; for example, during the 2014 initial hearings of the legal cases against Airbus and Air France (concerning the 2009 loss of Air France 447), heard in French courts, it was the SCM which was used by the families' representative lawyers to explain how the various levels within each organization contributed to the accident.

The SCM has been singularly successful in shifting the predominance of accident investigation focus away from the individual to the broader organizational and system influences which will be discussed in more depth in Part IV. It is the basis of numerous accident investigation and safety management models such as Bowtie (discussed later) and other barrier-based approaches to safety and risk management.

3.4.5   The Human Factors Analysis Classification System (HFACS)

The most famous and widely used approach linking human behavior to accident causation is the Human Factors Analysis Classification Systems (HFACS), developed by the American academics Scott Shapell and Douglas Weigmann.[33] This broad analysis tool was originally used by the US Navy as a means of investigating the human element in aviation accidents following a series of accidents where human failure was determined to be the underlying cause. The study sought to identify common behavioral patterns and establish a causal link to these accidents (Figure 3-5).

The HFACS system consists of four taxonomies, based on James Reason's SCM of accident causation: unsafe acts of operators, preconditions of unsafe acts, unsafe

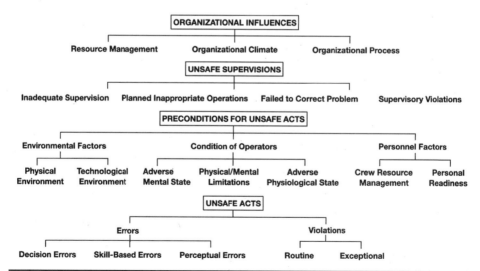

Figure 3-5   Based on the basic principles of Reason's SCM, HFACS attempts to link human behaviors with accident causation. (Source: Authors)

supervision, and organizational influences. Within each of these four groups, there are causal subcategories. The HFACS theory suggests that accident analysis will identify at least one causal condition within each of the four groups.

However, despite what might seem an obvious link between accident theory and human behavior, there has been little research that has established a link between the two areas.[34] Although there has been widespread use of HFACS, the lack of specific description of these causal categories has led to some criticism that the process is under-specified and therefore overly subjective.[35] However, the HFACS methodology has generated a vast database of human-centric accident information which has classified common behavioral traits in aviation accidents. Further validation of the methodology may well involve incorporation into systems-based approaches such as Leveson's STAMP model (discussed in Part IV). One proposal is for HFACS classifications to be applied to the control action analysis in the STAMP safety control structure thus bringing HFACS into line with a more structured approach to accident modeling.

3.4.6 Bowtie

Initially developed in the 1970s and widely used in the oil and gas industries in the 1990s, the Bowtie methodology has now had widespread use across many industries, including aviation. The model is largely based on the barrier approach to safety management promoted by Reason's SCM. Bowtie is effectively a merger of fault tree and event tree methodologies, or more simply put, how to avoid, trap, then mitigate risk. Bowtie is predominantly used to manage risk rather than define accident causation but its use has become so commonplace within aviation safety management and the oil and gas industry that it certainly deserves attention. It is also referred to in ICAO's SMM and ICAO Annex 19 and is a useful tool in the reactive classification of safety events (Figure 3-6).

The Bowtie model revolves around the hazard. Defining the hazard is a key element in the construction of an effective Bowtie. The hazard is something associated with an activity or organization that has the potential to cause damage or harm. The Top Event seen in the center of the model is the release or loss of control of the hazard which allows the system into an undesired state. Threats are potential causal routes

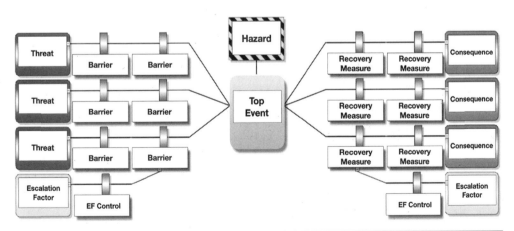

Figure 3-6 The widely used Bowtie approach is another safety risk management model based on the barrier approach proposed by James Reason. (Source: Authors)

to the Top Event ultimately resulting in potential consequences. On the left-hand side of the model are the preventative measures which can prevent the threat from triggering the Top Event or can eliminate the threat entirely. The right side of the model focuses on the Recovery measures which reduce the probability and/or severity of the Consequence once the Top Event is "live." The model also evaluates the Escalation Factors or reasons why barrier controls are sometimes ineffective.

The Bowtie methodology provides a useful tool for the management of safety and risk. However, while the use of linear causation makes the model intuitive, some have criticized the model in that it has done this at the expense of distorting the quantum of the relative risks. Another shortcoming suggests that although the Bowtie model encourages the active identification of hazards and their related controls, it does not determine nor indicate the adequacy of the system's control mechanisms.

3.5   Complex Causation Models

3.5.1   Normal Accident Theory

Describing an accident as something normal would seem paradoxical. Remember in our opening lines of this chapter, we considered the definition of an accident as something "unintended or unexpected," rather than "normal." The line of thought which evolved into Normal Accident Theory (NAT) emerged from a seminal book, *Man-Made Disasters*, written by a British academic, Barry Turner. The original book was first published in 1978[36] and challenged the common assumption that disasters were in some way unique events. The book was rewritten in 1995 and completed by Nick Pidgeon after Turner passed away during the project. Man-made accidents differentiated between natural catastrophes and the emergence of major sociotechnical disasters such as commercial aircraft crashes. Turner suggested that the surprise experienced in the initial aftermath was due to low levels of system knowledge and was a result of society's implicit trust in the resilience of safety-critical systems such as commercial aviation. Turner emphasized the influence of organizational culture and symbolism on the behavior of individuals involved in accidents rather than the more conventional focus on human error as a system anomaly.

Turner's concept was adapted and further developed by the Princeton-based academic, Professor Charles Perrow, and articulated in his famous book *Normal Accidents*[37] which analyzed the Three Mile Island nuclear disaster in 1979. In brief, Perrow's theory suggests that accidents are inevitable in complex safety-critical organizations when risk management practices become overly process-driven. Perrow labeled this tendency as tight coupling and describes how technology tends to drive organizations toward complex interactivity, so they progressively become increasingly more "tightly coupled." Perrow suggests that these systems can fail when they experience a combination of circumstances, but the overall driver of this state of vulnerability is the "normal" characteristics of the system itself rather than some unforeseen, random, or external influence. The tight coupling of complex and interdependent components is driven by an ongoing pursuit of productivity efficiency; the same efficiency is then transferred to the efficiency of the collapse of the system itself.

Perrow's work consolidated a new perspective on how large and often successful organizations with sophisticated risk management processes fail. However, NAT has been criticized for being more descriptive than analytical. While it provides a high-level explanation of many of the dynamics of organizational failure, it has made little progress toward practical utility as a safety risk management tool. Perrow's work has however been highly influential in shifting the focus of more traditional safety analysis away from an accident as an anomalous or random event and more of a function of inadequately managed system complexity.

3.5.2   High Reliability Theory (HRT)

In complete contrast to previous attempts to explain how and why accidents happen, HRT, posed a different question: why are some of the most potentially dangerous activities driving some of the highest safety performance? Predominantly based on work by a group of Berkeley scholars, the theory evolved from observations of highly reliable organizations (HROs) and their activities such as US Navy carrier operations (Figure 3-7), the nuclear power industry, and air traffic control. These academics noted that within these areas of activity, despite there being enormous potential for the release of catastrophic levels of energy or operating with fast-moving vehicles being in very close proximity, safety performance stayed high, often outperforming other human activities that held far less potential for disaster. What seemed to be the common trait or theme within these HROs was the extremely high level of safety-mindedness of operational personnel, perhaps generated by operating near these very obvious and physical threats.

What seemed to differentiate the way in which HROs dealt with emergent threats was the way operational authority was delegated. This "flat" managerial hierarchy meant that each area of the operation had its own subject matter experts. It is these experts who maintain operational control to bring any nonnormal situations back under control. These highly skilled individuals were not subjected to managerial or hierarchical overreach and maintained a strong sense of responsibility and local authority within their specialized field.

However, the observers of these highly skilled and highly motivated activities were at pains to point out the incredible amount of effort that it took to maintain these barriers and to achieve such a high level of resilience to disaster. The paradox of this field of safety research suggested that the obviously high level of threat posed by operating high-performance aircraft from navy carriers or producing energy from thermonuclear

Figure 3-7   US Navy carrier operations are a classic example of high safety performance despite the high potential for disaster. (Source: United States Navy)

reaction stimulated an authentic and intense safety-mindfulness. The immediate challenge that is presented by this conclusion is how to replicate such a safety culture in some of the more mundane of human activities. HRT also faced similar criticism to that of NAT; while it provided a novel and insightful perspective on organizational field, it provided little in the way of utility or predictive capability. In the words of Andrew Hopkins, the Australian safety writer, "HROs are very elusive creatures that inhabit the realm of theory rather than the real world."[38]

3.5.3   Practical Drift

After being wounded in a "friendly fire" incident by a US Airforce A7 fighter aircraft in Grenada in 1983, US Army Officer Scott Andrew Snook embarked on a research journey that culminated in the award of his Harvard PhD in 1996. In his thesis and subsequent book, Snook introduced the concept of "practical drift" which described how, in complex environments, a slow and steady divergence takes place between practice and written procedure. Practical drift describes the process over time where locally practical actions gradually drift away from the originally established procedure. When the system is suddenly brought back into close proximity (or using Perrow's NAT language, tightly coupled) there is a mismatch within the system which can have quite dramatic and disastrous implications.

> "Dramatic organizational failures and subsequent incident reviews open unique windows into the everyday lives of complex organizations. One such window opened on the 14th of April 1994, when two U.S. Air Force F-15 fighters accidentally shot down two U.S. Army Black Hawk helicopters in northern Iraq, killing all twenty-six people on board. After almost two years of extensive investigation, with virtually unlimited resources, no compelling explanation emerged."[39]

Although practical drift is a concept based on military operations in Iraq, in the mid-1990s, its relevance to the everyday mechanisms in complex systems such as commercial aviation is widely accepted. Like other theories linked to complex systems, it turns our attention away from the immediate and intuitive "fix" following an incident or accident and invites us to consider the broader emergent patterns or organizational behaviors. Similar to Normal Accident Theory, the concept of practical drift has been elaborated on by safety scientists such as Professor Sydney Dekker, in his book *Drift into Failure* who describes the cognitive shift away from looking for what is broken within a system and addresses the flaws that the system itself produces. Professor Eric Hollnagel elaborated on this concept of drift to explore how inherent human behavior traits such as "work arounds" from standard operating procedures are driven by a very natural trade-off between efficiency and thoroughness.

Practical drift again focuses attention on the emergent behavior of the system rather than the individual actors within it. One drawback to this approach is similar to some of the criticisms laid at the door of High Reliability Theory. That is that this approach to complex systems can become overly descriptive rather than usefully analytical. Although a systems manager may be encouraged to consider the dangers of organizational complacency, there is no roadmap to how that issue may be addressed. Having recognized some of the potential pitfalls of not addressing emergent system behavior, most systems-based theories leave us wanting for practical and implementable measures on how to address them.

3.5.4   Functional Resonance Analysis Method (FRAM)

FRAM is an accident analysis model developed by Professor Eric Hollnagel and published in 2012.[40] It is based on system theory which attempts to address some of the issues of practical application identified earlier in this chapter. FRAM decomposes a system into six functions in the form of a hexagon (six-sided shape). FRAM provides an assessment of system constraints that produce variability in system behavior; the model suggests that the "cause" of an accident is linked to variable performance within any given system. The model has been used across a number of safety-critical industries including commercial aviation. In one example it was used to explain the fatal mid-air collision between a Boeing 737-800 and an Embraer E-145, over the Amazon on the September 29, 2006, by assessing the "normal" performance level of the system rather than trying to identify anomalous events, or event sequences related to the accident.[41]

Although fluctuations are a normal characteristic of a complex system, abrupt changes (similar to Snook's abrupt realignment to tight coupling) can cause an accident. FRAM follows four steps to capture the level of variability in system behavior. The first step is to define basic functionality through traditional input/output parameters and adding time, control, precondition, and resource modules forming a hexagon. The second step is to characterize the potential variability of the system. Eleven common performance conditions (CPCs) are identified to determine potential variability. The third step is to define the possibility of functional resonance to find the possible connections under normal circumstances and unexpected connections under certain other conditions. This is achieved by creating a functional network; the output of one system may produce one or more inputs to other system functions. The fourth step is to establish control and protective barriers to initiate performance change. There are four types of control or protection barriers; physical, functional (or conditional), symbolic (signs designed to restrict some behaviors), and invisible (such as expert knowledge).

Hollnagel's FRAM model does provide us with a more practical application of systems theory. It is a qualitative method that effectively provides us with a consistent method of describing the interrelationships within complex systems. As with many complex systems models, it does however hit the same wall when it comes to model versus reality. Building a model of a complex system holds an inherent paradox: the more accurate the model, the less intuitive (and by implication useful) the model becomes. Hollnagel's FRAM (and indeed many of his writings) have been strongly criticized by Professor Nancy Leveson. The critical tone of the debate between two of the more prominent members of the safety science community emphasizes that we have much to resolve when it comes to managing the complexities of new and emerging technologies.

3.5.5   Deliberate Myths and Black Swans

Commercial aviation is a highly regulated and technically focused industry and perhaps for this reason the industry has tended toward scientific and engineering-oriented solutions to modeling accident causation. Nevertheless, commercial aviation safety has continued to use some models within the industry that have little or no theoretical or scientific grounding but do match the intuitive patterns of causation that attract human cognition. In this chapter, we have attempted to explain several of the better-known theories and models but also explain some of their inherent flaws.

Flawed models or theories do not necessarily suggest they are without value. There is a need for some level of mythology, or storytelling within aviation safety management.

Humans need these stories to understand and relate to each other about their environment; some of the less "scientific" theories simply tell a better story and their purpose is to articulate some of the high-level principles of safe operation. The value of these myths should not be underestimated. Relating to the earlier parts of this chapter, we considered how humans desperately need to understand events in their own way; not to do so is a type of pain according to Nietzsche.[xi]

This perspective is not an invitation to adopt a mindset of learned helplessness in the face of complexity, but to rationally appraise where the limits lie in our current knowledge of safety and risk management in commercial aviation. The status of safety management in commercial aviation is not one of "task complete," but one of a "work in progress." The highly regulated and technical focus of the industry perhaps creates inherent flaws in how we explain this complex environment. The very fact that we cannot fill all the gaps in providing an explanation of how and why accidents happen, leaves the door open to more intuitive explanations in other words, myths.

In his best-selling book *The Black Swan*, Nassim Nicholas Taleb suggested that some events are beyond the capabilities of our current predictive risk management models in science, finance, and medicine. The ultralow probability but high impact events that Taleb referred to as "Black Swans" are the product of the increasing levels of systems complexity driven by new technologies and increasingly tight coupling across different systems. When failure happens, the extent and level of impact has become almost impossible to predict using contemporary risk metrics, but as Taleb noted, humans are adept at explaining them after-the-fact, with the absolute clarity of hindsight.

Taleb's work stimulated a considerable amount of debate around the subject of risk and safety management. What some academic commentators have emphasized is that when we refer to a Black Swan, there is considerable difference between a truly random event, sometimes referred to as aleatory, and an event that takes us by surprise because of our lack of systems knowledge. This latter, epistemic limit is the challenge that commercial aviation, like many other complex industries, is facing today. The industry needs to start understanding and talking in terms of systems rather than component language. However, we need to establish what that common language actually is. The inevitable uncertainty produced by the interaction of complex systems is never going to be an easy story to tell, but it is one that we not only have to understand but, perhaps more importantly, learn how to explain. We may even need to sacrifice some degree of scientific accuracy to enable understanding. We may even need some new myths and better stories.

In the next section of this chapter, we will highlight a series of high-profile accidents in commercial aviation. We would invite the reader to attempt to understand these events using some of the models we have explained and to start to think about some of the broader issues we have described, that influence our understanding of why accidents happen.

3.6   Significant Commercial Aviation Accidents

In commercial aviation there is a history of significant accidents but also malicious acts such as terrorist attacks, or murder suicides, that do not really meet the definition of an accident. Some accidents and attacks have had such profound significance that

---

[xi] See endnote 17.

they have prompted industry, regulatory, and legal reform. It is these actions that have influenced the current shape and function of commercial aviation industry. The list of accidents that follow is extensive but is unfortunately by no means exhaustive. The list contains the most significant events that have helped create a safer world for air travel and transport that have occurred since the start of the 21st century.

Some of the events are referred to in conversations by industry professionals to this day and are regularly referred to during efforts to craft modern legislation in aviation, so the reader is urged to become familiar with the contents of the compilation. Each event has a very long investigation report associated with it which can be sourced through the online learning resources associated with this book. The database also gives links to other sources of information such as news articles and commentaries. As you read the reports, opinions, and commentaries please consider them in light of what we have discussed in this chapter. Think about the formal definitions, how the accident is being framed and how that is perceived by different interest groups. Consider the antecedent events, causes, and what differentiates them. Having read this book so far you have the basic tools to critically analyze these accidents, but most importantly, consider what we can still learn from these tragic events that will contribute to the future of commercial aviation safety.

- January 31, 2000: Alaska Airlines Flight 261, a McDonnell Douglas MD-83, en route from Puerto Vallarta, Mexico, to San Francisco, California, reported flight control problems with its horizontal stabilizer and a loss of stability before it crashed into the Pacific Ocean near Point Mugu, California. There were 88 fatalities. https://aviation-safety.net/database/record.php?id=20000131-0

- July 25, 2000: Air France Flight 4590, a Concorde supersonic aircraft, crashed on takeoff from Charles de Gaulle International airport near Paris after striking a titanium metal strip on the runway that had fallen from a previously departing Continental DC-10. The metal strip punctured a tire of the Concorde, and debris ruptured a fuel tank resulting in a fire. The aircraft lost lateral control after the engine shutdown procedure and crashed into a hotel. There were a total of 109 fatalities. https://aviation-safety.net/database/record.php?id=20000725-0

- October 31, 2000: Singapore Airlines Flight 006, a Boeing 747, crashed on takeoff in Taipei, Taiwan. The crew of this aircraft were attempting to takeoff on a closed parallel runway, hitting several large pieces of construction equipment. There were 83 fatalities. https://aviation-safety.net/database/record.php?id=20001031-0

- November 12, 2001: American Airlines Flight 587, an Airbus 300-600, experienced a loss of control upon initial climb after takeoff and crashed into a residential area in Queens, NY, killing 265. The investigation determined that excessive overuse of the rudder to counter a wake turbulence problem resulted in a separation of the vertical stabilizer from the aircraft. https://aviation-safety.net/database/record.php?id=20011112-0

- May 25, 2002: China Airlines Flight 611, a Boeing 747 from Taiwan to Hong Kong, broke up in flight and crashed due to a previous inadequate structural repair to the pressurized hull. There were 225 fatalities. https://aviation-safety.net/database/record.php?id=20020525-0

- January 3, 2004: Flash Airlines Flight 604, a Boeing 737 from Egypt to France, crashed in the Red Sea with 148 fatalities. The investigation considered spatial disorientation of the pilots resulted in the controlled flight into the sea. https://aviation-safety.net/database/record.php?id=20040103-0

- August 14, 2005: Helios Airways Flight 522, a Boeing 737 traveling from Cyprus to Greece, lost pressurization and crashed with 121 fatalities. An improper setting of the cabin pressurization switch incapacitated the crew and passengers due to hypoxia as the aircraft depressurized during the climb. The plane crashed near Athens after fuel starvation of the engines. https://aviation-safety.net/database/record.php?id=20050814-0

- September 5, 2005: Mandala Flight 091, a Boeing 737, crashed on takeoff in Indonesia. There were 117 fatalities. The accident report noted that the flight crew had failed to select an appropriate flap setting after omitting to use the checklist. https://aviation-safety.net/database/record.php?id=20050905-0

- May 3, 2006: Armavia Flight 967, an Airbus A320, crashed into the Black Sea at night on a missed approach in poor weather near Adler/Sochi airport in Russia. There were 113 fatalities. The official investigative report cited the psychoemotional stress level of the captain, poor cockpit resource management, and air traffic control problems; the aircraft was operating in below permitted weather conditions prior to the crash. https://aviation-safety.net/database/record.php?id=20060503-0

- July 9, 2006: An S7 Airlines Flight 778, an Airbus A310 from Moscow, crashed in Siberia with 125 fatalities. The aircraft failed to decelerate on landing due to inadvertent movement of one throttle to the forward thrust position which caused the plane to overrun the runway and crash into a concrete barricade. https://aviation-safety.net/database/record.php?id=20060709-0

- August 27, 2006: Comair Flight 191, a CRJ-100, crashed on takeoff from Lexington, Kentucky. The aircraft took off on the wrong (short) runway due to a low level of situational awareness from poor quality safety briefing, and the crew's nonpertinent conversations were in violation of the FAA's "Sterile Cockpit Rule." https://skybrary.aero/accidents-and-incidents/crj1-lexington-ky-usa-2006

- September 29, 2006: GTA Flight 1907, a new Boeing 737-800, was destroyed in a mid-air collision at Flight level 370 with an Embraer legacy business jet on a domestic flight over the Brazilian amazon jungle. There were 154 fatalities as the Boeing 737 crashed to the ground, while the Legacy continued flying and was able to safely land. The NTSB report cited a combination of ATC errors that placed both aircraft on the same airway at the same altitude. https://aviation-safety.net/database/record.php?id=20060929-0

- July 17, 2007: TAM Airlines Flight 3054, an Airbus A320, overran the runway and crashed into a factory on landing at Congonhas airport in São Paulo. There were 187 fatalities. The investigation indicated that the aircraft thrust reverser was deactivated, causing the plane to run off this short runway. https://aviation-safety.net/database/record.php?id=20070717-0

- August 20, 2008: SpanAir Flight JK 5022, a MD-82, crashed on takeoff in Madrid, Spain. There were 154 fatalities. The investigation indicated that following some hastily conducted maintenance that inhibited a warning device, the aircraft tried to take off with the flaps and slats retracted. https://aviation-safety.net/database/record.php?id=20080820-0

- January 15, 2009: A U.S. Airways Flight 1549, an Airbus A320, successfully ditched in the Hudson River near New York City after takeoff from LaGuardia airport (aka the "Miracle on the Hudson"). The aircraft lost thrust in one engine and suffered roll-back in the remaining engine during its initial climb out after striking a flock of Canadian Geese. All 155 occupants were safely evacuated from the airliner in the Hudson River. The NTSB cited the excellent crew resource management, safety equipment, and fast rescue response from ferry boat operators. https://aviation-safety.net/database/record.php?id=20090115-0

- February 12, 2009: Colgan Air Flight 3407, a Bombardier Dash 8, stalled and crashed on final approach to Buffalo, NY. There were 49 fatalities. The investigation established that the crew had failed to monitor the airspeed in icing conditions, made an inappropriate response to the stick shaker stall alarm system, and failed to comply with the FAA's Sterile Cockpit rule to minimize nonpertinent conversation. Following the crash, the Airline Safety and FAA Extension Act of 2010, which requires airline first officers to have an Airline Transport Pilot certificate. The rule essentially raises the experience level of new FAA first officers, requiring a minimum 1500 hours of total flight time experience unless certain strict educational conditions have been met. https://aviation-safety.net/database/record.php?id=20090212-0

- June 1, 2009: Air France Flight 447, an Airbus A330 from Rio de Janeiro to Paris, crashed into the Atlantic Ocean. There were 228 fatalities. The aircraft crashed in bad weather after receiving unusual airspeed indications from the plane's Pitot Static System. This accident is discussed in some depth in Chapter 1. https://aviation-safety.net/database/record.php?id=20090601-0

- November 4, 2010: Qantas Flight 32 had an uncontained engine failure on engine number two. The pilots received 54 computer messages alerting them of the failure. Pilots had to make an emergency landing at the Singapore Changi airport. The failure was the first of this kind in the Airbus A380, one of the largest passenger aircraft ever flown in commercial service. https://aviation-safety.net/database/record.php?id=20101104-1

- July 6, 2013: Air Asiana Flight 214, a Boeing 777, crashed on landing at the San Francisco International airport after the landing gear and the tail impacted the sea wall in the undershoot of the approach. NTSB experts found that the automation logic in the cockpit was not intuitive for the autothrottle system, among other factors. This accident is further discussed in Chapter 2. https://aviation-safety.net/database/record.php?id=20130706-0

- March 8, 2014: Malaysia Flight 370, a Boeing 777, departed Kuala Lumpur International airport in Malaysia headed toward Beijing Capital International airport. The flight deviated from its flight path and eventually fell off radar. As of 2022, the cause of this accident is yet to be established. https://aviation-safety.net/database/record.php?id=20140308-0

- July 17, 2014: While crossing over Ukraine enroute to Kuala Lumpur International airport in Malaysia, Malaysia Flight 17, a Boeing 777, was shot down with a surface-to-air missile. There were 298 fatalities. https://aviation-safety.net/database/record.php?id=20140717-0

- March 24, 2015: Flight Germanwings 9525, an Airbus A320, flying from Barcelona, Spain, to Dusseldorf, Germany, was deliberately crashed in the French Alps by one of the aircraft's pilots. There were 150 fatalities. This crash has raised significant questions about examining and treating mental health issues for commercial pilots. https://aviation-safety.net/database/record.php?id=20150324-0

- October 31, 2015: MetroJet Flight 7K9268, an Airbus A321-231, disintegrated during the climb through an altitude of 31,875 feet (9,410 meters). Investigators found evidence of a small explosive device which was thought to have initiated a catastrophic decompression and subsequent break-up of the airframe. https://aviation-safety.net/database/record.php?id=20151031-0

- March 19, 2016: During a second attempt at landing in poor weather at Rostov in southern Russia, a Boeing 737-800, Flight FZ-981, crashed into the airport. All 55 passengers and 7 crew were fatally injured. https://aviation-safety.net/database/record.php?id=20160319-0

- May 19, 2016: An Egypt Air, an Airbus A320-232, crashed into the Mediterranean Sea, approximately 200 miles north of the Egyptian coast without any radio transmission from the crew. Greek radar reported that the aircraft, Flight MS-804, suddenly turned sharply to the left then turned 360° in the opposite direction, before radar contact was lost. All 56 passengers and 10 crew were fatally injured. https://aviation-safety.net/database/record.php?id=20160519-0

- November 28, 2016: A LaMia Avro RJ85, Flight LMI2933, crashed into a wooded hillside close to Medellín Airport. The crew reported a low fuel state during the approach to Medellín. Subsequent analysis of the flight recorder showed the aircraft's engines shutdown sequentially during the final approach just before the crash. Of the 77 passengers and crew, only 6 survived the crash. https://aviation-safety.net/database/record.php?id=20161128-0

- December 7, 2016: A Pakistan International Airlines, ATR 42-500, crashed into terrain during approach to Islamabad airport. The aircraft had suffered the loss of control of one of its two engines. Possibly related to inappropriate maintenance schedules, the angle of the propeller pitch controller failed and created excessive drag, rolling the aircraft through 360° and subsequently reducing directional control and colliding with a mountain side. All 47 passengers and crew were fatally injured. The airline had faced repeated criticisms of its maintenance and safety procedures. https://aviation-safety.net/database/record.php?id=20161207-0

- February 11, 2018: An Antonov An-148-100B operated by Saratov Airlines crashed into a snowy field shortly after takeoff from Moscow's Domodedovo Airport. All 71 passengers and crew were fatally injured in the impact. The flight crew had not selected the pitot heaters after failing to complete a section of the checklist. https://aviation-safety.net/database/operator/airline.php?var=9248

- February 18, 2018: An ATR-72-212, operated by Iran Aseman Airlines, crashed into the summit of a 13,400-foot (4084 meters) mountain. The crew had lost control of the aircraft in mountain wave turbulence and been unable to maintain altitude. All 66 occupants were fatally injured in the impact. https://aviation-safety.net/database/record.php?id=20180218-0

- March 12, 2018: A de Havilland Canada DHC-8-402Q Dash 8, operated by US-Bangla Airlines, crashed after low-level maneuvering at low level during an attempted approach to Kathmandu-Tribhuvan Airport in Nepal. Of the 71 passengers and crew, only 20 survived the impact with the ground. https://aviation-safety.net/database/record.php?id=20180312-0

- May 18, 2018: A Cubana de Aviación Boeing 737-201 leased from Global Aviation crashed on takeoff from Havana-Jose Marti International Airport, Cuba. All 112 passengers and crew were fatally injured in the impact. The aircraft's trim setting was significantly different from the calculated center of gravity position of the aircraft. https://aviation-safety.net/database/record.php?id=20180518-0

- October 29, 2018: A Boeing 737 MAX 8 operated by Lion Air crashed into the sea shortly after takeoff from Jakarta-Soekarno-Hatta International Airport, Indonesia. All 189 passengers and crew were fatally injured on impact with the sea. The aircraft's maintenance log had recorded a series of previous occurrences related to the aircraft's airspeed system. Subsequent investigation found serious flaws in the aircraft's design and certification process, particularly relating to the aircraft's Maneuvering Characteristics Augmentation System, (MCAS). The MCAS system was found to have commanded an automatic trim force pitching the aircraft's nose down due to an erroneous AOA (Angle-of-Attack) reading from the aircraft's sensors. https://aviation-safety.net/database/record.php?id=20181029-0

- March 10, 2019: A second Boeing 737 MAX 8 operated by Ethiopian Airlines crashed six minutes after takeoff from Addis Ababa-Bole Airport. Under very similar circumstances to the Lion Air crash five months earlier, the aircraft produced an erroneous AOA signal which triggered the MCAS response to automatically pitch the aircraft's nose down. Despite repeated efforts by the flight crew to understand the emergency and bring the aircraft under control, the nose-down trim forces eventually resulted in an acceleration to approximately 500 knots before impacting the ground. All 157 passengers and crew died in the impact. This being the second of two very similar accidents involving the MCAS system resulted in the grounding of the global Boeing 737 MAX fleet. The aircraft manufacturer Boeing and the FAA who were the primary regulators, faced considerable criticism and a deep-rooted review of many of the certification practices in place at the time. https://aviation-safety.net/database/record.php?id=20190310-0

- May 22, 2020: An Airbus A320, Flight PK-8303, crashed into a residential area of the Pakistani city of Karachi killing all but two of the 99 occupants of the aircraft. One person on the ground was also killed in the accident. The aircraft, operated by Pakistan International Airlines, landed without the landing gear lowered. After touchdown, when both engines had contacted the ground, the landing was rejected, and a go-around initiated. The flight crew attempted a further landing, but the ground contact had significantly damaged the engines which both failed during the attempted return to the airfield. The aircraft crashed 1300 meters short of runway 25L. https://aviation-safety.net/database/record.php?id=20200522-0

- January 9, 2021: A Boeing 737-524 series operated by Sriwijaya Air crashed into the Java Sea shortly after takeoff from Jakarta Airport. All 62 people onboard

Flight SJ-182 were killed on impact with the sea. Low clouds and heavy rain were reported at the time of departure. The accident report noted that the aircraft's autothrottle system had been reported as faulty on three separate occasions during the week preceding the crash. The FDR (Flight Data Recorder) showed the faulty autothrottle had resulted in significant asymmetric thrust during the climb, before the autopilot eventually disengaged. The aircraft entered a steep dive and impacted the sea at over 10° nose down pitch angle. https://aviation-safety.net/database/record.php?id=20210109-0

- March 21, 2022: A China Eastern Airlines Boeing 737-89P series, Flight number MU-5735, crashed shortly after initiating a descent into Guangzhou Baiyun International Airport. All 132 passengers and crew died in the accident. The aircraft was filmed in a high-speed vertical dive to the ground and media speculation suggested sabotage. At the time of writing, the final accident report has not been released. https://aviation-safety.net/database/record.php?id=20220321-0

Key Terms

Antecedent Event

Black Swan Event

Causation

Flight Data Recorder

Functional Resonance Analysis Method

High Reliability Theory

Normal Accident Theory

Systems Theory

Swiss Cheese Model

Topics for Discussion

1. What is an accident and where are aviation accidents and incidents defined?
2. Give examples of thematic and episodic framing in media reporting of aviation accidents.
3. What do you think motivated some Air New Zealand employees to deliberately provide inaccurate information to the Flight TE-901 accident investigation? Would that happen today?
4. Define necessary and sufficient conditions of any one of the accidents listed in this chapter.
5. Explain the basic differences between linear causation and systems theory.
6. How could accident "myths" or storytelling be useful in aviation safety and risk management?
7. On reviewing the commercial accidents listed in this chapter, what themes or patterns can you identify?

References

1. ICAO. (2016). *Annex 13 to the Convention on International Civil Aviation. Aircraft Accident and Incident Investigation*, 11ed. Quebec, Canada: ICAO, Chapter One (Definitions).

2. Amalberti, R. (2001). The paradoxes of almost totally safe transportation systems. *Safety Science*, 37(2-3), 109–126

3. Lundberg, J., Rollenhagen, C., & Hollnagel, E. (2009). What-you-look-for-is-what-you-find—The consequences of underlying accident models in eight accident investigation manuals. *Safety Science*, 47(10), 1297–1311.

4. MacLennan, C. A. (1988). From accident to crash: The auto industry and the politics of injury. *Medical Anthropology Quarterly*, 2(3), 233–250.

5. Girasek, D. C. (2015). How members of the public interpret the word accident. *Injury Prevention*, 21(3), 205–210.

6. Goddard, T., Ralph, K., Thigpen, C. G., & Iacobucci, E. (2019). Does news coverage of traffic crashes affect perceived blame and preferred solutions? Evidence from an experiment. *Transportation Research Interdisciplinary Perspectives*, 3, 100073.

7. Cobb, R. W., & Primo, D. M. (2004). *The Plane Truth: Airline Crashes, the Media, and Transportation Policy*. Washington, DC: Brookings Institution Press.

8. Hollnagel, E. (2014). *Safety-1 and Safety-2: The Past & Future of Safety Management*. Farnham, Surrey: Ashgate.

9. Hopkins, A. (2006). Studying organisational cultures and their effects on safety. *Safety Science*, 44(10), 875–889.

10. Flight TE901 Accident Report. https://www.erebus.co.nz/Portals/4/Documents/Reports/Mahon/Mahon%20Report_web.pdf, p. 387.

11. Flight TE901 Accident Report. https://www.erebus.co.nz/Portals/4/Documents/Reports/Mahon/Mahon%20Report_web.pdf, p. 393.

12. Greenwood, M., & Woods, H. M. (1919). The incidence of industrial accidents upon individuals with special reference to multiple accidents. *Report no. 4. Industrial Fatigue Research Board*, London.

13. FAA. https://www.faa.gov/about/initiatives/maintenance_hf/library/documents/media/mx_faa_(formerly_hfskyway)/human_factors_issues/meeting_11/meeting11_7.0.pdf. Acessed September 2022.

14. ICAO. ICAO Circular 240-AN/144. https://skybrary.aero/bookshelf/books/2037.pdf. Accessed September 2022.

15. ICAO (International Civil Aviation Organisation). (2016). *Annex 13 to the Convention on International Civil Aviation. Aircraft Accident and Incident Investigation.* 11th ed. Quebec, Canada: ICAO, pp. 1–15.

16. Nietzsche, F. (1889). *The Twilight of the Idols*. English Translation. https://archive.org/details/TwilightOfTheIdolsOrHowToPhilosophizeWithAHammer/mode/2up, p. 5.

17. Russell, B. (1912). On the notion of cause. In: *Proceedings of the Aristotelian society, 13*, 1–26. Aristotelian Society, Wiley.

18. Mill, J.S., (1889), *A system of logic, ratiocinative and inductive: Being a connected view of the principles of evidence and the methods of scientific investigation.* Longmans, green, and Company, London.

19. Schein, E. H. (2017). *Organizational Culture and Leadership*. 5th ed. Hoboken, NJ: John Wiley & Sons.

20. Simon, H. A. (1965). Administrative decision making. *Public Administration Review*, 25(1), 31–37.

21. Simon, H. A. (1965). Administrative decision making. *Public Administration Review*, 25(1), xxvi.

22. Taylor, F. W. (1911). *The principles of scientific management.* New York: Harper Brothers.

23. Leveson, N. (2004). A new accident model for engineering safer systems. *Safety Science*, 42(4), 237–270.

24. Reason, J. (1997). *Managing the Risks of Organizational Accidents.* London: Routledge, p. 234.

25. Burnham, J. C. (2008). The syndrome of accident proneness (Unfallneigung): Why psychiatrists did not adopt and medicalize it. *History of Psychiatry*, 19(3), 251–274.

26. Leveson, N. (2011). *Engineering a Safer and More Secure World.* Cambridge, MA: MIT Press, p. 62.

27. Pasquini, A., Pozzi, S., & Save, L. (2011). A critical view of severity classification in risk assessment methods. *Reliability Engineering & System Safety*, 96(1), 53–63.

28. Nascimento, F., Majumdar, A., & Ochieng, W. (2013). Investigating the truth of Heinrich's pyramid in offshore helicopter transportation. *Transportation Research Record: Journal of the Transportation Research Board*, 2336(1), 105–116.

29. Dekker, S., & Pitzer, C. (2016). Examining the asymptote in safety progress: A literature review. *International Journal of Occupational Safety and Ergonomics*, 22(1), 57–65.

30. Manuele, F. A. (2011). Reviewing Heinrich: Dislodging two myths from the practice of safety. *Professional Safety*, 56(10), 52.

31. Leveson, N. (2011). Applying systems thinking to analyse and learn from events. *Safety Science*, 49(1), 55–64.

32. Dekker, S. (2006). Past the edge of chaos. Technical Report 2006–03. Lund University, School of Aviation, Sweden.

33. Shappell, S. A., & Wiegmann, D. A. (2000). The human factors analysis and classification system—HFACS. https://commons.erau.edu/publication/737. Accessed November 2022.

34. Dekker, S., & Hollnagel, E. (2004). Human factors and folk models. *Cognition, Technology & Work*, 6(2), 79–86.

35. Olsen, N. S., & Shorrock, S. T. (2010). Evaluation of the HFACS-ADF safety classification system: Inter-coder consensus and intra-coder consistency. *Accident Analysis & Prevention*, 42(2), 437–444.

36. Turner, B. A., & Pidgeon, N. F. (1997). *Man-Made Disasters.* Oxford, England: Butterworth-Heinemann.

37. Perrow, C. (1984). *Normal Accidents: Living with High Risk Technologies.* New York: Basic Books.

38. Hopkins, A. (2014). Issues in safety science. *Safety Science*, _67_, 6–14.

39. Snook, S. A. Practical drift: The friendly fire shootdown over northern Iraq. Harvard PhD Thesis Submission.

40. Hollnagel, E. (2012). *FRAM: The Functional Resonance Analysis Method. Modelling Complex Socio-technical Systems.* England: Ashgate.

41. De Carvalho, P. V. R. (2011). The use of Functional Resonance Analysis Method (FRAM) in a mid-air collision to understand some characteristics of the air traffic management system resilience. *Reliability Engineering & System Safety*, 96(11), 1482–1498.

CHAPTER 4

Investigating Accidents

Introduction

On first inspection, it is possible to think that accident investigation is simply "closing the stable door after the horse has bolted." Indeed, some may ask why it is worth expending significant time, money, and effort investigating an accident that has already happened? The simple answer is *to stop it happening again*. In fact, accident investigation is the backbone of the aviation safety system. Over the years, accident investigations have brought about some of the most significant changes in aviation, and elicited many lessons learned. Throughout this book you will see examples of where accidents have brought new procedures, systems, designs, and knowledge into the industry through the process of understanding: what happened, why it happened, and how to stop it happening again.

For some people, their first exposure to air accident investigation may be a television program recreating a particular investigation from history. The program may conjure images of an individual working alone (often late into the night, in a darkened room, while smoking a cigarette) to discover a single piece of evidence that reveals the entire chain of events—a eureka moment!

Although this may occur, it's a rarity; in reality, the process is generally very different. Accident investigation is a process of thorough, meticulous, and measured progress often undertaken by a large team of people, all looking for evidence, exploring possibilities, and challenging their conclusions trying to find the optimum way to improve safety. It can be challenging and frustrating, but it can also be immensely rewarding and crucial work. Following an accident, there is no shortage of "armchair investigators" willing to give interviews speculating on the cause and explaining their "theory" of what might have happened. Accident investigators do not have that luxury—when they release information about an accident it needs to be reliable and supported by evidence. There is no place for speculation.

The idea of learning from the worst outcomes is deeply ingrained in aviation and has become the gold standard for how to approach safety investigations. In 2005, Elaine Bromiley, the wife of Martin Bromiley, an airline pilot, died tragically and needlessly during a routine medical procedure. He was shocked to discover that there was to be no investigation into her death, and this prompted him to campaign for improved learning in healthcare and also to set up the Clinical Human Factors Group. His drive has contributed significantly to the setting up of a no-blame investigation unit in the UK medical sector, the Healthcare Safety Investigation Branch, which was built using the aviation model.

That is not to say that accident investigation in aviation is perfect—far from it—and later in the chapter we will look at examples where investigation, and the system of which it is a part, has failed to deliver the intended outcomes. However, what aviation accident investigation does have is a long history of learning from failure, achieved using a tried and tested framework. This chapter will focus on how that accident investigation framework, system, and process work.

4.1 Why and How to Investigate?

We defined what an aviation accident was in the previous chapter, but we shall now consider the why and how of aircraft accident investigation. The overarching ICAO Standards and Recommended Practices (SARPs) dealing with civil aviation accident investigation are known as "Annex 13 to the Chicago Convention."[1] Annex 13 describes the resources, protocols, and behaviors that should be in place in Contracting States, as to what constitutes an accident, who will lead or be involved with the investigation, and so on. Annex 13 requires an investigation to be conducted into all aviation accidents, and into serious incidents when the aircraft is of a maximum certified takeoff mass (MCTOM) of over 2250 kilograms, although national regulation may modify this requirement. In addition to this obligation to investigate accidents and serious incidents, Safety Investigation Agencies (SIAs) are also free to investigate other incidents, subject to their national legislation, when they expect to draw safety lessons for civil aviation from the accident investigation.

The sole purpose of the post-accident safety investigation is to prevent future accidents and incidents. By taking all the opportunities to see where things went wrong, we can learn what needs to be changed to improve safety, not just in identical situations, but for the whole aviation system where similar issues may arise and result in a different type of accident.

Annex 13 provides the definition of an accident that is used by most, if not all, accident investigation agencies and it is worth reproducing in full here:

> *An occurrence associated with the operation of an aircraft which, in the case of a manned aircraft, takes place between the time any person boards the aircraft with the intention of flight until such time as all such persons have disembarked, or in the case of an unmanned aircraft, takes place between the time the aircraft is ready to move with the purpose of flight until such time as it comes to rest at the end of the flight and the primary propulsion system is shut down, in which:*
>
> *a) a person is fatally or seriously injured as a result of:*
>
> > *— being in the aircraft, or*
> >
> > *— direct contact with any part of the aircraft, including parts which have become detached from the aircraft, or*
> >
> > *— direct exposure to jet blast,*
>
> *except when the injuries are from natural causes, self-inflicted or inflicted by other persons, or when the injuries are to stowaways hiding outside the areas normally available to the passengers and crew; or*
>
> *b) the aircraft sustains damage or structural failure which:*
>
> > *— adversely affects the structural strength, performance or flight characteristics of the aircraft, and*
> >
> > *— would normally require major repair or replacement of the affected component,*

except for engine failure or damage, when the damage is limited to a single engine (including its cowlings or accessories), to propellers, wing tips, antennas, probes, vanes, tires, brakes, wheels, fairings, panels, landing gear doors, windscreens, the aircraft skin (such as small dents or puncture holes), or for minor damages to main rotor blades, tail rotor blades, landing gear, and those resulting from hail or bird strike (including holes in the radome); or

c) the aircraft is missing or is completely inaccessible.

As with other ICAO Annexes, Annex 13 only provides a collection of SARPs—it is for each State to enact it in their local laws and regulations while notifying any "differences" (areas of the SARPs that are not adopted) with ICAO. However, subjectively at least, Annex 13 often seems to be widely accepted "as is," with relatively few registered differences. In the United States, Annex 13 is embodied as 49 CFR Part 831,[2] and in Europe as Regulation (EU) 996/2010.[3]

On the subject of why to investigate, Annex 13 states in Section 3.1, *Objective of the Investigation*:

The sole objective of the investigation of an accident or incident shall be the prevention of accidents and incidents. It is not the purpose of this activity to apportion blame or liability.

This phrase underpins all air accident investigations and is often reproduced on the front cover of reports to remind the reader of the purpose and approach.

4.1.1 Not-for-Blame Investigation

As described in the Annex 13 objective above, accident investigation in aviation is conducted using a "not-for-blame" approach. This is a different philosophy to, say, a criminal investigation where the aim is to find the person or persons responsible and bring them to justice. Some people are shocked when presented with the concept of not-for-blame as an investigation philosophy, so it is worth understanding the motivation and history for this approach.

As we will discuss in Part III, humans make mistakes. While it might be tempting to think that dismissing someone who makes a mistake will bring a replacement from a long line of pilots, maintainers, air traffic controllers, and so on, who don't make mistakes, it simply isn't true. So, we need to build safety systems that are resilient to mistakes.

The assumption is that most staff come to work to do a good job and that almost all incorrect actions are unintentional. Similarly, aviation is such a safety-critical industry, it is crucial that if, say, a baggage handler, in a rush to get the aircraft away on schedule, accidentally impacts the aircraft fuselage with a vehicle, they are able to draw attention to it without fear of punishment. In order to build the safest system possible, we need to understand exactly what happened during an accident and the information the personnel involved can provide is too valuable and too important to be lost because of fear of reprisals. The obvious outcome, if people fear losing their job for reporting mistakes, is that most will cover up all but the most visible mistakes in the hope of not being discovered.

One common confusion with the not-for-blame approach is to mistake it for "no responsibility," which is totally incorrect. You can't show up to work drunk, cut corners, falsify documentation, and expect to be treated with impunity by your company following the investigation into the almost inevitable accident that will result. This confusion gave rise to the concept of the Just Culture (which is expanded on in Part III). For now, it is sufficient to know that when an Annex 13 investigation is launched by a State SIA, it will invariably be a not-for-blame investigation.

Another reason for approaching accidents from a not-for-blame perspective is that, in many cases, an error, action, or inaction by an individual is merely the final outcome of an organizational or systemic problem, over which the individual at the frontline may have little or no control. We will return to this idea throughout the book. Aviation is not the only industry to use a not-for-blame approach—it is also followed in a number of other transportation modes including marine and railroad—but aviation was one of the first to adopt the concept.

4.1.2 Thinking Like an Investigator

Since each accident is unique, accident investigations face the challenge of an extraordinary range of potential scenarios and focal points ranging from, say, the failure of a single small component, such as a bearing in a complex system, through to large systemic failures of oversight and regulation. In some cases, these can appear in isolation or in the same accident—the challenge to the investigator is to address the full range of learning and prevention opportunities.

An investigation that "stops" when it finds someone did something wrong, rather than trying to understand why, has failed. In one extreme example, an investigator described investigating a fatal microlight accident in which a toxicological report found that the pilot had a high concentration of alcohol in their blood. An investigation that wasn't using a not-for-blame approach may have stopped at that point. However, the investigator continued the investigation and found, from talking to other people at the airfield where the accident pilot flew, that this pilot often flew under the effects of alcohol without incident. Therefore, there was possibly something else wrong with the aircraft that caused this accident which could help prevent recurrence, and so the investigation continued.

Aviation investigators will often have a professional specialism, such as experience as a pilot, engineer, or data recorder specialist. It is not a requirement to be an extraordinarily capable pilot or engineer to be an investigator—indeed that may actually be a hindrance as it may give you unrealistic expectations for what it is reasonable to expect of the "ordinary human"! However, anyone can adopt a not-for-blame investigator mindset—the key requirements are being inquisitive, keeping an open mind, avoiding judgment, and maintaining a focus on future safety. If an investigator (or someone reading a report for that matter) ever finds themselves thinking "*What they should have done is…*" or "*what would normally happen is…*" then they may have lost the necessary mindset and are not gaining the benefits of investigation.

Since safety levels in aviation are generally very high, and barriers have been developed to catch many mistakes, this can lead to an illusion that only the extremely unusual events turn into accidents. Take the example of:

- a maintainer who was distracted when called to an urgent job and on returning missed a step to lock the engine cowl doors, that might have been noticed by:
- a dispatcher who was tired having been woken intermittently by a noisy party next door to their house when trying to sleep, or

- a relief pilot who was hurrying an aircraft walk-around check due to delays in their arrival,
- but resulted in the aircraft taking off with the engine cowl doors unlatched.[i, 4]

Faced with this evidence during an investigation, one might ask, "What are the chances of all those things happening on the same day?" However, it is precisely because all these "unusual" things happened that we find ourselves investigating the accident. Had any of those factors not been present, the error would have been caught and there would have been no accident.

4.2 Role of Investigators and Regulators

Most countries have an agency which is nominated as being responsible for investigating civil aviation accidents. In many countries this takes the form of an independent SIA, including:

- The National Transportation Safety Board (NTSB) in the United States
- The Air Accidents Investigation Branch (AAIB) in the United Kingdom
- The Bureau d'Enquêtes et d'Analyses pour la sécurité de l'aviation civile (BEA) in France

ICAO maintains a list of accident investigation authorities on its website.[5]

In some countries, the function is performed by the aviation regulator. However, this can be considered less than ideal since an SIA needs to be independent, allowing it to investigate all matters, including the quality of regulation and oversight.

Annex 13 includes a Standard which states that:

A State shall establish an accident investigation authority that is independent from State aviation authorities and other entities that could interfere with the conduct or objectivity of an investigation.

4.2.1 Primacy for the Investigation

Annex 13 is very clear about the order of primacy in an investigation—in other words, which State will lead the investigation—and the relevant parties that should be involved. When an accident occurs (in an ICAO Contracting State[ii]), responsibility for its investigation falls to the State of Occurrence (the ICAO State in the territory of which an accident or incident occurs). This is for obvious practical and legal reasons (e.g., it is clear that granting powers to a SIA to investigate accidents in the territory of another State would be challenging). It is worth noting that different countries can have very different legal systems which will affect how they handle investigations (e.g., France uses the Civil or Napoleonic Code as the basis of its inquisitorial legal system, which demands very different processes to, say an adversarial common law system such as that in the United States or the United Kingdom). If appropriate, the

---

[i] This is an entirely fictional example, but there have been numerous instances of aircraft taking off with the engine cowl doors unlatched.[4]

[ii] An ICAO Contracting State is one that is a party to the Chicago Convention.

State of Occurrence can delegate responsibility for investigating a specific accident to another State by mutual agreement, although the State of Occurrence would remain as a participant.

Where the accident occurs in a noncontracting State, the State of Registry, Design or Manufacture should try to initiate an investigation, or conduct their own investigation with the information that is available. When an accident occurs in international waters or the location cannot definitely be established, responsibility for investigating falls to the State of Registry. Whichever agency takes primacy, it is then responsible for appointing an investigator-in-charge (IIC) and for notifying the other relevant parties, usually through their State SIAs. The IIC is responsible for coordinating and organizing the investigation, either on-site or remotely and is responsible for all aspects of the investigation.

4.2.2   Rights of Participation

Accident investigation is, by necessity, a collaborative process requiring information, analysis, and support from many different organizations, all of which have an interest in the investigation. For this reason, Annex 13 grants rights and obligations of participation to:

- The State of Registry (the State on whose register the aircraft is entered)
- The State of the Operator (the State in which the operator's principal place of business is located)
- The State of Design (the State having jurisdiction over the organization responsible for the type design)
- The State of Manufacture (the State having jurisdiction over the organization responsible for the final assembly of the aircraft)
- Any State which on request provides information, facilities, or experts to the State conducting the investigation

In this case, the relevant SIA in the State nominates a representative to participate in the investigation, designated an Accredited Representative (often abbreviated to AccRep). In addition to the Accredited Representatives, the interested parties (i.e., the operator, the type design organization, and the final assembly organization) shall nominate people to act as advisers to the Accredited Representatives.

In Europe, Regulation 996/2010 requires SIAs to invite EASA and the national civil aviation authorities of the concerned Member States to appoint a representative to act as an adviser to the investigation, provided that "the requirement of no conflict of interest is satisfied". Of course, deciding whether there is a conflict of interest before starting an investigation can be challenging.

4.2.3   The Party System

In the United States the participation of other States and their interested parties is formally managed through the "Party System" where advisers join functional investigation groups (e.g., powerplants, structures) usually under the supervision of NTSB staff—see 4.5.1.

Under the Party System, the NTSB may invite various qualified and interested organizations whose employees, functions, activities, or products were involved in the accident or incident, and who can provide suitable, qualified technical personnel, to assist in the investigation by participating as parties to the fact-finding phase of the investigation. Each participating party designates a party coordinator (spokesman) for its organization to act as the NTSB's direct and official point-of-contact for the party.

US Code of Federal Regulations 49 CFR § 831.21[6] grants the FAA party status to an NTSB investigation and Order 8020.11[7] describes the procedures and responsibilities for FAA involvement in an investigation. In brief, the FAA has nine responsibilities in an accident investigation, specifically to determine whether:

- Performance of FAA facilities or functions was a factor.
- Performance of non-FAA owned and operated air traffic control (ATC) facilities or a navigational aid was a factor.
- Airworthiness of FAA-certificated aircraft was a factor.
- Competency of FAA-certificated airmen, air agencies, commercial operators, or air carriers was involved.
- Federal Aviation Regulations were adequate.
- Airport certification safety standards or operations were involved.
- Airport security standards or operations were involved.
- Airman medical qualifications were involved.
- There was a violation of Federal Aviation Regulations.

The participation of the Accredited Representatives and the Party System serve multiple purposes. Firstly, it allows the investigating SIA direct access to those people and organizations able to supply the information and expertise needed to properly conduct the investigation. This, in turn, can bring significant resources to support an investigation—many SIAs may only be able to assign a few investigators to an accident and hence they may draw heavily on the expertise and resource of the aircraft design organization or others. In complex technical cases, thousands of person-hours may be spent performing analysis and testing.

Secondly, it grants rights to the organizations linked to an accident to be involved in an investigation, to have access to the evidence, and to contribute to the investigation outcomes.

It is important to note that under these participation systems, the investigation remains under the sole control of the investigating agency and their IIC, including any release of information.

4.2.4   Powers of an Investigator

In many countries, SIA investigators are granted significant powers to facilitate their investigation. In the United Kingdom, this includes granting investigators the power to enter premises and impound records or to compel a witness to attend an interview. However, these powers are considered a last resort, only to be used in extreme circumstances. Investigators understand that it is far more beneficial to explain that the point of investigation is to prevent recurrence, that the investigation is not for blame, and to allay any fears a witness may have about talking to an investigator, rather than simply compelling them to attend an interview.

4.2.5   Family Assistance

A more recent development in aviation safety has been a recognition of the need for SIAs, operators, and regulators to liaise with, include, and support the families of victims of air accidents as part of the process. Annex 13 provides for *The Participation of States Having Suffered Fatalities or Serious Injuries to Their Citizens:*

A State which has a special interest in an accident by virtue of fatalities or serious injuries to its citizens shall be entitled to appoint an expert to:

- *visit the scene of the accident;*
- *have access to relevant factual information which is approved for public release by the State conducting the investigation, and information on the progress of the investigation; and*
- *receive a copy of the Final Report.*

In the United States, the NTSB is designated (by the Aviation Disaster Family Assistance Act of 1996[8]) as the lead federal agency for liaison between the airline and the families. To this end, the NTSB has established the Transportation Disaster Assistance Division to keep clear distinction between family assistance and investigation. In the EU, Article 21 of Regulation 996/2010[3] defines the steps a Member State must take to support families. ICAO Doc 9973[9] provides guidance on the types of family assistance that may be provided to aircraft accident victims and their families and discusses what support should be provided, and by whom.

4.2.6   Approach to Investigation

While almost all SIAs work under the framework of Annex 13, there are different approaches to the processes involved such as gathering evidence or producing findings. Many SIAs outline their process on their website.[10–14] ICAO Doc 9756 *Manual of Aircraft Accident Investigation*[15] provides guidance for SIAs on the conduct of an investigation, in four parts:

Part I – Organization and Planning

Part II – Procedures and Checklists

Part III – Investigation

Part IV – Reporting

An additional complexity is that some agencies are responsible for additional modes of transport such as railroad and marine. These multimodal agencies deal with their broader mandate in different ways, which is beyond the scope of this chapter. Individual SIA websites give more details.

4.2.7   Inclusion of Human Factors

The acceptance and inclusion of Human Factors (HF—see Part III for more details) in accident investigation and analysis has varied greatly in the past and still does today. Agencies such as the Australian Transport Safety Bureau (ATSB) or its predecessor the Bureau of Air Safety Investigation (BASI) have embraced HF in investigations for many decades. Other Agencies have only recently started to include it, while others still have yet to make it part of their process.

One of the large challenges to including HF in investigations is accepting that it is a discipline in its own right, and not something that just anyone can do, simply by virtue of being human! HF is more than knowing that humans aren't perfect.

Historically, there has sometimes been a tendency to include an HF expert if any "HF issues" were uncovered. More modern thinking accepts that HF can pervade all aspects of the aviation system, be that engineering, operations, or even interpretation of flight data. All investigators need a grounding in HF, with the option to bring in an expert when the situation demands it. Many agencies employ investigators or support staff with a specialism in HF, Human Performance, or similar.

4.2.8   Specific Agencies and Their Structures

The National Transportation Safety Board (NTSB) is an independent, multimodal agency of the US government of about 400 personnel, with its headquarters in Washington DC. In 1974, through the *Independent Safety Board Act*,[16] Congress made the NTSB completely independent outside the Department of Transportation because "no Federal agency can properly perform such investigatory functions unless it is totally separate and independent from any other … agency of the United States."

Within the NTSB,[17] it is the *Office of Aviation Safety* that has the responsibility for investigating aviation accidents and incidents. The office has four regional offices:

- Eastern Region
- Central Region
- Western Pacific Region
- Alaska Region

In the United Kingdom, the relevant agency is the Air Accidents Investigation Branch (AAIB), headquartered in Farnborough. It is an independent unit within the Department for Transport, with the Chief Inspector reporting directly to the Secretary of State for Transport. The Branch[18] comprises around 35 investigators, organized into 6 teams of inspectors, each led by a principal inspector, who will often take the role of IIC in an investigation. Inspectors have a specialism of operations, engineering, or flight data recorders.

4.2.9   Board System

Some agencies, including the NTSB and the Transportation Safety Board of Canada (TSB) employ a Board system. Under this arrangement, the investigators bring forward a report to a public Board meeting (or to a delegated person). The NTSB and TSB Boards both have five Members. In the United States, these Members are nominated by the President and confirmed by the Senate to serve for 5 years.

NTSB Board Members perform multiple roles, including:

- Helping to guide the organizational goals and objectives
- Reviewing and approving accident reports and safety recommendations
- Acting as a spokesperson for the agency (often travelling to accident sites—see Figure 4-1)
- Testifying to Congress

Figure 4-1   NTSB investigator Eliott Simpson briefs NTSB Board Member Jennifer Homendy at the scene of a skydiving accident. (Source: NTSB)

4.2.10   Public Hearing

Following an accident, the NTSB may decide to hold a public hearing to collect additional information and to discuss in a public forum the issues involved in an accident. A hearing involves NTSB investigators, other parties to the investigation, and expert witnesses called to testify. At each hearing, a Board of Inquiry is established that is made up of senior safety board staff, chaired by the presiding NTSB Member. The Board of Inquiry is assisted by a technical panel and some of the NTSB investigators who have participated in the investigation serve on the technical panel. Depending on the topics to be addressed at the hearing, the panel often includes specialists in the areas of aircraft performance, powerplants, systems, structures, operations, ATC, weather, survival factors, and HF. Those involved in reading out the cockpit voice recorder and flight data recorder, and in reviewing witness and maintenance records may also participate in the hearing.

Expert witnesses are called to testify under oath about selected topics to assist the safety board in its investigation. The testimony is intended to expand the public record and to demonstrate to the public that a complete, open, and objective investigation is being conducted. The witnesses who are called to testify are selected because of their ability to provide the best available information on the issues related to the accident.

Following the hearing, investigators will gather additional needed information and conduct further tests identified as necessary during the hearing. After the investigation is complete and all parties have had an opportunity to review the factual record, a technical review meeting of all parties is convened. To better understand the overall processes of accident investigation, it will be useful to follow the lifecycle of an accident from notification to report.

4.3   Notification and Initial Response

Most SIAs have a specific notification phone number or address for notifying them of an accident. The event notification can come from many different people including the flight crew of the accident aircraft, the airline, the airport, ATC, members of the public, or many others.

Any notification process will aim to capture key information about the accident including:

- Aircraft type
- Aircraft registration
- Aircraft owner
- Crew and people onboard
- Date and time of accident
- Departure and destination
- Accident location
- Injuries
- Nature of the accident
- Aircraft damage

Once a notification is received it will be assessed by the agency and a decision made on the appropriate response, based on the description and circumstances. This may range from simply logging the event through to a full deployment. If the event is of significant potential safety interest, the agency will work straight away to preserve as much evidence as possible including securing the site, taking photographs, identifying witnesses, isolating cockpit voice recorders and flight data recorders, and preserving other recordings.

It is often the case that the information contained within an initial notification is inaccurate—sometimes to a very large extent. For the accident to G-YMMM (a Boeing 777 which crashed short of the runway while landing at London Heathrow), initial reports suggested it was a Boeing 757 that had departed the side of the runway on takeoff!

Of course, the notification may also be lacking information by simple practicalities. For example, if an aircraft is overdue then clearly an accident site location cannot be supplied. In this case, the SIA needs to adapt its response appropriately.

If the SIA initiates an investigation, an IIC will be appointed, and a notification of the accident will be sent to the relevant SIAs representing Interested Parties who will nominate Accredited Representatives.

4.4   Deployment and On-site

Accident sites fall into two main categories—those where the wreckage and/or accident site are accessible and those where it is not, which includes those in which the wreckage cannot be located. Each requires a different approach.

4.4.1   Inaccessible or Unknown Site

For some accidents, the first indication of a problem is an aircraft that is overdue, often prompting a search and rescue (SAR) effort based on available data such as radio traffic, radar, emergency locators,[iii] and so on. However, if the wreckage remains lost after SAR efforts have been stopped, usually at sea, it falls to the SIA to locate the wreckage. In this case, although the SIA may choose to move a team closer to the suspected accident site, if feasible, it is clearly not possible to deploy to a definitive site.

In the case of Airbus A330, F-GZCP operating flight Air France 447,[19] based on the information available to the BEA there was an initial search region of 17,000 km². In this accident, the vertical stabilizer floated on the sea surface (see Figure 1-1) and was spotted from the air a few days after the accident. However, it took nearly two years to locate the majority of the remaining wreckage on the seabed. The wreckage site when located was only about 600 meters by 200 meters in area (representative of the almost entirely vertical impact direction) and approximately 4 kilometers below the surface.

Underwater Locator Beacons (ULBs) are small devices attached to the aircraft that, on contact with water, emit a "ping" once a second for at least 90 days, allowing the recorders and the other wreckage to be located. However, the detection range of the signal depends on a variety of factors including depth and sea conditions but might be in the order of a few miles. Typically, an agency would use a sonar array towed behind a boat to conduct a wide search for the signal before then using a handheld hydrophone (an underwater microphone) to do more detailed direction/location finding. For Air France 447, extensive resources were deployed to try to detect the ULBs before their batteries expired, including a submarine and five ships (three carrying acoustic locating equipment). Unfortunately, this search was unsuccessful. However, the BEA were able to begin investigating the accident using floating debris recovered early in the investigation. In the case of the Boeing 777, 9M-MRO operating Malaysian Airlines Flight MH370 that disappeared in 2014, the accident site has never been identified despite extensive search operations and despite some debris washing up on shorelines thousands of miles from the intended route.

4.4.2   Accessible Site

If the accident site has been identified, and the SIA has decided to initiate a field investigation, they will launch a go-team to the site. The composition of this team will be based on the initial notification, the size and location of the accident, and so on. Most SIAs will operate a roster system with staff on call 24 hours a day, ready to deploy at short notice. Depending on the location, investigators will often drive or fly to the accident site, bearing in mind that the airport they are traveling to may be closed by the accident they are traveling to investigate!

Arriving on an accident site can be a daunting experience, particularly if it is a large site with significant disruption. You may be tired from traveling, there may be significant media interest, family members may have arrived at the site, there may still be rescue efforts ongoing (which must take priority), and you may be greeted gratefully as the "person in charge" or possibly with "who are you"? Whatever the scenario, the accident investigation team needs to establish control of the accident site. It is vital to

---

[iii] Most large aircraft carry an ELT (Emergency Locator Transmitter), using the COSPAS-SARSAT satellite network and uniquely coded, which is intended to allow location of the aircraft following an accident.

FIGURE 4-2   Fire and impact damaged throttle quadrant which still supplied useful information. (Source: NTSB)

remember that even in scenes of apparent total devastation, there can still be a great deal of useful evidence to the investigation. Figure 4-2 shows a badly damaged throttle quadrant from which useful information was still derived, despite the impact and fire damage.

4.4.3   Health and Safety

Before work on the site can start in earnest, the investigation team needs to ensure that it is a safe place to work. In addition to hazards from the location, weather, environment, and so on, working at aircraft accident sites has the potential to expose investigators to hazards specific to the accident ranging from sharp edges, through biological and chemical hazards to stored energy devices such as batteries, ballistic parachute systems, tensioned springs, and pressurized hydraulic systems. Cargo adds a new dimension to the possible hazards with the potential of Dangerous Goods being carried.

Most SIAs mitigate the hazards on the accident site through the use of Personal Protective Equipment (PPE) and dynamic risk assessments. ICAO Circular 315[20] discusses the nature and variety of occupational hazards and the management of risk associated with exposure to a wide range of health and safety hazards during the investigation of aircraft accidents.

4.5   Evidence Gathering

Once the investigation team has arrived on site, established control, and can work safely, the process of evidence gathering can begin. Some agencies, including the NTSB, choose to separate the evidence-gathering and analysis phases of the investigation. This has the advantage of avoiding any confirmation bias (looking for evidence that supports a theory) in the gathering of evidence and contributes to a common understanding, as all parties analyze the same evidence together. The downside of this approach is that it is impossible to ever gather *all* the evidence and so focusing on likely areas, identified by ongoing analysis, may help to ensure key evidence for this specific accident is gathered. As is often the case, a hybrid strategy may represent the most useful approach.

FIGURE 4-3   Accident site for G-YMMM at London Heathrow. (Source: AAIB)

Depending on the complexity of the site and the accident, investigators might spend up to one or two weeks on site. Investigators often face pressure to clear an accident site as quickly as possible, particularly if other infrastructure such as an airport or railway line is being affected. The skilled investigation team will work to allow as much of a return to normality as possible, as quickly as possible, while ensuring that all the required evidence is gathered as fully as necessary. This is not an easy balance to strike. Figure 4-3 shows the large complex site faced by investigators at London Heathrow airport following the accident to aircraft G-YMMM.

4.5.1   The Group System

For large investigations, some SIAs employ a "group" system (not to be confused with the Party System), forming working groups to focus on specific aspects such as power-plants, structures, operations, and so on.

In the United States, the IIC will assign and organize investigative groups to document specific aspects of the accident with individuals representing selected parties assigned to investigative groups and each group under the direction of an NTSB investigator who is designated as the Group Chairman. Not all parties will have members on every group and only those parties who can provide needed specific expertise relevant to the focus of the group will be considered for group assignments.

Under the direction of the Group Chairman, one or more sets of group notes, termed "field notes," will be developed collaboratively by each investigative group. Field notes should include all relevant factual information developed by the group and will typically also include appendices of supporting documentation, photographs, or other records collected by the group. Each NTSB Group Chairman will later prepare a Group Chairman Factual Report, drawing on the information in the field notes.

The benefits of this system include the provision of a focus area for investigators; creating clear points of contact and responsibility; and taking advantage of individual specialist knowledge. Disadvantages include the risk of investigative silos; the complexity of overlap or worse, missed areas; and the need for significant resources. Ultimately, how the system is implemented and how the team work together and communicate is more important than the specific details of any one approach.

4.5.2   Perishable Evidence

At the first opportunity, the investigative team will prioritize the collection of perishable evidence, that is, evidence that will not endure, for whatever reason. This will include more obviously perishable evidence such as fuel samples, witness marks in the ground, ice or soot on the fuselage, fracture surfaces, tire marks on the runway, ATC recordings, oil samples, cockpit switch, and display configurations. However, it will also include other perhaps more unusual perishable evidence such as CCTV recordings from surrounding buildings, a strong smell of fuel on the site, loose paper copies of flight plans or weather forecasts, personal electronics, and much more. This latter category would also include any part of the aircraft that is not on the accident site perhaps due to uncontained engine failure or midair collision. Therefore, investigators may take a quick inventory to ensure all major parts are accounted for. The investigative team needs to capture all of this evidence before it is lost.

Evidence can also be perishable for unexpected reasons. One investigator described arriving on site quite late in the evening, when it was dark. They debated as to whether to visit the site or wait until the morning after a good night's sleep. They elected to make a short visit and take some photos. This turned out to be fortuitous since it snowed, unexpectedly, overnight and the photos they had taken that night were the only record they had of some of the witness marks and other evidence. Another investigator described arriving on site to discover that despite the aircraft having overrun the end of the runway, the aircraft cabin had been serviced by a cleaning crew, removing significant evidence about the evacuation from the cabin.

Finally, a local authority may wish to clear a site as soon as possible (e.g., highway or train track) and so may clear wreckage without authorization from the investigation agency. Some evidence can be considered "vulnerable" rather than perishable, meaning it can be changed as opposed to destroyed completely. One example would be eyewitness evidence which can be easily corrupted by interactions such as conversations, questions, or watching news reports about the event. Another example could be the aircraft's recorders which if still powered, may overwrite the older accident recordings. The aircraft's recorders will be a high-priority item to locate.

4.5.3   Recorders

In a commercial aircraft accident, one priority for the on-site team is locating the flight data recorder (FDR) and the cockpit voice recorder (CVR) sometimes known colloquially as "black boxes" (despite being painted bright orange!). Some aircraft combine the two functions into a single combined voice and flight data recorder (CVFDR). The recorders are fitted with their own ULBs to aid location underwater.

As its name suggests, a digital flight data recorder (DFDR) records digital flight data onto nonvolatile memory (i.e., memory that endures after power is removed) to allow it to be extracted at a later date. This flight data includes speeds, altitudes, cautions and warnings, aircraft attitudes, engine data, control positions, and much more.

The binary data recorded to the DFDR is laid out in a specific way, inside frames, to make sure it "fits" into the memory (a hangover from the days when the medium was magnetic tape moving at a given speed). This organization map is called a Data Frame Layout or Logical Frame Layout (LFL) and is needed to allow the data to be decoded from binary into engineering units.[21] Since the LFL can be unique to an aircraft, it is the responsibility of the aircraft operator to be able to produce the LFL to the SIA in the case of an accident.

The CVR holds data in a more straightforward way, usually in the form of several audio files capturing different acoustic sources including the pilots' headset microphones and a Cockpit Area Microphone (CAM) to record speech, noises, and alarms in the cockpit. For new aircraft, both the CVR and DFDR must be capable of storing 25 hours of data.

Modern recorders might record many thousands of parameters from across the aircraft. ARINC Characteristic 767[22] (which, confusingly, is entirely unrelated to the Boeing aircraft of the same name) defines a standard for Enhanced Airborne Flight Recorders (EAFRs), and the Boeing 787 Dreamliner was the first aircraft to be equipped with an EAFR. Under the Characteristic, the EAFR can combine any or all of these functions: FDR, CVR, data link recorder, airborne image recorder, and integrated flight data acquisition. EAFRs also use the Flight Data Recorder Electronic Documentation protocol (FRED, defined by ARINC Characteristic 647[23]) whereby the LFL is embedded within the recorder, removing the need for it to be supplied separately.

Once the recorders have been located on an accident site, they will usually be transported immediately to a "read-out" facility (such as that shown in Figure 4-4). Most large SIAs have their own recorder capability, whereas others will seek support from other SIAs. Many agencies have dedicated recorder or laboratory staff who focus on data retrieval and analysis. As part of the investigation, some of these staff may also interrogate other specific aircraft systems and controllers, many of which contain some form of nonvolatile memory (e.g., Terrain Avoidance Warning System—TAWS box) or other electronic devices (e.g., tablets and phones) with some having facilities to repair and interrogate damaged chips.

4.5.4   Physical Evidence

Clearly, one of the most common pieces of physical evidence in an aircraft accident is the aircraft itself, whether damaged or not. The investigation team will conduct as much examination of the wreckage on-site as is necessary to avoid losing evidence. However, when practical, they will often remove and transport the wreckage and any other items of interest to their own storage facility or to a dedicated hangar to allow the investigation to continue.

As mentioned previously, investigators often face drastically different challenges both in terms of technical complexity and scale—from the overall structural integrity of a fuselage to whether an electrical pin properly engaged with a socket. As a result, SIAs maintain access, either internally or externally, to a wide range of laboratory facilities and expertise including metallurgy, corrosion, CT scanners, and more.

Annex 13 requires that an autopsy is performed on fatally injured flight crew and, if appropriate, any fatally injured passengers or cabin crew.

FIGURE 4-4   An NTSB recorder specialist prepares a CVR for read-out. (Source: NTSB)

4.5.5   People

In almost all investigations, there will be a need to gather information and evidence for the investigation from people who may be active participants in the operation and/or uninvolved witnesses. In the most benign incidents this might include flight crew, maintainers, passengers, ground crew, air traffic controllers, operator management, and more.

Many of the people interviewed in an investigation may feel nervous about talking about an accident, fearing repercussions. It is crucial that the investigator makes clear the purpose of the investigation and the protections they can offer regarding their evidence. It is often worth pausing and considering what the emotional state and concerns of a witness may be before commencing the interview.

Eyewitness evidence can be some of the most useful evidence for an investigator and yet it can also be some of the most unreliable. However, time is a factor in accurate recollection—memory is not like a recording that can be reliably viewed repeatedly without it changing. As a result, it is preferable to talk to eyewitnesses as soon as possible after the event.

Even when talking to them soon after an event, a wide variation in perceptions and recollections can be expected from eyewitnesses. One investigator described tabulating the different observations made by 10 different eyewitnesses for one particular accident, with some very marked differences between descriptions of the same event.

To help gather the best possible evidence, some investigators have adopted specific techniques such as one used by the police called Cognitive Interviewing. This technique is aimed at maximizing recall by stimulating as many cues as possible.

Whatever the approach, interviewing is a skill that requires training, practice, and preparation, particularly for key witnesses. The investigator will almost inevitably have specific questions that they will want answers to, but they must resist the urge to go straight to these closed questions as this will limit the information they retrieve. The witness may have useful information about which the investigator currently knows nothing.

4.5.6 Documents

In a highly regulated industry such as aviation, there will inevitably be a large amount of documentation associated with the operation connected with an accident. Company Operating Manuals, aircraft technical records, crew records, and licensing will often be of interest to investigators as they relate to the direct operation of the aircraft. However, there may also be other more detached documents of interest that may relate more to the organizations and cultures involved, including briefing sheets, emails, and company audits. Also of relevance will be any documents used by the crew for flight planning and delivery, whether in paper format or as part of an Electronic Flight Bag (EFB). Although it is possible to describe some typical evidence that will usually be of interest, it is impossible to create a checklist that will cover every eventuality.

Under a group system, each group chair (the senior investigator overseeing a specific area of the investigation) will complete a factual report on their area of responsibility. The reports might also include proposed safety recommendations. Under the US system, all factual material is placed in the public docket that is open and available for public review, but this is not commonly done in other countries. Once sufficient evidence has been gathered, analysis can begin in earnest, although there will invariably be a need to supplement the evidence as the analysis progresses.

4.6 Analysis

As described above, the first stage of accident investigation is discovering *what* happened. However, often far more challenging, but arguably more important, is the issue of *why* it happened. Analysis is the process of taking evidence and ordering it, examining it, and supplementing it, in order to draw conclusions.

It can be tempting to confuse the act of "writing opinions down" with analysis. The difference between these two is evidence. Investigators must work with evidence and data since, as the statistician W. Edwards Demming said, "Without data, you're just another person with an opinion." An investigation report that descends into unsupported opinion and supposition will quickly damage the credibility of the SIA and ultimately undermine aviation safety.

This doesn't mean that every single detail in an investigation must be supported by independent evidence or be discarded—if a person normally sleeps for 8 hours a night and the night before an accident it is established that they had no sleep at all, it is reasonable to assume that they may have been tired without a written statement from that person or an independent assessment from the day! Investigations are often forced to rely on reasonable deductions, but the analysis process must stay mindful that these are deductions and avoid inadvertently turning them into fact.

4.6.1 Analysis Models

As described in other chapters, there are many models which have been used to explain accident causation. These are variously used by accident investigators, but like all tools, there is no single model that will help in all circumstances. Indeed, in most investigations a range of tools will be used to help the investigator form and test their hypotheses against the available evidence.

Chapter 3 describes a number of models of causation ranging from simple, linear models to more complex systemic approaches such as STAMP and FRAM. Both extremes can be applicable in the same investigation depending on what is being examined and what level of analysis is being applied. For example, linear causal sequences may be useful in describing the cause and effect of component failures, whereas systemic models may be needed to examine the influence of less tangible, nonbinary influences such as cultural or management factors. A major challenge for investigators becomes the "burden of proof" which differs from that which might be expected in litigation or criminal prosecution.

The ATSB and its predecessor BASI have been notable pioneers in their application of a more systemic approach to accident investigation. Their work in the 1990s using Professor James Reason's Organizational Accident "Swiss Cheese Model" eventually led to changes within ICAO's guidance material for accident investigation. By looking beyond proximal causes of accidents, they were able to identify organizational and regulatory failures in a range of accidents. In some cases, this resulted in deeper systemic findings that were less well-evidenced than, for example, technical failures. This was justified on the basis that "findings" did not mean the same as "causes"—in fact, BASI ceased to use the term *cause* in the 1980s to avoid the implication of blame or liability.

For example, following a serious incident where a Boeing 747 landed at Sydney Airport with its nosewheel retracted, the investigation identified areas of concern deep within the airline and the regulatory authority. For example, in describing the operating airline, Ansett Australia, BASI found: "Commercial imperatives resulted in the accelerated introduction of the B747 to the operator's fleet which, in turn, contributed to deficiencies in the management of manuals, procedures and line training" and "The development team leader did not recognize the need to delay the start of B747 operations when it became apparent that some requirements would not be met." Within the aviation safety regulator, BASI found: "The Civil Aviation Authority's project manager issued an air operators certificate to Ansett before all the regulated requirements were met and without Ansett having developed and/or put in place all the necessary procedures" and "Real or imagined pressure, caused by a seemingly inflexible starting date for Ansett's B747 operations, probably influenced some of the actions taken by Civil Aviation Authority staff."

However, as other investigation agencies around the world followed suit in choosing to go deeper into organizational and systemic factors, a tension developed over the evidence that was needed to justify some of the conclusions that were being drawn. Investigators work in a complex world where physical evidence and hard data are complemented by witness recollections and hard-to-quantify elements such as safety culture. Legal and quasilegal processes ranging from Coroners and Medical Examiners to civil and criminal litigation tend to demand a much higher burden of proof and yet often have less access to the evidence that comes from an aircraft accident investigation.

A similar tension seems to exist between academics and investigators where the former promotes a sometimes bewildering array of often complex analytical models and tools, and the latter crave for practical, straightforward tools that can help them to unravel an investigation. Underwood and Waterson[24] reviewed over 100 such tools as part of their work to understand why there appears to be a gap between academia and

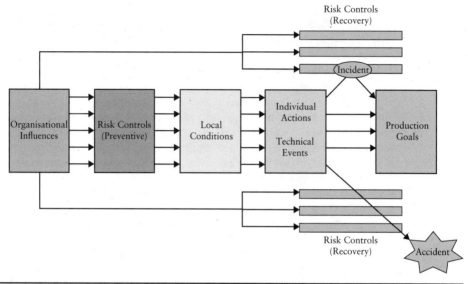

FIGURE 4-5   ATSB Adaptation of the Reason Model. (Source: ATSB)

the practitioner community. The answer essentially comes down to usability, especially in the heat of an accident investigation where evidence may be absent, incomplete, inaccurate, or contradictory, as well as not necessarily being identified in the correct order.

The ATSB's approach was challenged as a result of a clash with the South Australian Coroner over a fatal accident involving a double engine failure on a Piper PA31-350 Chieftain in the Spencer Gulf. The coroner was of the view that the two engines failed independently of each other, whereas the ATSB investigators concluded that, based on the limited evidence available, the second engine failed as a result of the first. This led the ATSB to review its analytical approach in a publication titled *Analysis, Causation and Proof in Safety Investigations*.[25] As the authors note,

> ...*safety investigations require analysis of complex sets of data and situations where the available data can be vague, incomplete and misleading. Despite its importance, complexity, and reliance on investigators' judgements, analysis has been a neglected area in terms of standards, guidance and training of investigators in most organisations that conduct safety investigations.*

The ATSB analytical approach is now based on an adapted Reason model (see Figure 4-5) which adds risk controls and acknowledges that in most cases, despite the limitations of any system, there is usually a successful outcome and the "production goal" such as a safe landing is achieved.

The investigation path is described using a simple diagram (Figure 4-6) which although correlating with the colors in the adapted Reason model also works for most tools. In simple terms, it characterizes the imperative to start with the bad outcome and work backwards, asking "why" questions for as long as there is evidence available.

For most investigators, while running out of evidence becomes the point that they stop their analysis, it doesn't necessarily mean that this "root cause" is where they make their recommendations. These can be identified at all layers of the system, and it is for

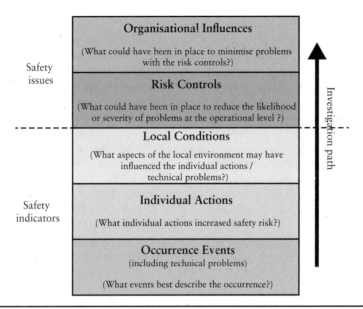

FIGURE 4-6   ATSB investigation analysis model. (Source: ATSB)

the safety regulator to evaluate what is likely to have the greatest effect within reasonable cost constraints.

New analytical tools create opportunities to consider increasingly complex socio-technical systems in different ways. As investigators gain confidence in some of the more academic models, we may see greater adoption "in the field." Until then, it is worth remembering that no single tool will be suitable for every facet of an investigation. Nothing replaces the curiosity of an investigator and the desire to make sure every hypothesis is thoroughly tested by the evidence.

4.7   Reporting

The primary output of most accident investigations is one or more written accident reports. Some agencies supplement these reports with explanatory videos or animations but, on the whole, these are used to support, explain, and clarify the evidence, analysis, and conclusions of the investigation as described in the accident report.

4.7.1   Interim Reports

For large, complex, or long-running investigations, an agency may choose to issue interim reports. These can vary from short statements of fact (often very soon after an accident occurs) through annual updates on the progress of an accident (as required by Annex 13) to very large reports presenting what is known to date (as in the case of MH370).

Another reason for issuing interim reports may be to alert the industry to relevant findings and promulgate recommendations requiring action before the investigation is complete. Clearly, it is beneficial to release such information in as timely a manner as possible.

4.7.2   Final Reports

Appendix 1 of Annex 13 describes the format of the final accident report and Doc 8756 offers further guidance. One significant distinction is between the factual information (Section 1) and analysis (Section 2) of the report. The first presents the facts established and evidence that has been gathered during the investigation, without commenting on its significance. The analysis section presents an evaluation of the evidence given in the earlier section and should not introduce new facts.

Accident investigators are forced to tread a fine line when reporting—they have an obligation to protect the evidence given to them and the people who supplied it, but they must also provide enough detail and context to assure the reader that an inference, conclusion, or recommendation is justified.

For the NTSB, under the Board system, the IIC and the NTSB senior staff create a final draft report, called the notation draft, for presentation to the Board Members. Following a period for review of the draft report, a public meeting of the Board Members is held in Washington DC (see Figure 4-7). The NTSB staff will present and comment on the draft report; party representatives are permitted to attend but may not make any kind of presentation or comment. At this meeting, the Board Members may vote to adopt this draft, in its entirety, as the final accident report; may require further investigation or revisions; or may adopt the final accident report with changes that are discussed during the meeting.

Annex 13 mandates that once a draft Final Report is ready, it should be sent to all of the participating States for their comments, which should be "significant and substantiated" within 60 days. If it receives comments, the State conducting the investigation

FIGURE 4-7   An NTSB investigative hearing in progress. (Source: NTSB)

shall either amend the draft Final Report to include the substance of the comments, or if they are rejected append them to the Final Report (if requested by the State making the comments). However, these latter comments must be restricted to technical aspects of the Final Report on which no agreement could be reached. As such, this is a relatively unusual occurrence and is regarded by all as a last resort.

4.8   Findings, Causes, Recommendations

Having discovered what happened, and why it happened through the process of investigation, evidence gathering, and analysis, the final stage of the process can be summarized as—*so what?*

Although finding out what happened, and why, is a necessary part of the investigation process, it is not sufficient to prevent an accident from recurring. For that, changes need to be made to the aviation system and this is done through findings and causes leading to recommendations.

4.8.1   Findings

The ICAO *Manual of Aircraft Accident and Incident Investigation*[15] gives the following definition:

> *The findings are statements of all significant conditions, events or circumstances in the accident sequence. The findings are significant steps in the accident sequence, but they are not always causal or indicate deficiencies. Some findings point out the conditions that pre-existed the accident sequence, but they are usually essential to the understanding of the occurrence. The findings should be listed in a logical sequence, usually in a chronological order.*

The findings tell the abridged story of the relevant parts of the accident, including those elements that were found to not be part of the sequence leading to the accident, for example, *There was no evidence of airframe failure or system malfunction prior to the accident*, and those that were investigated but eliminated, for example, *Flight crew fatigue was not a factor in the accident*. The findings give a useful insight into which elements of the factual information and analysis are worthy of focus.

4.8.2   Causes and/or Contributing Factors

ICAO[15] defines causes as "those events which alone, or in combination with others, resulted in injuries or damage. Causes are defined as actions, omissions, events, conditions, or a combination thereof, which led to the accident or incident."

In contrast, they define contributing factors as "actions, omissions, events, conditions, or a combination thereof, which, if eliminated, avoided or absent, would have reduced the probability of the accident or incident occurring, or mitigated the severity of the consequences of the accident or incident." Some SIAs use the term *contributory cause*.

SIAs can use causes, contributing factors, or both in the conclusions of a report. Some States choose not to prioritize their causes and/or contributing factors, usually listing them sequentially. Others choose to prioritize particular elements, calling them primary causes or similar. The modern view of causation is that most accidents are the result of multiple causes or factors and therefore, identifying a single or primary cause runs the risk of confusing that message.

Where the SIA is reasonably sure, but not certain, of a cause or factor, they will use a term such as *probable* or *likely*. This is also true if the cause draws on an event or occurrence that is not certain.

4.8.3   Recommendations

Annex 13 defines a safety recommendation as:

A proposal of an accident investigation authority based on information derived from an investigation, made with the intention of preventing accident or incidents.

Safety recommendations are the ultimate mechanism for SIAs to improve the aviation safety system. However, they need to be carefully developed to ensure that, among other things, they:

- are supported by the evidence and analysis;
- are aimed at the appropriate organization;
- are likely to be accepted; and
- bring the greatest likely safety benefit.

Safety recommendations can be issued at any point during an investigation and should be issued as soon as possible to give maximum benefit.

In general, recommendations will be phrased to describe the safety outcome the nominated party should achieve, rather than *how* they should achieve it. It is not for the agency to prescribe the method of achieving an effect, as there may be other factors, not visible to the SIA, that dictate how best to achieve the aim. In this respect, recommendations follow a similar style to certification regulation—see Chapter 6.

A well-written recommendation, supported by evidence, is hard to refuse and, on the whole, all parties want to see the safety of the system improved and will work toward that end. Equally, an SIA is allowed, and it is generally beneficial, to consult on draft recommendations. However, in general, the SIA is not a regulatory authority and is therefore not empowered to enforce or mandate its recommendations. Annex 13 requires that "A State that receives safety recommendations shall inform the proposing State, within ninety days...of the preventative action taken or under consideration, or the reasons why no action will be taken." This means that at least a refusal to act must be documented and justified. To help reinforce this system, agencies keep track and publish the status of their recommendations in Safety Reviews[26] or similar.

If during the investigation, a recommendation has been acted upon or an organization has taken a safety action voluntarily, that will usually be recorded in the report as an "agreed action." Different SIAs record these in different parts of the report.

If a recommendation is not acted upon, in addition to logging the lack of acceptable response, the agency may restate the recommendation in the reports following future accident investigations. For example, multiple agencies, including the BEA and the NTSB, have made multiple recommendations for the installation of cockpit cameras to aid investigation, with little success to date.

However, sometimes a single investigation can produce major change within the industry. One such example is that of Bombardier Q400 N200WQ, operating Colgan Air Flight 3407, a commuter airline fatal accident that occurred on February 12, 2009, near

FIGURE **4-8**   The wreckage of Colgan Flight 3407. (Source: NTSB)

Buffalo, New York. Figure 4-8 shows the wreckage from Flight 3407. The NTSB investigation very quickly produced recommendations in a wide range of areas, including:

- Effect of icing on aircraft performance
- Cold weather operations
- The "sterile cockpit" (limiting unrelated discussions between pilots in critical phases of flight)
- Flight crew experience and training
- Human fatigue management
- Stall recovery

4.8.4   Most Wanted/Significant Seven

Many regulators and agencies choose to prioritize or highlight specific recommendations or safety focus areas. This can help to prevent issues becoming stale and add emphasis to those items they consider to provide greatest benefit. The NTSB scheme is called the "Most Wanted List" and can be found on their website.[27] The UK CAA has run similar schemes in the past, such as the Significant Seven, focusing more on accident types rather than specific prevention strategies, in order to make sure that effort is being focused in the right areas.

For 2021-2022, the two most wanted improvements for aviation from the NTSB were:

- Require and verify the effectiveness of Safety Management Systems in all revenue passenger-carrying aviation operations.
- Install crash-resistant recorders and establish Flight Data Monitoring programs.

The latter item included a call for cockpit image recorders to be installed, joining calls from other agencies including the BEA, as described above. The NTSB cites the 737 MAX accidents, among others, as an example of where an image recorder would have been "extremely helpful."

4.9   When the System Fails

Given the comprehensive process and protocols described above, it would be tempting to think that after each accident there is a rapid, thorough investigation by an independent body, revealing all possible lessons to be learned, and resulting in recommendations that are acted upon in their entirety by the industry, which prevent that type of accident from ever happening again. Sadly, in reality, this is not the case; not all accidents are reported, not all investigations are perfect, not all parties implement all recommendations, and all for a wide variety of reasons. Many of these issues can and will be fixed by attitudes, organizations, and resources developing over time. However, there are also situations where an accident is investigated, good recommendations are made and may be acted upon and yet, despite this, the system fails, and an accident recurs.

On April 1, 2009, in the North Sea off the United Kingdom, a Eurocopter AS332 L2 "Super Puma"[iv] helicopter, G-REDL, suffered a catastrophic failure of the main rotor gearbox (MGB) while cruising at 2000 ft above mean sea level. As a result, the helicopter departed cruise flight and struck the surface with high vertical speed. The helicopter was destroyed and all 16 people onboard were killed.

The UK AAIB conducted a thorough investigation of the accident which took more than two years and resulted in a report more than 200 pages in length.[28] The investigation identified the following causal factor:

> *The catastrophic failure of the Main Rotor Gearbox was a result of a fatigue fracture of a second stage planet gear in the epicyclic module.*[v]

and three contributory factors. It is worth noting that one section (about one-third) of the second stage planet gear, containing the origin of the crack, was never recovered, which restricted the analysis and conclusions that could be made.

The AAIB investigation also highlighted a 1980 accident in Brunei to another variant of the helicopter (an SA330 Puma, registration 9M-SSC) which showed similar signs of second-stage planet gear failure within the MGB and in which all 12 persons onboard were killed.

[iv] The designation "Super Puma" is applied to several aircraft including the AS332, AS332 L1, AS332 L2, EC225 and H225. The MGBs of the latter three aircraft share the same main module and epicyclic module reduction gears.
[v] Similar to a car gearbox, the MGB takes power from the fast-turning engines, reduces the speed and transmits it to the rotor. A planet gear is one of the gears in that reduction / transmission chain.

The investigation into G-REDL issued 17 safety recommendations (6 shortly after the accident in 2009 and 11 in the final report in 2011), including:

Safety Recommendation 2011-036: It is recommended that the European Aviation Safety Agency (EASA) re-evaluate the continued airworthiness of the main rotor gearbox fitted to the AS332 L2 and EC225 helicopters to ensure that it satisfies the requirements of Certification Specification (CS) 29.571 [which relates to fatigue tolerance evaluation of metallic structure] *and EASA Notice of Proposed Amendment 2010-06* [which relates to damage tolerance and fatigue evaluation of metallic rotorcraft structures].

Following the accident to G-REDL, a number of activities were undertaken by the Original Equipment Manufacturer (OEM) and regulator, including an Airworthiness Directive (see Chapter 6) to remove a ring of magnets installed in the gearbox and designed to collect any metal "chips" suspended in the gearbox oil, but not register a warning. There was also a testing campaign launched inside Airbus Helicopters, but only part of this was completed by 2016. In the years following the G-REDL accident there were also a number of issues related to a separate part of the MGB (referred to as the bevel gear vertical shaft) that were unrelated.

On April 29, 2016, while returning from a platform off the coast of Norway, an accident occurred to an EC 225, registration LN-OJF. The aircraft was established in the cruise at 2000 ft at 140 kt when the main rotor suddenly detached. The aircraft departed cruise flight and impacted a small island, destroying the aircraft and killing all 13 people onboard. Figure 4-9 shows the main wreckage of the aircraft being recovered.

The subsequent comprehensive investigation[29] by the Accident Investigation Board Norway (AIBN)[vi] revealed that:

the accident was a result of a fatigue fracture in one of the eight second stage planet gears in the epicyclic module of the main rotor gearbox.

and that there were "clear similarities" with the accident to G-REDL, but that:

the post-investigation actions were not sufficient to prevent another main rotor loss.

The investigation into LN-OJF produced a further 12 recommendations including recommendations to change certification specifications and the type design.

The investigations into G-REDL and LN-OJF were both highly technical, complex investigations focusing heavily on the material properties of a single component and are beyond the scope of this chapter. However, the question that remains is how, given the initial investigation into G-REDL (and the earlier accident to 9M-SSC), the performance of the overall safety system was not sufficient to stop this almost-identical accident to LN-OJF from occurring. Given the similarities between the two accidents, part of the AIBN investigation into the accident to LN-OJF focused on the follow-up to the recommendations from the G-REDL investigation.

Taking the example of one recommendation, in the case of G-REDL, around 36 hours before the accident, a magnetic particle had been discovered.[vii] Unfortunately, miscommunication between the operator and the OEM meant that the "chip" was misinterpreted. After this point, the detection methods in place did not provide any further indication.

[vi] Renamed the Statens havarikommisjon (SHK) or Norwegian Safety Investigation Authority (NSIA) in 2020 with an expanded remit to include the defense sector.
[vii] The degradation process of spalling involves liberating small chips of metal from the bearing surface. By collecting these chips on magnetic chip detectors, operators can be alerted to their presence.

FIGURE 4-9   The wreckage of LN-OJF being recovered. (Source: AIBN)

Therefore, following the G-REDL accident, the AAIB introduced a recommendation (SR 2011-032) to "introduce further means of identifying in-service gearbox component degradation." Airbus Helicopters" response to that recommendation was that magnetic plug detectors and/or "chip" detectors "are sufficient to ensure flight safety." The AAIB assessed this response as "Not adequate" and "Closed," meaning they were expecting no further response. Unfortunately, in the case of LN-OJF, there were no indications of the impending failure as the crack in the planet gear had propagated from an origin below the bearing surface of the planet gear without any spalling to release detectable "chips." This suggests a weakness in the recommendation system, but one that is not easily resolved.

There were many other subtle technical aspects to these investigations, but viewed as a whole, these accidents are a tragic example of missed opportunities by the aviation system. The AIBN investigation identified that the action taken by the manufacturer could have been more effective and that the regulator could also have been more effective in its implementation of the safety recommendations and follow-up, saying:

> *In summary, Airbus Helicopters and EASA did not successfully manage to realise the safety potential from the G-REDL accident report.*

The commercial aviation system failed to prevent recurrence and so it is the system that needs to learn and improve.

4.10 Conclusion

The importance of accident investigation in the aviation safety system cannot be over-stated. Even where data safety programs claim to be predictive and forward-looking, they invariably still draw on historic investigation results or current investigation efforts to inform the analysis. In this way, investigations, and the recommendations they produce run through the entirety of the aviation system.

However, investigation is only useful when it is done "properly." An investigation that succumbs to bias, misses key evidence, or fails to deliver meaningful recommendations can have an overall detrimental effect on the aviation safety system. Investigation is a difficult skill requiring training and practice, but it is essential to get the process of accident investigation right and ensure that the system continues to learn from every opportunity.

Key Terms

Accredited Representative

Air Accidents Investigation Branch (AAIB)

Board of Inquiry

Bureau d'Enquêtes et d'Analyses pour la sécurité de l'aviation civile (BEA)

Cause

Cockpit Voice Recorder (CVR)

Contributory Factor

Field Investigation

Final Accident Report

Findings

Flight Data Recorder (FDR)

Go-team

Group System

ICAO Annex 13

National Transportation Safety Board (NTSB)

Party System

Recommendations

Safety Investigation Authority (SIA)

Technical Adviser

Topics for Discussion

1. What is the ICAO document that describes how accident investigations should be conducted?
2. Why does aviation use a no-blame approach to accident investigation?
3. Does no-blame investigation remove responsibility from those connected with an accident?

4. Describe the "party system" and the role of an accredited representative.
5. List five pieces of evidence that might be gathered at an accident site.
6. What is the purpose of the NTSB Most Wanted List?
7. What power does an SIA have to mandate safety recommendations?
8. Why does accident investigation not always prevent future accidents?

References

1. ICAO. *Annex 13: Aircraft accident and incident investigation*. https://elibrary.icao.int/home
2. US Code of Federal Regulations, 49 CFR Part 831. https://www.ecfr.gov/current/title-49/subtitle-B/chapter-VIII/part-831?toc=1
3. Regulation (EU) No 996/2010. https://www.easa.europa.eu/en/document-library/regulations/regulation-eu-no-9962010
4. AAIB` Report on the accident to Airbus A319-131, G-EUOE London Heathrow Airport, May 24, 2013. https://www.gov.uk/aaib-reports/aircraft-accident-report-1-2015-airbus-a319-131-g-euoe-24-may-2013
5. ICAO list of Safety Investigation Agencies. https://www.icao.int/safety/aia/Pages/default.aspx
6. US Code of Federal Regulations, 49 CFR Part 831.21. https://www.ecfr.gov/current/title-49/subtitle-B/chapter-VIII/part-831/subpart-B/section-831.21
7. FAA Order 8020.11. Aircraft Accident and Incident Notification, Investigation, and Reporting. https://www.faa.gov/regulations_policies/orders_notices/index.cfm/go/document.information/documentID/1033315
8. Aviation Disaster Family Assistance Act, 1996. https://www.ntsb.gov/tda/er/Pages/tda-fa-aviation.aspx
9. ICAO Doc 9973. *Manual on assistance to aircraft accident victims and their families.* https://store.icao.int/en/manual-on-assistance-to-aircraft-accident-victims-and-their-families-doc-9973
10. NTSB Investigation Process. https://www.ntsb.gov/investigations/process/Pages/default.aspx
11. TSB Investigation Process. https://www.tsb.gc.ca/eng/enquetes-investigations/index.html
12. AIBN Investigation Process. https://www.nsia.no/About-us/Methodology
13. AAIB Investigation Process. https://www.gov.uk/government/publications/how-we-investigate/how-we-investigate
14. ATSB Investigation Process. http://www.atsb.gov.au/about_atsb/investigation-process/
15. ICAO Doc 9756 Manual of Aircraft Accident and Incident Investigation Part I–IV. https://www.icao.int/safety/airnavigation/aig/pages/documents.aspx
16. Independent Safety Board Act, 1974. https://www.ntsb.gov/legal/Pages/ntsb_statute.aspx
17. NTSB organization. https://www.ntsb.gov/about/organization/Pages/default.aspx
18. The UK AAIB. https://www.gov.uk/government/organisations/air-accidents-investigation-branch/about
19. Final Report on the accident on June 1, 2009 to the Airbus A330-203 registered F-GZCP. https://bea.aero/docspa/2009/f-cp090601.en/pdf/f-cp090601.en.pdf

20. ICAO Circular 315: Hazards at Aircraft Accident Sites. https://www.icao.int/safety/airnavigation/aig/pages/documents.aspx

21. BEA, Flight Data Recorder Read-out. https://bea.aero/uploads/tx_scalaetudessecurite/use.of.fdr_01.pdf

22. ARINC Characteristic 767 Enhanced Airborne Flight Recorder. https://aviation-ia.sae-itc.com/standards/arinc767-1-767-1-enhanced-airborne-flight-recorder

23. ARINC Characteristic 647 Flight Recorder Electronic Documentation (FRED). https://aviation-ia.sae-itc.com/standards/arinc647a-1-647a-1-flight-recorder-electronic-documentation-fred

24. Underwood, P., & Waterson, P. (2013). *Accident analysis models and methods: Guidance for safety professionals.* Loughborough: Loughborough University. https://repository.lboro.ac.uk/articles/report/Accident_analysis_models_and_methods_guidance_for_safety_professionals/9354404

25. Australian Transport Safety Bureau. (2008). *Analysis, causality and proof in safety investigations.* Aviation Research and Analysis Report AR-2007-053. Canberra City: Australian Transport Safety Bureau. https://www.atsb.gov.au/media/27767/ar2007053.pdf

26. AAIB Annual Safety review 2021. https://www.gov.uk/government/news/annual-safety-review-2021

27. NTSB Most Wanted List. https://www.ntsb.gov/Advocacy/mwl/Pages/default.aspx

28. AAIB Report on the accident to Aerospatiale (Eurocopter) AS332 L2 Super Puma, registration G-REDL, 11 nm NE of Peterhead, Scotland on April 1, 2009. https://www.gov.uk/aaib-reports/2-2011-aerospatiale-eurocopter-as332-l2-super-puma-g-redl-1-april-2009

29. AIBN Report on the air accident near Turøy, Øygarden municipality, Hordaland county, Norway, April 29, 2016 with Airbus Helicopters EC 225 LP, LN-OJF, operated by CHC Helikopter Service AS. https://www.nsia.no/Aviation/Published-reports/2018-04

PART II

Technical Safety

CHAPTER 5

An Introduction to
Technical Safety

Introduction

The purpose of this chapter is to introduce the concept of Technical Safety and prepare the reader for the upcoming technical chapters. It will explore how technical safety has evolved in aviation to reach the enviable levels it has today. It will also examine how learning from failure is embedded within the aviation system and fundamental to the continued improvement of the safety of the aviation system, particularly for technical aspects. Finally, it will look at some of the technologies, both evolutionary and revolutionary, currently under development.

5.1   Fundamentals of Flight

Before going any further, it will be useful to briefly discuss the fundamental principles of flight, which apply to all aircraft regardless of size or complexity. The concept of "lift" is at the heart of all aviation, whether dealing with hot air balloons, gliders, powered aircraft, rotorcraft, or any other machine—"defying gravity" is what distinguishes an aircraft from a car or similar powered vehicles. We will generally use the example of powered aircraft from this point, but the principles are universal.

When air flows over and around a suitably shaped surface (usually an airfoil) a lift force is generated "upwards" (in relation to the surface). The size of that force is related to the relative speed of the airflow, the shape of the airfoil, and the density of the air.[i]

By moving the airfoil through the air,[ii] lift is generated that can be used to overcome the force generated by gravity (Figure 5-1).

When thinking about lift, it is important to distinguish between airspeed (the speed of the aircraft relative to the air around it) and groundspeed (the speed of the aircraft relative to the ground below it)—it is airspeed, the movement of the air over the airfoil, that generates lift. That flow of air relative to the airfoil can be generated by moving the

---

[i] The equation defining lift is $L = \dfrac{1}{2}C_L \rho v^2 A$ where C_L is the lift coefficient, ρ is air density, v is flow velocity, and A is the wing area. However, these dependencies can be grouped as airspeed, airfoil shape, and air density without any loss of understanding.

[ii] By towing in the case of gliders, under its own power in the case of powered aircraft, or by rotation in the case of rotary wing aircraft.

125

LIFT

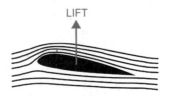

FIGURE 5-1   Generation of lift due to airflow over an airfoil. (Source: Authors)

airfoil through still air; by wind passing over a stationary airfoil; or, as is usually the case, by a combination of the two.

It is for this reason that airplanes take off and land into wind; the relative airflow from the wind generates additional lift. The flipside of this means that landing with a tailwind (wind coming from behind the aircraft) is usually avoided, since maintaining the airspeed needed to generate the required lift results in a higher groundspeed (the very thing you're trying to reduce to zero on landing!). Similarly, on takeoff, a tailwind necessitates a higher groundspeed to achieve the necessary relative airspeed. Of course, once airborne in the cruise phase of the flight, a tailwind is desirable since, for a given airspeed, you reach your destination faster.

To help visualize this difference between airspeed and groundspeed, imagine riding a bicycle along a slow-moving walkway (aka moving sidewalk, travelator, etc.) at an airport.[iii] To balance, you need to keep pedaling the bicycle moving relative to the walkway, but you have a choice of whether to ride the bicycle "with" or "against" the direction of the moving walkway. Riding with the walkway will get you through the airport fastest. However, let's imagine you have to check your gate on a screen fixed to the wall at the side of the walkway, riding with the walkway would send you sailing past the screen at high speed. Instead, for that task, you might prefer to be riding *against* the walkway since you can still pedal to keep your balance while moving more slowly relative to the screen. In this analogy, your speed relative to the screen is your groundspeed and the speed you pedal relative to the walkway is your airspeed. Let's leave that analogy and return to the real world.

On takeoff, as well as increasing airspeed, another way to increase lift is to increase the "angle of attack" of the airfoil (the angle between the incoming airflow and the airfoil). This can be achieved by raising the nose of the aircraft ("pitching up"). It is this "rotation" at a suitable speed on takeoff that causes an airplane to lift off the ground. Lift is broadly proportional to the angle of attack up to a limit, beyond which the airflow detaches itself (separates) from the airfoil and lift is lost resulting in a "stall" (Figure 5-2).

Changes to aircraft attitude are brought about by using "control surfaces" which are smaller aerodynamic surfaces usually attached to the wings or tailplane. Moving these surfaces in a coordinated way can adjust the aircraft in pitch (nose up or down), roll (e.g., left wing down and right wing up), and yaw (rotation about a vertical axis, e.g., nose to the left, tail to the right) as described in Figure 5-3.

On the flight deck, these surfaces are normally controlled using a center stick, a yoke, or a sidestick (forward to pitch down, left to roll left) and pedals (left pedal to yaw the nose left). In smaller and older aircraft there is a direct connection from the controls

---

[iii] Please don't try this at your local airport—it is heavily frowned upon!

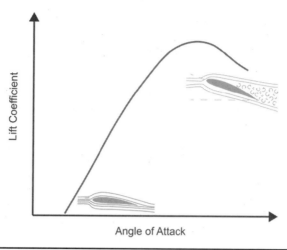

FIGURE 5-2   Change in lift with angle of attack. (Source: Authors)

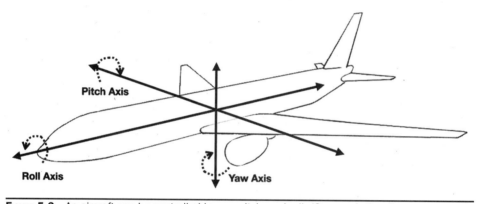

FIGURE 5-3   An aircraft can be controlled in yaw, pitch, and roll. (Source: Authors)

to the surfaces, whereas in modern aircraft the inputs are interpreted by algorithms which decide how to move the control surfaces in response to that input.

While lift counteracts the force due to gravity in the vertical direction, the forces acting in the horizontal direction are thrust (pushing forward either through jet engines or propellers) and drag acting backwards (Figure 5-4).

In general, the cost of gaining additional lift is additional drag. Therefore, most aircraft have a system for changing the shape of the wing (such as flaps) to allow the same lift to be generated at lower speed (e.g., for approach and landing) and then a lower drag, lower lift configuration for higher speed cruise. We will return to this subject in Chapter 6.

5.2   Aircraft Evolution

In the 120 or so years that have passed since the Wright Brothers' first powered flight, aircraft have changed almost beyond recognition, but the fundamental physical principles

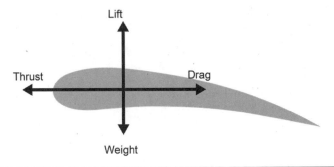

FIGURE 5-4 Simplified forces on an aircraft. (Source: Authors)

of flight remain unchanged. To appreciate the existing state of technical safety in the aviation industry, it is helpful to have a very brief history of the development of aircraft over the last 100 or so years. We will deal with airframe, engine, and flight controls separately.

5.2.1 Airframe Development

Following the First World War and the expedited development of aircraft for military purposes, the first commercial passenger service started in 1914 and the first transatlantic flight in 1919. The first aircraft to achieve successful and profitable flights around this time was the Douglas DC-3, a twin-engine aircraft with radial piston engines driving propellers, launched in 1936 albeit with a capacity of just 32 passengers.

The next major technical advance was the introduction of the jet engine, developed during the Second World War by both British inventor Frank Whittle and the German physicist Hans von Ohain, independently of each other. The first passenger jet aircraft was the British-designed and built de Havilland Comet which entered service in 1952. While it has become reasonably common to expect "teething" problems with new aircraft technologies, the Comet certainly had a poor start by any standards. Within the first twelve months of entering service, three aircraft experienced catastrophic inflight breakups, discussed later. Although only 114 Comet variants (including prototypes) were ever produced, 25 of these resulted in hull-losses, 13 of which were fatal accidents.

The Comet had a number of competitors, including the Tupolev Tu-104, the Boeing 707, and the Douglas DC-8. Of these, the Boeing 707, Boeing's first passenger jet aircraft and the first of the 7-series commercial jetliners, was the most successful, and as such is sometimes seen as the originator of the Jet Age.

The 707 was capable of a wider speed range than the previous generation of aircraft, longer duration, and higher altitude. This presents a technical challenge since as aircraft fly higher, the outside pressure reduces as does the level of oxygen. Therefore, for comfort and safety, the cabin is pressurized so that conditions in the cabin stay similar[iv] to those at sea level. As a result, there is a pressure difference between the higher-pressure

---

iv Similar, but not identical. Typically, the pressure in the cabin is maintained to match the conditions at about 8,000 ft altitude. You may have noticed that if you open an empty plastic water bottle in the cruise phase of flight and then seal it, when you land, the bottle will be crushed by the greater cabin pressure on the ground than in cruise.

air inside the cabin and the lower-pressure atmosphere outside. This pressure difference, which increases with altitude, must be contained by the aircraft fuselage.

The Boeing 707 featured a "swept wing" (i.e., angled back from the fuselage) rather
than "straight wing" (i.e., perpendicular to the fuselage). The swept wing provides
aerodynamic benefits at higher cruising speeds and altitudes (which are also beneficial
for jet engine performance) but adds difficulties at lower speeds requiring additional
aerodynamic systems to address the problems. These will be discussed in more detail
in Chapter 6. The 707 also featured pod-mounted engines below the wings compared
with the Comet which featured engines mounted inside the wing profile. This allowed
greater diameter engines to be used.

There have been fewer fundamental structural changes since the 1950s—there have
been incremental changes in aircraft (e.g., winglets/sharklets—the small "flick-ups" at
the wingtips) and engine design, but today's aircraft are fundamentally similar. In 1967,
Boeing introduced the 737 and variants of that aircraft are still on sale today, currently
the 737 MAX Series, and the Airbus A320, introduced in 1987, still has variants in production, including the A320 neo.

One clear area of aviation evolution has been the progressive introduction of larger
aircraft such as the Airbus A300, introduced in 1974 as the first twinjet widebody (or
twin-aisle) airliner, the Boeing 747, Boeing 777, and Airbus A380—but even these aircraft can be viewed as developments and evolutions of existing designs, rather than
fundamentally different aircraft.

One obvious exception to this evolution was Concorde, the supersonic airliner
launched in 1976. Concorde required more power, being the only commercial aircraft to use reheat technology, and a revised wing design, adopting a "delta wing" to
provide an acceptable compromise between subsonic and supersonic performance.
Concorde was also notable for being the first fly-by-wire airliner. Until this innovation,
there was a mechanical connection between the pilot and the control surface. In the
new arrangement, sensors measure the pilot input and then send an electrical signal
to an actuator which moves the control surface. Rather confusingly the traditional
mechanical connections to control surfaces were often made using metal control wires
rather than electrical wires carrying signals! We mean the latter when we talk about
fly-by-wire.

However, despite its innovation, Concorde was retired; following a fatal crash after
taking off from Paris, Charles De Gaulle Airport in 2000, the Anglo-French program
then struggled with the reduction in flying following the 2001 Terrorist attacks in the
US and increasing maintenance costs.

5.2.2 Jet Engine Development

Over the years, the jet engine has evolved to improve performance in a range of areas
including fuel consumption, noise, reliability, durability, stability, and thrust. This in
turn allowed aircraft manufacturers to achieve greater ranges, greater payloads, and
improved efficiency. This evolution was even more impressive as it occurred concur-
rently with enhancements in jet engine reliability and safety. The inflight engine shutdown rate has continued to decrease to the point where a modern airline pilot may
never experience an engine failure in their entire career. Such a claim would have been
inconceivable half a century ago.

Most modern jet aircraft use a type of jet engine called a turbofan. This has a large
fan on the front drawing in air, some of which is fed into the core of the engine and

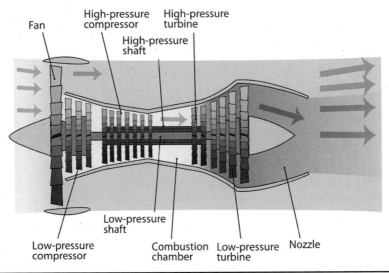

High-pressure   High-pressure
compressor   turbine

Fan

High-pressure
shaft

Low-pressure
shaft

Low-pressure   Combustion   Low-pressure   Nozzle
compressor   chamber   turbine

Figure 5-5   Main components of a turbofan engine. (Source: K. Aainsqatsi[1])

some of which "bypasses" the core. On entering the core, the air is compressed (by the aptly named compressor), mixed with fuel, and ignited (in the combustor). The hot gas then has energy removed from it by the turbine and finally the gas is expelled from the nozzle along with the bypass air. Figure 5-5 shows the main components of a turbofan engine.

The jet engine has seen numerous innovations such as:

- Single crystal superalloy materials, giving longer thermal and fatigue life, greater corrosion resistance, reduced weight, and higher melting temperatures
- Large composite swept fan blades to accommodate higher bypass ratios
- Chevron/serrated edge nacelles to improve noise performance

One of the more remarkable innovations is that by passing relatively cool air (around 750°C) through holes in the turbine blade, the nickel blades which would normally have a melting temperature of around 1300°C are able to work in hot gases with a temperature of 1600°C.[2] However, despite all this innovation and optimization, the fundamental physics of jet engine operation remains the same as the original design.

5.2.3 Flight Deck Development

The Boeing 707, Boeing 727, DC-8, early Boeing 747, DC-10, and L-1011 flight decks used a three-crew arrangement consisting of a captain, first officer, and flight engineer. The airplane systems were designed for the flight engineer to be the systems operator. Large instrument panels were mounted behind and to the right of the two pilots. The flight engineer was also expected to monitor and operate the vital hydraulic, electrical, fuel, air conditioning, and pressurization systems with minimal supervision. The design of the short-range Boeing 737 was changed radically to provide a flight deck that only needed a two-person flight crew. To accomplish this crew reduction, the airplane control systems were simplified.

Automatic operation of a system was also provided for selected equipment failure cases to avoid the necessity of crew intervention at a critical time in flight. The Boeing 737 electrical system load-shedding feature is an example. In the event of a single generator failure on the two-generator non-paralleled electrical system, the remaining generator picks up the essential load, and non-essential loads are shed. The galley power and other similar loads are shed so that the remaining generator can provide all essential electrical loads without the need for crew attention or intervention. Later, when the crew has time to restore a second generator, the shed loads can be recalled. This same design philosophy has prevailed through the other two-crew designs on the Boeing 757, Boeing 767, Boeing 747-400, and the Boeing 777 series of aircraft.

One of the largest changes to all flight deck design was the introduction of the "glass cockpit," a term which describes the use of digital instruments and multifunction displays (MFDs) rather than traditional individual ("steam gauge") instruments. Figure 5-6 shows the difference between the two types of cockpit.

The glass cockpit technology gives an opportunity to present information in a different way and allow flight crew to configure displays to best suit the phase of flight and tasks at hand. However, as more information is concentrated onto a single screen, the risk associated with that screen failing increases. As a result, most glass cockpits provide independent standby instruments.

In addition to the flight instruments, the flight controls have also changed through the years. The first Airbus aircraft, the A300 and the A310, used a conventional cockpit design with a yoke. The A320, introduced in 1987, used a sidestick to command a full fly-by-wire (FBW) system. The sidestick acts as an "input" to the flight control law which then "decides" what control surfaces to move in response to that input. This allows different behavior to be embodied. For example, in a conventional aircraft, to maintain a left turn, the yoke needs to be turned to the left and held. Under the Airbus philosophy, in normal circumstances, the pilot is commanding a roll rate, meaning that to maintain a left turn, a left input is made to the sidestick until the desired roll angle is reached and then the stick is returned to the center—this will result in a constant turn being held. This is only made possible through the use of FBW. Indeed, there are other control laws (Alternate and Direct) in the Airbus control philosophy that can be reverted to if needed, for example, due to sensor issues such as those seen in the crash of flight AF447, operated by an Airbus A330, as discussed in Chapter 1. This ability to provide processing between input and actuation enables multiple possibilities. For example, the Eurofighter Typhoon is inherently unstable to provide agility, with the FBW adding the necessary stability. Similarly, the sidestick, FBW philosophy has also been adopted on the forthcoming Bell 525 helicopter.

One outcome of using a passive sidestick is the lack of feedback (either through movement or force) to the pilot, or between the pilots, with the associated potential impact on situational awareness, as already discussed in the description of the loss of Air France 447 in Chapter 1. However, active sidesticks have recently been introduced in production aircraft such as the Gulfstream G500/600. The all-Russian designed and built Irkut MC-21 received type-certification in December 2021 and became the first commercial airliner with active sidestick control.

Returning to general evolution, even in aircraft not using FBW technology, the development of automated systems in the last half century has been remarkable. Systems such as the flight director, autopilot, automatic flap configuration, autothrust, and autobrakes have developed to a point where their use is commonplace. Similarly, autoland

Figure 5-6   Boeing 737 traditional cockpit versus a Boeing 777 glass cockpit. (Source: B Larkins[3], C Watts[4])

systems, which make use of the radio altimeter and the airport Instrument Landing System (ILS) to allow landings in zero visibility, have been in operation for many decades. In this type of operation, the pilots become a type of "system supervisor," monitoring the aircraft behavior and standing ready to take over if needed.

Modern pilots are increasingly described as system operators who control aircraft mostly through complex interfaces consisting of knobs, dials, and typing on computers and who supervise the work of automated systems, versus directly controlling aircraft through manual inputs on controls. As such, one "partly intended" consequence of improving automation is the reduction in time spent hand-flying that can occur. During a typical sector, pilots will typically connect automation at 400 to 1500 feet, and fly the complete sector on automation before retaking control a few hundred feet before touchdown. This amounts to only a few minutes of manual flying per sector.

However, automation and its use have yet to reach a point where it is reliable beyond question, meaning that the human is still needed to monitor and occasionally intervene. The challenge for automation is to provide the greatest support to pilots, particularly in areas where humans are naturally less capable, such as monitoring and repetitive tasks. This must be balanced with the need to keep the human sufficiently engaged to a level where their skills are available when needed.

Finally, there is a common assumption that the natural extension to ever-improving automation will be an eventual switch to autonomous, pilotless operation. However, there are numerous technical and social hurdles to be overcome before that switch occurs,[v] not least acceptance by the traveling public.

5.3　The Importance of Evolution

Aviation is a high-reliability industry that has depended on continual improvement to develop and maintain its enviable safety record; by investigating accidents, learning from the findings, and adapting the system to prevent them from recurring, we can continuously improve the system of aviation.

The corollary to this is that, contained within the current aviation system are countless developments, modifications, and lessons learned from history. That is not to say that nothing in the aviation system should be changed—it can, and must, be improved— but in order to make that improvement, and certainly before we throw away any parts of the existing system, we must understand the history behind them and the purpose they serve. Chapter 6 will look in more detail at the technical aviation safety system as it exists today. However, before that, to illustrate the point, let us look at two examples of design changes brought about through learning from accidents.

5.3.1　The Comet Accidents

As described above, the de Havilland Comet was the first commercial jet airliner. Work on the design began after the Second World War in 1946, with the first prototype flying in 1949 and a Certificate of Airworthiness issued in 1952 with passenger service shortly thereafter. The aircraft operated at a cruise height of more than 35,000 feet, twice that of the existing airliners giving a pressure difference about 50% greater than in other

---

[v] As our hero is told in the Top Gun sequel, *"The end is inevitable Maverick. Your kind is heading to extinction"* to which Maverick pauses, then replies *"Maybe so Sir, but not today"*…which is actually one of the more accurate parts of the film!

aircraft, accompanied by a greater temperature difference. The designers were aware of this additional requirement and gave it "special attention." They adopted a conservative design pressure of 2.5 times the expected difference, rather than the 2 times required by regulation. This was done to guard against failure due to material fatigue (the process of material weakening due to repeated loading and unloading), and to improve the safety margin around openings (doors, windows, and hatches). They also gave attention to the wings where there was a growing understanding of the importance of repeated loading in shortening the life of the fuselage.[5]

Before the middle of 1952, the importance of fatigue resistance for the pressure cabin had not been realized. In the middle of 1952, the test requirements changed to include repeated application of 15,000 pressure differences, in excess of expected levels, and requiring greater design loads for riveted joints, doors, windows, etc. As a result of these changes, de Havilland conducted additional pressure cycle testing to assure the airworthiness of the pressure hull. These tests were ended when the skin at the corner of a window failed due to fatigue, but the number of pressurizations that had been applied meant that the safety margin was considered sufficient.

On May 2, 1953, Comet G-ALYV crashed in an exceptionally severe tropical storm near Calcutta, India, killing all 43 on board. The subsequent Inquiry concluded that the accident was due to a structural failure due to overstress caused by either severe weather, or overcontrolling and/or loss of control by the flight crew. Fatigue failure of the cabin was not suspected in this Inquiry.

Only 8 months later, on January 10, 1954, an accident to Comet G-ALYP occurred over the Mediterranean with the loss of 6 crew and 29 passengers. Following this accident, the operator British Overseas Air Corporation (BOAC) suspended its Comet passenger services to carry out detailed examination of its fleet. The subsequent review with de Havilland considered many possible causes including fire and structural failure due to gust loads. Fatigue was considered, but mostly in relation to wind gusts on the wing structure (which occurs many times each flight) since the cycles are much higher than the pressure hull due to cabin pressurization (which has one cycle per flight). After a review, in which Lord Brabazon, the Chair of the Air Registration Board, wrote that "Although no definite reason for the accident has been established, modifications are being embodied to cover every possibility that imagination has suggested as a likely cause of the disaster. When these modifications are completed and have been satisfactorily flight-tested, the Board sees no reason why passenger services should not be resumed." Comet operations resumed on March 23, 1954, with the agreement of the Minister of Transport and Civil Aviation.

A little over 2 weeks later, Comet G-ALYY crashed near Naples, in a similar manner to G-ALYP, with the death of all 21 people on board. The operator, BOAC, immediately suspended all Comet services and the Certificate of Airworthiness for the Comet was withdrawn. When the fleet was grounded, the type had accumulated nearly 25,000 hours.

A Court of Inquiry was convened and a complete investigation of the problem by the Royal Aircraft Establishment (RAE) was ordered. As part of this work, a repeated loading test of the whole cabin was ordered. This was achieved by placing an aircraft in a water tank and filling both the tank and the cabin with water. More water was then pumped into the sealed cabin to raise the pressure in the cabin. The pressure was then reduced and increased again to create a fluctuating pressure cycle. Using this method, the aircraft was subjected to loading representative of repeated flight cycles including

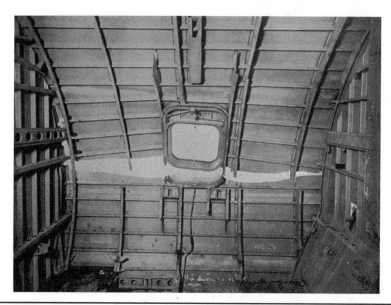

FIGURE 5-7   View of failure from inside G-ALYU. (Source: Inquiry report)

both cabin pressurizations and fluctuating loads to simulate gusts on the wings. Every 1000 flight cycles it was subjected to a high-pressure proof test of 1.3 times the expected pressure difference.

The airframe under test (G-ALYU) had already made 1230 pressurized flights and the cabin structure failed in testing after a further 1830 flight cycles (total 3060). The starting point of the failure was the corner of one of the cabin windows, shown in Figure 5-7, where fatigue was evident. Examination of the distortion in the wreckage from G-ALYP confirmed that this was the likely failure mode in that accident.

Further tank testing on G-ALYU showed that the stress at the corners of the windows was higher than had been believed by the designers. The Court of Inquiry noted that "No suggestion was made that any party wilfully disregarded any point which ought to have been considered or wilfully took unnecessary risks" and that "de Havillands were proceeding in accordance with what was then regarded as good engineering practice." The designers believed that the static testing they conducted was sufficient to assure the airworthiness of the pressure hull for the life of the aircraft. This was supported by repeated loading tests conducted in 1952–1953 in which a test section survived 16,000 applications of just over the working pressure.

One of the modifications made by de Havilland was to "strengthen and redesign windows and cutouts, and so lower the general stress," giving rise to the rounded windows found in today's modern aircraft.

The Comet is an excellent example of the aviation industry learning from its failings and introducing changes to ensure those failings are not repeated. At the time, not enough was understood about the potential of fatigue to weaken the pressure hull of an aircraft. A significant amount of effort was expended in understanding the cause of the accident and it is worth remembering the Comet was not fitted with a Flight Data Recorder or Cockpit Voice Recorder. In fact, the Comet accident helped inspire David

Warren, who is widely accepted as the inventor of the Flight Data Recorder,[6] to start developing what has become a fundamental component of commercial aviation safety.

It might be tempting to think that more should have been done to prevent the second and third accidents to the Comet. However, this "hindsight bias"—the temptation to think, looking back, that a situation was predictable, now the outcome is known—is a trap that is all too common in aviation. However, that doesn't mean the situation cannot be improved. Unfortunately, effectively preventing "repeat" accidents is an area where the aviation industry still has much work to do—we will return to this subject in Chapter 6.

The Comet accident occurred nearly 70 years ago, in an industry that was effectively only a few years old, and only 50 years after the first Wright Brothers flight. Yet, despite the relative youth of the industry, those lessons learned are still with us today. Our second case study gives a different perspective on the practice of learning from failure in the field of technical safety.

5.3.2   The Manchester Fire

The fatal accident to a Boeing 737, often referred to as "The Manchester Fire," was a seminal accident in aviation history. The UK Air Accidents Investigation Branch (AAIB) investigated the accident, and their report is comprehensive.[7] A summary of events is presented below but, as with all accidents, those interested in the details should take the time to read the full report.

On August 22, 1985, a Boeing 737 operating a flight to Corfu, with 131 passengers and 4 cabin crew, began a takeoff from runway 24 of Manchester Airport. About 36 seconds later, as the airspeed passed 125 knots, the No. 1 (left) engine suffered an uncontained failure (meaning parts of the engine were ejected from the side of the engine rather than being contained inside). This in turn punctured part of the fuel tank, causing a fuel leak which ignited.

On hearing a "thud" (which they took to be a burst tire or bird strike), the flight crew aborted the takeoff. Because they suspected a failed tire, they did not use full braking effort. They then received a fire warning for No 1 engine, and ATC confirmed "a lot of fire." On asking whether they should evacuate,[vi] ATC suggested they should evacuate from the right side of the aircraft.

As the aircraft turned off the runway onto a link taxiway, the commander ordered an evacuation. The aircraft stopped and the commander ordered the No 1 (left) engine fire drill to be carried out and the right engine to be shut down (to enable the evacuation from the right side). However, before that was complete, the commander saw fuel and fire spreading forward on the left side of the aircraft, so he ordered the copilot to evacuate through the right sliding flight deck window (using a fabric escape strap) and the commander followed.

In the cabin, those toward the rear of the plane were immediately aware of the intense fire from the heat, smoke, and "crackling and melting" of the windows although this was not obvious to everyone in the cabin. The purser, situated at the front of the cabin, then issued the call to evacuate a number of times over the PA (Public Address) system. Then, as the aircraft was coming to a halt he opened the right, front door. The door unlocked normally but the lid of the slide container jammed on the doorframe.

[vi] There are significant risks associated with conducting an evacuation, so if a fire can be contained it may be safer to keep passengers onboard. The decision to evacuate is made by the aircraft commander.

After a short time trying to clear it, he abandoned that effort and crossed to the left front door. He opened it a little to confirm the fire had not yet spread too far forward to prevent evacuation, then opened it fully to confirm the inflation of the slide. It was now 25 seconds since the aircraft had come to a stop, and it was at this point that the first fire vehicle to arrive started discharging foam. Passengers started evacuating from the front left door with the supervision and help of a member of the cabin crew, who had to pull free some passengers who had become jammed. Seventeen passengers eventually escaped through this exit.

The purser then returned to the right front door, succeeded in opening it and inflating the slide, and supervised the exit of passengers. By this time, dense smoke was starting to reach the front of the aircraft and when it began to threaten the cabin crew, they vacated the aircraft, along with 34 passengers who used the same exit.

In the center of the cabin, as the aircraft came to a stop, the passenger seated beside the right overwing exit attempted to open it by pulling on the armrest, which was attached to the door. Their companion, seated in the center of the row, then stood up and pulled the "Emergency Pull" handle on the hatch which then fell on top of the passenger closest to the door. The hatch, which weighed around 22 kg (48 pounds), pinned them in their seat. With the assistance of a passenger from the row behind, the hatch was lifted and moved to a vacant seat. This exit was opened about 45 seconds after the aircraft stopped and 27 passengers, including 2 children, evacuated through this exit.

As the aircraft turned onto the taxiway link, the right rear door was seen to be open with the slide deployed and inflated. A member of the cabin crew was initially visible in the doorway but was obscured by thick black smoke as the aircraft stopped. No one escaped through this door. The left rear door was opened by fire crew after the fire had been extinguished. Neither rear cabin crew survived. During the fire, the tail section and fuselage behind the wings collapsed, as seen in Figure 5-8.

Figure 5-8 The remains of the British Airtours Boeing 737 G-BGJL. (Source: AAIB report)

The AAIB report notes that:

Many of the factors which affected this accident should have biased events towards a favourable outcome. The cabin was initially intact, the aircraft remained mobile and controllable and no one had been injured during the abandoned take-off. The volume of fuel involved, although capable of producing an extremely serious fire, was relatively small compared with the volume typically carried at take-off, the accident occurred on a well equipped major airport with fire cover considerably in excess of that required for the size of aircraft and the fire service was in attendance within 30 seconds of the aircraft stopping. However, 55 lives were lost.

In addition to the 55 of the 137 people on board the aircraft who lost their lives in this accident, 15 people suffered serious injuries. There is much more detail about the accident and detailed analysis of the causal factors in the AAIB report. While a full discussion is beyond the scope of this introduction, it is informative to look at just some of the changes in the aviation system that stemmed from the investigation of this accident and the 31 safety recommendations it produced.

The flight crew, unaware of the fire, elected to try to clear the runway. The AAIB recommended that operators direct crews "on any rejected takeoff or emergency landing to stop on the runway and review the situation before a decision on clearing the runway is made." Stopping as soon as possible, regardless of potentially blocking the runway, is now standard practice for much of the industry.

The AAIB concluded that "The primary reason for the majority of the fatalities was rapid incapacitation due to inhalation of the toxic smoke atmosphere." This created a far greater focus on the materials used in the cabin and their by-products in a fire. Fire-resistant wall and ceiling panels and seat materials are now standard on many aircraft. The AAIB observed that "The narrow gap of 10½ inches [27 cm] between row 9 and 10 seats impeded passengers' access to the right overwing exit." Modern aircraft have wider emergency exit rows that need to be clear of baggage for takeoff and landing. Passengers are briefed on how to operate the overwing emergency exit, and floor lighting for guiding passengers to exits is now also standard. These beneficial changes represent just a few of the those introduced in response to the accident that are still present in the industry today.

Similar activity was also underway in the United States. In 1985, the NTSB addressed the issue of cabin safety on several fronts. It issued a study on emergency equipment and procedures relating to in-water air carrier crashes. The study found that equipment and procedures were either inadequate or designed for *ditching* (emergency landing on water where there is time to prepare and that involve relatively little aircraft damage) rather than for the more common short (or no warning) in-water crashes. The NTSB recommended improvements in life preservers, passenger briefings, emergency-evacuation slides, flotation devices for infants, and crew postcrash survival training.

Later in 1985, the NTSB completed a study on airline passenger safety briefings. The NTSB concluded that in the past, "The survival of passengers has been jeopardized" because they did not know enough about cabin safety and evacuation. The study also found wide variances, and sometimes inaccuracies, in oral briefings and in information on seatback-stored safety cards.

In October 1985, the FAA issued a notice of proposed rulemaking to require protective breathing equipment for flight crews and cabin attendants. The NTSB had recommended making this equipment available following its investigation of the fatal fire on the Air Canada DC-9 at Cincinnati International Airport in June 1983.

The recommendation repeated one made a decade earlier after the NTSB participated in the investigation of a foreign accident involving a cabin fire. The NTSB believed that without protective breathing equipment, flight attendants could easily be incapacitated by fire and smoke, could not make effective use of fire extinguishers, and could be of no use during an evacuation.

5.3.3   Conclusions

These two case studies help to show the importance of continuous improvement to provide many of the achievements we take for granted in the modern aviation system. It might be tempting to think that, as they are from the 1960s and the 1980s, these accidents belong to a different time and have no relevance today. While it is certainly true that they are from a different time (the Comet accidents occurred closer to the Wright Brothers' first flight than to present day), the lessons learned from them and their impact on the aviation system are as relevant today as they have ever been. "Those who cannot remember the past are condemned to repeat it," wrote George Santayana, and nowhere is this truer than in aviation. We forget these hard-won lessons at our cost.

As evidence for that point, on takeoff from Las Vegas airport on September 8, 2015, a Boeing 777 suffered an uncontained engine failure.[8] The crew rejected the takeoff and despite a significant fire breaking out on the left side of the aircraft which rendered only 2 of the 8 exits usable, all 170 onboard were able to evacuate with only 1 serious injury. The similarities with the Manchester Fire are striking, but thankfully the outcome was very different.

Similarly, on 27 June 2016, a Boeing 777 departed Singapore for Milan, Italy.[9] About two hours into the flight there was a failure in the right engine and the flight crew elected to turn back. As they landed back at Changi Airport, a fire broke out in the vicinity of the right engine, of which the flight crew were initially unaware. The aircraft stopped on the runway and a fire developed. Airport rescue and firefighting were able to extinguish the fire and all passengers disembarked safely.

Every accident is different, and there were certainly more lessons to be learned from these two accidents, but it is clear that the successful outcome in these events is owed in large part to the system adaptations put in place following the investigation of the Manchester Fire. Let us now move forward to look at the form of the aviation safety system as it stands today.

5.4   The Modern Safety System

The safe operation of the global aviation system relies on each of the different groups involved playing their part effectively. Many of these organizations will be described in greater detail in Chapter 12—The Role of Government, but for now it will be useful to take an overview of the key players involved in the design, certification, operation, and maintenance of an aircraft.

5.4.1   ICAO

The International Civil Aviation Organization, known as ICAO (usually pronounced *eye-kay-oh*), is a branch of the United Nations.[10] In 1944, toward the end of the Second World War, the Convention on International Civil Aviation was drafted by 54 nations to promote cooperation and "create and preserve friendship and understanding among the nations and peoples of the world." This agreement is known today as the Chicago Convention and established some of the fundamental principles of international air travel. It also led to the creation of ICAO which is today funded and directed by 193 national governments. ICAO's core function is to "maintain an administrative and expert bureaucracy...and to research new air transport policy and standardization innovations as directed and endorsed by governments."

In practice, ICAO produces Standards and Recommended Practices (SARPS) some of which are produced as Annexes to the Chicago Convention. Examples include:

- Annex 6—Operation of Aircraft
- Annex 8—Airworthiness of Aircraft
- Annex 13—Aircraft Accident and Incident Investigation
- Annex 19—Safety Management Systems

These SARPS are not law, and do not supersede the primacy of national regulatory requirements; it is for the nations to implement the SARPS in local and national regulations. An example would be:

Regulation (EU) No 996/2010 of the European Parliament and of the Council of 20 October 2010 on the investigation and prevention of accidents and incidents in civil aviation which embodies the requirements of Annex 13 into European Union law.

5.4.2   Government Regulators

Aviation regulators are government authorities that work at a national level to approve and regulate civil aviation. These regulators are often referred to as a National Aviation Authority (NAA) or a Civil Aviation Authority. In the United States, this function is performed by the Federal Aviation Administration (FAA) with a mission "to provide the safest, most efficient aerospace system in the world."[11].

In 2002, the European Union created the European Union Aviation Safety Agency (EASA), representing 31 EASA Member States and Associated Countries. Over time, EASA adopted more responsibility from the NAAs. Part of EASA's current role is to certify and approve products and organizations in fields where they have exclusive competence (e.g., airworthiness) and provide oversight and support in fields where they have shared competence (e.g., Air Operations, Air Traffic Management). The FAA and EASA are two of the largest aviation authorities, in part because of the number of aircraft designed, built, and certified under those agencies.

5.4.3   Design Organizations

Design Organizations are often referred to as the aircraft Original Equipment Manufacturer (OEM), for example, Boeing, Airbus, Sikorsky, and Leonardo. The regulator holds the responsibility for certifying an aircraft design, including issuing the Certificate of Airworthiness, and they also issue approvals for entities to operate as a design organization. EASA calls this a Design Organization Approval (DOA) and the FAA an Organization Designation Authorization (ODA). The Design Organization is responsible for demonstrating to the Regulator the compliance of their products with Certification Standards. The approvals granted by a Design Organization under delegated authority, in the development and support of their aircraft products can be far-reaching. Chapter 6 looks at this important subject in greater detail.

5.4.4   Aircraft Operator

In commercial transport, the aircraft operator is the entity that the public usually interacts with, examples include United Airlines, British Airways, and American Airlines. Operators need to be approved by the regulator, usually holding an Air Carrier Operating Certificate (ACOC) in the US, or an Air Operators Certificate (AOC) in Europe. The requirements for an ACOC/AOC include having suitable staff, training, and aircraft maintenance. Depending on the size and approvals of the operator they may outsource some or all, of these tasks. Equally a large operator may be authorized to perform their own training, maintenance, and approvals.

5.4.5   Maintenance Organization

Maintenance organizations are those performing different levels of aircraft maintenance, sometimes referred to as a Maintenance and Repair Organization (MRO). An Approved Maintenance Organization (AMO) must also be approved by the government regulator. Some aircraft operators choose to embed an MRO within their organization and may even sell those maintenance services to external customers. A more thorough description of the complexity surrounding maintenance processes and approvals is provided in Chapter 6.

5.4.6   Air Navigation Service Provider

Air Navigation Service Provider (ANSP),[vii] as the name suggests, is a wide-ranging term covering organizations playing a part in assisting operators in the navigation part of the aviation transportation system, including:

- Air Traffic Service (ATS)
- Aeronautical Information Service (AIS)
- Meteorological Services (MET)
- Communication, Navigation, Surveillance (CNS)

ANSPs are also licensed by the government regulator. More detail is provided in Chapter 8.

5.4.7   Airport or Aerodrome Operator

The airport or aerodrome operator is responsible for the safe operation of the airport. In order to be certified, there are countless design requirements that must be complied with for all aspects of an airport including taxiways, markings, visual aids, electrical power distribution, and much more.

In operation, the airport operator is responsible for all aspects of ground handling, including:

- Maintenance of the runways, navigation equipment, lighting
- Wildlife hazard management
- Airport rescue and firefighting

[vii] Do not confuse with Aircraft Network Security Program in the United States.

- De-icing, refueling, and aircraft towing
- Provision of airport information
- Passenger, baggage, and cargo handling

Chapter 7 discusses airport safety in more detail.

5.4.8 Safety Investigation Agencies

The work of an Accident Investigation Body, now more commonly referred to collectively as Safety Investigation Agencies (SIAs), was discussed in Chapter 4. They're mentioned here as a reminder of the role they play in the aviation safety system. Even though they don't necessarily become involved in "normal operations," SIAs play a key role in the continuous improvement of the safety system.

5.4.9 Summary

The description above is a vastly simplified representation of the overall system which ignores the countless other organizations and personnel working in and around the industry. However, it does outline many of the key stakeholders and organizations involved in delivering safe aviation which will be key for the upcoming chapters.

5.5 Interactions and Complexity in the Modern Aviation System

The complexity of large modern aircraft means that no single person can know the detailed design and operation of an entire aircraft. This presents potential issues for designers, regulators, and flight and maintenance crew and increases the necessity for checklists, processes, procedures, documentation, and so on, so that the system can avoid an incomplete understanding becoming the source of an issue. Inevitably, this complexity also increases the difficulty of avoiding unanticipated interactions, particularly when considering system failures.

5.5.1 Incident to an Airbus A340 in 2005 (G-VATL)

A very clear example of this issue of complexity is the 2005 incident involving an Airbus A340 en route from Hong Kong to London Heathrow.[12] The A340 has four engines, numbered 1 to 4 with Engine 1 the outer engine on the left wing and Engine 4 the outer engine on the right wing. (This "left to right from the cockpit" numbering scheme is a standard to minimize errors in identification.) The A340 also has multiple fuel tanks, with remaining fuel being moved between them as it is burned in flight.

Around 11 hours after takeoff, at Flight Level 380 (a pressure altitude of 38,000 feet), the No. 1 engine lost power and ran down. Initially the pilots suspected a leak had emptied the contents of the fuel tank feeding No. 1 engine, but a few minutes later the No. 4 engine started to lose power. You don't have to be a pilot to understand what a troubling situation this is. Losing two engines on a four-engine aircraft has already reduced operational margins significantly. Further to that, when the engines have run down (slowed and stopped producing useful thrust) rather than, say, failing catastrophically, one might suspect a system problem and so losing another engine may be considerably more likely than raw engine failure rates may suggest.

As a result, the flight crew elected to manually open the crossfeed valves, to start transferring fuel between tanks, and No. 4 engine recovered to normal operation.

This coupled with an empty indication on one tank convinced the crew they had a fuel management problem. Fuel had not been transferring from the center, trim, and outer wing tanks to the inner wing tanks, so the pilots attempted to transfer fuel manually. Although this was partially achieved, they did not get the expected indications of fuel transfer and so the commander decided to divert to Amsterdam Schiphol Airport where the aircraft landed safely on three engines.

So, ultimately there were no injuries, and no damage was done to the aircraft, which is why this was classified as an incident rather than an accident. However, clearly there was potential for a much worse outcome; hence, the AAIB elected to investigate. This was a highly complex investigation involving multiple computer systems, failure modes, alerting systems, redundancy logic, and more described in a report of more than 150 pages. The description below presents some of the key points.

The A340 has two fuel control and management computers (FCMCs) which control the fuel system and conduct tasks such as fuel transfers and center of gravity management. The FCMCs send outputs to the various valves and pumps to achieve fuel transfer and receive feedback from those components. These FCMCs determine their own health (graded 0 to 7) through continuous monitoring and the healthiest is designated as the master. If a failure occurs in the master, it should degrade its health level and the slave should take over as master.

In this incident, both FCMCs were suffering degraded health and failures (a common occurrence at the time) and had been reset by the flight crew. At some point in the flight, FCMC2 lost its ability to relay information through its outputs and hence to transfer fuel. At this point FCMC1 should have taken over as master, but its health was considered to be lower than FCMC2 so did not take over master status. In addition, this meant that fuel warnings for the flight crew were not able to be issued. If both FCMCs fail, then there is a redundant route by which the warnings would have been issued; however, because the slave FCMC was still providing outputs, this was not triggered. So, a situation developed where the master FCMC was unable to issue commands to transfer fuel, but was healthy enough to make those decisions, while the slave was in a more degraded state but was still able to issue commands which it didn't do due to its slave status.

This was a highly complex investigation into a situation that was very difficult to predict. It is presented here as example of the complexity in ensuring safety in modern systems. The designers of the FCMC had already attempted to mitigate the possibility of any degradation (through the "health comparison" protocol) or failure (through redundancy) and yet, despite those efforts, a situation arose that had the potential to threaten the safety of the aircraft.

As with almost every accident or incident, there were other factors involved, including possible low arousal of the crew, a failed attempt to relight No. 1 engine above the maximum guaranteed altitude for a relight, and the flight crew's decision to manually open the crossfeed valves. We will return to the issue of the human as both problem and solution in Chapters 10 and 11. However, it clearly shows the types of issues that can emerge when unexpected events take place in complex systems.

5.5.2   Qantas Airbus A380 (QF 32)

A related, but very different, example can be seen in the accident that occurred to Qantas Flight QF32, investigated by the Australian Transport Safety Bureau (ATSB). The comprehensive final report is more than 300 pages in length. As previously, a synopsis extracted from the report is presented below, but the report is the true document of record.

On November 4, 2010, an Airbus A380 aircraft (a large four-engine, twin-passenger deck aircraft) being operated as QF32 departed from Changi Airport, Singapore, for Sydney, Australia. On board were 440 passengers, 24 cabin crew, and 5 flight crew. The flight crew comprised the normal operating crew (captain and first officer) with the addition of a second officer for crew relief purposes. This flight also carried a check captain[viii] under training and a supervising check captain, who was supervising the check captain under training. All flight crew were located in the cockpit at the time of the incident.

Following a normal takeoff, the crew retracted the landing gear and flaps and changed the thrust setting to climb. Around 5 minutes into the flight, while maintaining 250 knots in the climb and passing 7000 feet above mean sea level (AMSL), the crew heard two, almost coincident "loud bangs." The captain immediately selected altitude and heading hold mode on the autoflight system control panel.

The crew reported that there was a slight yaw and the aircraft leveled off in accordance with the selection of altitude hold. The captain stated that he expected the aircraft's autothrust system to reduce thrust on the engines to maintain 250 knots as the aircraft leveled off. However, the autothrust system was no longer active, so he manually retarded the thrust levers to control the aircraft's speed.

The A380 is fitted with an electronic centralized aircraft monitoring (ECAM) system which provides information to the crew on the status of the aircraft and its systems. It also presents the required steps of applicable procedures when an abnormal condition has been detected by the monitoring system. The captain took control of the aircraft and instructed the first officer to commence the procedure on the ECAM. The ECAM displayed a number of messages about the degraded health of Engine No. 2, including a fire warning for approximately 1 or 2 seconds.

The crew transmitted a PAN call (announcing an urgent situation but not requiring immediate assistance) to Singapore Air Traffic Control (ATC). As part of the procedures, the crew elected to shut down No. 2 engine. As they were doing this, the ECAM reported a No. 2 engine failure. Since the damage to No. 2 engine was assessed as serious, the crew discharged one of the engine's two fire extinguisher bottles, but no confirmation was received. This was then repeated for the first bottle and twice for the second bottle, but no confirmations were received. They then returned to the engine failure procedure, including transferring fuel. At this time, engines 1 and 4 were in degraded mode, engine 2 was in a failed mode, and engine 3 was in an alternate mode.

After discussing options, the crew elected to maintain their present altitude and continue processing the ECAM messages, a decision they frequently reviewed (which is good practice in a complex emergency). The aircraft then took up a holding pattern to the East of Changi Airport.

The second officer then went into the cabin to visually assess the damage. Through the cabin windows, he could see damage to the wing and fuel leaking from the wing. The crew then elected not to transfer any more fuel as they were uncertain of the fuel system integrity. They were also unable to jettison fuel due to damage to the fuel management system. It took the flight crew about 50 minutes to complete all of the initial procedures associated with the ECAM messages.

[viii] A check captain is a captain appointed by the operator to conduct flight proficiency tests on other pilots.

The crew then assessed the aircraft's systems and one of the check captains attempted to calculate the required landing distance based on reduced braking, inoperative leading edge lift devices, reduced spoilers, and inactive left-side thrust reverse (see Chapter 6 for system explanations). The A380's landing performance calculation assessed the required distance to land by taking into account all of the aircraft's failures; however, for each individual failures, it added a safety margin. The resulting calculation included all the cumulative safety margins and meant that the initial performance calculation suggested the A380 in this configuration could not land back at Changi Airport. Eventually the crew used pragmatism. Rationalizing that if this aircraft couldn't safety land and stop on any of the 4-kilometer runways at Changi, the longest runways available in that region of the world, then where was it going to land? The flight crew's experience and rationality determined that their damaged aircraft would be capable of landing and decelerating; however, the number of interrelated failures prompted some concerns.

The crew progressively configured the aircraft for landing and continued to check for controllability of the aircraft in each new configuration. The gear had to be lowered using an alternative procedure employing gravity to extend it. Engines No. 1 and 4 were set to provide symmetrical thrust and speed was controlled using Engine No. 3. The autopilot was functional but disconnected twice on the approach, after which the captain elected to fly the approach manually. The aircraft touched down just under 2 hours from takeoff and despite reduced retardation, the aircraft was able to stop before the end of the runway.

The flight crew shut down the remaining engines but the No.1 engine continued to run, even after selecting the fire extinguisher bottles. Although there was concern about fuel leaking onto hot brakes (>900°C after absorbing all the kinetic energy from the massive aircraft) it was being managed by the fire crew who were laying a foam blanket over the fuel. Therefore, the flight crew elected to disembark, rather than evacuate, the passengers. They chose to use a single set of stairs on the right side of the aircraft thereby keeping the possibility of an emergency evacuation by slides on the other doors. The disembarkation was completed approximately two hours from after the aircraft had landed.

Discussions and efforts to shut down No.1 engine by various means continued. Finally, a decision to "drown" the engine with water then firefighting foam was taken. The engine was finally shut down around 3 hours after the aircraft landed.

The initiating event for this accident was the failure of the wall of an oil feed pipe in the No. 2 engine, due to a nonconformance in the manufacture of the pipe resulting in reduced wall thickness. Following the investigation, numerous other engines were found to have a similar nonconformance. This resulted in an uncontained engine rotor failure causing damage which exceeded the modeling used in the safety analysis. Much of that damage was to wiring looms and either as part of that damage, or as a result of the actions taken by the crew as part of the ECAM procedures, the following systems were affected:

- Hydraulic power
- Electrical power
- Flight controls
- Engine control

- Landing gear
- Braking
- Bleed air system
- Fuel system
- Engine fire protection

Most of these systems were affected by damage to the respective wiring loom. In addition, a large engine fragment passed through the left inner fuel tank which gave rise to a low-intensity fire. Fortunately, conditions within the fuel tank were not sufficient to sustain combustion.

The sophistication and complexity of a modern aircraft are such that no one person can know and understand all aspects of it. As a result, automated logic in systems, detection of degradations, and unforeseen damage and interactions can sometimes lead to situations such as seen with the ECAM in QF32. The ECAM system is intended to aid crews by monitoring parameters and presenting prioritized warning or caution messages. However, the ECAM philosophy meant that high-priority messages (as judged by the ECAM logic) had to be completed before lower-priority messages.

It also highlights the fragility of electrical systems in the face of mechanical failures. There are countless ways in which the accident could have ended in tragedy including inflight fire, loss of control, and fuel exhaustion. Clearly, a single initiating engine failure should not be allowed to lead to such a level of system failure. However, ensuring an acceptable level of safety performance in such a complex aircraft is not a trivial challenge.

Finally, the successful outcome of this accident is another example of the power of human decision-making when facing an unpredicted set of complicated and inter-related failures.

5.5.3   Conclusions

These case studies highlight some of the different challenges encountered with modern aircraft. While overall mechanical understanding and reliability have improved since the early days of the Comet aircraft, the introduction of multiple computer systems presents new and more complex possible routes to failures. However, possibly in contrast to the early days of aviation, in the modern aircraft it is not feasible for flight crew, or any single person, to understand, let alone diagnose, every system failure that could occur. Which leaves a question—how do we effectively assure safety in the modern aviation system? We will attempt to address this question in the remainder of Part II of this book.

Key Terms

Airspeed

Aircraft Operator

Airfoil

Comet Aircraft

Drag

ECAM

FCMC

Fly-by-Wire

Glass Cockpit

Government Regulator

Groundspeed

ICAO

Lift

Sidestick

Stall

Turbofan

Yoke

Topics for Discussion

1. What produces lift in an airfoil, airspeed, or groundspeed?
2. List the key components of a turbofan engine.
3. What was the technical cause of the Comet airplane crashes?
4. Describe hindsight bias. How does it affect aviation safety?
5. List 3 changes that resulted from the Manchester Fire accident.
6. What caused the two engines to run down in the incident to A340 G-VATL?
7. What was the initiating event in the accident to the A380 operating flight QF32 and what was the result?
8. Describe three future technical challenges for the aviation industry.

References

1. Aainsqatsi, K. CC BY-SA 3.0 via Wikimedia Commons.
2. https://www.thenakedscientists.com/articles/interviews/how-do-you-stop-jet-engine-melting
3. Larkins, B. CC BY-SA 2.0 via Wikimedia Commons.
4. Watts, C. B., from Madison, WI. CC BY-SA 2.0 via Wikimedia Commons.
5. https://lessonslearned.faa.gov/Comet1/G-ALYP_Report.pdf
6. https://web.archive.org/web/20130514175925/http://www.abc.net.au/dimensions/dimensions_future/Transcripts/s642790.htm
7. https://www.gov.uk/aaib-reports/8-1988-boeing-737-236-g-bgjl-22-august-1985
8. https://reports.aviation-safety.net/2015/20150908-0_B772_G-VIIO.pdf
9. https://www.mot.gov.sg/docs/default-source/about-mot/investigation-report/b773er-(9v-swb)-engine-fire-27-jun-16-final-report.pdf
10. https://www.icao.int/about-icao/Pages/default.aspx
11. https://www.faa.gov/about/mission
12. https://www.gov.uk/aaib-reports/aar-4-2007-airbus-a340-642-g-vatl-9-february-2005

CHAPTER 6

Aircraft Safety Systems

Introduction

In this chapter we will discuss technical aspects of safety, specifically as they relate to the aircraft. Aviation is a high-reliability industry and much of that reliability is related to the fact that it is also a highly regulated industry. As you read this chapter, many of the acronyms and much of the language may seem obscure or impenetrable—phrases and acronyms such as Design Organization, AOC,[i] Continuing Airworthiness, and CAMO[ii] are commonplace and may seem unnecessarily confusing. However, they represent and codify a system of regulation that has been built and refined over decades that, although far from perfect, has helped to develop the enviable safety record the industry has today. A solid understanding of this system, and how safety is assured in the roots of aviation, is fundamental to playing an effective part of that industry; the complete system only works when everyone is clear about their responsibilities and those are assured by effective oversight.

This part of the aviation safety system may seem mundane, particularly when compared with higher level concepts such as safety management and human factors. However, we are only given the freedom to concentrate on those higher level, secondary safety areas by assuring the lower level, primary safety that is embedded deep in the system.

This chapter will:

- Explain how aircraft are designed and certified for safety, including the roles of the Design Organization and the Regulator.
- Describe the processes around maintaining an aircraft and ensuring its continuing airworthiness.
- Outline the fundamentals of operating an aircraft safely.
- Discuss some of the safety systems installed.
- Look at some examples of when this part of the aviation safety system has failed.

---

[i] Air Operator Certificate.
[ii] Continuing Airworthiness Management Organization.

6.1   Aircraft Design for Safety

The process of designing a new aircraft is a hugely complex, time-consuming, and expensive undertaking. Aircraft OEMs[iii] effectively "bet the company" whenever they choose to design a new aircraft. For example, the A380 development program took Airbus nearly 7 years and is estimated to have cost up to $20 billion. However, it remains a source of debate whether, with production ending in 2021, they have ever recouped those development costs. In 2015, Airbus celebrated the (modest) success that they were no longer producing each A380 airframe at a loss! In comparison, Boeing spent somewhere between $17 and $23 billion developing the 787, which gave a break-even point at around 1000 aircraft. Ultimately the success or failure of a type is decided by commercial sales and operational longevity. However, the majority of the development cost of a new aircraft is taken by the design and certification process, and it is that process which this section aims to outline.

6.1.1   Regulation

It is useful at this point to distinguish between the concepts of *Airworthiness* and *Continuing Airworthiness*. Airworthiness is the measure of being Airworthy, which ICAO Annex 8,[1] the ICAO Standards and Recommended Practices (SARPs) relating to Airworthiness, is defined as:

> The status of an aircraft, engine, propeller or part when it conforms to its approved design and is in a condition for safe operation.

Continuing Airworthiness on the other hand is defined as:

> The set of processes by which an aircraft, engine, propeller or part complies with the applicable airworthiness requirements and remains in a condition for safe operation throughout its operating life.

Put a different way, continuing airworthiness is the route by which the airworthiness of an aircraft, engine, propeller, or part is upheld. The focus of this section is airworthiness, and we will return to the issue of continuing airworthiness in Section 6.2.

Annex 8,[1] Section 1.2.3 states that:

> The design shall not have any features or characteristics that render it unsafe under the anticipated operating conditions.

This simple statement of intent underpins all the subsequent requirements of the Annex which are aimed at ensuring this is the case. In addition to Annex 8, ICAO has published Doc 9760,[2] an Airworthiness Manual which provides guidance to States on the suggested content of various airworthiness regulations.

In the United States, Annex 8 is implemented into US domestic law through 14 CFR Chapter 1, Subchapter C, Part 21 (or 14 CFR Part-21[3]) which gives the *procedural requirements for issuing and changing: design approvals; production approvals; airworthiness certificates; and airworthiness approvals*. In Europe, the implementing regulation is Regulation (EU) No 748/2012,[4] Annex I of which describes *the requirements and procedures for the*

[iii] OEM, or Original Equipment Manufacturer, is an abbreviation for a company manufacturing components or aircraft.

certification of aircraft and related products, parts and appliances, and of design and production organizations. Both the EU and US regulations are commonly referred to as "Part 21." This is the starting point for the design and certification of a new type. In addition to Annex 8 covering Airworthiness, Annex 6 *Operation of Aircraft,*[5] defines specifications to ensure a minimum level of safety around the operation of the aircraft, minimum aircraft equipment fits, maintenance, etc.

6.1.2 Organization Designation Authorization and Design/Production Organization

In order to have a new design certified, an organization needs to be recognized as being competent and having the necessary systems and processes in place. In the United States, the Organization Designation Authorization (ODA) program is the means by which the FAA grants designee authority to organizations or companies as described in 14 CFR Part 183, Subpart D. This authorization allows organizations to conduct the types of functions which they would normally seek from the FAA. For example, approved aircraft manufacturers may be authorized to approve design changes in their products and repair stations may be authorized to approve repair and alteration data. Regular FAA oversight of an ODA is required to ensure the organization holding ODA approval functions properly and that any approvals or certificates issued meet FAA safety standards.

In Europe, Subpart J of Part 21 describes the process for approving a Design Organization (DO). It also describes the requirements for maintaining that approval including oversight (through documentation, audits, assessments, etc.), maintaining a Design Organization handbook, and maintaining a design management and assurance system. Design Organizations are granted individual approvals which define the tasks that the DO is permitted to perform (e.g., designing minor changes to type design or minor repairs). A similar process exists for Production Organization approval, defined in Subpart G. Perhaps unsurprisingly, it is common for the same organization to hold both Design Organization and Production Organization approvals. Granting of Design or Production Organization approval can bring significant changes to the certification and production process, often granting substantial delegated authority from the regulator. We will return to this issue in Section 6.1.6.

6.1.3 Type Certification

Type Certification is the process by which the regulator approves the design and manufacturing standards of a new aircraft type and is a collaborative process between the regulator and the Design Organization. A Type Certificate is issued for the first aircraft of the type, thereby defining the design and build standard that must be met by subsequent aircraft, which are each issued an individual aircraft Certificate of Airworthiness. A Type Certificate is issued for unlimited duration, unless otherwise revoked, suspended, or surrendered. The full process of Type Certification is well beyond the scope of this book, but more resources are available in the reference list at the end of the chapter.[6-9] However, it is important to understand the various regulations with which an aircraft must comply to achieve Type Certification. The Type Certification process has four main steps.[7-9] Although they are named and grouped differently between FAA and EASA, as shown in Table 6-1, they broadly achieve the same aims.

	FAA	EASA
Step 0	Conceptual Design	Definition and Working Methods
Step 1	Requirements Definition	Technical Familiarization and Initial Certification Basis
Step 2	Compliance Planning	Certification Program
Step 3	Compliance Demonstration	Compliance Determination
Step 4	Postcertification Activities	Technical Closure and Issue of the Approval

TABLE 6-1   Stages of the Type Certification Process

The technical requirements for certifying a large fixed-wing aircraft are stated in "Part 25" (14 CFR Part-25[10] in the United States and Certification Specification CS-25[11] in Europe). For large helicopters the equivalent standards are 14 CFR Part-29[12] and CS-29.[13] As the name suggests, the certification specification contains the rules that must be followed in order to achieve type certification, which can often be quite concise. However, the standards also contain or refer to supporting material such as Advisory Circulars (AC), Orders and Notices, Acceptable Means of Compliance (AMC), or Guidance Material (GM) which are intended to clarify and explain the standard and give examples of how an OEM might show compliance. The methods described in the AMC represent one or more acceptable ways of demonstrating compliance, but a Design Organization is free to offer alternative means to the regulator. Put another way, the Part/CS describes what must be complied with, while the AC, AMC, and the GM describe how to implement and demonstrate compliance. For example, the requirement of CS-25.1309 (provided in Section 6.1.3.1) fills about a page, but the corresponding AMC that follows fills nearly 40 pages.

CS-25 is an extensive document, running to nearly 1500 pages and divided into the following main sections:

Flight	Covering characteristics such as performance, stability, stall behavior, etc.
Structure	This section deals with aspects such as flight loads, control surface loads, fatigue evaluation, etc.
Design and Construction	Dealing with areas such as control surfaces and systems, landing gear, ventilation and heating, pressurization, fire protection, etc.
Powerplant	Divided into areas such as fuel and oil system, cooling, intake and exhaust, fire protection, etc.
Equipment	This section covers items such as flight instruments, lights, and safety equipment.
Electrical Wiring Interconnection Systems	Regulations covering all aspects of wiring including system separation and safety, fire protection, bonding, accessibility, etc.
Auxiliary Power Unit Installations	This section has similar considerations to the powerplant sections.

If the reader is new to type certification regulations, it is highly valuable to choose a section of interest and read the regulation and supporting material in full, to get a real flavor of the requirements for certifying an aircraft.

There is a similar Type Certification process for other major components such as engines, auxiliary power units (APUs) and propellers. These can be certified independently and then "added" to different aircraft. For engines, 14 CFR Part-33[14] covers the relevant requirements in the United States (with AC 33-2B, *Aircraft Engine Type Certification Handbook*[15] providing additional information) whereas in Europe, CS-E is the relevant regulation defining the initial airworthiness requirements for engines. These documents also specify the extensive engine testing procedures including endurance, rain and hail ingestion, and containment tests for compressor and turbine blade failure. (If you have never seen the latter, it is worth searching the internet for "blade-off test" videos![16])

You will notice that Certification Specifications are written in a manner intended to produce safe outcomes without dictating precisely how a Design Organization achieves those outcomes. One of the clearest examples of a difference in approach can be seen between the Boeing and Airbus approaches to cockpit design. Both of the cockpit designs are aimed at solving the same problem (that of controlling and operating an aircraft effectively) and both are certified under the same regulations, and yet they look and behave completely differently.

Between them, the FAA and EASA have "determined that the aircraft certification systems of each Authority for the design approval, production approval, airworthiness approval, and continuing airworthiness of the civil aeronautical products and articles identified in this document, are sufficiently compatible in structure and performance." to allow mutual recognition.[17] In practice, this means that each recognizes authorizations, Type Certificates, etc., issued by the other, as if they had been issued by them. They also maintain Significant Standards Differences (SSD) documents to highlight significant differences between, say, specific versions of CS-25 and 14 CFR Part-25.[18] Similarly, there are other bilateral arrangements involving other countries such as Canada and Brazil.

Other regulators, e.g., Russia, have aligned some of their certification regulations with the FAA and EASA (e.g., AP-25 harmonized with Part-25 and CS-25). Before the invasion of Ukraine, the FAA and EASA had recognized the type certificates of certain Russian aircraft certified to this standard (e.g., Tupolev Tu-204). However, these were revoked following the invasion in 2022.

6.1.3.1   Safety Assessment

Returning to the previous example of CS-25.1309, the section of the regulation is shown in full below.

CS 25.1309 Equipment, systems and installations

The requirements of this paragraph, except as identified below, are applicable, in addition to specific design requirements of CS-25, to any equipment or system as installed in the aeroplane. Although this paragraph does not apply to the performance and flight characteristic requirements of Subpart B and the structural requirements of Subparts C and D, it does apply to any system on which compliance with any of those requirements is dependent. Jams of flight control surfaces or pilot controls covered by CS 25.671(c)(3) are excepted from the requirements of CS 25.1309(b)(1)(ii). Certain single failures covered by CS 25.735(b) are excepted from the requirements of CS 25.1309(b). The failure conditions covered by CS 25.810 and CS 25.812 are excepted from the requirements of CS 25.1309(b). The requirements of CS 25.1309(b) apply to powerplant installations as specified in CS 25.901(c).

(a) The aeroplane equipment and systems must be designed and installed so that:

    (1) Those required for type certification or by operating rules, or whose improper functioning would reduce safety, perform as intended under the aeroplane operating and environmental conditions.

    (2) Other equipment and systems are not a source of danger in themselves and do not adversely affect the proper functioning of those covered by sub-paragraph (a)(1) of this paragraph.

(b) The aeroplane systems and associated components, considered separately and in relation to other systems, must be designed so that -

    (1) Any catastrophic failure condition

        (i) is extremely improbable; and

        (ii) does not result from a single failure; and

    (2) Any hazardous failure condition is extremely remote; and

    (3) Any major failure condition is remote; and

    (4) Any significant latent failure is eliminated as far as practical, or, if not practical to eliminate, the latency of the significant latent failure is minimised; and

    (5) For each catastrophic failure condition that results from two failures, either one of which is latent for more than one flight, it must be shown that:

        (i) it is impractical to provide additional redundancy; and

        (ii) given that a single latent failure has occurred on a given flight, the failure condition is remote; and

        (iii) the sum of the probabilities of the latent failures which are combined with each evident failure does not exceed $1/1\ 000$.

(c) Information concerning unsafe system operating conditions must be provided to the flight crew to enable them to take appropriate corrective action in a timely manner. Installed systems and equipment for use by the flight crew, including flight deck controls and information, must be designed to minimise flight crew errors which could create additional hazards.

(d) Electrical wiring interconnection systems must be assessed in accordance with the requirements of CS 25.1709.

(e) Certification Maintenance Requirements must be established to prevent the development of the failure conditions described in CS 25.1309(b), and must be included in the Airworthiness Limitations Section of the Instructions for Continued Airworthiness required by CS 25.1529.

CS-25.1309 (and its FAA equivalent FAA AC 25.1309-1A) is a fundamental part of the type certification process, and again entire books have been written about the process.[19] However, as can be seen from above, there is a need to assess *any equipment or system installed in the aeroplane* for safety, including ensuring that a catastrophic failure condition does not result from a single failure. There is also a need to calculate and justify failure probabilities, with a range of tools available including: probability methods; fault tree analysis; functional hazard assessment; failure modes and effects analysis, zonal safety analysis, and more. This type of analysis can be particularly challenging when assessing innovations for which there is little operational experience for example, the Boeing 787 introduced a new use of batteries and adhesive bonding of the composite wings to the fuselage.

A further challenge is raised when considering designing and assuring software. We saw in the serious incident involving G-VATL in Chapter 5, the potential negative impact that unexpected or coincident failures can have when considering automated logic. Standards exist[20,21] that can guide development and provide a route to certification. Clearly the risk is multiplied greatly when considering fly-by-wire aircraft and the associated flight laws. This difference in potential risk leads to the concept of Design Assurance Level (DAL). This "rating," which is determined from the safety assessment process described above, gives a measure of the potential impact of a failure in that system, with DAL A implying *Catastrophic* through to DAL E implying *No Safety Effect*. This rating can then be used to decide on the necessary assurance level ensuring that software is developed and assessed in proportion to its importance.

Given recent advances in computing techniques such as machine learning and neural networks it is worth noting that at present, regulators have no accepted route to certify any "Artificial Intelligence" (AI) which is stochastic or nondeterministic (meaning the same input will not always produce the same output and outputs can evolve). However, EASA has published an AI Roadmap[22] and concept paper[23] anticipating that "the first certifications of assistance to pilots are expected in 2025, with a gradual ramping up towards full autonomy in around 2035."

6.1.4   Requirements Beyond the Aircraft Design

As part of the type certification process, the Design Organization is also under an obligation to help and support the operator to introduce the type into their operation, including providing a range of documentation. As such, they will supply a Maintenance Planning Document, intended to help operators to devise an appropriate maintenance program. This will draw on a Maintenance Review Board Report (MRBR) which is an acceptable means of compliance for developing a maintenance program. The MRBR can be generated using the Maintenance Steering Group (MSG-3)[24] process which is a decision-logic process for determining initial scheduled maintenance requirements for new aircraft and is a means for developing acceptable maintenance tasks and intervals.

The Design Organization will also supply a Flight Crew Operations Manual (FCOM) explaining how the aircraft systems work and how the designers envisaged the aircraft to be operated. The DO must also prepare a master minimum equipment list (MMEL) for a particular aircraft type, which identifies those items which (individually) are permitted to be unserviceable at the start of a flight. The MMEL may be associated with special operating conditions, limitations, or procedures. The MMEL is particularly important since, as aircraft become more complex, engineers and flight crew cannot be aware of all the possible interactions between systems. For example, in isolation, the failure of a radio altimeter (RadAlt) may seem to be a manageable problem not worthy of cancelling a flight. However, loss of the RadAlt may affect the autopilot, terrain awareness warning systems (TAWS), and more. This was part of the reason for the concern around the introduction of 5G towers in the United States that might interfere with proper RadAlt operation.[25] It is also worth noting that the Type Certificate Holder (TCH) must collect and review operational information concerning safety of the type design and correct any potentially unsafe condition. We will return to these points in Section 6.2.

6.1.5 "Grandfather" Rights

Given the cost and timescales involved in "clean sheet" designs, some Design Organizations opt to modify, develop, and update an existing design. Where the design of an aircraft or component is the same as, or similar to, an already-certified product, it can be possible to certify it through so-called "Grandfather Rights." This process allows variants of an initial type design to be manufactured under variations to the original type certificate. This can mean that developments and improvements brought into later version of the Certification Specification are not reflected in the certification of a "similar" design, although an authority can invoke the new standard (as was the case for the Boeing 737NG). The standard against which the type is assessed is known as the Certification Basis.

Examples of the "grandfathering" process include the Airbus A321, A319, and A318, all of which were developed as variations of the A320. The Boeing 737 was first certified in 1967 and every model since then has been approved as a combination of new requirements and original requirements, the latter benefiting from grandfather rights. The Super Puma family described in Chapter 4 was another beneficiary of this arrangement. The grandfather process avoids much of the cost and complexity involved in gaining completely new certification. Clearly this can be very attractive to aircraft OEMs since, as well as reducing costs, it can also de-risk the process. Equally, if a product is similar enough in operation to a previous type, it may not be necessary for crew to undergo retraining. However, this process can produce some unintended outcomes. For example, the grandfathering process has produced a modern Boeing 737 that is almost unrecognizable compared with the original design. As well as visual differences, this includes handling differences, avionics differences, and multiple system differences, of which MCAS is one (see Chapter 13).

6.1.6 Certification Pressures and Delegation

One by-product of the increasing complexity of modern aircraft is the increasing certification pressure on regulators. Aircraft manufacturers are huge organizations with thousands of employees involved in the design of enormously complex aircraft. For example, Airbus employs somewhere around 125,000 people worldwide, whereas EASA employs approximately 800. Regulators simply do not have the resources, and in some cases the expertise, necessary to perform detailed oversight of all the certification activities that are asked of them. FAA Order 8110.4C acknowledges that:

> The heavy workloads for FAA personnel limit involvement in certification activities to a small fraction of the whole. FAA type certification team members must review the applicant's design descriptions and project plans, determine where their attention will derive the most benefit, and coordinate their intentions with the applicant.

For this reason, a process exists for delegating certain aspects and functions of the design and type certification process to an approved person or organization. However, this process is not without issue. The most obvious recent example of a problem in this area is the 737 MAX accidents (see Chapter 13), where questions were raised about the FAA's oversight of Boeing. The US House of Representatives report stated that:

> *The Committee's investigation has also found that the FAA's certification review of Boeing's 737 MAX was grossly insufficient and that the FAA failed in its duty to identify key safety problems and to ensure that they were adequately addressed during the certification process.*[26]

It is worth noting that questions around the delegation of authority are not new. Even in the Comet Inquiry (see Chapter 5), the question of whether it was appropriate for aircraft manufacturers to have delegated authority for inspections was raised, with the Court of Inquiry concluding that the system was *"satisfactory."* In addition to the workload of certification, a new design can risk the continued existence of an OEM, as noted previously. This can increase pressures on all parties to make a success of a new type certification.

6.1.7   Modifications and Additions

Clearly, the effort involved in designing and certifying an aircraft to ensure safety is considerable. Therefore, it is unsurprising that there are restrictions on how the aircraft may be modified once in operation; there is no point in carrying out detailed aerodynamic modeling and flight testing if the first thing an operator does is bolt a new aerodynamic surface onto the wing! To avoid this problem, any changes to an aircraft need to be subject to an approval process and documented. The size of the change will dictate the type of approval needed with the least significant being approved in normal maintenance and the most significant requiring a Supplementary Type Certificate (STC).

6.1.8   When the Certification and Initial Airworthiness System Fails

As mentioned above, the high reliability of the complete aviation system is built on the best efforts of all involved to deliver safety, including by learning from mistakes, occurrences, and accidents. However, it is of course totally apparent that at times these processes and systems fail to deliver the safety outcomes we desire. In Chapter 4, we have already seen examples where the investigation and recommendation process has failed to prevent the recurrence of accidents. In this section we will look at examples where the design of the aircraft was flawed, but for different reasons, using the example of the Boeing 777, G-YMMM[27] which crashed at London Heathrow airport on January 17, 2008. The following summary is based on the AAIB[iv] report[28]:

The aircraft was operating a scheduled flight from Beijing to London. In cruise, the aircraft stepped through altitudes of around 35,000 ft, 38,000 ft, and 40,000 ft. The flight was uneventful until the aircraft was stabilized on the ILS to Runway 27L (the left-hand runway, heading due West) with autopilot and autothrottle engaged. At around 800 ft above the airfield level, the copilot took control intending to disconnect the autopilot at 600 ft and land manually. At that point, the autothrottles commanded an increase in thrust to both engines, but after both initially responding, the thrust on the right engine reduced, followed by the left around 7 seconds later. The engines did not shut down but continued to produce thrust, albeit less than commanded. A thrust differential was noted by the copilot at this point, about 48 seconds before touchdown.

[iv] AAIB—Air Accidents Investigation Branch investigates civil aircraft accidents and serious incidents within the United Kingdom, its overseas territories, and crown dependencies.

About 27 seconds before touchdown, the copilot noticed that the airspeed was reducing below the expected value of 135 knot and the crew discussed the lack of thrust being produced. The engines failed to respond to further demands from the autothrottle and manual movement of the thrust levers. At 115 knot, the AIRSPEED LOW warning and a master caution sounded, and the airspeed stabilized briefly, so the commander retracted the flaps from FLAP 30 to FLAP 25 to reduce drag. At 200 ft the airspeed had reduced to 108 knot.

Ten seconds before touchdown, the stick-shaker indicated an imminent stall, so the copilot pushed the control column forward, thereby disconnecting the autopilot and reducing the nose-high pitch. As the aircraft approached the ground, the copilot pulled back on the control column, but the aircraft struck the ground short of the runway around 330 m short of the start of the runway and 100 m inside the perimeter fence. During the impact and short ground roll, the nose landing gear (NLG) and both the main landing gears (MLG) collapsed. The right MLG separated from the aircraft, but the left MLG remained attached. See Figure 4-3.

6.1.8.1   When We Don't Know Enough...

Some accidents result from new and unforeseen situations, perhaps where foresight or engineering experience did not extend far enough to prevent a bad outcome. We saw in the previous chapter in the example of the Comet how insufficient understanding (in that case, relating to material fatigue) can lead to catastrophic outcomes. The accident to G-YMMM provides a more contemporary example. After detailed investigation, simulation, data-mining, and testing, the UK AAIB concluded that the reason the engines failed to respond to the demand for increased thrust on approach was a restriction in fuel flow at the fuel oil heat exchanger (FOHE—a device which allows heat energy to be transferred from the hot engine oil into the cold fuel). The combination of the low outside temperature and low fuel flow rates had allowed the fuel to form a type of "sticky" ice that coated the inside of pipes. When increased thrust was demanded, the increased fuel flow caused the ice to dislodge and restrict flow through the two FOHEs to both engines.

Given everything discussed so far in this chapter around the safety measures embedded in the certification process, it is relevant to ask how this situation could have arisen? Perhaps the clearest explanation comes from the final two (of four) causal factors identified in the report:

3) *The FOHE, although compliant with the applicable certification requirements, was shown to be susceptible to restriction when presented with soft ice in a high concentration, with a fuel temperature that is below −10°C and a fuel flow above flight idle.*

4) *Certification requirements, with which the aircraft and engine fuel systems had to comply, did not take account of this phenomenon as the risk was unrecognised at that time.*

Simply put, the aviation system did not "know enough" to recognize the risk or enough to effectively mitigate against it. In situations such as this, we can only accept the shortcoming and be thankful that no one was fatally injured. Of course, that is no reason not to strive to know more, predict more, and mitigate every conceivable risk, particularly when considering design requirements. Similarly, design and certification standards should be constantly evolving to learn the lessons of past accidents, incidents, and errors. With that in mind, the AAIB report made 18 safety recommendations, of which 3 are of particular interest in this discussion:

2008-049: It is recommended that the Federal Aviation Administration and the European Aviation Safety Agency review the current certification requirements to ensure that aircraft and engine fuel systems are tolerant to the potential build up and sudden release of ice in the fuel feed systems.

2009-031: It is recommended that the Federal Aviation Administration and the European Aviation Safety Agency jointly conduct research into ice formation in aviation turbine fuels.

2009-032: It is recommended that the Federal Aviation Administration and the European Aviation Safety Agency jointly conduct research into ice accumulation and subsequent release mechanisms within aircraft and engine fuel systems.

It is in this way that the system improves.

6.1.8.2   When We Think We Know Enough...

As well as being an example of not knowing enough, the accident to G-YMMM is also an example of where the system appeared to "know enough,", but in reality that understanding (and the mitigations it produced) was based on flawed assumptions. As part of the accident sequence, the aircraft contacted the soft ground short of runway 27L at Heathrow airport with a vertical rate of descent of 25 ft/s. Both MLG partially separated at the initial impact and the NLG collapsed due to the high vertical and lateral loads.

The left gear behaved as it was designed to do, with six "fuse pins" (designed weak points intended to be the point of failure in extreme conditions) fracturing and allowing the fuel tank to remain intact. However, on the right landing gear, the two upper fuse pins did not fracture. This resulted in a piece of the gear separating and impacting the fuselage, causing injuries in the cabin. It also caused the spar web to rupture causing a breach in the center fuel tank (with the release of nearly 15,000 lb / 7,000 kg of fuel), and the disruption of the passenger oxygen bottles which all vented to the atmosphere. Had there been a postimpact fire, there was huge potential for greater injury or fatalities.

Analysis conducted as part of the investigation showed that the progression of failures can be very different when comparing between landing on soft ground as opposed to hard runway. In the case of soft ground, a delayed build-up of shear forces in the fuse pins when compared to impact with hard ground prevented most of them from reaching their failure loads. (This is analogous to dropping an egg onto a mattress compared with hard concrete.) With some fuse pins still intact, the design breakaway scenario is changed.

Emergency landings can occur both onto the runway itself and soft surfaces outside the airfield boundary, and yet the certification requirements in place at the time did not distinguish between landings on different surfaces. As a result, the AAIB issued the following safety recommendation:

2009-096: It is recommended that the Federal Aviation Administration in conjunction with the European Aviation Safety Agency review the requirements for landing gear failures to include the effects of landing on different types of surface.

The regulations were written with an understanding of the importance of the failure sequence of the landing gear, and yet despite that, a detail had been overlooked that was only brought to light through experience and the investigation process.

6.1.8.3 Concluding Remarks

The case of G-YMMM highlights two ways in which the certification process can fail to meet its intended objective, either as a result of insufficient knowledge or through the use of flawed assumptions. In both areas, thorough investigation yielded recommendations that will hopefully inform and improve the design and certification process to prevent a repeat of the situation, although, as we have seen in Chapter 4, that feedback system is not always perfect and there are a number of examples where opportunities to prevent accidents existed and yet were not taken. These examples highlight not only the importance of continual improvement but also the importance of effective investigation.

6.2   Maintaining Aircraft for Safety

Clearly, after investing the time and effort to design a safe, type-certificated product, it is essential that it is maintained in that same airworthy state—this is the role of continuing airworthiness. As given above, Continuing Airworthiness is defined as "The set of processes by which an aircraft, engine, propeller or part complies with the applicable airworthiness requirements and remains in a condition for safe operation throughout its operating life." The responsibility for continued airworthiness is shared between the regulator, the type-certificate holder, the operator, their maintenance organization, and their certifying staff, with each having a role to play.

6.2.1   The Regulator

It is the regulator that issues Type Certificates and each individual Certificate of Airworthiness (C of A). Then, as a type enters service, and as experience grows throughout its operational life, issues will be discovered that require correction in order to ensure the continuing airworthiness of the aircraft. Therefore, throughout the life of the aircraft, the regulator and TCH will issue publications detailing work that must be carried out by anyone operating that type.

The primary publication of this type is an Airworthiness Directive (AD). These are legally enforceable rules issued by the regulator to correct an unsafe condition. Compliance with an AD is mandatory and if they are not embodied, then the C of A is not valid, and the aircraft cannot legally fly. Emergency Airworthiness Directives (EADs) are ADs that must be applied urgently. ADs and EADs are publicly available from regulators,[29,30] although the Alert Service Bulletins or Service Bulletins (described in Section 6.2.2) to which they refer are usually only available to operators of the aircraft. The FAA and EASA will usually emulate the other's E/ADs however this can sometimes take time as each follows their own process.

When first entering the world of continuing airworthiness, it can be shocking to see the number of EADs and ADs being issued—it is not unusual to see 2, 3, or 4 issued in a working day. However, this is how the system is intended to work; types are under continual review by regulators and TCHs, to eliminate unsafe conditions or identify potential improvements.

6.2.2   The Type Certificate Holder

The TCH provides technical support on the type to the regulator and operators of the aircraft. This includes support for ADs, issuing Service Bulletins, and advising on maintenance and repairs. Service Bulletins (SBs) are notices issued to an aircraft operator by the TCH describing a product change or improvement. SBs are not mandatory (only an E/AD can be mandatory), but the relevant airworthiness organization (described in Sections 6.2.3 and 6.2.4) is responsible for developing an assessment policy and determining the appropriate/necessary action or response. Implementing an SB will often be at a cost to the operator. Alert Service Bulletins (ASBs), also referred to as Mandatory or Emergency, are issued when an unsafe condition is discovered by the TCH. Like SBs, they are not mandatory, but they are encouraged to be carried out within the specified timeframe.

The TCH may also make use of Service Letters (called various other names including In-Service Information) to convey information to operators, but they do not contain change information for action. As part of the type certification process, the applicant is responsible for providing documentation that will allow an operator to maintain the airworthiness of the aircraft. In the United States, this is described as Instructions for Continued Airworthiness (ICA) described in FAA Order 8110.54A[31] or the Maintenance Planning Document (MPD). These instructions will include airworthiness limitations (including mandatory replacement and inspection times), maintenance manual, maintenance instructions, aircraft engine instructions, and more. They will also draw on the MRBR generated during type certification.

6.2.3   The Operator

When an operator starts operating an aircraft type, the operator must ensure that the C of A remains valid and they must have an Aircraft Maintenance Program (AMP) and an Approved Maintenance Schedule (AMS) in place, which has been approved by the regulator. These will be bespoke to each operator, but will be based on the TCH's MPD.

To help satisfy these requirements, EASA requires that there must be an approved Continuing Airworthiness Management Organization (CAMO), detailed as part of the AOC and regulated under Part-CAMO. There must also be periodic review of the C of A, producing an Airworthiness Review Certificate (ARC) which is typically valid for 1 year.[32] The CAMO maintains the airworthiness to the regulations defined in Part-M of the EASA regulations. The FAA does not use the CAMO system. Instead, the operator is authorized, through its AOC, to develop its own Continuous Airworthiness Maintenance Program (CAMP), which does not require approval from the regulator. The CAMP contains specific maintenance and inspection tasks with guidelines available in AC 120-16F *Air Carrier Maintenance Programs*.[33]

In Europe, the CAMO develops a Continuing Airworthiness Management Exposition (CAME) describing the responsibilities, procedures, and methods of the CAMO.[34] The CAME is the equivalent of the CAMP in FAA terms, except that the CAME is subject to approval by the regulator. The CAMO/CAMP ensures that ADs and AMP tasks are performed, along with overseeing any Modifications or Repairs, with compliance demonstrated by the Certificate of Release to Service (CRS). The work must be carried out by an approved FAR Part-145 MRO.

If an aircraft suffers major damage, any repairs that are not described in the approved maintenance plan, such as damage to the airframe, need to be approved, usually by the authorized Design Organization or TCH. It is worth noting that any major repairs may require a change in the continuing airworthiness requirements of that specific aircraft.

6.2.4 FAR and EASA Part-145 Maintenance Repair and Overhaul

The aircraft operator assigns an organization to carry out work in accordance with the AMP and AMS, normally termed a Maintenance, Repair and Overhaul (MRO) organization. The MRO will need to be approved under Part-145 of the regulation, with an approval for the specific aircraft type in question. A Part-145 organization must meet specific requirements particularly in its staffing, tooling, premises, and capacity, including using licensed maintainers.

The Part-145 MRO can be a separate, standalone organization or may be part of the same company—the regulatory requirements are the same in both cases. Many operators find it advantageous to have their own MRO operation which can also offer services to other operators. It is important to note that even when subcontracting maintenance work, the continuing airworthiness of the aircraft remains the responsibility of the CAMO, regardless of any approvals held by the organization performing the maintenance including Part-145 or CAMO. This means the CAMO management must remain directly engaged with safety assurance.

6.2.5 Maintainer Licensing

Maintenance staff that certify work carried out must hold an Aircraft Maintenance License (AML), approved for the particular aircraft type (called a Type Rating). There are also limitations on the type of work a specific engineer can carry out, reflected in the type of license issued. FAA maintainer licenses are issued under Part-65[35] and EASA licenses under Part-66 which "defines the aircraft maintenance licence and establishes the requirements for application, issue and continuation of its validity." Organizations which provide training for Part-65/-66 licensed engineers must themselves have approvals from the regulator. The requirements for this approval are defined in EASA/FAA Part-147 covering training and examination.

6.2.6 Aging Aircraft and Changes to Certification

Aircraft are designed for many years of service, if maintained properly. However, service experience has shown that aging aircraft can require modified maintenance and inspection plans to retain their airworthiness, due to factors such as environmental deterioration and fatigue. Extended service lives mean that the continued airworthiness requirements must consider virtually every component. In addition, as an aircraft grows older, any slightly flawed assumptions that have been made about its usage such as sector lengths, operating altitudes, etc., are amplified.

A significant accident, highlighting the importance of this area of regulations occurred in 1988. On April 28, a Boeing 737-200 suffered an explosive decompression in flight which required an emergency landing on the Hawaiian island of Maui. The explosion killed one stewardess but remarkably, considering the extent of damage (see Figure 6-1), the remaining 94 passengers and crew survived the ordeal. The NTSB investigation report concluded that the accident was caused by a failure of the maintenance program to "detect the presence of significant disbonding and fatigue damage." The age of the aircraft became a key issue (it was 19 years old), and the 737 had sustained a remarkable number of takeoff-landing cycles, 89,090, the second most such cycles for a plane in the world at that time. The Aloha Airlines accident focused public and congressional attention on *aging aircraft*.

In response to this accident, in 1991 the US Congress passed the *Aging Aircraft Safety Act* which required air carriers to demonstrate that maintenance of an airplane's

Figure 6-1 Damage to the Aloha Airlines 737 following explosive decompression. (Source: NTSB)

age-sensitive parts has been *"adequate and timely enough to ensure the highest degree of safety."* This legislation was codified by the FAA as final rule entitled *Aging Airplane Safety* and requires airlines to ensure that repairs or modifications made to their airplanes are damage-tolerant and that repetitive inspections and record reviews are conducted every 7 years for all transport airplanes with more than 14 years in service.[36] Both the FAA[37] and EASA[38] require a *Supplemental Structural Inspection Program* (SSIP) to be incorporated into the AMP as part of a damage-tolerance-based inspection program. The FAA give guidance on the SSIP,[39] as do EASA.[40] The SSIP should be implemented before any indication that a significant increase in inspections is required to maintain airworthiness. Since the TCH is aware of the effects of corrosion on the aircraft, the FAA and EASA consider that objectives of a Corrosion Prevention and Control Program (CCPC) are addressed through the MSG-3 process (see Section 6.1.4).

6.2.7 Loss of Airworthiness

If an aircraft does not meet, or cannot be shown to meet, the applicable airworthiness requirements it is considered no longer airworthy and hence cannot fly. This can happen for a variety of reasons including overrun of component inspections or maintenance and failure to embody ADs. However, if the aircraft is considered safe to fly in certain conditions, the regulator may issue a Permit to Fly, for specific purposes such as development, delivery, exhibition, etc.

If the regulator has reason to doubt the continued airworthiness of an aircraft or type, they have the ability to "ground" the aircraft by suspending its certificate of airworthiness (C of A) or type certificate (TC) or issuing an AD prohibiting flight. This can often be in response to an accident or some other failure of the airworthiness system. Of course, operators are free to choose to stop flying a particular aircraft type at any point if they have lost confidence in it, but this is not the same as a regulatory suspension of C of A or TC.

For example, following the accident to LN-OJF (described in Chapter 4) EASA issued an Emergency AD[41] instructing operators not to operate the aircraft except for a single ferry flight with no passengers to a maintenance or storage location. One major consideration for regulators when taking a decision to suspend operation of an aircraft, is how they will "unground" the aircraft; what tests, evidence, experience will they require to reinstate its airworthiness. This can be challenging, but it is not a reason to delay action if the aircraft is considered unairworthy.

If a TCH no longer wishes to support a type, they can apply to surrender its TC. If no new suitable TC holder is found, the regulator can accept the surrender and take the decision to revoke the TC, effectively removing support for the aircraft and invalidating its Certificate of Airworthiness. This was the process applied to Concorde in the early 2000s.

6.2.8 When the Continuing Airworthiness System Fails

We have already seen the example of G-REDL in Chapter 4 which highlighted the importance of effective continuing airworthiness. A more commonly known example is that of the BAC One-Eleven G-BJRT operating as British Airways flight 5390 from Birmingham to Malaga on June 10, 1990.[42] While climbing through 17,300 ft, the left windscreen, which had recently been replaced, was blown out by the differential cabin pressure. The commander was sucked halfway out of the windscreen aperture and blown back over the flight deck roof where he was restrained by cabin crew while the copilot flew the aircraft to a safe landing. The subsequent investigation identified that 84 of the 90 bolts used to fit the replacement windscreen had a diameter 0.7 mm smaller than specified, but that this was symptomatic of larger problems in British Airway's maintenance system. Eight Safety Recommendations were made as a result of the investigation.

Continued airworthiness was also a significant factor in the accident to Flight JAL 123, an accident that is sadly still recognized as having the highest number of fatalities for a single aircraft accident. A Boeing 747 SR-100 (a short-range derivative of the 747-100), registration JA8119 was operating Japan Airlines Flight 123 from Tokyo to Osaka on August 12, 1985. Around 12 minutes after takeoff at nearly 24,000 ft, the aircraft suffered a rapid decompression damaging the fuselage and the vertical stabilizer. This caused the rudder to separate from the aircraft and caused damage to the hydraulic system. The aircraft became uncontrollable, and the effects of hypoxia (caused by the lack of oxygen at altitude and which can impair judgement) hindered the efforts of the crew to control the aircraft. About 30 minutes into the flight, the aircraft crashed killing all but 4 of the 524 passengers and crew.

The subsequent investigation by Japan's Aircraft Accident Investigation Commission revealed that the aircraft had been involved in a tailstrike during a bounced landing 7 years prior. The repair that was made by Boeing to the damaged pressure bulkhead after the tailstrike was not in accordance with their approved repair method, which compromised the resistance of the part to fatigue cracking. (Remember the effects of pressurization on metal fatigue in the case of the Comet discussed in Chapter 5). During the investigation, it was calculated that this incorrect repair would fail after around 10,000 pressurization cycles. This failure occurred after around 12,300 cycles.

This accident highlights the vital importance of accurate maintenance of continued airworthiness. The erroneous installation in this case was that a splice, or joining plate, simply was too small, meaning that it was not joined by the intended number of rivets. The issue went undetected for 7 years, until the catastrophic accident.

6.3 Operating an Aircraft Safely

In addition to maintaining the aircraft in an airworthy state as described above, the operator of an aircraft must also operate that aircraft in an approved way. The approval given by the regulator to operate an aircraft commercially is an Air Operator's Certificate (AOC) or Air Carrier Certificate (ACC). The regulation governing this process for typical commercial aviation in the United States is 14 CFR Part 121[43] and in Europe Part-ORO. AOC operating to Part-CAT and Part-SPA. The operating certificate lists the approved types of operation and any limitations (in the Operations Specifications or Ops Spec).

The requirements to hold an AOC/ACC are unsurprisingly extensive, including the right accountable staff, safety management system, an appropriate quality system, airworthy aircraft, and sufficient funds. In this section we will focus on some of the safety-relevant aspects of operation.

6.3.1 Flight Crew Licensing and Training

The licensing of flight crew started at the very beginning of aviation. The very first flight crew licenses (FCLs) were issued in 1909 followed by international licensing standards in 1919. The process is normally the function of the state National Aviation Authority (NAA). As we will discuss in Chapter 16, EASA broke the standard NAA regulatory mold with the introduction of an international flight crew licensing system which could then be implemented by each participating NAA. Like all international regulatory standards, flight crew licensing originates with ICAO. Chapter 2 of Annex 1[44] covers the requisite qualification and issue of license ratings for pilots. Chapter 3 describes the requirements for flight navigators and flight engineers. Typically, FCL regulation covers the approval of flight crew training, flight simulators, and appropriate arrangements for the certification of medical fitness for flight crew.

Professional pilots will hold a minimum qualification of a Commercial Pilot's License (CPL). The award of a CPL typically follows an 18-month (minimum) structured program of aerodynamic theory, navigation, meteorology, technical and human factors–related subjects. These courses are expensive (in the region of at least US$150,000) and often paid for by the student pilot themselves through commercial or family loans. This has raised the issue of whether the system is producing the best candidates as the talent pool is restricted by the ability to pay for the training. Early flying training is conducted on single-engine light aircraft followed by progression to more capable twin-engine aircraft which are required to attain multiengine competence ratings. The CPL training culminates in a series of ground and flying examinations which must be passed to a suitably high standard in order to graduate.

This is not the end of the road as the recently graduated pilot now needs to find a job. Very few commercial operators will operate the type of aircraft that was used for training, so now the potential commercial pilot needs to obtain a type-rating on a commercial aircraft. The choice at this point is again determined by the resources of the individual pilot. They may be willing and able to pay thousands of dollars to gain a type rating on a popular commercial aircraft type such as an Airbus A319 or a Boeing 737. If not, they then must find a willing employer who will pay for the type rating. In considering a potential hire, the employing airline faces two main risks. Is the student sufficiently capable of obtaining the type rating with the minimal amount of expensive training and, will they then stay with the airline for a sufficient length of time to recoup the investment in the pilot's training? To mitigate this risk, airlines will often impose a training bond on the pilot. In this way the pilot will have to reimburse the airline for the costs incurred in type rating training should they take alternative employment.

The new pilot will now enter the airline's induction training which is both a way of the airline assessing the candidate's capabilities and for the pilot to assess whether the airline is the company they want to work for. The training doesn't ever really end in a pilot's career path but the objective for most is now to obtain sufficient hours and experience to gain a command. When undertaking their early examinations most pilots opt for the Airline Transport Pilot's License or ATPL rather than the simpler CPL. In this way they effectively hold a "frozen" ATPL until they achieve sufficient hours to qualify for a full ATPL required to become the aircraft captain. This is not just a case of accumulating time on the aircraft. Setting the training pattern that will last their entire career, a new pilot will have to pass competency checks in the simulator every 6 months which will assess their technical and human factors skills. To establish their ability to fly in the real world, that is out on the line, a line check is conducted normally every year. The airline's AOC requires that the holder must suitably maintain qualified and licensed flight crew as described in Part-61 in the United States and Part-FCL in Europe. The checks and training described above will invariably be undertaken by the airline's own personnel, but they will technically be acting as representatives of the regulator, checking under the privileges of their training license and qualification.

6.3.2   Operations Manual

As part of their AOC, operators must maintain an operations manual, the relevant sections of which must be made available to all crew. The structure and content of the manual are prescribed in regulation (Part-ORO.MLR.101) and it includes information such as the (safety) management system, flight time limitations, Standard Operating Procedures (SOPs), emergency procedures, performance operating limitations, and dangerous goods. In Europe, the manual is divided into four sections:

- **Part A:** General/Basic
- **Part B:** Aircraft operating matters—type related
- **Part C:** Route/role/area and aerodrome/operating site instructions and information; and
- **Part D:** Training

These parts are often abbreviated as OMA, OMB, etc. Any changes need to be approved by the regulator and notified to all personnel affected.

6.3.3   Extended Range Operations

In certain circumstances, regulators are able to give operators an approval under ETOPS[45] approval. This originally meant Extended-range Twin-engine Operations Performance Specifications (ETOPS) but is now broadened by ICAO to simply mean "extended operations" or *Extended Diversion Time Operations* (EDTO). This approval enables twin-engine airplanes to operate on routes that were previously only accessible to aircraft with three or four engines. Without an ETOPS approval, twin-engine aircraft must remain within 1-hour flying time of a diversion airport at the approved one-engine inoperative cruise speed. This means that if a twin-engine aircraft suffers an engine failure, it would be at most 1 hour from a diversion airport, thereby minimizing the exposure to the potential of a double-engine failure.

To obtain ETOPS certification, an operator needs an AOC to demonstrate that the necessary airworthiness, maintenance, and operational requirements are in place. ETOPS is a special AOC authorization that will be detailed in the Ops Spec. ETOPS operation is regulated in the United States under Part-121[46] with guidance[47] and under Part-SPA[48] (short for SPecific Approvals) in Europe with guidance.[49]

The operator also needs to be operating an airplane-engine combination that has been type-design-approved for ETOPS. The challenge of ETOPS is essentially one of engine and system reliability and so the approval is underpinned by aircraft and engine reliability data. In addition, ETOPS operations require different maintenance procedures. The operator must produce a different CAMP,[50] and the CAMO must produce a different CAME.[51] The flight crew, maintenance, and dispatch personnel must also have appropriate training around ETOPS. Various ETOPS approvals can be given including 180 and 240 minutes' diversion time. The A350XWB currently has an ETOPS-370 approval meaning it can operate more than 6 hours (at single-engine speed) from a diversion airport.

6.3.4 Minimum Equipment List

An operator is responsible for preparing a Minimum Equipment List (MEL) which is as restrictive, or more restrictive than the MMEL—see 6.1.4) produced for the aircraft type as part of the Type Certificate. This MEL is included in the Operations Manual and is approved by the regulator.[52,53] A separate MEL may be required to satisfy any ETOPS approvals.

6.3.5 Leasing Aircraft

Airlines often choose to lease aircraft for a variety of financial reasons, including short-term contracts, financial risk management, or unwillingness to commit to purchase. In this case, it is important to distinguish between wet lease and dry lease and the resulting impact on continuing airworthiness. In a wet lease, the organization providing the aircraft also supplies the crew, insurance, and maintenance and takes operational control of the flight under their AOC. In a dry lease the aircraft is supplied but without crew, maintenance, or insurance, and the operation is carried out by the organization taking the aircraft under their AOC. An analogy could be drawn with car rentals, where wet lease is the equivalent of hiring a taxi and dry lease is similar to having a long-term company car on lease. The latter requires much more involvement on the part of the person leasing the car but also benefits from more control. The regulator needs to be made aware of, and approve, any aircraft leasing agreements.

6.3.6 Power by the Hour

Power by the Hour (PBH) is an approach to aircraft leasing that can form part of an aircraft management plan. Under this model the operator can agree a rate based on flight hours and/or calendar time. This could apply to the complete aircraft or just components such as engines. Under this agreement, the supplier agrees to provide parts, maintenance, overhaul, etc., and the operator pays an ongoing fee (or in some cases pays into "pot" from which future costs are drawn). This provides the supplier with an ongoing income and minimizes significant outlays for the operator on overhaul or replacement.

6.3.7   Conclusions

To this point, this chapter has been heavy with regulation which is not always the most fascinating experience. However, what has been included here is only a small portion of the regulation that covers commercial aircraft operations. The regulatory complexity in designing, maintaining, and operating a modern commercial aircraft should not be underestimated but neither should its importance. It is this highly regulated system which, although far from perfect, has allowed aviation to achieve remarkable safety levels.

6.4   Aircraft Safety Systems

We move now from the areas of design, initial, and continued airworthiness, all of which aim to deliver a reliable aircraft to the frontline, to the area of frontline operations and the flight deck Human-Machine Interface (HMI). Modern aircraft employ multiple systems to support flight crew in the piloting task. Some of these systems address controllability of the aircraft or reducing workload on pilots, while others aim to enhance the situational awareness of the flight crew or alert them to potentially dangerous situations. We will divide them broadly into those two categories.

6.4.1   Operational Aids

Aircraft systems have developed beyond recognition from the start of powered flight, and yet the piloting task remains broadly the same. The main systems developments have focused on helping the flight crew, expanding safety and flight envelope margins, and reducing workload.

6.4.1.1   *High-Lift Systems*

As described in Chapter 5, the lift generated by an airfoil is dependent on its shape. Different shapes can yield higher lift, but usually at the expense of greater drag. Therefore, it is desirable to be able to change the shape of the airfoil dependent on the phase of flight. High lift is required at lower speeds for takeoff and landing to give better controllability and margin before stall, and low drag at higher speeds during the cruise phase of flight. This ability to effectively change the airfoil shape on command is the function of flaps and slats. Slats are moving surfaces at the front of the wing, and flaps are extending surfaces at the rear of the wing. Figure 6-2 shows the outer slats deployed on an A350 on approach to land.

On modern aircraft, flap and slat operation is achieved using a single control lever with different settings corresponding to different combinations of slats and flaps ranging from the "cleanest" or low drag setting (Flaps 0 or "flaps up") to the highest lift/highest drag setting ("flaps full"). Figure 6-4 shows examples of different flap configurations.

6.4.1.2   *Speed Brakes/Spoilers*

While main wheel brakes remain the primary method of stopping aircraft on the runway, speed brakes and spoilers are designed to assist them. Some dedicated speed brakes (or airbrakes) are attached to the fuselage as shown in Figure 6-3 or as seen on military aircraft.

Spoilers can serve a double purpose; they can reduce lift as well as increasing drag and they can operate in a number of ways. In operation as a ground spoiler, they deploy during the ground roll while landing, or during a rejected takeoff. They reduce lift,

FIGURE 6-2   Outer slats deployed on an A350 on approach to land. (Source: Author)

FIGURE 6-3   Speed brakes deployed on BAe146. (Source: A Pingstone[54])

thereby increasing the load on the landing gear to give better braking, and they increase drag. As speed brakes they can increase drag during flight. Finally, as roll spoilers they can act in tandem with ailerons to help give better roll authority. Figure 6-4 shows combinations of flaps and spoilers in a variety of flight configurations. On many aircraft, the deployment of spoilers during landing is now automatic on landing.

a) Flaps only

b) Spoilers only

c) Flaps and spoilers

d) Flaps, spoilers/speed brakes, and ground spoilers

FIGURE 6-4 Flaps and spoilers in different configurations in flight. (Source: Author)

Some aircraft are fitted with a configuration warning system such as Take-Off Warning System (TOWS). Sadly, it is this system that failed to produce an alert for the pilots of SpanAir Flight 5022 that attempted takeoff without flaps and slats and crashed just after takeoff, killing 154 of the 172 on board.[55]

6.4.1.3   Autobrakes

The autobrake system enables automatic brake application on landing or during a rejected takeoff (RTO). The landing autobrake system controls brake pressure to maintain aircraft deceleration at one of several pilot-selected values, provided that sufficient runway friction is available to maintain this level. The RTO autobrake system applies full braking upon closing throttles above a fixed speed (e.g., 85 knot). Using autobrakes frees the pilot to concentrate on other activities, such as applying reverse thrust and directional control of the airplane.

6.4.1.4   Brake-to-Vacate and Runway Overrun Prevention

Brake-to-Vacate (BTV) is an Airbus system that extends the autobrake concept. During approach preparation, the BTV system displays to the pilots the landing distance that can reasonably be achieved under normal conditions. The crew can then select which runway exit they wish to use, and the aircraft will optimize deceleration for passenger comfort, thrust reverser usage, brake wear, and temperature and runway occupancy, to achieve taxi speed by that exit.

Motivated by the significant issue of runway excursions, the Runway Overrun Protection Scheme (ROPS) is a development of BTV and consists of two parts, the Runway Overrun Warning (ROW) and the Runway Overrun Protection (ROP). The ROW calculates and displays the reasonable landing distance in dry and wet conditions. If either distance is calculated to be beyond the available runway length the PFD caution "IF WET: RWY TOO SHORT" or the warning "RWY TOO SHORT" is displayed. During the landing roll, the ROP scheme calculates the remaining distance, selects maximum autobrake if necessary, and gives the warning "MAX REVERSE" if necessary.

6.4.1.5   Antiskid Braking

For the majority of operations, the antiskid feature of brakes is not required, especially if the runway is dry and long, since the brakes are not applied hard enough to lock the wheels. However, when the runway is slippery and short, the ability of the antiskid system to maximize braking effectiveness becomes very important. The early antiskid systems for the four-wheel truck main gears controlled tandem pairs of wheels. Later the technology to control each wheel independently was incorporated to maximize the effectiveness of the antiskid system.

6.4.1.6   Thrust Reversal

Thrust reversal involves diverting the thrust (usually acting rearwards) from the engine to act forwards and decelerate the aircraft; the core of the engine still operates in the same way. Many modern aircraft do this by blocking bypass air and diverting it forwards through "doors" on the side of the engine. Other solutions include external clamshells that close in the exhaust path and divert it forwards. Thrust reversers can be extremely effective and are useful when landing on short runways (such as regional airports) or when the braking surface is slippery. The Boeing 737-200 has such an effective reverser that the airplane can stop within its required distance on a wet runway using only thrust reversers. Thrust reverse is more effective at high speed since thrust generation is related to the speed of the aircraft.

Although the vast majority of decelerating force is from the aircraft's braking system, on virtually every landing, at least some reverse is used to help minimize brake wear and to keep the brake temperatures low. This often becomes the critical reason when deciding whether to use full reverse, typically discussed during the approach briefing. When schedules dictate a quick turnaround (perhaps under an hour until the aircraft has to take off again), there is sometimes insufficient time to allow the brakes to cool to a safe level to operate. The use of full reverse can help to keep the brakes cooler and therefore minimize turnaround times.

6.4.1.7 Autoland

Autoland is a system that automates the landing phase of flight with the pilots in a monitoring role. It was designed to allow landings in poor meteorological conditions but is not restricted to poor visibility. Typically, the system will combine various aircraft systems such as RadAlt, ILS localizer and glideslope guidance, autothrottle, and autobrake. At the limit of operations, a "fail operational" autoland system can allow landings with no decision height, meaning the pilot is not required to see the runway prior to touchdown. Conversely, Airbus has recently demonstrated automatic takeoff using an image recognition-based system rather than an ILS-based system as part of its autonomous taxi, takeoff, and landing aspirations.

6.4.1.8 Aircraft Communications Addressing and Reporting System

The Aircraft Communications Addressing and Reporting System (ACARS) is a communication datalink system that sends messages between an airplane and the airline ground base. The operational features of ACARS equipment and the ways in which ACARS is used in service vary widely from airline to airline.

There is nothing new about sending messages between the airplane and the ground. What makes ACARS unique is that messages can be sent, including fuel quantity, subsystem faults, and air traffic clearances, in a fraction of the time it takes using voice communications, in many cases without involving the flight crew. The ACARS relieves the crew of having to send many of the routine operational voice radio messages by instead downlinking preformatted messages at specific times in the flight. These may include the time the airplane left the gate, liftoff time, touchdown time, and time of arrival at the gate. ACARS is a very useful operational tool, allowing the gathering of important information such as weather and NOTAM information of the flight crew. In addition, ACARS can be asked by the airline ground operations base to collect data from airplane systems and downlink the requested information to the ground.

6.4.2   Situational Awareness and Crew Alerting

Whereas the technologies described in the previous section are aiming to assist the flight crew by reducing their workload, the technologies described below are intended to assist the flight crew's situational awareness and alert them to any potential problems. Once an alerting condition is detected the aircraft may offer a combination of tones, warning messages, lights, and tactile alerts (such as stick-shaker) to attract the crew's attention.

6.4.2.1 Head-up Displays

The use of head-up displays is becoming commonplace in modern aircraft. For example, the Boeing 787 and Airbus A350 both incorporate head-up screens onto which a

FIGURE 6-5 The view through a HUD while in cruise. (Source: Authors)

PFD-type display is projected. Head-up displays present vital aircraft performance and navigation information to the pilot's forward field of view. The information on the HUD is collimated so that symbols appear to the human eye to be at infinity and overlaid on the actual outside scene. Figure 6-5 shows how symbols appear to a pilot looking through the HUD.

Also, use of the new *enhanced flight vision system (EFVS)*, a display that allows pilots to see through low-visibility weather, could become commonplace in the cockpit of the future. *Forward looking infrared (FLIR)* thermal cameras have been developed to display a real-world visual image to the cockpit allowing the pilot to "see" in very low-visibility conditions. Additionally, HUD systems are also being developed to display new *synthetic vision system (SVS)* images. In contrast to the EFVS, SVS produces a computer-generated artificial image using moving map and terrain databases to provide a more realistic view of the outside environment. Recently, display researchers have even combined EFVS with SVS to improve the visual picture. Combining the real-world reliability of the EFVS FLIR with the terrain and situational awareness of SVS could dramatically reduce issues associated with low-visibility approaches in the future.

6.4.2.2 Electronic Flight Bags

Electronic flight bags (EFBs) are commonly used in modern operations. These are electronic devices (typically tablets or laptops) that contain charts, flight plans, and approach plates, which can perform weight and balance calculations, and plan flights, among other functions.

6.4.2.3 Additional Cockpit Screens

Today, all aircraft modern feature large format LCD displays to provide information to the pilot. Most screens (except those offering Primary Flight Displays (PFD)) will be Multifunction Displays (MFDs) which bring the added benefit of being configurable to the pilot's needs. Modern aircraft tend to benefit from even more displays. For example, on the A350, in addition to the pilot having access to the PFD, Nav display, System and warning display, and MFD, they also have a large Onboard Information System. This can fulfill the function of an EFB system displaying items like airport charts and airway information as well as providing documentation such as FCOM, Flight Manual, and MMEL.

6.4.2.4 Weather Detection

Weather remains a major cause of delays and accidents. Real-time, accurate knowledge of the weather ahead is required. Aircraft must be able to detect weather conditions; alert the flight crew to avoid dangerous situations; and effectively cope with turbulence, precipitation, and wind shear. Modern aircraft are fitted with airborne weather radar allowing flight crew to detect and avoid severe weather and turbulence.

Onboard system enhancements to aid the pilot in unexpected wind-shear encounters are incorporated on current aircraft. These wind-shear alert and guidance systems are combined with crew training, (where the emphasis is on detection and avoidance of hazards occurring along the flight path), improved ground-based alerting, and weather radar systems. If wind shear is unavoidable, the wind-shear system quickly alerts the crew and provides pitch attitude guidance to effect an optimal escape maneuver. The huge increase in Apps and connectivity has prompted the development of weather Apps such as *eWAS* and *MyRadar* which are able to give detailed real-time information and forecasts to give flight crew unprecedented planning data.

6.4.2.5 Terrain and Ground Proximity Warning (TAWS/EGPWS)

Since the advent of powered flight, inadvertent ground or water contact has been a worldwide problem. While much early effort went into avoiding such accidents, no major advance occurred until introduction of the ground-proximity warning system (GPWS) in the early 1970s. Although there has been a marked reduction in controlled flight into terrain (CFIT) accidents since then, they still occur. In the period 2012–2021, 4 of the 36 fatal accidents (11%) identified in the Boeing Statistical Summary[56] were attributed to CFIT. IATA[57] identified CFIT as the second highest fatal accident category.

The *Enhanced* GPWS, termed EGPWS, is an advanced terrain warning system used in most modern commercial airline fleets. The EGPWS improves situation awareness and increases warning times by using a terrain database that compares actual location of the aircraft to terrain in the database and uses a "look ahead" feature to detect what terrain ahead of the flight path may pose a risk. Depending on the mode triggered, the aircraft will provide an aural caution (such as "CAUTION TERRAIN") or warning (such as "TERRAIN TERRAIN, PULL UP") alert, sometimes accompanied by visual alerts and tactile alerts such as stick-shaker.

6.4.2.6 Traffic Collision Avoidance System (TCAS)

TCAS (pronounced *tee-cas)* is an electronic aircraft collision warning system to help prevent midair collision accidents. Typically, TCAS requires mutual "Mode S" transponder technology which automatically communicates with other aircraft independent of ATC systems to display the relative position of other such transponder-equipped aircraft.

Upon detecting a potential conflict aircraft, TCAS issues a coordinated traffic advisory (TA) or resolution advisory (RA) to the flight crew to climb or descend as necessary to avoid the other aircraft. The advisories are coordinated between aircraft, meaning that both aircraft will be given appropriate avoiding instructions rather than, say, both being instructed to climb.

A 2021 study undertaken by Eurocontrol[58] suggested that only approximately a third of RAs were correctly followed by flight crew. The response to low compliance rates from the human pilot has prompted Airbus to introduce their AP/FD TCAS[59] Mode Concept. Ostensibly this is an auto-TCAS, where the aircraft automatically flies the RA, rather than relying on the correct response form the pilot.

6.4.2.7 *Engine Alerting (EICAS and ECAM)*

The Engine Indicating and Crew Alerting System (EICAS) or Electronic Centralized Aircraft Monitor (ECAM) is a digital computer system which monitors and indicates engine and aircraft subsystem information. It will also alert the crew to aircraft configuration and system faults and failures. It can prioritize errors and alert messages as well as display the appropriate Quick Reference Handbook (QRH) checklist. New Airbus and Boeing aircraft models have improved EICAS/ECAM to enhance functionality and reduce display footprint on the flight deck. Each manufacturer has its own strategies for engine indications and crew alerts, but all perform essentially the same basic functions.

6.5   Conclusion

It may seem obvious, but commercial aviation safety is intimately connected to the design of the various aircraft that form the backbone of the transportation system. Certification, airworthiness, and maintenance are all systems and processes that are extremely easy to take for granted and yet are well-hidden from the traveling public. However, maintainers can have as much impact on the safety of a flight as the flight crew as, arguably, can the aircraft designers and regulators. The system only works when everyone plays their part, but sadly this often only shows when something goes wrong.

However, for the most part, this system delivers remarkable safety levels, day in and day out. The systems for achieving type certification, continuing airworthiness, etc., are complex, but necessarily so. Ensuring that the lessons learned over time get carried through to every new design is challenging, but vital to ensure that the system retains, and improves, its enviable safety record.

Key Terms

Aging Airplane Safety Rule (AASR)

Air Operator Certificate (AOC)

Aircraft Communications Addressing and Reporting System (ACARS)

Airworthiness

Anti-Skid System

Cockpit Display of Traffic Information (CDTI)

Cockpit Voice Recorder (CVR)

Continuing Airworthiness Management Organization (CAMO)

Controller Pilot Data Link Communications (CPDLC)

Design Organization (DO)

Electronic Flight Bags (EFBs)

Engine-Indicating and Crew-Alerting System (EICAS)

Extended-Range Twin-Engine Operations (ETOPS)

Flaps

Flight Management System (FMS)

Fly-by-Wire

Ground-Proximity Warning System (GPWS)

Head-Up Display (HUD)

High-Lift Systems

Liquid Crystal Display (LCD)

Multifunction Displays (MFDs)

Organization Designation Authorization (ODA)

Primary Flight Display (PFD)

Speed Brakes

Spoilers

Stick-Shaker

Thrust Reversers

Traffic Collision Avoidance System (TCAS)

Type Certification

Topics for Discussion

1. Describe the process of Type Certification.
2. How does Airworthiness differ from Continuing Airworthiness?
3. Describe the function and position of flaps and slats.
4. Name one advantage and one drawback of the "Grandfather" system.
5. Explain the process of "grounding" an aircraft.
6. Why did the accident to the 777 G-YMMM at Heathrow occur?
7. What is the purpose of Service Bulletins and how do they relate to Airworthiness Directives?
8. Name two aids to situational awareness for pilots and describe how they work.

References

1. ICAO. (2022). *Annex 8: Airworthiness of aircraft*. 13th ed.
2. ICAO Doc 9760. (2020). *Airworthiness manual*. 4th ed. https://store.icao.int/en/airworthiness-manual-doc-9760
3. 14 CFR Part-21. https://www.ecfr.gov/current/title-14/chapter-I/subchapter-C/part-21?toc=1

4. EASA Part-21, EU Regulation 748/2012. https://www.easa.europa.eu/en/downloads/20143/en
5. ICAO. (2022). *Annex 6: Operation of aircraft*. 12th ed.
6. De Florio, F. (2016). *Airworthiness: An introduction aircraft certification and operations*. 3rd ed. St. Louis, MO: Elsevier.
7. FAA Type Certification process. https://www.faa.gov/regulations_policies/orders_notices/index.cfm/go/document.information/documentid/15172
8. EASA Type certification process. https://www.easa.europa.eu/en/document-library/certification-procedures/airworthiness-type-design
9. Skybrary, Certification of Aircraft. https://skybrary.aero/articles/certification-aircraft-design-and-production
10. 14 CFR Part-25. https://www.ecfr.gov/current/title-14/chapter-I/subchapter-C/part-25?toc=1
11. EASA CS-25. https://www.easa.europa.eu/en/document-library/easy-access-rules/easy-access-rules-large-aeroplanes-cs-25
12. 14 CFR Part-29. https://www.ecfr.gov/current/title-14/chapter-I/subchapter-C/part-29
13. EASA CS-29. https://www.easa.europa.eu/en/document-library/easy-access-rules/easy-access-rules-large-rotorcraft-cs-29
14. 14 CFR Part-33. https://www.ecfr.gov/current/title-14/chapter-I/subchapter-C/part-33
15. FAA AC 33-2B. *Aircraft engine type certification handbook*. https://www.faa.gov/documentlibrary/media/advisory_circular/ac%2033-2b.pdf
16. A380 Blade-off test. https://youtu.be/j973645y5AA
17. FAA and EASA mutual recognition. https://www.faa.gov/aircraft/air_cert/international/bilateral_agreements/baa_basa_listing/media/EUTIP_Rev6_w_amdt1_amdt2.pdf
18. EASA Significant Standards Differences. https://www.easa.europa.eu/en/document-library/bilateral-agreements/eu-usa/easa-significant-standards-differences-ssd-between-cs-codes-and-faa-14-cfr-codes
19. Lloyd, E., & Tye, W. (1982). *Systematic safety*. 2nd ed. Taylor Young Ltd.
20. RTCA DO-178C. https://my.rtca.org/productdetails?id=a1B36000001IcmqEAC
21. SAE ARP4754A. https://www.sae.org/standards/content/arp4754a/
22. EASA Artificial Intelligence Roadmap. https://www.easa.europa.eu/downloads/109668/en
23. EASA Level 1 Machine Learning Concept Paper. https://www.easa.europa.eu/downloads/136368/en
24. Skybrary MSG-3. https://www.skybrary.aero/articles/maintenance-steering-group-3-msg-3
25. FAA, 5G and Aviation Safety. https://www.faa.gov/5g
26. House of Representatives Boeing 737 MAX report. https://transportation.house.gov/imo/media/doc/2020.09.15%20FINAL%20737%20MAX%20Report%20for%20Public%20Release.pdf
27. Aviation Safety Network, Accident G-YMMM. https://aviation-safety.net/database/record.php?id=20080117-0
28. AAIB report into G-YMMM. https://www.gov.uk/aaib-reports/1-2010-boeing-777-236er-g-ymmm-17-january-2008
29. FAA Dynamic Regulatory System. https://drs.faa.gov

30. EASA Safety Publications Tool. https://ad.easa.europa.eu

31. FAA Order 8110.54A. *Instructions for continued airworthiness.* https://www.faa.gov/regulations_policies/orders_notices/index.cfm/go/document.information/documentid/638511

32. EASA Part-CAMO.A.320. *Airworthiness review.* https://www.easa.europa.eu/en/document-library/easy-access-rules/easy-access-rules-continuing-airworthiness-0

33. FAA, AC 120-16F. *Air carrier maintenance programs.* https://www.faa.gov/documenTLibrary/media/Advisory_Circular/AC%20120-16F.pdf

34. EASA Part-CAMO.A.300 Continuing airworthiness management exposition (CAME). https://www.easa.europa.eu/en/document-library/easy-access-rules/easy-access-rules-continuing-airworthiness-0

35. 14 CFR 65.71-65.95. https://www.ecfr.gov/current/title-14/chapter-I/subchapter-D/part-65/subpart-D?toc=1

36. 14 CFR 121.1105. https://www.ecfr.gov/current/title-14/chapter-I/subchapter-G/part-121/subpart-AA/section-121.1105

37. 14 CFR 121.1109. https://www.ecfr.gov/current/title-14/chapter-I/subchapter-G/part-121/subpart-AA/section-121.1109

38. Appendix I to AMC M.A.302. https://www.easa.europa.eu/en/document-library/easy-access-rules/easy-access-rules-continuing-airworthiness-0

39. FAA AC 91-56B. *Continuing structural integrity program for airplanes.* https://www.faa.gov/regulations_policies/advisory_circulars/index.cfm/go/document.information/documentID/73486

40. EASA AMC 20-20A. https://www.easa.europa.eu/en/document-library/certification-specifications/amc-20-amendment-20

41. EAD 2016-0104-E. https://ad.easa.europa.eu/blob/EASA_AD_2016_0104_superseded.pdf/EAD_2016-0104-E_1

42. AAIB report in G-BJRT. https://www.gov.uk/aaib-reports/1-1992-bac-one-eleven-g-bjrt-10-june-1990

43. 14 CFR Part 121. https://www.ecfr.gov/current/title-14/chapter-I/subchapter-G/part-121?toc=1

44. ICAO. *Annex 1: Personnel licensing.*

45. Skybrary ETOPS. https://www.skybrary.aero/articles/extended-range-operations

46. Appendix P to Part-121. https://www.ecfr.gov/current/title-14/chapter-I/subchapter-G/part-121/appendix-Appendix%20P%20to%20Part%20121

47. FAA AC 120-42B. https://www.faa.gov/regulations_policies/advisory_circulars/index.cfm/go/document.information/documentid/73587

48. EASA Part-SPA. https://www.easa.europa.eu/en/document-library/easy-access-rules/easy-access-rules-air-operations-regulation-eu-no-9652012

49. EASA AMC 20-6 rev.2. https://www.easa.europa.eu/en/downloads/1696/en

50. 14 CFR 121.374. https://www.ecfr.gov/current/title-14/chapter-I/subchapter-G/part-121/subpart-L/section-121.374

51. EASA AMC 20-6 Appendix 8 https://www.easa.europa.eu/en/document-library/easy-access-rules/online-publications/easy-access-rules-acceptable-means?page=8#_Toc256000026

52. 14 CFR 121.628. https://www.ecfr.gov/current/title-14/chapter-I/subchapter-G/part-121/subpart-U/section-121.628

53. EASA Part-ORO.MLR.105. https://www.easa.europa.eu/en/document-library/easy-access-rules/easy-access-rules-air-operations-regulation-eu-no-9652012

54. By Adrian Pingstone—Own work, Public Domain. https://commons.wikimedia. org/w/index.php?curid=4043632

55. CIAIAC, Spanair DC-9 Report. https://www.mitma.gob.es/recursos_mfom/pdf/ EC47A855-B098-409E-B4C8-9A6DD0D0969F/107087/2008_032_A_ENG.pdf

56. Boeing Statistical Summary to 2021. https://www.boeing.com/resources/boeing- dotcom/company/about_bca/pdf/statsum.pdf

57. IATA CFIT Report. https://www.iata.org/contentassets/06377898f60c46028a4dd3 8f13f979ad/cfit-report.pdf

58. https://www.skybrary.aero/sites/default/files/bookshelf/5842.pdf. Accessed November 2022.

59. https://safetyfirst.airbus.com/app/themes/mh_newsdesk/documents/archives/ airbus-ap-fd-tcas-mode.pdf. Accessed November 2022.

CHAPTER 7

Airport Safety

Introduction

Although it is not true for all types of aviation, the commercial aviation operations discussed in this book predominantly commence and terminate at airports. The design of airports and the operations that support commercial flights must all be structured around safety. Airports play host to a diverse collection of complex operations each with its own set of safety issues which will be explored in this chapter. It starts by providing an overview of the certification process and design constraints for airports. It then explores the hazards presented in ground-based operations, before addressing the potential risks faced by flight operations.

7.1 The Passenger Experience

However, before discussing the technical aspects of the airport, it is worth reflecting on the passenger's interaction with the airport. For most commercial airlines, passengers are the reason that the service exists and yet most passengers will see very little of what goes on behind the scenes, and the importance of safety in those operations.

Airports have changed significantly over the years. For example, in the early 1960s, passengers for British European Airways (BEA) flights from Heathrow airport would check in at the West London Air Terminal in Kensington, run by the airline, and hand over their luggage, before being taken by bus to Heathrow airport. Today, airports are the "public interface" for commercial aviation, and they have their own business models. Clearly, airports charge airlines for the services they provide (partly by charging "landing fees") but with regard to passengers, airport terminals are an essential opportunity to raise revenue. Many travelers will have disposable income and so many airport terminals have become a home to luxury brands, meaning airports work hard to maximize footfall and revenue. Having passed through security (to be "airside"), some airports gently "force" passengers to walk through Duty Free to get to their gate. Some have signage on which the "Gate will be announced at…" serves two purposes—it allows operational flexibility on allocating a gate, and it keeps passengers in the main concourse, beside the shops, for longer. This means that many passengers are kept separated from the airport processes. On arriving at the gate, passengers may see some of the work going on to prepare the aircraft but to understand the complexity involved, it is important to appreciate the numerous different processes that are underway.

7.2 "Turning Round" an Aircraft

The reality of aviation economics is that aircraft are only making money when they are flying. This means that, within some constraints, aircraft are generally "turned around" (the process of completing one flight and preparing for another) as quickly as possible to get them back flying. This is particularly true in the low-cost, short-haul market where some airlines have perfected a 20-minute turnaround, helped by keeping the same cabin and flight crew for the next leg. Many of us will have seen a low-cost airline arrive, the passengers deplane, their (minimal) hold baggage is unloaded, maybe some fuel is added, and after a quick security check and tidy-up, the new passengers are welcomed onboard. Some low-cost carriers prefer to use two sets of mobile steps to avoid the cost, complexity, and potential delay of waiting for ground staff to position an air bridge.

Long-haul flights generally take longer to turnaround (typically 2 hours minimum) as there is more that must be done every time including: deplaning passengers (on the left of the aircraft); security checks; offloading cargo and bags; performing any necessary maintenance; cleaning; emptying toilets; refueling; recatering and loading baggage and cargo (on the right of the aircraft); loading passengers; and (perhaps) de-icing. However, long-haul flights are also generally more impacted by scheduling than short-haul flights are. It may be feasible to turn a flight around in 2 hours or less, but if that then has the flight arriving at a transatlantic destination at say, 0300 (assuming the airport is even able to accept flights at that time), it may not be a popular flight!

7.3 Airport Design and Certification

In commercial fixed-wing aviation, it is a given that the flight will originate from, and travel to, an airport of some kind. The Boeing Statistical Summary[1] covering accidents from 2011 to 2020 suggests that 13% of accidents occur on takeoff or initial climb, and 54% on final approach or during landing. Including the 8% of fatal accidents that occur during ground operations (taxi, tow, etc.) suggests that 75% of accidents occur at, or close to, airports. For this reason, it is imperative that as much as possible is done to entrench safety in the design and operation of the airport.

7.3.1 ICAO Standards and Recommended Practices

The underlying ICAO SARPS related to airports are contained in Annex 14.[2] Volume I deals with *Aerodrome Design and Operations* with guidance in areas including aerodrome data, physical characteristics, obstacle restriction, visual aids, and aerodrome services. Volume II concerns *Heliports*. The Annex is supplemented by ICAO Doc 9157[3] *Aerodrome Design Manual* which comes in six parts dealing with:

- Part 1 – Runways
- Part 2 – Taxiways, Aprons, and Holding Bays
- Part 3 – Pavements
- Part 4 – Visual Aids
- Part 5 – Electrical Systems
- Part 6 – Frangibility

7.3.2 Airport Certification

The regulations around airport certification perhaps have greater diversity than those for aircraft certification, due in part to the fact that airports (unlike aircraft) can be assured to remain in their country of certification. We will consider certification requirements in the United States and in Europe.

7.3.2.1 Certification in the United States

The Federal Aviation Act of 1958 was broadened in 1970 to authorize the FAA administrator to issue operating certificates to certain categories of airports serving air carrier aircraft and also barred any person from operating an airport, or any air carrier from operating in an airport, that did not possess a certificate if it was required.

The relevant certification requirements are defined in 14 CFR Part-139.[4] To be certified by the FAA, airports are required to meet certain standards for airport design, construction, maintenance, and operations as well as firefighting and rescue equipment, runway and taxiway guidance signs, control of vehicles, management of wildlife hazards, and record-keeping. Under FAR Part-139, the FAA classifies US airport as follows:

- **Class I Airport**—An airport certificated to serve scheduled operations of large air carrier aircraft that can also serve unscheduled passenger operations of large air carrier aircraft and/or scheduled operations of small air carrier aircraft.

- **Class II Airport**—An airport certificated to serve scheduled operations of small air carrier aircraft and the unscheduled passenger operations of large air carrier aircraft. A Class II airport cannot serve scheduled large air carrier aircraft.

- **Class III Airport**—An airport certificated to serve scheduled operations of small air carrier aircraft. A Class III airport cannot serve scheduled or unscheduled large air carrier aircraft.

- **Class IV Airport**—An airport certificated to serve unscheduled passenger operations of large air carrier aircraft. A Class IV airport cannot serve scheduled large or small air carrier aircraft.

- **Joint-Use Airport**—Means an airport owned by the US Department of Defense, at which both military and civilian aircraft make shared use of the airfield.

Scheduled large air carrier aircraft require a Class I certified airport, of which there are currently more than 400 in the United States.[5] An airport that meets FAR Part-139 criteria is issued an Airport Operating Certificate (AOC).

Certificated airports must maintain an *Airport Certification Manual* (ACM) that details operating procedures, facilities, and equipment and other appropriate information. 14 CFR 139.203, supported by FAA Advisory Circular AC 150/5210-22,[6] defines the contents that the ACM of a Class I certified airport must contain:

1. Lines of succession of airport operational responsibility

2. Each current exemption issued to the airport from the requirements of this part

3. Any limitations imposed by the Administrator

4. A grid map or other means of identifying locations and terrain features on and around the airport that are significant to emergency operations

5. The location of each obstruction required to be lighted or marked within the airport's area of authority

6. A description of each movement area available for air carriers and its safety areas, and each road described in § 139.319(k) that serves it

7. Procedures for avoidance of interruption or failure during construction work of utilities serving facilities or NAVAIDS that support air carrier operations

8. A description of the system for maintaining records, as required under § 139.301

9. A description of personnel training, as required under § 139.303

10. Procedures for maintaining the paved areas, as required under § 139.305

11. Procedures for maintaining the unpaved areas, as required under § 139.307

12. Procedures for maintaining the safety areas, as required under § 139.309

13. A plan showing the runway and taxiway identification system, including the location and inscription of signs, runway markings, and holding position markings, as required under § 139.311

14. A description of, and procedures for maintaining, the marking, signs, and lighting systems, as required under § 139.311

15. A snow and ice control plan, as required under § 139.313

16. A description of the facilities, equipment, personnel, and procedures for meeting the aircraft rescue and firefighting requirements, in accordance with § 139.315, § 139.317 and § 139.319

17. A description of any approved exemption to aircraft rescue and firefighting requirements, as authorized under § 139.111

18. Procedures for protecting persons and property during the storing, dispensing, and handling of fuel and other hazardous substances and materials, as required under § 139.321

19. A description of, and procedures for maintaining, the traffic and wind direction indicators, as required under § 139.323

20. An emergency plan as required under § 139.325

21. Procedures for conducting the self-inspection program, as required under § 139.327

22. Procedures for controlling pedestrians and ground vehicles in movement areas and safety areas, as required under § 139.329

23. Procedures for obstruction removal, marking, or lighting, as required under § 139.331

24. Procedures for protection of NAVAIDS, as required under § 139.333

25. A description of public protection, as required under § 139.335

26. Procedures for wildlife hazard management, as required under § 139.337

27. Procedures for airport condition reporting, as required under § 139.339

28. Procedures for identifying, marking, and lighting construction and other unserviceable areas, as required under § 139.341

29. Any other item that the FAA Administrator finds is necessary to ensure safety in air transportation

The Airport Emergency Plan (point 20 above), is an integral part of the ACM although it may be published and distributed separately to each user or airport tenant. The plan must contain sufficient detail and provide sufficient guidance on the procedures for prompt response to a range of emergencies including *inter alia* aircraft accidents; bomb incidents; sabotage; hijackings; major fires; natural disasters such as floods, tornadoes, and earthquakes; and power failures. The Airport Emergency Plan must be reviewed at least annually, and for Class I airports there must be a full-scale exercise of the plan every 3 years.

See FAA AC 150/5200-31C for more details.

7.3.2.2   Airport Certification in Europe

In Europe the relevant regulation is contained in CS-ADR-DSN *Certification Specifications and Guidance Material for Aerodrome Design*.[7] Part-ADR.OPS[8] describes the organizational requirements and Part-ADR.OP the operational requirements for organizations running an airport. Many of the requirements in US regulation are mirrored in the European regulation, but with different detail. For example, in ADR.OR.E.005 there is a requirement to have the equivalent of an ACM described above but it is called an Aerodrome Manual with the contents of that manual defined in AMC3. That content has some similarities but it is not identical.

7.3.3   Airport Site Selection

Site selection for a new airport requires a careful balance between a range of aeronautical, air-transport/commercial, practical, and environmental requirements. This section examines some of these constraints.

7.3.3.1   Environmental and Meteorological Considerations

Part of the consideration when designing an airport is meteorological. The typical prevailing weather such as winds, including any natural turbulence, gusting, or windshear; fog; snow; ice and disposition to microbursts will all need to be accounted for. Once a site is selected, these factors will also affect runway placement. Environmental considerations require that the airport should minimize the destruction of fauna and flora, and ecology of the surroundings.

Restrictions on maximum allowable slopes for runways, which are dictated by aircraft performance on landing and takeoff, require flat or very gently undulating terrain. Grading, drainage, and soil characteristics play an important role in the construction and maintenance of airports which in turn influences the site selection. Runoff from rain, snow, mixed with de-icing chemicals used on aircraft and airfield pavements and cleaning chemicals used in aircraft maintenance pose contamination issues for groundwater and surface water bodies like lakes and rivers. As a result, some airports may be required to provide at least primary treatment of all runoff discharges and can incur discharge restrictions on the nature of the chemicals that can be used.

Most governments require environmental impact assessments including evaluations of population relocation, changes in employment patterns, and distortion of existing regional land use and transportation planning before granting permission to build airports.

7.3.3.2   *Noise Restrictions*

Aircraft noise at the airport represents a significant environmental impact. Airports manage aircraft noise in a number of ways that include limiting hours of operation through the use of night curfews, restricting or banning noisy aircraft, selecting runways to limit or distribute noise more evenly over the community, designating approach and departure routes over less-populated areas and encouraging airlines to modify approach and departure gradients and operating procedures to reduce engine noise over highly populated areas. The airport in Nice, France is highly constrained by the need to minimize noise so as not to upset its very influential neighbors. These noise restrictions can mean that crews must delay configuring the aircraft for landing, which in turn can contribute to the risk of flying unstable approaches.

7.3.3.3   *Surrounding Area*

From an aeronautical standpoint, an airport should have land that is flat and large enough to accommodate the runways and other facilities and that the surrounding area be free from obstructions to aircraft such as mountains, powerlines, and tall buildings. To satisfy passenger requirements, airport sites must be sufficiently close, or sufficiently well served by transport links, to be accessible to their users. However, there is an almost inevitable tension between the convenience of an airport being accessible from a large city and the risk posed by an airport being close to that city. Sadly, the new trend of naming airports with their city, regardless of how far from that city they are, doesn't really satisfy the problem of convenience—the renamed *Frankfurt Hahn* airport is a former military base situated 75 miles from Frankfurt!

It is worth noting that, when Heathrow first started operating as a civil airport in 1946, London did not extend in the way it does now, but during the expansion that followed, it was important to remember the reason for retaining clear approach and departure paths. In the G-YMMM (Boeing 777 Shanghai to Heathrow) accident described in Section 6.1.8, the aircraft landed short of the runway inside the airport perimeter. Had that perimeter been closer to the runway, had the aircraft landed even shorter, or had the surrounding obstacles been higher, there was the potential for significant casualties, both on the aircraft and on the ground.

Unfortunately, even keeping the immediate area clear is not sufficient. On October 4, 1992, a Boeing 747 operating as El Al Cargo Flight 1862 had made a scheduled stop in Amsterdam. After taking off from Schiphol Airport, climbing through 6500 feet engines 3 and 4 separated from the right wing with their pylons. The crew declared an emergency and started to return to Schiphol. However, given the enormous asymmetric forces created with the loss of both engines on the right wing, the aircraft became uncontrollable and crashed into an 11-floor apartment building in the Bijlmermeer, a suburb of Amsterdam, approximately 13 kilometers east of Schiphol Airport. The 4-flight crew and 39 people on the ground were killed. Although the accident didn't occur particularly close to the airport, it is an inevitable (albeit small) risk of having airports near to large cities that they will be regularly overflown by aircraft.

7.3.3.4   *Unusual Airports*

Some airports have unusual siting or configurations, often due to necessity, which can in turn lead to unusual approach or departure requirements, often exacerbated by noise and emission restrictions. The examples below show what airport certification can mean when mixed with a range of physical, environmental, economic, and political requirements.

Saint Helena	For a significant number of days per year, the airport can be almost unusable due to extreme low-level turbulence/windshear.
Gibraltar	Has similar traits, albeit presenting a less severe problem than St Helena. However, when the wind blows over the Rock of Gibraltar, approaches are suspended.
Kai Tak, Hong Kong	Now closed, this airport had a challenging approach to Runway 13 that had to be flown manually over the city. Pilots would aim for "Chequerboard Hill,' make a 47-degree right turn then line up for the runway with 2 miles to go at 650 feet.
St Maarten	The runway is positioned beside the sea with aircraft landing and taking-off meters from the beach, meaning jet blast is an issue. Figure 2-6 shows an approach to St. Maarten.
Narita, Japan	This airport has a home and farm in the middle after the owners refused to move out when the airport expanded.
JFK, New York	The "Canarsie approach" requires a steep turn around 800 feet above ground level which makes meeting stable approach criteria difficult.

7.4   Ground Operations Risk

Airport ground operations are complex and diverse, with various hazards across different activities. These activities will be discussed under the following headings:

- Ramp operations
- Specialized airport services
  - Aviation fuel handling
  - Aircraft rescue and firefighting (ARFF)
  - De-icing and anti-icing
  - Wildlife and Foreign Object Debris (FOD)
- Hangars and maintenance
- Airport terminal buildings

It is interesting to note that ICAO Doc 9859, the *Safety Management Manual*[9] which helps aircraft operators to implement Safety Management Systems in their organization, is explicitly considered to apply equally well to aerodrome operators.

7.4.1   Occupational Safety

Airports are the place of work for large numbers of people and, as such, need to comply with local occupational safety regulations and requirements from bodies such as The Occupational Safety and Health Administration (OSHA) in the United States, their European counterpart EU-OSHA, or the Health and Safety Executive (HSE) in the United Kingdom. There are numerous areas of focus that are relevant to airports including:

- Fire
- Manual handling

- Lighting
- Working at height/fall protection
- Compressed gas
- Vehicle safety
- Hot work (welding, cutting brazing)
- Noise
- Vibration
- Ventilation
- Chemical/hazardous/toxic materials handling
- Electrical installation
- Heat and cold stress

Occupational health requirements are far-reaching and location specific, and as such are beyond the scope of this book. The OSHA regulations[10] and the EU-OSHA regulations[11] give further details. However, it is important to note that although airports and aviation represent a very different working environment (compared to an office) this does not absolve the airport operator of their occupational health requirement. In fact, the opposite is true—the larger potential hazards that are encountered call for greater safety management.

7.4.2   Ramp Operations

The ramp or apron of the airport is the area designated for parking, loading, and unloading aircraft. As described in Section 7.2, an aircraft "turnaround" can be a hectic operation and the ramp area sees a diverse collection of high-paced activities that involve aircraft, vehicles, and individuals working in close proximity to one another. Some of these activities include:

- Moving the aircraft—by marshaling (guiding) the aircraft in under its own power, chocking to stop it moving, or towing / pushing the aircraft back from the gate
- Aircraft refueling (see Section 7.4.3)
- Aircraft servicing—restocking catering, cleaning, emptying lavatories, etc.
- Baggage and cargo loading and unloading
- Connecting and disconnecting from ground power
- Routine maintenance and checks
- Aircraft de-icing (see Section 7.4.5)

This is a busy area, often working on tight timescales. Figure 7-1 shows an A380 being serviced—the double decks call for two catering trucks and two airbridges. When introduced, the A380 required adaptations to be made at airports, with some building dedicated gates for the type. The orange trolley and folding cable run is providing ground power to the aircraft. Figure 7-2 shows this connection on a different aircraft type.

FIGURE 7-1   An A380 with two catering trucks, a cargo truck, and two airbridges.
(Source: M Salam[12])

FIGURE 7-2   A ramp worker connects an aircraft to ground power. (Source: J Goh[13])

7.4.3 Aviation Fuel Handling

Fuel handling is an important safety issue not only to the fuelers but also to other airport personnel, the traveling public, and the operation of the aircraft. Failure to adhere to safe operating procedures when fueling aircraft and/or transporting fuel from one location to another on the airport can result in major disasters. This potential for harm is well recognized even though millions of gallons of aircraft fuel are handled each year without major incidents for the most part. The basic types of aviation fuel are aviation gasoline (avgas for general aviation piston operations), Jet A or Jet B (commercial jet uses) fuels, and JP series (military jet uses) fuels. Figure 7-3 shows a ramp refueling operation. Fueling operations present a number of hazards including:

- Health hazards to fuelers
- Fuel contamination
- Explosions and fires during fueling or fuel transfer, paying attention to sources of ignition such as:
  - Static electricity
  - Sparks
  - Explosions and fires during fuel tank repair
  - Miscellaneous including smoking
- Hazards from spills

7.4.4 Aircraft Rescue and Firefighting

As we have seen in Chapters 5 and 6, the effective operation of Airport Rescue and Fire Fighting (ARFF) can be crucial to the outcome of any accident on or near an airport.

FIGURE 7-3 Aircraft refueling operations. (Source: Wikimedia Commons)

Figure 7-4 Typical ARFF vehicle. (Source: Wikimedia Commons)

Aircraft rescue services and firefighting equipment are based on an index as determined by the length of the air carrier aircraft and the frequency of departures. Figure 7-4 shows an example of a typical ARFF vehicle. Part 139 (14 CFR 139.319) requires that within 3 minutes of the alarm, at least one required ARFF vehicle must be able to reach the midpoint of the farthest runway serving air carrier aircraft (or a comparable point) and begin applying an extinguishing agent. Within 4 minutes from the time of alarm, all other required vehicles must be able to reach the same point and also begin applying extinguishing agent.

There are a large number of FAA Advisory Circulars describing ARFF operations[14] covering aspects from vehicle markings to driver-enhanced vision systems. However, as aircraft energy storage changes, so ARFF will have to change to keep step. The most obvious examples are the introduction of high-energy density batteries, which can ignite fires that can be extremely difficult to control, and the introduction of hydrogen as a fuel. Although it is less energy dense than current kerosene fuels, hydrogen burns differently, tending to evaporate, rise and burn off very quickly in a flash fire as well as burning with a very pale flame that can be almost invisible in daylight.

7.4.5 De-icing and Anti-icing

Ice and snow on airfoil, control, and sensor surfaces can have serious repercussions on the safe operation of the aircraft, including loss of lift, difficulty controlling the aircraft, and erroneous aircraft data. Hence, when freezing or near-freezing conditions are likely to be present, the aircraft should not be allowed to take off before being sprayed with de-icing fluid, if contamination is present, or anti-icing fluid, if contamination may occur prior to becoming airborne. Once in the air, commercial aircraft have systems that prevent the accumulation of most types or icing, but those systems

FIGURE 7-5   De-icing operations prior to takeoff. (Source: P Cerutti[15])

are inadequate on the ground due to several unavoidable design criteria. Figure 7-5 shows a de-icing or anti-icing operation underway, spraying protective fluid onto the wing. Sometimes a de-icing operation is performed first to remove contaminants followed by an anti-icing operation if precipitation is still occurring that could cause future airframe contamination. Alternatively, sometimes de-icing occurs without anti-icing and vice versa. Often both operations are described under the term "de-icing" for simplicity. The amount of time that de-icing/anti-icing fluid will remain effective is termed the Holdover Time (HOT) and depends on the fluid being used and the weather conditions.

Numerous accidents have occurred due to attempted takeoffs with contaminated surfaces, and it is important to realize that such accidents can happen outside of winter. For aircraft performing long flights at altitude, unused fuel can be cooled to low temperatures causing moisture to form ice on the wing during turnaround, even at temperatures above freezing.

Application crews typically work on elevated platforms (baskets at the end of booms) which presents a working at height hazard that must be mitigated. The fluid also represents a hazard if inhaled or ingested. Finally, damaging the aircraft, either by contact or application of fluid to inappropriate areas such as static ports, pitot heads, etc., is also a significant risk during de-icing operations.

The FAA has recently adopted the European de-icing standard called the Standardized International Aircraft Ground Deice Program (SIAGDP). FAA Advisory Circular 120-60B[16] provides primary FAA guidance for the development of a ground de-icing program. Likewise, in Europe, CAT.OP.MPA.250 *Ice and other contaminants— ground procedures*[17] provides guidance.

7.4.6   Hangars and Maintenance Shops

Hangars present a range of hazards to the airport, most of which fall under fire, building, and occupational health regulations. They can be broadly grouped as follows:

Category	Examples
Walkways and Working Surfaces	Working at height, slips, trips, and falls
Electrical Hazards	Adequate earthing
Compressed Gas, Flammables, and Hazardous and Toxic Substances	Hydraulic systems, oleos, oxygen generators, tires, chemicals
Fire	Extinguisher placement, flammable and combustible material storage
Material Handling Equipment	Cranes, hoists, slings, fork lift trucks
Miscellaneous Hazards	Noise, welding, cutting, vibration from tools

The considerations are little different from a hangar or maintenance shop situated away from the airport, although some issues, such as noise, may be magnified compared to other more remote locations.

7.4.7   Wildlife and Foreign Object Debris (FOD)

Airport operators have responsibility for wildlife management and for establishing means and procedures to minimize the risk of collisions between wildlife and aircraft. This can include active control measures such as bird-scaring as well as broader passive measures such as grass-cutting and planting by local landowners. Airport operators are also required to issue Notice to Airmen (NOTAMs) on hazards including the presence of wildlife on airport property. The management of wildlife can also create a tension between what is best for safety and what is best in terms of an airport's environmental sustainability.

FOD[i] describes objects which can represent a hazard to an aircraft. As such, FOD can represent a significant risk to all aircraft and can come from multiple sources. In July 2000, a Concorde aircraft crashed shortly after takeoff after running over a piece of debris on the runway, causing a tire to disintegrate and rupture a fuel tank and a fire to break out. All 109 people on board were killed as well as 4 people on the ground. Airport operators are obliged to have a system of FOD control as part of their oversight and safety programs. Airport operators are also obliged to inspect the movement area for the presence of FOD and the status of wildlife as well as the status of visual aids and current surface conditions, with the specific frequency dependent on the category of airport.

7.4.8   Airport Terminal Buildings

FAA planning and design guidance for airport terminals are included in AC 150/5360-13[18] including gate and apron planning. Much of the planning involved is around efficiency of movement whether for vehicle traffic arriving at the airport, flow of people in the terminal building or gate access for aircraft. However, from a safety perspective, all of the usual occupational and public health and safety considerations apply as in other buildings serving the public.

The big difference is that airports are designed to control passenger flows and are often quite unfamiliar buildings for their users, especially in fires or in the event of a

---

[i] Usually pronounced as a word to rhyme with "rod" rather than as the individual letters.

terrorist attack. This can create problems for evacuation as was highlighted by the fire at Dusseldorf Airport in 1996 where 17 people died including 9 passengers who became trapped in a self-service lounge. They were unfamiliar with the escape route and when they telephoned staff for help, the people they spoke to were unaware of the lounge location and could not provide any assistance. Construction work is near constant in large airport buildings and their work often needs to be conducted while the terminal is still in operation. This requires careful management from both a safety and security point of view.

7.5   Flight Operational Risks

Until this point, the risks we have described have mostly been related to the operation of the airport and the provision of services. Of course, the purpose of an airport is to support air operations and so, in this section, we shall consider some of the risks presented to, and by, flight operations.

7.5.1   Runway Incursions

A runway incursion can be defined as "Any occurrence at an aerodrome involving the incorrect presence of an aircraft, vehicle or person on the protected area of a surface designated for the landing and take-off of aircraft." The "incorrect presence" described here might be an intentional or unintentional failure to comply with an ATC clearance, an incorrectly issued clearance, or possibly miscommunication. A runway incursion is a very serious safety issue and in 2019, the FAA reported[19] more than 1750 reported runway incursions.

The typical airport surface environment is a complex system of markings, lighting, and signage coupled with layouts that vary by airport. Airport ground operations are managed by air traffic controllers, pilots, and airport personnel. These groups jointly manage aircraft and vehicular traffic, and monitor and maintain conditions of runways, taxiways, aprons, signs, markings, and lighting. As we will see in Chapter 8, controllers serve as the hub for the entire operation by issuing clearances, instructions, and advisories to pilots and other ground vehicle operators. In this way, controllers guide all traffic that operates on the aprons, taxiways, and runways.

Pilots and controllers rely primarily on visual feedback to maintain situational awareness and separation, hence the need for changed safety margins during reduced visibility operations or low visibility operations (RVOP and LVOP respectively). Vehicle operators require specific airside driving permits and are required to follow controllers' instructions to maneuver, especially if they have to cross runways. It is within this environment that large numbers of individuals with varying levels of experience, training, and language proficiency must concentrate on performing their tasks against a backdrop of intense radio communications for coordinated actions and procedures that are needed for smooth and safe operations. Specialized signs and surface markings together with pilots new to a particular airport pose additional difficulties. Given the complexity and intensity of the operation, it is easy to understand how even well-trained, highly conscientious individuals can make errors, especially when faced with unusual or unexpected situations.

7.5.2   Classification and Prevention

The FAA[20] has produced guidance material that classifies the source of runway incursions into three categories as indicated below:

- *Operational Incidents*—Action of an Air Traffic Controller that results in: Less than required minimum separation between two or more aircraft, or between an aircraft and obstacles (vehicles, equipment, personnel on runways) or ATC clearing an aircraft to take off or land on a closed runway.

- *Pilot Deviations*—Action of a pilot that violates any Federal Aviation Regulation. Example: A pilot crosses a runway without a clearance while en route to an airport gate.

- *Vehicle/Pedestrian Deviations*—Pedestrians or vehicles entering any portion of the airport movement areas (runways/taxiways) without authorization from air traffic control.

It is important to highlight that these "deviations" do not necessarily imply intent. For example, a pilot deviation might include crossing a runway when the flight crew are unaware that is what they are doing. This is clearly an undesirable situation, and no less dangerous for the lack of intent, but it highlights the hazard posed by, say, loss of situational awareness rather than attributing it to "bad actors."

In order to promote global harmonization and effective data sharing, ICAO and the FAA have established a standard classification scheme to measure the severity of runway incursions, outlined in Figure 7-6.

ICAO Doc 9870,[21] the *Manual on the Prevention of Runway Incursions* highlights contributory factors and recommendations for prevention. This was followed in 2017 when ICAO held a second Global Runway Safety Symposium (GRSS) and issued a Global Runway Safety Action Plan (GRSAP).[22] The GRSAP presents a comprehensive list of contributing factors (grouped as latent conditions, threats and active human performance) and a list of actions for mitigation of those factors.

The FAA have produced a National Runway Safety Plan (NRSP) covering the period 2021–2023.[23] This describes a number of initiatives to address the problem of runway incursions, an issue also addressed through the FAA Runway Incursion Reduction Program (RIRP). One measure was the Runway Incursion Mitigation (RIM) program which identified airport "hotspots" where more incursions were occurring or likely. Mitigation is then made by *site-specific enhancements including taxiway reconfigurations and changes to lighting, markings, and aircraft operations.*

Another approach to addressing the issue is through the use of technology. The FAA has laid out a technology roadmap that includes:

ASDE-X	Airport Surface Detection Equipment, Model X (described in Chapter 8)
RWSL	Runway Status Lights, red lights embedded in the pavement to signal potentially unsafe situations
ASSC	Airport Surface Surveillance Capability, using data fusion to show aircraft and vehicle locations
RIPSA	Runway Incursion Protection by Situational Awareness using noncooperative surveillance and indicator lights to the pilot

Available Reaction Time	Evasive of Corrective Action	Environmental Conditions	Speed of Aircraft and/or Vehicle	Proximity of Aircraft and/or Vehicle

Runway Incursion categories

Category D	Category C	Category B	Category A	Accident
Incident that meets the definition of runway incursion such as incorrect presence of a single vehicle/person/aircraft on the protected area of a surface designated for the landing and take-off of aircraft but with no immediate safety consequences.	An incident characterized by ample time and/or distance to avoid a collision.	An incident in which separation decreases and there is a significant potential for collision, which may result in a time critical corrective/ evasive response to avoid a collision.	A serious incident in which a collision was narrowly avoided.	An incursion that resulted in a collision

FIGURE 7-6 Runway incursion severity classification. (Source: FAA)

SAFRE	Situational Awareness for Runway Entrances, combining speech and cooperative surveillance
RIDS	Runway Incursion Devices

7.5.3   Runway Incursion Case Studies

As described above, runway incursions have taken a variety of forms. In this section we will look at some case studies highlighting different types of runway incursions. These are very brief descriptions showing how the airport can form part of the accident sequence—interested readers are encouraged to read the full report for the details of each event.

7.5.3.1   The "Tenerife" Accident

On March 27, 1977, a KLM flight from Amsterdam and a PanAm flight from America, along with others, were diverted to Tenerife due to a bombing at their intended destination (Las Palmas). Las Palmas reopened a few hours later and the PanAm flight was able to leave straight away. However, congestion at the airport (due to the diversions) meant that the PanAm flight had to wait for 2 hours behind the KLM aircraft while its passengers reboarded and it was refueled.

The KLM flight was then cleared to backtrack (taxi against the direction of take-off/landing) to the end of runway 12 and turn 180 degrees to prepare for takeoff. The PanAm flight was cleared to follow the KLM flight backtracking down runway 12, but to leave at the third taxiway (to line up behind the KLM aircraft) and report leaving.

The following exchange then took place between the KLM aircraft and the tower[24]:

KLM 4805:	*The KLM four eight zero five is now ready for takeoff and we are waiting for our ATC clearance.*
Tower:	*KLM eight seven zero five you are cleared to the Papa Beacon, climb to and maintain flight level nine zero, right turn after takeoff, proceed with heading four zero until intercepting the three two five radial from Las Palmas VOR*
KLM 4805:	*Ah—Roger sir, we are cleared to the Papa Beacon, flight level nine zero until intercepting the three two five. We are now (at takeoff)*
Tower:	*OK…Stand by for takeoff, I will call you*

In this exchange, the KLM flight was now ready for takeoff and were given instructions for their departure but were *not* given a clearance to takeoff because the PanAm flight had not yet vacated the runway. However, the KLM flight misinterpreted the exchange, they repeated the instructions and announced, "We are now at takeoff" and began their takeoff roll. This was the point at which the KLM flight unintentionally breached their clearance and committed the runway incursion.

At that point the tower, knowing the PanAm flight was still on the runway, instructed "OK…" and then added "Stand by for takeoff…I will call you" but this partially coincided with a message from the PanAm aircraft "No, uh…and we're still taxiing down the runway" which only caused a shrill noise in the KLM cockpit as the two radio transmissions merged. Further communication between the PanAm aircraft and the tower ("report runway when clear," "ok will report when we're clear") was heard by the KLM flight engineer who questioned whether the PanAm flight was clear of the runway, but the captain was certain they were. The KLM flight saw the PanAm aircraft on the runway and tried to climb over it as it turned and applied full power. However, the KLM aircraft impacted the PanAm fuselage, before crashing 150 m further down the runway.

Reading the exchange between the tower and the aircraft, it is possible to see where the confusion occurred. There is ambiguity around the term "clearance" and the controller's use of the word "ok." It is for this reason that aviation uses standard phraseology, and following this accident, the use of the word "clearance" was changed to only refer to takeoff. All 248 people onboard the KLM flight were killed, and 335 people onboard the PanAm aircraft died, totaling 583 lives lost. This disaster still represents the greatest loss of life in a single aviation accident.

7.5.3.2   Boeing 747 in Taiwan

On October 31, 2000, a Boeing 747 registered 9V-SPK, which was operating Singapore Airlines flight SQ006, crashed on takeoff from Chiang Kai-Shek airport in Taiwan with 20 crew and 159 passengers on board. The aircraft had attempted to take off from a partially closed runway and was destroyed by its collision with construction equipment including concrete barriers, and the subsequent fire. The accident resulted in 83 fatalities, 39 serious injuries, and 32 minor injuries.

In the report by the ASC,[25] the Taiwanese investigation agency, observed that at the time of the accident, heavy rains and strong winds from a typhoon prevailed. The flight crew were aware that the Runway 05R was partially closed but was available for taxi and that their clearance was for runway 05L. However, they did not review the layout of the airport sufficiently, and they lined up on runway 05R in visibility of about 800 meters. Exacerbating factors included nonstandard airfield lighting and the fact that the runway lights for the closed runway remained on, due to them being on the same circuit as the taxiways.

7.5.3.3   Boeing 737 at Schiphol

It may be tempting to assume from the case study above that confusion over runways and taxiways only occurs when the layout is modified in some way, or that this problem has been solved. This case study shows that neither is the case. The Dutch Safety Board (DSB) reported a serious incident of a similar, but different, type.[26] While not technically a runway incursion, it has many similar characteristics, including loss of situational awareness.

On September 6, 2019, a Boeing 737-800, registration PH-HSJ, was operating a passenger flight from Schiphol Airport in the Netherlands to Chania, Greece. The layout of the relevant portion of Schiphol Airport is shown in Figure 7-7. The crew was taxiing the aircraft in the dark in a northerly direction on Taxiway C, on the way to Runway 18C in order to take off. While taxiing, the crew received a takeoff clearance for runway 18C.

The flight crew then turned left twice onto Taxiway D (mistakenly) and commenced the takeoff. ATC noticed the aircraft starting to accelerate on the taxiway and instructed the crew to stop immediately. The crew rejected the takeoff and taxied back to the beginning of Runway 18C, after which the airplane took off uneventfully.

The DSB report identified that the flight crew failed to recognize that they were on a taxiway from the cues supplied to them by the centerline lights and markings. Other cues, such as signs indicating Runway 18C enhanced their (incorrect) perception. The flight crew's decision to continue the flight after the rejected take off also meant that the CVR recording of the event could not be recovered. This serious incident bears a strong resemblance to the accident to SQ006—it was fortunate for everyone concerned that the controller noticed and was able to stop the crew from continuing.

Figure 7-7 The taxi path and rejected take-off path taken by the aircraft (Source: Authors; Adapted from the DSB report.)

7.5.3.4 *Runway Incursion by an ARFF Vehicle*

As this book was being prepared, on November 18, 2022, an Airbus A320 operated by LATAM which was taking off from Lima-Jorge Chávez airport in Peru struck an Airport Rescue and Firefighting vehicle which had entered the runway.[27] The landing gear appears to have collapsed and a fire soon broke out. The 108 people onboard the aircraft evacuated successfully but two firefighters who were in the vehicle were killed and a third was reported to be seriously injured. Initial indications are that the response was part of a preplanned exercise but there was a temporary loss of situational awareness. Regardless of the cause, the accident shows the potential devastation that even a single-vehicle runway incursion can cause.

7.5.4 Runway Excursions

A different form of risk in flight operations is that of a runway excursion. The term "runway excursion" describes a situation where the aircraft leaves the runway surface in one of two ways. The first is overrunning the end of the runway, either by failing to become airborne on takeoff, or failing to stop before the end of the runway when landing. The second is by departing the side of the runway on takeoff or landing, commonly known as "veer off.' Many of the documents described for runway incursion, such as the FAA NRSP 2021-2023, also deal with the issue of runway excursion. The Flight Safety Foundation has also produced a Runway Excursion Risk Reduction (RERR) toolkit[28] incorporating their report on *Reducing the Risk of Runway Excursions*.

7.5.4.1 *Runway Overrun*

Runway overruns on landing usually occur either by the aircraft carrying too much energy into a landing (including tailwind), insufficient available runway due to long landing, system failure such as loss of thrust reverser, or insufficient retardation due either to braking action or flight crew actions. The condition of the runway surface plays an important role, especially if it becomes contaminated by rubber (from tire skid marks), water, ice, or snow. Airports need to routinely monitor the friction coefficient and may choose to groove or scrub the runway to aid braking. We have already discussed the scenario of excess energy in Chapter 6 which described the Runway Overrun Protection System which aims to alert pilots to carrying too much energy into a landing, or insufficient runway length. "Floated" landings, (where aircraft touch down well past the threshold), reduce the available runway for retardation.

Failure to become airborne, causing runway overrun on takeoff can occur for a range of reasons, including insufficient thrust; incorrect aircraft configuration (e.g., flaps and slats); incorrect performance calculations; tire failures; and weight/balance being outside operational limits. Similarly, a late initiation of a rejected takeoff (RTO) at greater than V_1 (the maximum speed at which a rejected takeoff can be initiated in the event of an emergency) can also result in a takeoff overrun.

7.5.4.2 Runway Veer-off

Runway veer-offs can occur on both landing and takeoff, although previous research in Europe has suggested that they occur three times as often on landing as they do on takeoff.[29] Veer-offs can occur for a number of reasons including wind gusts, asymmetric thrust, and system failures, although further research[30] in Europe identified that the three most commonly occurring causal factors in veer-off accidents were: inaccurate crew performance; wet/contaminated runway; and strong crosswinds.

7.5.5   Runway Excursion Case Studies

In this section we will look at some case studies highlighting different types of runway excursions.

7.5.5.1 Air France A340 Overrun at Toronto

One of the most well-known and dramatic runway overruns occurred to an Air France A340 in Toronto. The Canadian Transport Safety Board (the National Safety Investigation Agency) investigated this accident and published[31] the following synopsis.

On 2 August 2005, Air France 358, an Airbus A340-313 aircraft, left Paris, France, for Lester B. Pearson International Airport in Toronto, Ontario, with 297 passengers and 12 crewmembers on board. The pre-flight inspection had been performed, and the crew received the Toronto weather before departure. Thunderstorms were forecast for arrival.

On final approach, the crew received a report from an aircraft ahead of them: heavy precipitation was causing the runway to have poor traction. Approximately 320 feet above the ground, the pilots retook manual control of the aircraft for landing. The aircraft crossed the threshold of Runway 24L 40 feet above the glide slope. Due to poor visibility, they could not visually gauge where they were over the runway. They landed 3800 feet down the 9000-foot runway, well past the 1000-foot intended touchdown zone. Unable to stop in time, the aircraft slid off the far end at a speed of 80 knots, and skidded into a ravine just beyond the runway. A fire broke out, forcing the passengers to evacuate the aircraft as quickly as possible. Two crewmembers and 10 passengers were seriously injured, and the plane was destroyed, engulfed by the flames.

The accident report[32] describes the detail behind the accident which, as always, is far more complex than any simple synopsis can convey. Figure 7-8 shows the post-accident fire in progress and Figure 7-9 shows the accident site and wreckage. It is remarkable, looking at these images, that no one was killed in this accident due, in part,

FIGURE 7-8   The post-accident fire to flight AF358. (Source: TSB)

FIGURE 7-9   AF358 accident site and subsequent wreckage. (Source: TSB)

to the progress in fire resistance materials and evacuation considerations described in Chapters 5 and 6.

In 2019, the TSB published a follow-up to the investigation[33] and noted that since this accident, there have been 140 occurrences in Canada involving runway overruns, 19 of which have been the subject of a comprehensive TSB investigation. They also noted that to mitigate the risk of overruns, operators of airports with runways longer than 1800 meters must conduct formal runway-specific risk assessments and take appropriate action. They also observed the need for ICAO standard Runway End Safety Areas (RESAs) or equivalent, such as the Engineered Materials Arrestor System (EMAS) seen in the next case study.

7.5.5.2   *PSA Airlines Flight 2495*

On January 19, 2010, a PSA Airlines Bombardier CRJ 200 jet was scheduled to depart from Yeager Airport in Charleston, West Virginia, for Charlotte, North Carolina. Everything leading up to the takeoff on Runway 23 seemed normal. However, when the pilots reached the 80 knot callout, the captain realized that the flaps were not configured properly on the plane. He tried to correct the settings of the flaps, but this prompted an airplane warning. As a result, the captain decided to abort the takeoff. The plane was unable to stop before the end of the runway and was forcibly stopped 130 feet into the 455 feet long Engineered Materials Arrestor System, also known as an EMAS.

EMAS is a safety measure often used as part of a safety case for reduced RESAs. It is made of crushable concrete that collapses as the aircraft crosses it, thus stopping the aircraft's motion to prevent a crash or running off the runway. Figure 7-10 shows the PSA Airlines in the EMAS area.

All 34 passengers were uninjured, and the aircraft sustained only minor damage. Had it not been for the EMAS, the severity of the accident would have been much worse. The EMAS was constructed as part of a runway extension; Yeager airport is built on the top of a cleared mountain and there was not enough room to extend the runway the required distance since there was a steep slope that consisted of loose sediment rock.

Figure 7-10   EMAS after arresting PSA 2495. (Source: NTSB)

Figure 7-11   EMAS arrest of Boeing 737 at LaGuardia in 2016. (Source: Unknown)

In 2022, the FAA reported 18 incidents to date where EMAS systems have safely arrested an aircraft.[34] A high-profile example of EMAS in action was seen in 2016 when a Boeing 737 aircraft carrying a vice presidential candidate overran the runway at New York LaGuardia airport. Figure 7-11 shows the site after the aircraft crossed the EMAS. There had been a tailwind, an early flare, and a delay in reducing the throttles, meaning the aircraft had touched down more than halfway down the runway (4200 feet beyond the threshold). A further delay in deploying the speed brake resulted in the aircraft reaching the end of the runway at around 40 knots groundspeed.

7.5.5.3   *Veer-off Accident*

On July 13, 2012, a Gulfstream G-IV suffered a veer-off accident at Le Castellet airport in France. The accident report[35] by the BEA found that while conducting a visual approach the crew omitted to arm the ground spoilers and on touchdown, they did not deploy. The nondeployment generated a low load on the landing gear causing a temporary loss of on-ground condition, which inhibited the deployment of the thrust reversers. The crew applied a nose-down input which resulted in unusually heavy loading of the nose gear. The aircraft then deviated to the left due to a left orientation of the nose gear for an unknown reason. The crew were unable to maintain control, in part because they were untrained on the relevant procedure, and the aircraft veered off the runway, hit some trees, and caught fire. The crew were unable to evacuate the aircraft and the single firefighter was unable to bring the fire under control. The required fire protection was not ensured because one firefighter was late arriving at the airport. All three crewmembers were killed. The runway safety area for this runway was 75 meters on either side of the runway, and the trees that the aircraft struck were located 95 meters or more from the center line i.e. outside the runway safety area.

7.5.5.4   *Runway Incursion Resulting in Veer-off*

On October 8, 2001, a Boeing MD-87 was taking off in fog from Linate airport in Milan Italy with 6 crew and 104 passengers on board. Part way through its takeoff, after rotation, the MD-87 collided with a Cessna 525 with 4 people on board which had inadvertently entered the runway. The MD-87 was airborne for a short while but veered off the runway and impacted a baggage handling building, located 50 meters from the runway centerline, at around 140 knots. All 114 people aboard both aircraft were killed in the accident as well as 4 on the ground. One other sustained serious injury.

The subsequent investigation report[36] by the ANSV (the Italian NSIA) found that there was confusion between the Cessna crew and the air traffic controller as to the aircraft's position and that the "crew were not aided properly with correct publications, lights and signs to enhance their positional awareness."

The report also highlighted numerous failings around the airport operations. These included:

- No functional Safety Management System was in operation.
- The aerodrome standard did not comply with ICAO Annex 14.
- No aerodrome Operations Manual was established.
- No Quality System was established for all activity sectors.

The ANSV summarized the systemic failings saying, *"the system in place at Milano Linate airport was not geared to trap misunderstandings, let alone inadequate procedures, blatant human errors and faulty airport layout."*

7.5.6   Parked Aircraft

In some cases, even a parked aircraft can represent a hazard to an airport. On July 12, 2013, a parked and unoccupied Ethiopian Airlines Boeing 787 suffered a ground fire while on a remote stand at London Heathrow Airport. This event was not officially an accident, based on the Annex 13 definition. However, clearly the risk of an in-flight fire, in this relatively recently introduced aircraft type, justified an investigation. The fire was initiated by a fault in the lithium-metal battery in the Emergency Locator Transmitter. The aircraft suffered extensive heat damage in the upper portion of the aircraft's rear fuselage.

7.6   Conclusion

As this chapter has illustrated, there are diverse and complex activities happening simultaneously at any commercial airport. The role of the modern airport as part of the aviation safety system is now fully acknowledged with airports implementing their own SMS and hazard management across all areas of the operation. Yet, despite that maturity, we still see a significant number of incidents, such as runway incursions, every year. This reflects, in part, the point made at the start that all the fixed-wing commercial operations described here start and end at an airport. The severity of some of the accidents also highlights the inherent risk associated with takeoff and landing. However, despite this, airports still function extremely well as part of the high-reliability aviation system for the majority of the time. Improved technology, greater awareness, and systemic improvements are all part of the efforts to improve airport safety performance even further.

Key Terms

Airport Certification Manual

Airport Emergency Plan

Airport Operating Certificate

Aircraft Rescue and Firefighting (ARFF)

Airport Surface Detection Equipment (ASDE)—X Series

Grounding

Engineered Materials Arrestor System (EMAS)

FAA Advisory Circular (AC)

Final Approach Runway Occupancy Signal (FAROS)

Operational Incident

Pilot Deviation

Runway Excursion

Runway Incursion

Runway Status Lights (RWSL)

Runway End Safety Area (RESA)

Veer-off

Vehicle/Pedestrian Deviation

Topics for Discussion

1. What class of airport must large air carriers operate in the United States?
2. Name five categories of information the Airport Certification Manual must contain.
3. List three considerations when selecting an airport site.
4. Describe five occupational safety risks encountered in airport.
5. Name three services that are involved in an aircraft turnaround.
6. Describe the events leading up to the Tenerife accident.
7. What are the two types of deviations in the category of runway excursion?
8. Describe the function of EMAS.

References

1. Boeing Statistical Summary, https://www.boeing.com/resources/boeingdotcom/company/about_bca/pdf/statsum.pdf
2. ICAO Annex 14 Volume I.
3. ICAO Doc 9157 Aerodrome Design Manual.
4. 14 CFR Part-139, https://www.ecfr.gov/current/title-14/chapter-I/subchapter-G/part-139
5. FAA Part-139 Certification List, https://www.faa.gov/airports/airport_safety/part139_cert/part_139_airport_certification_status_list

6. FAA Advisory Circular AC 150/5210-22, https://www.faa.gov/documentLibrary/media/Advisory_Circular/150_5210_22.pdf
7. EASA CS-ADR-DSN, https://www.easa.europa.eu/en/document-library/easy-access-rules/easy-access-rules-aerodromes-regulation-eu-no-1392014
8. EASA Part-ADR, https://www.easa.europa.eu/en/document-library/easy-access-rules/easy-access-rules-aerodromes-regulation-eu-no-1392014
9. ICAO Doc 9859 Safety Management Manual, https://www.icao.int/safety/safety-management/pages/guidancematerial.aspx
10. US OSHA regulations, https://www.osha.gov/laws-regs/regulations/standardnumber/1910
11. EU-OSHA regulations, https://osha.europa.eu/en/safety-and-health-legislation/european-directives
12. Mohammed Tawsif Salam—Own work, CC BY-SA 3.0
13. Image Jason Goh from Pixabay.
14. FAA ARFF. https://www.faa.gov/airports/airport_safety/aircraft_rescue_fire_fighting
15. Paolo Cerutti, CC BY 2.0 via Wikimedia Commons
16. FAA AC 120-60B. *Ground Deicing and Anti-icing Program.* https://www.faa.gov/documentLibrary/media/Advisory_Circular/AC120-60B.pdf
17. EASA Air Operations. https://www.easa.europa.eu/en/downloads/20342/en
18. FAA AC 150/5360-13A. *Airport Terminal Planning.* https://www.faa.gov/documentLibrary/media/Advisory_Circular/AC-150-5360-13A-Airport-Terminal-Planning.pdf
19. FAA *Runway Safety Statistics.* https://www.faa.gov/airports/runway_safety/statistics/
20. FAA *Runway Incursions.* https://www.faa.gov/airports/runway_safety/resources/runway_incursions
21. ICAO Doc 9870 *Manual on the Prevention of Runway Incursions.* https://www.icao.int/safety/RunwaySafety/Documents%20and%20Toolkits/ICAO_manual_prev_RI.pdf
22. ICAO *Global Runway Safety Action Plan (GRSAP).* https://www.icao.int/safety/RunwaySafety/Documents%20and%20Toolkits/GRSAP_Final_Edition01_2017-11-27.pdf
23. FAA *National Runway Safety Plan.* https://www.faa.gov/sites/faa.gov/files/airports/runway_safety/statistics/NRSP_2021-2023.pdf
24. NASB Final Report on Tenerife Accident. http://www.project-tenerife.com/nederlands/PDF/finaldutchreport.pdf
25. ASC Taiwan *Report into Flight SQ006.* https://www.ttsb.gov.tw/media/4252/sq006-executive-summary.pdf
26. DSB *Report into Aborted Takeoff.* https://www.onderzoeksraad.nl/en/page/15141/aborted-takeoff-from-taxiway-boeing-737-800-amsterdam-airport
27. Aviation Safety Network. https://aviation-safety.net/database/record.php?id=20221118-1
28. Flight Safety Foundation. *Runway Excursion Risk Reduction Toolkit.* https://flightsafety.org/toolkits-resources/past-safety-initiatives/runway-excursion-risk-reduction-rerr-toolkit/
29. Van Es (2010) Van Es, G.W.H, "A Study of Runway Excursions from a European Perspective," NLR-CR-2010-259, NLR, 2010 https://skybrary.aero/sites/default/files/bookshelf/2069.pdf

30. Future Sky Safety. *Identification and Analysis of Veer-off Risk Factors in Accidents/ Incidents.* https://www.futuresky-safety.eu/wp-content/uploads/2016/01/ FSS_P3_NLR_D3.4_v2.0.pdf

31. TSB *Beyond Air France 358.* https://www.tsb.gc.ca/eng/medias-media/articles/ aviation/2019/20191203.html#fn2

32. TSB Canada AF358 Accident Report. https://www.tsb.gc.ca/eng/rapports-reports/ aviation/2005/a05h0002/a05h0002.pdf

33. TSB *Beyond Air France 358.* https://www.tsb.gc.ca/eng/medias-media/articles/ aviation/2019/20191203.html#fn2

34. FAA *Engineered Material Arresting System.* https://www.faa.gov/newsroom/ engineered-material-arresting-system-emas-0?newsId=13754

35. BEA N823GA Accident Report. https://bea.aero/en/investigation-reports/ notified-events/detail/accident-on-13-july-2012-at-le-castellet-aerodrome-83-to- the-gulfstream-g-iv-aeroplane-registered-n823ga-operated-by-universal-jet-avia- tion-ujt/

36. ANSV MD-87 Accident Report. https://reports.aviation-safety.net/2001/20011008-0_ MD87_SE-DMA_C25A_D-IEVX.pdf

CHAPTER 8

Air Traffic Safety Systems

Introduction

While some of the previous chapters have focused on how aircraft and airports are designed for safety, once commercial flights become airborne, they navigate through what is often highly complex airspace and rely on an intricate air traffic system and human controllers for separation and sequencing. However, changes are needed to efficiently accommodate the expected growth in air traffic, improve efficiency, and safely incorporate Unmanned Aircraft Systems (UAS).

ICAO statistics provide some idea of the vast scale of global air traffic.[1] In 2018, more than 4.3 billion passengers were carried, with around 1 billion carried in North America and 1 billion in Europe, totalling around nearly 50% combined. Asia and Pacific region flights accounted for 1.6 billion passengers (37%). Those passengers were carried on nearly 38 million departures (11 million in the United States and 9 million in Europe). In the United States, more than 14,000 air traffic controllers are responsible for handling 45,000 flights per day, with more than 5000 aircraft in the sky at one time.[2]

The mission of the global Air Traffic Management (ATM) system is to promote safe, orderly, and expeditious flow of aircraft. Once a commercial airliner is airborne and into the airspace, crews constantly assess factors such as aircraft weight, fuel burn, winds aloft, cloud buildups, and turbulence in order to optimize their route and altitude. This can result in frequent requests to Air Traffic Control (ATC) for changes which must be balanced against potential conflicts from other traffic in the immediate airspace. In some cases, the results are of short duration, such as pilots requesting to deviate laterally to avoid a cloud buildup that may contain severe turbulence, and then returning to the pre-agreed flight path once past the buildup. In other cases, the requested deviations may be significant, such as a descent to a lower altitude due to the need to operate anti-ice systems that reduce the available thrust to engines and preclude maintaining the current altitude. This chapter brings together some of the ATM protocols and technologies aimed at keeping aircraft safe and looks at some of the challenges the system faces.

The information in this chapter has been broken up as follows. The first part of the chapter gives:

- A brief history of ATC, links to the regulations involved and some of the stakeholders
- A description of the ATC system and its components, including tracking and communicating with aircraft

- The steps and technologies involved in aircraft navigation
- A description of a typical commercial flight, bringing together the different parts described above

The second part of the chapter will look at:

- The challenge of UAS integration
- The future of ATC and new technologies
- Accident case studies in which ATC has played a significant role

8.1   History, Regulation, and Stakeholders

Before exploring the details of the ATC system, it will be useful to give a short history of the ATC system, a reference to the appropriate regulations, and a brief outline of some of the stakeholders involved.

8.1.1   History of Air Traffic Control

Some of the notable milestones in ATC history are presented below:

June 1956	Midair collision over the Grand Canyon in Arizona, Unites States, killing all 128 occupants of the two aircraft. The accident highlighted that much of the airspace in the United States was uncontrolled without radar above 20,000 feet. It also spawned the overhaul of ATC operations and the development of the FAA.
Aug 1958	FAA established in the United States.
June 1960	Britain, France, Belgium, Italy, Luxembourg, West Germany, and Holland decided to coordinate air traffic control.
Dec 1960	Another disastrous midair collision occurred over New York City. A TWA Super Constellation and a United Air Lines DC-8, both in a holding pattern under instrument flight rules, collided killing all aboard. The United Air Lines aircraft had navigational problems and excessive speed, which could not be detected by the New York ATC towers that lacked proper surveillance radar equipment. As a result of the accident, there were equipment upgrades and a new regulation mandating the limit of 250 knots airspeed when within 30 nautical miles of the airport and below 10,000 feet altitude.
Dec 1962	The United Kingdom announced the new National Air Traffic Control Services (NATCS).
Mar 1964	EUROCONTROL was established.
Dec 1968	An agreement was signed to build Europe's first international control center at Maastricht, to open in 1972, called the Maastricht Automatic Data Processing system or MADAP, which is now called the Maastricht Upper Area Control Centre.
Apr 1972	The UK NATCS became NATS when it joined with the CAA.
1975	Computer flight plans were first implemented in the United Kingdom.
1981	In 1981, the first computer data link was established, between London Terminal Control Centre and EUROCONTROL.

Aug 1981	Over 12,000 US air traffic controllers went on strike over labor conditions and were fired by the FAA. It took almost 10 years before overall ATC staffing levels returned to normal.
Jan 1982	The FAA released the first National Airspace System Plan, a comprehensive 20-year blueprint to modernize ATC and air navigation systems in the United States.
Apr 2000	Creation of the FAA Air Traffic Organization (ATO), a performance-based department focusing on efficient operation of the US ATC system.
Dec 2003	The Next Generation Air Transportation System (NextGen) concept was authorized, a new, ambitious, multi-agency effort to develop a modern air transportation system for the 21st century.

8.1.2 Regulation

The relevant ICAO SARPs are contained in Annex 11, *Air Traffic Services*[3] and the procedures are documented in ICAO Doc 4444, the *Procedures for Air Navigation Services—Air Traffic Management* (PANS-ATM).[4]

Air Navigation Service Providers (ANSPs) are regulated bodies, giving one or more of the following services[5]:

Air Traffic Service (ATS)	Such as an Air Traffic Control (ATC) service, Flight Information Service (FIS), or alerting service
Aeronautical Information Service (AIS)	Including Aeronautical Information Publication (AIP), aeronautical charts, and Notices to Airmen (NOTAM)
Meteorological (MET)	Such as Meteorological Aerodrome Reports (METARS) and Terminal Aerodrome Forecasts (TAFs)
Communication, Navigation, Surveillance (CNS) systems	Focusing on the infrastructure to provide these services

We will focus predominantly on ATS in this chapter. As with manufacturers, operators, and maintenance organizations, ANSPs are required to have proper organizational processes in place, including staff licensing, a Safety Management System, and accident and incident reporting. In the United States, ATSs are impacted by a range of CFR Parts including FAR Part-71 around designation of airspace, Part-73 on Special Use Airspace, and Part-97 on Standard Instrument Approach Procedures, In Europe, EASA terms these services Air Traffic Management / Air Navigation Services or ATM/ANS and they are covered by the corresponding ATM/ANS regulations[6] and the Standardized European Rules of the Air (SERA).[7]

8.1.3 Stakeholders

In the United States, the FAA is responsible for the National Airspace System (NAS) which is a network of both controlled and uncontrolled airspace, both domestic and oceanic.[8] It also includes air navigation facilities, equipment, and services; airports and landing areas; aeronautical charts, information, and services; rules and regulations; procedures and technical information; and manpower and material. The FAA's ATO is the operational arm of the FAA[9] relating to ATM. It is responsible for providing safe and efficient air navigation services to the US airspace, including all of the United States and large portions of the Atlantic and Pacific Oceans and the Gulf of Mexico. In Europe, EUROCONTROL is an international organization dedicated to safe ATM across Europe with many separate ANSPs operating in the different countries. In UK airspace, ANSP services are predominantly provided by NATS.

8.2 The ATC System and Its Components

The global ATC system is made up of a vast array of components and describing all of them is beyond the scope of this chapter. However, there are some key aspects of the system and an understanding of those can help to provide a functional understanding. This chapter will attempt to lay them out clearly to aid that understanding.

8.2.1 Airspace Division

Global airspace is divided into 9 ICAO air navigation regions, which are then further divided into Flight Information Regions (FIRs). Each FIR is managed by a controlling authority that has responsibility for ensuring that air traffic services are provided to the aircraft flying within it. As Figure 8-1 shows, FIRs vary in size—smaller countries may have one FIR in the airspace above them and larger countries may have several. Airspace over the ocean is typically divided into two or more FIRs and delegated to controlling authorities within bordering countries.

In some cases, FIRs are split into lower and upper sections. The lower section remains referred to as a FIR, but the upper portion is referred to as an Upper Information Region (or UIR). The European UIRs are shown in Figure 8-1.

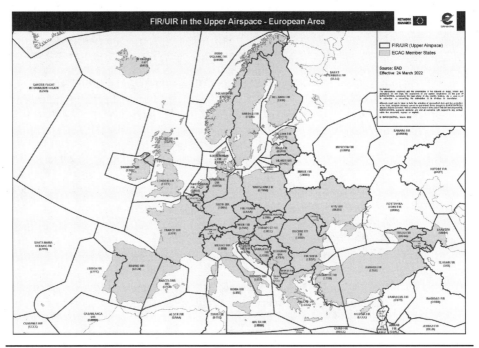

Figure 8-1 European upper flight information regions. (Source: EUROCONTROL)

8.2.2   FIR Subdivisions

Within each FIR, there are additional regions relating to the facilities below. NATS, the UK ANSP describes airspaces in the United Kingdom as follows[10]:

- **Control Zones:** Around airports are Control Zones, typically 2 miles around the airport and up to around 2000 feet. This zone is restricted to aircraft using the airport. Airports sometimes extend this region with an additional control zone (also called a Control Traffic Region—CTR) which can extend up to 10 miles. In the United States, the term control zone has been replaced with Class D airspace.

- **Control Area:** Above the control zone is usually a Control Area (CTA) to protect aircraft climbing out of an airport to join an airway or descending from an airway to the airport. These areas can be combined for multiple airports in close proximity, known as Terminal CTAs.

- **Airways:** Airways are typically 10 NM wide with a defined lower base of 1200 feet above ground level in the United States or between 7000 and 10,000 feet in Europe, extending up as far as 19,500 feet.

- **Upper Airspace:** All airspace above a given height, varying by country FL180 and above in the United States and 19,500 feet AMSL and above in Europe. It is controlled airspace and here aircraft follow defined routes (paths which connect defined points). Routes in the 19,500–24,000 feet range in Europe are Lower Air Routes and those above are Upper Air Routes.

8.2.3   Airspace Classification

Airspace within a FIR (and UIR) is usually divided into pieces that vary in role, size, and classification. ICAO Annex 11, Section 2.6 describes the different classifications of regions into which airspace is divided.[11] Each is assigned a classification from A to G. Classes A-E are known as controlled airspace and classes F and G are uncontrolled airspace. Aircraft flying in controlled airspace must follow instructions from air traffic controllers. Aircraft flying in uncontrolled airspace are not mandated to take air traffic control services but can call on them if and when required (e.g., flight information, flight following, alerting and distress services).

The highest level of control is applied to Class A airspace, where all operations must be under Instrument Flight Rules (IFR), aircraft are subject to ATC clearance, and separation is provided by ATC. At the other end of the scale is Class G or uncontrolled airspace where pilots are responsible for their own separation. Although some traffic information and separation services are available, the use of these ATC services are optional and their provision discretionary. These ICAO standard classifications are not universally applied with some countries modifying the rules to fit their existing system before there was standardization. The UK-based ANSP NATS provides an excellent explanation of airspace classification on their website.[12]

8.2.4   Air Traffic Services

There are two fundamental types of air traffic service: Flight Information Service (FIS) and Air Traffic Control (ATC) service. ICAO Annex 11 defines a Flight Information Service as "a service provided for the purpose of giving advice and information useful for the safe and efficient conduct of flights." An FIS is available to any aircraft in the

FIR and is typically used by General Aviation aircraft. Information from an FIS can include weather conditions, availability of radio navigation services, changes in aerodrome conditions etc. FIS can also be asked for emergency support and assistance. In the United States, a FIS also offers a "flight following" service. The FIS may be provided by an air traffic controller; however, their primary ATC duties always take precedence. The FIS does not provide separation information—pilots remain solely responsible for the safe conduct of their flight.[13]

In contrast, an air traffic control service is defined in Annex 11 as

A service provided for the purpose of

  a) preventing collisions:

    1) between aircraft, and

    2) on the maneuvering area between aircraft and obstructions; and

  b) expediting and maintaining an orderly flow of air traffic.

Under an air traffic control service, the controller will be responsible for maintaining appropriate separation of aircraft, including taking into account any wake turbulence considerations from larger aircraft. However, the Pilot in Command (PIC) is ultimately responsible for the safe operation of the aircraft and hence, in extreme conditions, may choose to deviate from ATC instructions to maintain the safety of the aircraft.

8.2.5 Types of ATC Facilities

Air traffic controllers operate primarily from three types of facilities. These basic ATC facilities are as follows (starting with the airport, then moving away):

- **Airport traffic control towers (ATCTs)**—usually the tower at an airport (and what many people would picture when talking about air traffic control). There are 520 ATCTs in the United States.
- **Terminal radar approach control (TRACON) facilities**—also known as terminal control centers, these are facilities hosting air traffic controllers who guide aircraft approaching and departing airports. There are 147 TRACONs in the United States.
- **Air route traffic control centers (ARTCCs)**—also known as area control centers or enroute centers. There are 21 ARTCCs in the United States. In some cases, area control is combined with terminal control, as is the case for London at Swanwick in the United Kingdom.

8.2.6 Flight Plans

A flight plan is required for a range of flights, including international flights, flights to be provided with air traffic control service, and flights under IFR. This means that since all flights in Category A airspace must be IFR, they must all file a flight plan. A flight plan includes information such as aircraft identification, departure aerodrome, cruising speed, route, fuel endurance, and more. Figure 8-2 shows the ICAO model flight plan form.

8.2.7 Radar

Radar is still the predominant means of tracking and maintaining aircraft separation, particularly around airports. However, radar can take a number of forms.

1. **ICAO model flight plan form**

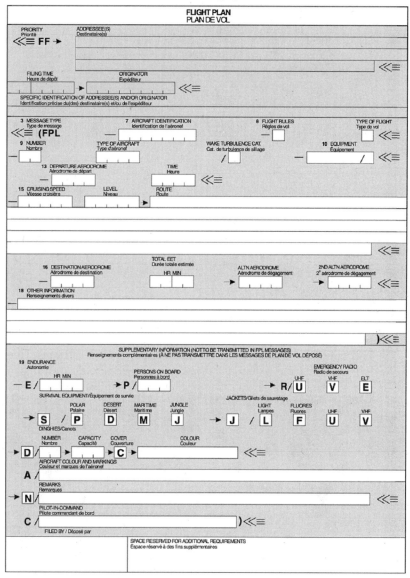

FIGURE 8-2 ICAO model flight plan. (Source: ICAO)

Primary radar is the classical form of radar in which a radio wave is sent out from the measurement point, and reflections are bounced back and detected. The bearing and distance can then be calculated by measuring the time of flight of the signal roundtrip and the angle at which the reflection was detected. This system requires only that the target reflects radio waves, meaning that as well as aircraft, birds, clouds, and land may also be detected. This also means that there is no form of identification of the target. Primary radar is still used in ATC today, often as a backup or additional system.

Secondary Surveillance Radar (SSR) is the most common form of radar used in ATC and requires the aircraft to be fitted with a transponder. This transponder replies to an interrogation signal by transmitting data dependent on the particular mode being used. SSR is limited to line-of-sight operations and provides range and bearing information. In addition, an SSR ground station transmits interrogation pulses for Modes A, C and S at 1030 MHz and the transponder replies with data at 1090 MHz. The spacing between the transmitted pulses identifies which mode is being requested. For a Mode A request, the transponder replies with a 12-pulse response representing the code configured in the transponder, sometimes known as the "squawk code." This code can be used to indicate certain states including failed radio (7600) or emergency (7700). In addition to the squawk code, a transponder capable of responding to a Mode C request will also provide the aircraft's pressure altitude to a resolution of 100 feet. Mode S (for Selective) enhances this service giving altitude to a resolution of 25 feet and expansion of the identification field to a 24-bit address, allowing unambiguous identification of aircraft. Traffic alert and Collision Avoidance System (TCAS) uses Mode S interrogations of surrounding aircraft to get range, bearing and altitude in order to compute any necessary avoiding action (Resolution Advisory, or RA). Mode A/C and Mode S equipment are interoperable. Airport Surveillance Radar (ASR) combines primary and secondary surveillance radar to monitor aircraft within the immediate vicinity of an airport.

The NATS Airspace Explorer app, shown in Figure 8-3, predominantly uses radar data for its display. However, for popular aircraft tracking sites such as adsbExchange and Flightradar24, ADS-B Out is the primary data source.

Automatic Dependent Surveillance Broadcast (ADS-B) is an extension of the SSR Mode S technology, sometimes called 1090ES (Extended Squitter). ADS-B is actually two separate services, ADS-B Out and ADS-B In. Using ADS-B Out, the enabled aircraft

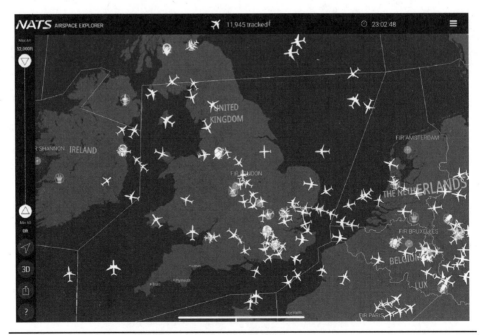

FIGURE 8-3   NATS Airspace Explorer app. (Source: Author)

broadcasts a range of data, including identification, altitude, position (usually derived from GNSS—see Section 8.3.2) and speed. ADS-B differs from SSR in that no interrogation signal is necessary.

Since 2020, the United States has required ADS-B Out when operating in Class A airspace and other designated classes of airspace.[14,15] Traditionally ADS-B monitoring is achieved through ground-based systems. However, the Aireon company is enabling ADS-B Out monitoring through Low Earth Orbit satellites giving nearly 100% global coverage. NATS and NAVCANADA have been early adopters of this technology.

ADS-B In allows receivers to ingest information from other ADS-B equipped aircraft. One of the technologies that is enabled by ADS-B In is Cockpit Display of Traffic Information (CDTI). There are multiple systems available using this technology, reviewed by the FAA in 2016,[16] including on the Boeing 787.

8.2.8 Data Link Communications

Data Link is an overarching term for several systems, networks, and protocols used for transmitting and receiving digital information. Data link can use a number of communication methods including VHF Data Link (VDL), HF Data Link (HFDL) and satellite (SATCOM) data link. VDL requires line of sight and SATCOM is limited in polar regions, and so HFDL is used for polar coverage and remote areas.

ACARS messages (see Section 6.4.1.8) were originally sent over VHF channels, but they now also use data link. The data link system also allows operators to transmit Airline Operational Control (AOC) messages between flight crew and airline operational centers. FAA AC 90-117[17] contains more information on data link. The table in that document which describes interoperability between old and new technologies highlights the challenge of introducing any new technology into the aviation system. The level of technology installed on the fleet varies very slowly and hence there is a long period where legacy technologies need to be supported in parallel with the new technology.

8.2.8.1 Controller Pilot Data Link Communications

Controller Pilot Data Link Communications (CPDLC) is a means of communication between controller and pilot, using the data link technology. It can be thought of as a secure messaging service for nonurgent messages between the controller and flight crew and when a message is received in the aircraft it is displayed on the FMS or similar. Figure 8-4 shows example messages as they appear in the aircraft's cockpit.

The messages exchanged between the two parties are selected from a subset of messages, in general reproducing agreed aviation phraseology. Controllers are provided with the capability to issue ATC clearances such as level assignments, lateral deviations/vectoring, speed assignments, radio frequency assignments, and various requests for information. There is a range of services within CPDLC to permit different operations, including Data Link Initiation Capability (DLIC), ATC Clearance Service (ACL), and Departure Clearance (DCL).

This system overcomes several problems inherent to human voice transmission such as message deformation or poor pronunciation, and to the transmission or reception of messages due to frequency band saturation or the poor propagation of radio waves.

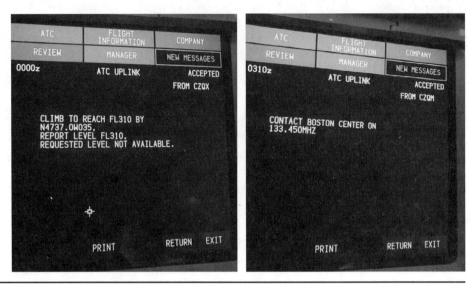

FIGURE 8-4   Example CPDLC messages. (Source: Authors)

A number of ICAO documents,[18,19] as well as some useful websites,[20] give more detail on CPDLC and data link. CPDLC is being globally implemented and currently is in different implementation stages around the world.

8.2.8.2   *Automatic Dependent Surveillance—Contract Services*

Automatic Dependent Surveillance—Contract Services (abbreviated as ADS-C), are another protocol provided over data link. ADS-C uses various systems on board the aircraft to automatically transmit aircraft position, altitude, speed, meteorological data, and more to an air traffic service unit. This data can then be used for surveillance and automated route conformance checking. ADS-C can also take the place of voice position reports.

The type of information to send and when (known as the "ADS contract"), is agreed between the aircraft and the air traffic systems. Some types of information are included in every report, while others are only provided if requested. A connection can support multiple simultaneous contracts with a ground unit and an aircraft can connect with up to five separate ground units simultaneously. The aircraft can also send unsolicited emergency messages to any connected unit. Although they are similar terms, ADS-C should not be confused with ADS-B (see Section 8.2.7 above).

ADS-C provides better surveillance performance and CPDLC gives better communication performance. In combination, they allow operation with reduced ATC minima.[21] The CPDLC and ADS-C system were a peripheral feature of the accident to Air France Flight 447—see Section 8.7.2.

While much of what is described in this section is the current state of the technology, it is by no means a global standard. There are still some parts of the world where communication and coordination are made through High Frequency (HF) radio (with highly variable performance), and even sometimes coordinated between aircraft on a "self-help" basis. Across substantial parts of the African continent (the IATA in-flight broadcast—AFI Region), aircraft assist in the safe procedural separation from each other by transmitting on a common frequency—126.9 MHz.[22]

8.3 Aircraft Navigation

Inherently coupled with the ATC task of routing and separating aircraft is the flight crew task to navigate the aircraft accurately. Traditionally this was achieved using radio navigation aids. However, the availability and reliability of Global Navigation Satellite Systems (GNSS) such as GPS means their use for navigation is increasing significantly. This section discusses some of the key aspects of how this function is performed.

8.3.1 Traditional Navigation

It was not that many years ago that aircraft navigated by the stars. Up until the 1970s it was still reasonably common for some aircraft to use a "star-shot" system. The aircraft's position could be determined by the azimuth of celestial objects and referring them to an almanac (a published guide to the expected position of heavenly bodies as they changed by time and geodetic position). However, predominantly, air navigation has been facilitated by ground-based systems (NAVAIDS) whose signals were used through aircraft avionics for enroute navigation and landing guidance, although this didn't always give full coverage of airports and airspace. Some of the basic NAVAIDS and radar surveillance equipment used in this system are described below.

A VHF Omnidirectional Range (VOR) is a short-range (up to c. 200 NM) navigation device that transmits signals allowing the aircraft receiver to calculate a bearing ("radial") to the beacon. Tracking two VORs simultaneously allows a position to be calculated.

A nondirectional beacon (NDB) is a radio beacon with no direction information encoded. However, automatic direction finding (ADF) equipment will indicate the direction of the NDB relative to the aircraft; the distance to the NDB is not known directly. For this reason, Distance Measuring Equipment (DME) stations are often collocated with NDBs. DME works by calculating the "time of flight" of a radio signal sent in response to an interrogation signal to calculate the distance of the aircraft from the beacon.

As GNSS (satellite) navigation becomes the dominant system, these navigation devices are being progressively decommissioned. However, in some cases they are being repurposed. For example, as the FAA moves to Performance Based Navigation (see Section 8.3.5 below) the VOR infrastructure is being repurposed to provide a backup system for the case of GPS outages.[23] Navigating by VOR and the other radio navigation aids above can facilitate a nonprecision approach. A precision approach requires an Instrument Landing System (ILS) consisting of a localizer (providing horizontal guidance), a glideslope (providing vertical guidance), a marker beacon, and an approach light system.

8.3.2 Global Navigation Satellite Systems

GNSS refers to a constellation of satellites that transmit signals from space carrying satellite position and time data to receivers. A GNSS receiver such as those on an aircraft or in your phone, can then receive the signals and calculate, by how long the signal took to arrive, how far they are from that satellite. By repeating this process with a number of satellites, the receiver can calculate its precise three-dimensional position in space.

The NAVSTAR Global Positioning System (GPS) is the most well-known GNSS system. It is owned by the US government and was originally designed for the US Department of Defense, before being made available for civilian use in the 1980s. GPS positioning is now ubiquitous with phones, cars, watches, and even dog collars using the technology. The NAVSTAR network consists of 24 satellites orbiting the earth and providing global coverage. The control segment consists of a master control station in Colorado, an alternative master control in California, 11 command and control sites and 16 monitoring stations.

The equivalent GNSS system in Russia is called Global'naya Navigatsionnaya Sputnikovaya Sistema or GLONASS, if your Russian is a little rusty. China has its own system, known as the BeiDou Navigation Satellite System (BDS). In Europe, the equivalent system is called Galileo. It consists of two systems: one higher precision system open to government-authorized users, and a lower accuracy service that is free and open to all users. It was created in part for redundancy and independence from the GPS and GLONASS systems and it went live in 2016.

It is important to remember that GNSS is a "listen-only" system—receivers capture the signals being transmitted by satellites to calculate their own location, similar to an analog radio. There is no return communication with the satellite. This is particularly important when considering cases such as Malaysian Airlines Flight MH370, the Boeing 777 that diverted from filed flight plan and was lost on 8 March 2014. Some have questioned whether the flight could be "tracked through GPS," which it couldn't; the GPS receiver on its own allows the aircraft to calculate its own position accurately, but it cannot be traced by third parties (see Section 8.7.3).

8.3.3   Satellite-Based Augmentation Systems

The performance of GNSS systems can be improved by regional satellite-based augmentation systems (SBAS). In brief, accurately located reference stations track the GNSS signals and report the errors to a centralized computer, which calculates the appropriate differential correction and transmits it to the receiver by geostationary (i.e., fixed relative to the earth) satellite, thereby improving the accuracy of receivers able to apply the correction.

In the United States, the SBAS system is named Wide Area Augmentation System (WAAS) and provides service for all classes of aircraft in all phases of flight including enroute navigation, airport departures, and airport arrivals. This includes vertically guided landing approaches in instrument meteorological conditions at all qualified locations. In Europe, the SBAS is named the European Geostationary Navigation Overlay Service (EGNOS). There are numerous other systems across the globe including SDCM in Russia, GAGAN in India, and KASS in South Korea. All of these systems comply with a common standard and so are compatible and interoperable. SBAS has become an invaluable asset in commercial aviation since, for example, basic GPS does not satisfy ICAO operational requirements for critical flight stages such as final approach.[24] However, with the addition of SBAS, the performance standards can be met.

8.3.4   Advantages of Satellite-Based Navigation

Satellite-based navigation provides significant operational and safety benefits over past methods that relied on ground-based navigation aids. The new mode of navigation meets the needs of growing operations because pilots are able to navigate virtually anywhere in the system, including to airports that currently lack ground navigation and landing signal coverage. Satellite-based navigation will also support Performance-Based Navigation (PBN).

With satellite navigation, the number of published precision approaches has also greatly increased so that in addition to finding airports, pilots can also successfully land there during inclement weather conditions. In addition, combining GNSS with flight deck electronic terrain maps and ground-proximity warning systems can help pilots reduce flight profiles that expose them to the risk of controlled flight into terrain (CFIT). Satellite-based navigation also decreases the number of ground-based navigation systems necessary, thereby reducing infrastructure costs. Today, using PBN, GPS augmented by WAAS can provide precision approach guidance for most of the runways in the United States that would otherwise require a dedicated ILS.

8.3.5   Performance-Based Navigation

Modern aircraft performance and functional capabilities mean that future airspace designs can save time and fuel, thereby reducing emissions. A key component of this is PBN. ICAO Doc 8613[25] describes the requirements for PBN in terms of accuracy, integrity, and continuity of service. The FAA[26] and EASA[27] both provide guidance on PBN.

As described above, traditional navigation relies on navigating from a position relative to one ground-based navigation aid (e.g., VOR, DME or NDB) to a position relative to another. This method prevents aircraft from flying the most direct possible route and is inefficient. PBN, made up of Area Navigation (RNAV) and Required Navigation Performance (RNP), aims to improve this situation as well as consolidating and standardizing disparate implementations of the same concept.

RNAV involves navigating from one point, waypoint, or fix, to another, with these points often defined by latitude and longitude. This separates the flightpath from ground-based navigation stations. RNP is an extension of RNAV in which the aircraft flies a specific path between two 3D points in space. Under RNP, the navigation system must be able to monitor the navigation performance being achieved and alert any failures to achieve the required standard—this is the "performance-based" aspect of the technology. Flight crew require an endorsement to conduct PBN operations, which requires a theoretical course, flying training, and a skills test.

Under standard operations, in oceanic airspace, aircraft follow "tracks" that are optimized dependent on the prevailing winds. The lack of radar surveillance and direct controller–pilot communications often require oceanic separation standards to be 20 times greater than those in domestic airspace and these large separations limit the number of available tracks. Therefore, some flights are assigned a less-than-optimum altitude in order to still transit the airspace without undue delay, and there is insufficient opportunity to adjust altitudes to conserve fuel. PBN offers an opportunity to reduce separations and increase efficiencies.

8.4   A Typical Commercial Flight

Having learned about the various components, stakeholders, protocols, and technologies involved in getting an aircraft safely from A to B, it may be helpful to describe the relevant aspects of ATC and navigation of a typical commercial flight from start to finish, to see how the parts work together in practice. This is a typical generic description, but of course the details will vary from flight to flight.

8.4.1   Flight Planning and Aircraft Preparation

Before the flight, a flight plan will be generated and filed (usually by a flight planning department in the airline) which will include routing, fuel planning, diversion airports etc. The plan will take account of expected weather and winds, airport restrictions, NOTAMs etc. Once filed with the appropriate authority, the plan will either get accepted as submitted, or will give a modified route, consisting of waypoints and altitudes, to the same destination.

On arrival at their normal place of reporting the flight crew will review the plans and look for any obvious errors. They will then assess the serviceability state of the aircraft for the intended route, the enroute, and destination weather and any other considerations before deciding on the required fuel load. Once at the aircraft, the flight crew will load the Flight Management System (FMS) with the designated route either automatically from the published clearance transmission from CPDLC or manually. They will also calculate the takeoff and climb performance of the aircraft considering the aircrafts' serviceability state, the ambient conditions, and the length of the intended departure runway and climb-out routing. They will assess the published departure routes and how to avoid terrain, particularly should they experience an engine failure. Much of this time is spent briefing on what threats they think they will encounter during departure from the airfield and what they would actually do during any non-normal event.

8.4.2 Start-Up, Pushback, and Taxi

Once the aircraft is ready to depart (passengers, baggage, fuel, catering, etc., loaded), the flight crew will notify the ground controller at the airport, and wait for ATC and pushback clearance. Once cleared, the aircraft will be pushed back into the controlled movement area of the airport. The flight crew will start the engines, re-asses the serviceability of the aircraft, and when cleared, taxi toward the runway, following the route given by the ground controller, eventually stopping short of the runway at a holding point.

8.4.3 Takeoff and Departure

When prompted or at a predetermined location, the aircraft will then be handed over to the tower controller on a different radio frequency. The tower controller is responsible for the runway and the airspace proximate to the runway under their control. The tower controller issues a takeoff clearance, granting permission for the crew to, either enter the runway and hold position or enter and take off. After takeoff, the tower controller will pass the aircraft to the departure controller, who issues heading and level instructions to the aircraft as it climbs to its cruise altitude, avoiding other traffic in the area. Normally an aircraft will have been precleared to follow a Standard Instrument Departure (SID) but to optimize airspace use, the controller often gives intermittent clearances. The departure controller will then deliver the aircraft to an agreed point and altitude and pass the aircraft to an enroute centre, known as an Air Route Traffic Control Centre (ARTCC).

8.4.4 Enroute

The aircraft will then cross the various sectors of airspace, with each controller handing the aircraft to the next. The controllers will coordinate with the pilots, via radio or CPDLC, issuing instructions on speed, route, and altitude to maintain separation. The flight crew may also request routing changes to avoid weather and to shorten the route if there is limited traffic. The cruise altitude will usually rise as the flight continues as the weight of the aircraft decreases due to fuel burn. The aircraft requires less lift to support the aircraft and so it can climb to thinner air, where there is also less drag.

8.4.5   Descent, Approach, and Landing

As the aircraft approaches its destination, the enroute controller will begin the aircraft's descent. The enroute controller will deliver the aircraft to a waypoint which often marks the start of a Standard Instrument Arrival or STAR. The aircraft may be given clearance to "descend via" the standard arrival and follow a prepublished vertical and horizontal path. However, just as with the SID, a controller may well intervene to optimize airspace use. The approach controller may issue speed, altitude, and flight path instructions to the pilot. Depending on the density of air traffic at the destination airport, the aircraft may be cleared into a holding pattern to await its turn to approach and land. These are often predefined hold patterns or "stacks" where the aircraft enters at the top and flies a descending "racetrack" or oval pattern before exiting the bottom to approach the airport.

Once cleared to approach, they will be handed to the approach and then tower controller at the destination airport. The approach controller ensures the arriving aircraft are adequately spaced using time as the metric. TBS or time-based separation helps to maximize the use of airspace and to mitigate congestion and delays. The tower controller ensures that there is enough spacing between departures and arrivals, both in the air and on the runways, and gives the pilot clearance to land. It remains the pilot's decision whether the aircraft is suitably stable to continue to land, that is: properly configured, on profile, at the right speed and position, and that the weather continues to be safe to land. Sometimes, unusual weather (as in Figure 8-5) or unexpected events such as a vehicle taxiing onto the runway, make the safest course of action a rejected landing or "go-around."

8.4.6   Taxi-in and Shutdown

After landing, the tower controller passes responsibility for the flight to the ground controller who guides the aircraft through the airport taxiways to the gate, where the flight crew park and shut down the aircraft. Just as they started their journey, the crew will discuss how safe and effective the flight was and whether there is more to learn from the flight.

FIGURE 8-5   In unusual weather conditions, such as a low bank of cloud or fog, as seen here at London's Heathrow Airport in March 2022, the crew of this Virgin Airlines Boeing 787 elected to do the safe thing and go-around from their approach to runway 09L. (Source: Authors).

8.5   Integration of Other Air Vehicles

8.5.1   Background

A major change in the area of airspace safety has been the rapid growth in the development of new types of air vehicle, whether Unmanned Aircraft / Aerial Systems (UAS), often referred to as *drones* or UAVs, or the new generation of Urban Air Mobility (UAM) vehicles under development. This section will treat the two categories separately although some of the solutions apply to both platforms.

8.5.2   Small Unmanned Aerial Systems

Improvements in materials and computing mean that drones are accessible to the general public in an unprecedented way. They have, however, been around for many years in military applications, albeit in different forms. There are innumerable commercial uses for drones including law enforcement, package delivery, filmmaking, agriculture, search and rescue, security inspections, photography, real estate listings etc. as well as recreational use by amateurs and hobbyists.

14 CFR Part-107[28] allows for the routine civil operation of small Unmanned Aircraft Systems (sUAS), defined as unmanned aircraft weighing less than 55 pounds. Part-107 defines a range of rules for sUAS registration, licensing for commercial pilots, and much more. However, a key provision of this rule is a waiver mechanism to allow individual operations to deviate from many of the operational restrictions of Part-107 if the FAA administrator finds that the proposed operation can safely be conducted under the terms of a certificate of waiver. For example, waivers have been granted to allow operations Beyond Visual Line Of Sight (BVLOS). EASA has similar rules,[29] categorizing drones as either "open" (<25 kg or 55 lbs), "specific," or "certified."[30]

8.5.3   Urban Air Mobility

One of the most disruptive technologies in aviation at present is the emergence of UAM solutions. According to the Vertical Flight Society, there are currently 700 projects being run by around 300 companies trying to produce a viable "air taxi." However, while much of the focus is unsurprisingly on the technology around propulsion, lift generation and control into which huge sums of money are being invested ($8 billion in 2021 according to the *Financial Times* January 31, 2022), the issue of airspace integration is also very real. As we have discussed, the high-reliability aviation system is built on change through evolution with aircraft, and the technology they incorporate, surviving decades. Existing systems are not well-suited to radical change in a short time period and so require a new approach.

UAS Traffic Management (UTM)[i] is the term for the traffic management system aimed at autonomous or unmanned operations. It is often considered as separate from, but complementary to, existing ATM. The FAA has established a research team and has laid out a vision[31] and in 2022 published a revised Concept of Operations (ConOps) for UTM.[32] EASA has also been working to develop the regulatory framework around UAM in airworthiness (with a special condition for VTOL), operations and pilot licensing, and airspace integration.[33]

---

[i] UTM is also referred to as Unified Traffic Management or sometimes Uncrewed Traffic Management. All these are really the same thing—air traffic control that deals with drones or UAVs.

One of the challenges facing UTM development for UAM, and for the airspace integration of unmanned aerial systems in general, is the lack of a shared vision for the "end point." For example, at one end of the spectrum, some imagine small, hobby-size drones delivering small packages, always under a few hundred feet in permanent contact with a base no more than a few miles away. However, others envision fully autonomous, independent aerial transport solutions able to move many people hundreds of miles at more than 100 knots. Clearly these two scenarios (and all of the combinations between them) present very different challenges in terms of airspace integration.

Altitude limits (e.g., 400 feet) and geofencing (restricting drone operations around sensitive areas such as airports) can go some way to limiting the risk of a drone-aircraft conflict for large commercial aviation. However, this does little to protect smaller aircraft including helicopters which may face much greater exposure.

To address this issue in Europe, the U-space concept has been developed. U-space is a set of digital, automated services inside a volume of airspace. The four basic services inside the space will be: network e-identification, geo-awareness, UAS flight authorization and traffic information. A recent SESAR paper entitled *Demonstrating the Everyday Benefits of U-Space*[34] described a number of different possible use cases including air taxis, drone deliveries, and public safety and security. The concept is due to begin in 2023.

In early 2022, the UK government has funded a $5 million feasibility study named ALIAS (Agile and Integrated Airspace System) as part of an Innovate UK Future Flight Phase 3 program to develop new air transport technologies. The ALIAS project aims to reconcile many of the interoperability issues, by working closely with the US RTCA and the primary ANSP, Ports of Jersey, in the Channel Islands, the project will explore how crewed and uncrewed aircraft can safely operate within the same airspace during flight trials planned for 2024.

8.5.4　UAS Safety Events and Investigations

There have been a number of accidents where aircraft have come into conflict with UAS or "drone" aircraft. On September 21, 2017, a US Army Blackhawk collided with a DJI Phantom in New York.[35] In this case, the sUAS remote pilot intentionally flew the aircraft 2.5 miles away, well beyond their visual line of sight. On October 12, 2017, a Beechcraft King Air collided with a drone on approach to a Quebec airport.[36] Fortunately, in both of these accidents there were no injuries. There have also been numerous near-misses, with Wikipedia listing more than 30 specific cases.[37] The NTSB has recently amended the definition of "Unmanned aircraft accident" in 14 CFR 830 by replacing the weight categorization (previously 300 lb/150 kg) with an airworthiness certificate requirement.[38]. The UK AAIB began investigating accidents to UAS in 2015.[39] At that time, like the NTSB, they were only required to investigate accidents and serious incidents if the UAS was over 150 kg (330 lb). They have since reported on 68 UAS events.

An example of the threat posed by UAS is shown in the UK AAIB investigation into an Alauda Airspeeder Mk II[40] summarized below. While performing a demonstration flight, the remote pilot lost control of the 95 kg (210 lb) UAS. After the loss of control had been confirmed, the safety "kill switch" was operated but had no effect. The Unmanned Aircraft then climbed to approximately 8000 feet, entering controlled airspace at a holding point for flights arriving at Gatwick Airport, before its battery depleted and it fell to the ground. It crashed in a field of crops approximately 40 m from occupied houses and 700 m outside of its designated operating area. There were no injuries.

The AAIB found that the aircraft was not designed, built, or tested to any recognizable standards and that its design and build quality were of a poor standard. The operator's Operating Safety Case contained several statements that were shown to be untrue. The report made 15 Safety Recommendations regarding the operator's procedures, airworthiness standards, and the regulatory oversight. The risk presented in this case is obvious, both to other aircraft and also to the public on the ground. The need to bring this new technology into the stringent safety regime mainstream commercial aviation is the long-term objective of regulators.

8.6   Future Developments

In the ideal *Next Generation* (NextGen) environment, pilots would be released from the rigid discipline of being spaced in nose-to-tail time blocks, along less-than-optimal routes, and often at inefficient altitudes. Extended to its ultimate application, NextGen would embrace the airplane's complete operation, from start-up at the originating airport to shut down at the destination after having flown a direct, nondeviating course, flying the optimal performance figures straight out of the airframe manufacturer's operating handbook. This vision is a major goal of ATC technology improvements.

8.6.1   Aircraft 4D Trajectory Concept

The 4D trajectory[41] of an aircraft consists of its three-dimensional position in space, plus time as the fourth dimension. The implication of considering the desired trajectory in this way is that a delay is considered to be a trajectory deviation in the same way as a route deviation would be. Accurate location of an aircraft in space *and* time allows for greatly improved efficiency through time-based management and the assignment of crossing times at specific points in space.

8.6.2   FAA NextGen Program

NextGen[42] is the FAA's ongoing program to modernize the US NAS. It is a multibillion-dollar project and arguably the most ambitious infrastructure project in US history, the ultimate goal of which is to deliver Trajectory Based Operations (TBO). NextGen is a collection of different technologies, many of which are described elsewhere in the chapter, aimed at enhancing all aspects of the air traffic system. Some of the technologies involved include:

- **ADS-B**—see 8.2.7
- **Established on RNP (EoR)**—to shorten flightpaths by allowing compliant aircraft to turn earlier (see also 8.3.5)
- **Performance Based Navigation (PBN)**—airports operating RNAV SIDS, STARS, and Approaches. See 8.3.5.
- **Data Communications (DataComm)**—including Datalink, CPDLC, and ADS-C. See 8.2.8.

- **Wide Area Augmentation System (WAAS)**—see 8.3.3
- **System Wide Information Management (SWIM)** is the network backbone structure that will carry NextGen digital information. SWIM will enable cost-effective, real-time data exchange and sharing among users of the NAS.
- **Integrated Terminal Weather System (ITWS)** provides improved integration of weather data into timely, accurate aviation weather information.
- **NextGen Weather** will help reduce weather impacts by producing and delivering tailored aviation weather products via SWIM, helping controllers and operators develop reliable flight plans, make better decisions, and improve on-time performance.
- **Terminal Flight Data Manager (TFDM)** modernizes ATC tower equipment and processes. Using SWIM capabilities, TFDM will share real-time data among controllers, aircraft operators, and airports so they can better stage arrivals and departures for greater efficiency on the airport surface.
- **Remote Tower Services (RTS)**—see 8.6.6

NextGen is discussed at more length in Chapter 17, it is an ongoing project, with progress updates available from the FAA at its website.[39]

8.6.3   Single European Sky ATM Research

Single European Sky ATM Research (SESAR)[43] is a joint undertaking between European institutions, responsible for all ATM-relevant R&D efforts in the European Union. It is the technology pillar of the Single European Sky project. Projects range from exploratory research, through industrial research to very large-scale demonstrations. SESAR has funding of around €1.6 billion for the period 2016–2024. In 2021, SESAR produced its *Solutions Catalogue*[44] detailing the projects under development. The groupings and some of the delivered projects are given below:

High-performing airport operations

- Precision approaches using GBAS Cat II/III (rather than ILS)
- Time-based separation
- Minimum pair separations based on required surveillance performance (RSP)
- Ground traffic alerts for pilots for airport operations
- Independent rotorcraft operations at the airport

Advanced air traffic services

- Extended arrival management (AMAN) horizon
- Point merge in complex terminal airspace
- Optimized route network using advanced RNP
- User-preferred routing
- Enhanced short-term conflict alerts (STCA) with downlinked parameters
- Enhanced ACAS using autoflight systems

Optimized ATM network services

- Automated support for dynamic sectorization
- Automated support for traffic complexity detection and resolution
- Calculated takeoff time and target time of arrival
- Reactive flight delay criticality indicator

Enabling aviation infrastructure

- Initial SWIM technology solution
- Initial ground-ground interoperability
- Extended flight plan
- ACAS ground monitoring and presentation system

As the above shows, there is a huge number of developments taking place in the ATM system—far more than could be described here. However, two more specific study areas are worth noting. The first is SAFELAND, a concept for single pilot operations in which a ground-based operator, working from a remote ground station can interact with onboard automation and air traffic controllers to support a single pilot who becomes incapacitated at any point. In the second subject area, in October 2022, SESAR released a collection of results on the subject of Artificial Intelligence (AI) in ATM from 15 different projects.[45] The report envisages multiple uses for AI, including:

- Reducing delays
- Maintaining controller situational awareness
- Route planning
- Speech recognition
- Weather management

The report also highlights ATM as an ideal candidate for AI automation given the repetitive nature of the task, the clearly defined goals and the significant data generation involved. SESAR has also reported on a project around building trust and confidence in AI by making it explainable.

8.6.4   NextGen–SESAR Harmonization

In 2018, the FAA published the third edition of the *NextGen—SESAR, State of Harmonization*[46] document. The document gives the following areas of progress:

- **Coordination in support of ICAO**—The United States and Europe are supporting ICAO in developing the next edition of the GANP.
- **Harmonization risk management**—A new joint harmonization risk issue opportunity management (HRIOM) framework has been established
- **Air/ground data communications**—The first joint US-EU air/ground data communication strategy supported by industry from both sides of the Atlantic was delivered.

- **SWIM**—An initial joint US-EU SWIM strategy has been developed.
- **Navigation**—There has been progress in developing a joint US-EU navigation systems roadmap.
- **UAS/Drones**—Collaboration has been established on the harmonization and interoperability aspects related to the integration of unmanned aerial systems, or drones.
- **Cybersecurity**—Harmonization activities on cybersecurity have been initiated.

8.6.5   Airport Surface Detection Equipment

Airport Surface Detection System—Model X (ASDE-X) is a surveillance system (first mentioned in Chapter 7) that allows air traffic controllers to track surface movement of aircraft and vehicles. Part of the purpose of the system is to help reduce critical runway incursions. By providing detailed monitoring of movement on runways and taxiways, ASDE-X can alert air traffic controllers of potential runway conflicts. By combining data from multiple sources, the system is able to track and identify vehicles without transponders, as well as transponder-equipped vehicles and aircraft. In the United States, 35 airports have currently been equipped with the technology including Atlanta (ATL), Dallas (DFW), Denver (DEN), Chicago (ORD) and Los Angeles (LAX).[47]

In addition, there are new products[48] in the market making use of millimeter-wave radar sensor units combined with near-infrared optical sensors to detect and alert ground staff to the presence of Foreign Object Debris (FOD) on the runway or airport surfaces, reducing the need for inefficient inspections and the risk of FOD being missed.

8.6.6   Remote Towers

Remote Tower Service (RTS) is a system which allows aerodrome ATC or FIS to be provided from a location other than the aerodrome. The system is designed to maintain a level of operational safety which is equivalent to that achievable using a manned tower, located at the aerodrome, to oversee both air and ground movements.[49] In 2016, Sweden was the first country in the world to certify a remote tower, controlling air traffic at Örnsköldsvik from a remote tower hub 93 miles away at Sundsvall.

Remote towers are part of NextGen and in June 2022, Selma Alabama announced plans to build the United States' first Remote Tower Air Traffic Control Center.[50] The center will also act as a training center. In Europe there are a number of operational remote towers, mostly in Sweden and Norway. EASA revised its guidance material on Remote Aerodrome ATSs in 2020.[51]

8.6.7   Holographic Radar

One of the challenges of future ATC will be to incorporate small, low-level UASs into a traffic control system designed for large, high-level aircraft. One technology aimed at addressing this problem is holographic radar.[52] Holographic radar aims to provide a full, digital, 3D picture of the sky.

Rather than using a traditional "scanning" beam to survey an area, holographic radar uses a "staring array" (similar to some imaging satellites). As the name suggests, it uses a phase difference approach, similar to the way in which holograms are created. However, this approach takes significant computing power, hence its relatively recent introduction. The system is part of a study region in the United Kingdom, the National Beyond Visual Line of Sight Experimental Corridor (NBEC), a safe, segregated airspace initiative for testing drone infrastructure, airspace structures and unmanned traffic management (UTM) concepts. Holographic radar can produce higher resolution target signatures by combining the returns from multiple sensors in the array. Because it is not scanning, it can constantly "illuminate" the targets, aiding resolution. The system can also remove reflections generated by the rotating blades of wind turbines.

8.7   ATC Related Case Studies

Although millions of flights move safely through the ATC system every day, there are still times when the system does not perform as it should. It is important to understand these issues in order to prevent future accidents. You will see echoes of many of these accidents in the technology improvements described above.

8.7.1   Überlingen

Possibly the most notorious accident related to ATC is the midair collision that occurred over Überlingen in Germany on 1 July 2002. In this accident, a Tupolev Tu-154 collided with a Boeing 757 cargo aircraft while both were under ATC control by Zurich ACC. All of the passengers and crew from both aircraft, 71 in total, were killed.

On the night of the accident, maintenance work was underway at the ATC center meaning that the radar system was in "fallback mode" and hence visual Short Term Conflict Alerting (STCA) was not available. There were two controllers working, but one was resting as was unofficial common practice at quiet times. This meant that at the time of the accident, one controller was left covering three aircraft, but across two screens and two frequencies meaning that messages could be lost or transmitted simultaneously.

The two aircraft were approaching each other at FL360, creating a conflict. A controller from another facility did receive and notice an STCA alert, but the maintenance work meant that controller could not reach Zurich ACC by phone. In the final moments of the accident, the controller, who had not become aware of the conflict in time, instructed the TU-154 to descend. This instruction was too late to maintain the required minimum separation, however the controller did not impart the urgency of the situation. The TU-154 initiated a descent and the controller turned their attention to the third aircraft (an A320) on the second screen.

Both the TU-154 and B757 were equipped with TCAS II and 7 seconds later, the B757 received a TCAS collision RA to "descend" and the TU-154 received an RA to "climb." However, the TU-154 continued the descent. The copilot questioned this continuation, but this was not addressed in the cockpit. The 757 reported the TCAS descent to ATC, but this was not noticed by the controller, in part due to cross-traffic from the A320 on the second frequency. The ATC computer in ACC Zurich issued an aural Short Term Conflict Alert (STCA) warning, but this was not noticed in the control room. Both TCAS RAs were extended in the cockpit ("increase descent" and "increase climb" respectively) but both aircraft continued to descend. The aircraft collided at just below 35,000 feet.

While it may be tempting to attribute the cause of this accident as a simple failure by ATC to maintain separation, the report[53] by the BFU, the German investigation agency, gives an excellent demonstration of how multiple factors can contribute to an accident. Many organizational issues were identified including CRM, regulations and procedures, safety culture, and risk assessment. Again, this brief summary does not do justice to the complexity of the accident and the investigation, and the report is highly recommended. As we discussed in Chapter 3, there were multiple barriers in the system to prevent this accident, and yet all were ineffective.

8.7.2 Air France Flight 447

We have already discussed the Air France Flight 447 accident a number of times in the book, however, there is a specific element related to ATC that is worth highlighting. Over the course of just over an hour, the aircraft was traversing three FIRs, specifically RECIFE, ATLANTICO, and DAKAR. Radar coverage ended just outside the RECIFE FIR as shown in Figure 8-6. The crew were in HF contact with ATC but radio communications over HF were reported as being poor quality that evening.

CPDLC was not in operation in the ATLANTICO FIR on the day of the accident and DAKAR were trialing CPDLC communications at the time of the accident, with a requirement to confirm all instructions on HF radio. In addition, Senegal was trialing a new ATC system and the specifics were not well understood by the controllers. Due to a formatting error in the flight plan, the DAKAR ATC system was not able to accept attempts by the crew to connect via ADS-C or CPDLC.

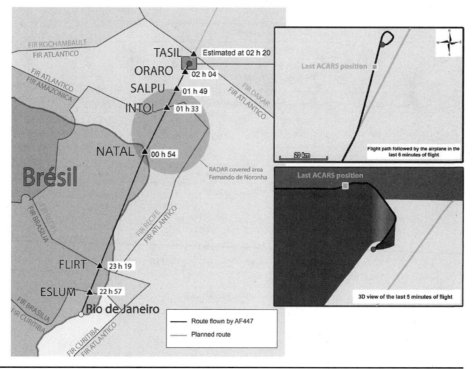

Figure 8-6 Flightpath of AF447. (Source: BEA[51])

The extract below summarizes some key parts of the ATC exchanges.

0h 36m 40	The RECIFE controller instructed the crew to maintain FL350 and contact ATLANTICO when they were over INTOL
01h 14m 58	The RECIFE controller coordinated with the ATLANTICO controller that AF447 was estimated at INTOL at 01h 32 at FL350
01h 31m 44	The RECIFE controller gave the crew the ATLANTICO and DAKAR frequencies and told them to contact DAKAR only after TASIL (since TASIL is on the boundary between ATLANTICO and DAKAR)
01h 33m 25	The crew contacted ATLANTICO saying they had passed INTOL and giving their estimates for the next two waypoints
01h 35m 26	The ATLANTICO controller started coordination to handover the flight to the DAKAR controller, but the DAKAR controller interrupted this, saying they would call back
01h 35m 38	The ATLANTICO controller sent a SELCAL[ii] call, whose completion the crew confirmed and thanked the controller
01h 35m 46	The ATLANTICO controller asked the crew to maintain FL350 and give an estimate for TASIL. Over the next 40 seconds, the controller asked the crew three times to give its estimated time at TASIL.
01h 46	The DAKAR controller asked the ATLANTICO controller for more information, since they had no flight plan. The DAKAR control center then created a virtual flight plan and trajectory through their FIR. There was no radio communication or ADS-C connection with the flight.
01h 49	The radar showed the aircraft passing SALPU
c. 02h 08	The crew changed course 12 degrees to the left (most likely to avoid the weather ahead)
02h 10m 05	The initial incorrect speed indications were logged and there was a subsequent loss of control.
02h 14m 28	The aircraft impacted the sea
02h 47m 00	The DAKAR controller coordinated the flight with the SAL (Cape Verde) controller (the next FIR) that AF447 was estimated to leave the DAKAR FIR (at point POMAT) at 03h 45, but that it had not yet established contact with them (02h 48m 7)
3h 54m 30	The SAL controller called the DAKAR controller to confirm the expected time at POMAT. *[100 minutes after the time we now know the aircraft crashed]*
	There were then attempts to clarify the expected times and to contact AF447 by controllers and nearby flights, including discussions with the ATLANTICO controller and another nearby FIR.
05h 23	The disappearance of the flight was registered

TABLE 8-1   Key ATC interactions of AF447. (Source: BEA[54])

Had an ADS-C and CPDLC connection been made, the loss of altitude by AF447 would have been alerted to the DAKAR controller. This would not have prevented the accident, but it may have been possible to raise the alarm sooner. The BEA made two safety recommendations relating to the ATC aspects of the accident:

> *The Brazilian and Senegalese authorities make mandatory the utilization by aeroplanes so equipped, of ADS-C and CPDLC functions in the zones in question.*

[ii] SELCAL is a way of addressing a radio voice message to a particular aircraft—almost like a phone number for the aircraft.

and

> *ICAO request the involved States to accelerate the operational implementation of air traffic control and communication systems that allow a permanent and reliable link to be made between ground and aeroplane in all areas where HF remains the only means of communication between the ground and aeroplanes.*

There is no indication that the ATC issues were contributory to the accident, however it is conceivable that large delays in obtaining a clearance in oceanic airspace, particularly without a CPDLC connection, may have discouraged the pilots from seeking a larger deviation around the bad weather ahead. Regardless, it is evident that clear and timely communication with ATC is fundamental to aviation safety.

Irrespective of the above, the delay in recognizing that the flight was lost was clearly undesirable. In addition, part of the reason it was so difficult to locate the aircraft wreckage was that, in the absence of any contradictory evidence, the investigators assumed that the flight had continued with significant forward groundspeed, giving them a likely impact area well ahead of the last known aircraft position. As we know now, the aircraft entered a very severe stall (see Figure 8-6), meaning that forward speed was reduced significantly, coupled with very high vertical speed. Despite the difficulties in locating the aircraft wreckage, attempts made in the following years by the investigation agencies to introduce methods of locating lost aircraft, including fitting deployable recorders were not successful.

8.7.3   Malaysian Airline Flight MH370

On March 8, 2014, a Boeing 777, registration 9M-MRO was operating Malaysian Airlines Flight MH370 with 239 persons on board. After a routine handover from Malaysian ATC to Viet Nam ATC the aircraft went missing and the Mode S transponder stopped responding. The aircraft was tracked by civilian and military radar for a period before disappearing over the sea.

Several underwater searches were conducted, coordinated by the ATSB, with the location informed by SATCOM "handshakes" (short connections) between the aircraft and a satellite. The main wreckage of the aircraft remains undiscovered, but parts identified as almost certainly having come from the accident aircraft have washed ashore. Away from the events of the accident, what MH370 highlighted was that, although aircraft can accurately calculate their own position, without primary radar it is very difficult to locate an aircraft that does not have its tracking systems (Mode S, ADS-B, etc.) working.

The combined occurrences of AF447 and MH370 convinced ICAO to introduce the Global Aeronautical Distress & Safety System (GADSS).[55] This initiative has generated activity around aircraft tracking, autonomous distress tracking and postflight localization and recovery. The aircraft tracking phase is already a requirement, with aircraft needing to report their position every 15 minutes. A requirement for one-minute tracking when an aircraft is in distress is scheduled for 2023.

8.7.4   Midair Collision in Brazil

On September 29, 2006, one of the deadliest accidents in Brazilian aviation history occurred. The crash involved a Boeing 737 from Gol Transportes Airlines operating as GOL Flight 1907 and an Embrarer-136 with US registration number N600XL, operating as an executive aircraft under the company ExcelAire Services.

While both flights were on their respective routes, they collided at Flight Level 370. The left wing of each aircraft made contact as they passed with a supersonic closure rate,

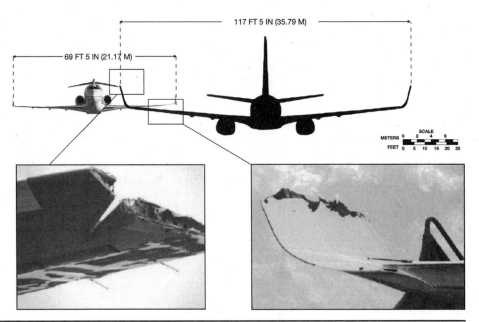

FIGURE 8-7   The geometry of the Brazil midair collision. (Source: Centro de Investigação e Prevenção de Acidentes Aeronáuticos[56])

as depicted in Figure 8-7. The Embraer jet lost part of the left winglet and sustained damage in the left stabilizer and left elevator. It did however remain controllable in flight and made an emergency landing at a military facility. Conversely, the Boeing aircraft lost about one-third of the left wing in the collision, leaving it uncontrollable. The ensuing abrupt spiral dive caused the aircraft to come apart in flight before violently impacting the ground in the middle of the thick rainforest. There were no injuries onboard the Embraer jet, but there were no survivors on the Boeing aircraft due to the impact forces.

The investigation report[53] presented several contributing factors related to ATC, and these highlight how ineffective attention can erode safety:

- There was an incomplete flight clearance issued by the assistant controller in Sao Paulo and the ground controller at São José dos Campos. This caused the Embraer jet pilots to believe they could maintain Flight Level 370 all the way to their destination despite a discrepancy with their flight plan.

- The controller of sectors 5 and 6 in the Brasilia airspace did not provide the controller of sectors 7, 8, and 9 with the necessary information for handing off the Embraer jet.

- The air traffic controller for sectors 7, 8, and 9 did not make radio contact with the Embraer pilots to change the aircraft's flight level and switch the frequency, thus losing contact with the aircraft.

- The air traffic controller of sectors 7, 8, and 9 did not perform the necessary procedures for the loss of transponder and loss of radar. This enabled the Embraer jet to continue flying at the incorrect flight level.

- There was a lack of communication between sectors and control centers during the coordination and handoff.

- There was also a lack of monitoring, advisory, and guidance associated with individual ATC decisions at the Brasilia area control center.

- Standard procedures were not followed during the hand-offs of both aircraft between controllers and also when radar contact was lost with the Embraer jet.

- The collision avoidance technology aboard the aircraft did not function likely due to inadvertent inactivation of the transponder on N600XL. (NTSB comment)

As with the Überlingen accident, there were multiple factors involved in this accident and multiple barriers that were ineffective. In response to the accident, the investigation team made more than 60 safety recommendations. It is a sad irony that one of the factors in this accident was the ability of both aircraft to track a course at a given altitude with great precision. Had that not been the case, the aircraft may not have collided. Today, the Strategic Lateral Offset Procedure (SLOP) aims to address that risk.

8.8   Conclusion

It is easy to overlook the role that air traffic controllers play in the aviation field because they are an invisible force. In general, the traveling public never sees the role that air traffic control plays from the start to the end of a flight. However, as we have seen, the systems, technologies, and procedures designed to allow safe passage are complex, in order to accommodate all airspace users safely. The chapter has also shown that when those systems are ineffective, the results can often be catastrophic.

The ultimate challenge for the worldwide air traffic system is how to modernize the system, by making best use of new technology to increase efficiency and incorporate new air vehicles and missions, while maintaining and improving current levels of safety.

Key Terms

Air Route Traffic Control Centers (ARTCCs)

Airport Surface Detection Equipment, Model X (ASDE-X)

Airport Surveillance Radar (ASR)

Airport Traffic Control Towers (ATCTs)

Area Navigation (RNAV)

Automatic Dependent Surveillance—Broadcast (ADS-B)

Controlled Airspace

Controller Pilot Data Link Communication (CPDLC)

Data Communications (Data Comm)

EGNOS

Galileo

Global Positioning System (GPS)

Instrument Flight Rules (IFR)

Instrument Landing System (ILS)

Navigational Aids (NAVAIDS)

NextGen Weather

Performance-Based Navigation (PBN)

Reduced Vertical Separation Minimums (RVSM)

Required Navigation Performance (RNP)

Satellite-based Augmentation System (SBAS)

Small Unmanned Aircraft Systems (sUAS)

System Wide Information Management (SWIM)

Terminal Automation Modernization and Replacement (TAMR)

Terminal Flight Data Manager (TFDM)

Terminal Radar Approach Control (TRACON)

Traffic Flow Management (TFM)

Uncontrolled Airspace

Unmanned Aircraft Systems (UAS)

Uncrewed Traffic Management (UTM)

VHF Omnidirectional Range (VOR)

Visual Flight Rules (VFR)

Wide Area Augmentation System (WAAS)

Topics for Discussion
1. Describe how air traffic is controlled close to an airport.
2. Briefly describe how GPS works.
3. What are some of the advantages of satellite-based navigation?
4. What is automatic dependent surveillance—broadcast (ADS-B)?
5. Distinguish between ADS-B and ASDE-X.
6. What is NextGen and what are its goals?
7. Discuss the basic provisions of the FAA UAS Rule (FAR Part-107).
8. Outline some of the factors in the Überlingen accident.

References
1. ICAO *2018 Annual Report*. https://www.icao.int/annual-report-2018/Documents/Annual.Report.2018_Air%20Transport%20Statistics.pdf
2. FAA *Traffic by the Numbers*. https://www.faa.gov/air_traffic/by_the_numbers
3. ICAO Annex 11 *Air Traffic Services*.
4. ICAO Doc 4444 *Procedures for Air Navigation Services (PANS)-ATM*. https://store.icao.int/en/procedures-for-air-navigation-services-air-traffic-management-doc-4444
5. UK CAA *ANSP Certification and designation*. https://www.caa.co.uk/commercial-industry/airspace/air-traffic-management-and-air-navigational-services/air-navigation-services/ansp-certification-and-designation/ansp-certification-and-designation/
6. EASA ATM/ANS Regulations. https://www.easa.europa.eu/en/document-library/easy-access-rules/easy-access-rules-air-traffic-managementair-navigation-services

7. EASA Standardised European Rules of the Air (SERA). https://www.easa.europa.eu/en/downloads/68174/en

8. FAA *Air Traffic.* https://www.faa.gov/air_traffic

9. FAA *Air Traffic Organization.* https://www.faa.gov/about/office_org/headquarters_offices/ato

10. NATS *Introduction to Airspace.* https://www.nats.aero/ae-home/introduction-to-airspace/

11. ICAO *Annex 11, Section 2.6.* https://elibrary.icao.int/explore;searchText=annex%2011

12. https://www.nats.aero/ae-home/introduction-to-airspace/. Accessed November 2022.

13. EGAST *Flight Information Service.* https://skybrary.aero/sites/default/files/bookshelf/2700.pdf

14. CFR 91.225 and 91.227. https://www.ecfr.gov/current/title-14/chapter-I/subchapter-F/part-91/subpart-C?toc=1

15. FAA AC 90-114B *Automatic Dependent Surveillance-Broadcast Operations.* https://www.faa.gov/documentLibrary/media/Advisory_Circular/AC_90-114B.pdf

16. FAA Report DOT-VNTSC-FAA-16-14 *Cockpit Display of Traffic Information (CDTI) and Airport Moving Map Industry Survey.* https://rosap.ntl.bts.gov/view/dot/12357

17. FAA AC 90-117 *Data Link Communications.* https://www.faa.gov/documentLibrary/media/Advisory_Circular/AC_90-117_(E-update).pdf

18. ICAO *Annex 10, Volume II, Chapter 8.* https://elibrary.icao.int/explore;searchText=Annex%2010

19. ICAO Doc 10037 *Global Operational Data Link (GOLD) Manual.* https://store.icao.int/en/global-operational-data-link-gold-manual-doc-10037

20. Code 7700 *CPDLC.* https://code7700.com/cpdlc.htm#gsc.tab=0

21. Code 7700 *ADS-C.* https://code7700.com/ads-c.htm#gsc.tab=0

22. IATA IFB Region. https://www.skybrary.aero/sites/default/files/bookshelf/2976.pdf

23. FAA *Very High Frequency Omnidirectional Range Minimum Operational Network (VOR MON).* https://www.faa.gov/about/office_org/headquarters_offices/ato/service_units/techops/navservices/gbng/vormon

24. EUSPA, *What is SBAS?* https://www.euspa.europa.eu/european-space/eu-space-programme/what-sbas

25. ICAO Doc 9613 *Performance-based Navigation (PBN) Manual.* https://www.icao.int/sam/documents/2009/samig3/pbn%20manual%20-%20doc%209613%20final%205%2010%2008%20with%20bookmarks1.pdf

26. FAA Performance Based Navigation (PBN) Guidance & Approval. https://www.faa.gov/about/office_org/headquarters_offices/avs/offices/afx/afs/afs400/afs410/pbn

27. EASA *Performance Based Navigation Guide.* https://www.easa.europa.eu/community/system/files/2021-03/PBN%20Guide.pdf

28. FAA *Drones.* https://www.faa.gov/uas

29. EASA *Easy Access Rules for Unmanned Aircraft Systems.* https://www.easa.europa.eu/en/downloads/110913/en

30. EASA *Civil Drones (unmanned aircraft).* https://www.easa.europa.eu/en/domains/civil-drones

31. FAA Unmanned Aircraft System Traffic Management (UTM). https://www.faa.gov/uas/research_development/traffic_management

32. FAA *UAS UTM Concept of Operations v2.0.* https://www.faa.gov/sites/faa.gov/files/2022-08/UTM_ConOps_v2.pdf

33. EASA *Urban Air Mobility (UAM).* https://www.easa.europa.eu/en/domains/urban-air-mobility-uam

34. SESAR *Demonstrating the Everyday Benefits of U-Space.* https://www.sesarju.eu/U-space_everyday_benefits

35. NTSB *Report into UH-60M Collision with DJI Phantom.* https://data.ntsb.gov/carol-repgen/api/Aviation/ReportMain/GenerateNewestReport/96058/pdf

36. TSB *In-flight collision with drone.* https://www.tsb.gc.ca/eng/enquetes-investigations/aviation/2017/a17q0162/a17q0162.html

37. Wikipedia *List of UAV-related incidents.* https://en.wikipedia.org/wiki/List_of_unmanned_aerial_vehicles-related_incidents

38. NTSB *Amendment to the Definition of Unmanned Aircraft Accident.* https://www.federalregister.gov/documents/2022/07/14/2022-14872/amendment-to-the-definition-of-unmanned-aircraft-accident

39. AAIB *Investigating accidents to Unmanned Aircraft Systems.* https://www.gov.uk/government/publications/investigating-accidents-to-unmanned-aircraft-systems/investigating-accidents-to-unmanned-aircraft-systems

40. AAIB *Investigation to Alauda Airspeeder Mk II.* https://www.gov.uk/aaib-reports/aaib-investigation-to-alauda-airspeeder-mk-ii-uas-registration-n-slash-a-040719

41. Skybrary *4D Trajectory Concept.* https://www.skybrary.aero/articles/4d-trajectory-concept

42. FAA *Next Generation Air Transportation System (NextGen).* https://www.faa.gov/nextgen

43. SESAR website. https://www.sesarju.eu/

44. SESAR *Solutions Catalogue 2021.* https://www.sesarju.eu/sites/default/files/documents/reports/SESAR%20Solutions%20Catalogue%202021%20small.pdf

45. CORDIS *AI in Air Traffic Management.* https://www.sesarju.eu/sites/default/files/documents/AI%20in%20air%20traffic%20management%20brochure.pdf

46. FAA *NextGen—SESAR State of Harmonisation.* https://www.faa.gov/nextgen/nextgen-sesar-state-harmonisation

47. FAA *Airport Surface Detection Equipment, Model X (ASDE-X).* https://www.faa.gov/air_traffic/technology/asde-x

48. Thales *FOD Detection System.* https://www.thalesgroup.com/sites/default/files/database/d7/asset/document/fodetect_datasheet.pdf

49. Skybrary *Remote Tower Service.* https://www.skybrary.aero/articles/remote-tower-service

50. Made in Alabama *Remote Tower.* https://www.madeinalabama.com/2022/06/pioneering-remote-aviation-control-tower-landing-at-selmas-craig-field/

51. EASA *Guidance on Remote Tower Operations.* https://www.easa.europa.eu/en/document-library/easy-access-rules/easy-access-rules-guidance-material-remote-aerodrome-air-traffic

52. Thales *Holographic Radar.* https://www.thalesgroup.com/en/holographic-radar

53. Ueberlingen *Final Accident Report.* https://www.bfu-web.de/EN/Publications/Investigation%20Report/2002/Report_02_AX001-1-2_Ueberlingen_Report.pdf?__blob=publicationFile

54. EA *AF447 Final Report.* https://bea.aero/docspa/2009/f-cp090601.en/pdf/f-cp090601.en.pdf

55. ICAO *GADSS Concept of Operations.* https://www.icao.int/safety/globaltracking/Documents/GADSS%20Concept%20of%20Operations%20-%20Version%206.0%20-%2007%20June%202017.pdf

56. CENIPA *PR-GTD / N600XL Accident Report.* https://sistema.cenipa.aer.mil.br/cenipa/paginas/relatorios/rf/pt/PR_GTD_N600XL_29_09_06.pdf

PART III

Human Safety

CHAPTER 9

An Introduction to Human Safety

Introduction

Aviation is often described as a complex sociotechnical system, one in which humans and technology work closely and where many of the connections and interactions are not well understood. Imagine how many humans have been involved by the time a commercial aircraft lands at its destination. In the broadest sense, these include the aircraft designers, manufacturers, maintainers, pilots, cabin crew, trainers, ground handlers, airport operations, air traffic control, and even passengers. Designing a system that makes sense to all the different personalities, competencies, and experiences is enormously challenging.

When things go wrong, media reports seem to focus on human frailty far more frequently than the technical failures or vulnerabilities that often precede an error or violation. The terms "pilot error" or "human error" are often used to simplify complex sets of causal and contributory factors with the result that safety measures can be focused on individual actions rather than higher systemic factors. It is easy to forget the very positive contribution that humans make to the aviation safety system or to put them in roles that fail to exploit their relative strengths. Rather than try and "engineer-out" the humans from aviation, successful safety management depends on achieving an appropriate blend.

This section of the book looks at the impact of the human on commercial aviation safety. Having introduced what is a vast topic, Chapter 10 considers the threat from humans to commercial aviation safety, Chapter 11 offers a balancing view of the positive contribution. Retaining a sense of balance is imperative—the human is always going to be part of the system, either as a designer, operator, maintainer, or customer. However, just as those roles have evolved into the modern civil aviation system, so too they will continue to do so as new technologies and constraints become apparent. Human performance will remain an important topic for commercial aviation safety.

9.1 Asiana 214

Chapter 2 introduced the accident which happened on July 6, 2013, in which a Boeing 777 aircraft operated by the Korean airline, Asiana, crashed during final approach to San Francisco Airport, USA (Figure 9-1). Three people were killed and 49 were seriously injured in a preventable accident. The National Transportation Safety Board

FIGURE 9-1 The main fuselage of Asiana Flight 214 after it crash-landed in San Francisco. (Source: NTSB. Public Domain, https://en.wikipedia.org/wiki/Asiana_Airlines_Flight_214#/media/File:NTSBAsiana214Fuselage2.jpg)

investigation into the accident identified many safety issues including flight crew adherence to standard operation procedures (SOPs), design complexity, automation training, and crashworthiness. While these are all topics which have appeared in many accident reports, their presence in 2013 is a stark reminder of the importance of human-centered design and effective training.

On the day of the accident, the weather was good, the aircraft was serviceable, but the glideslope for runway 28L was not functioning. The crew were vectored for a visual approach, intercepting the final approach course about 14 miles from the runway. Due to the length of the sector, there were three pilots on flight deck—Commander, First Officer, and Relief First Officer with the First Officer performing as Pilot Flying (PF) with the Commander as Pilot Monitoring (PM). At 11:23:05.2, the aircraft intercepted the localizer and at 11:23:17.3, ATC requested the 777 to reduce speed to 180 knots because of nearby traffic. Over the next four and half minutes, the crew struggled to maintain a stable approach to the airport but did not decide to conduct a go-around—the ultimate defense when landing an aircraft, which effectively allows the crew to climb, circle, and have another go.

As the aircraft conducted its final approach, airspeed was reducing, and the aircraft drifted below the glideslope—the imaginary trajectory that an aircraft should follow during this critical stage of flight. As the Enhanced Ground Proximity Warning System (EGPWS) alerted the crew that they were 200 feet above the runway, the PM commented, "it's low," which the PF acknowledged. Nearly 5 seconds later, the EGPWS called "100" and the PM stated, "speed." At this point, he advanced the thrust levers, but within 2 seconds the stick shaker activated to alert the crew to an impending stall.

The aircraft was too low and slow and at 1127:50, the aircraft hit the sea wall at a speed of 106 knots. Airport surveillance cameras captured the terrifying 16 seconds that followed until the wreckage of the aircraft came to a stop 2400 feet from its initial impact.

Television viewers around the world were soon bombarded with images from the scene including video of the aircraft's impact with the seawall and subsequent breakup. These were supplemented by passenger imagery as the survivors escaped burning wreckage. Sadly, in the chaos, one passenger who had survived the accident was run over and killed by a fire truck. This was a tragic and shocking event in which three people lost their lives and many more were injured. However, there are positive as well as negatives to this case study which illustrate advances that have been made in commercial aviation safety. The Boeing 777 is known as a third-generation commercial jet airplane. Accident statistics clearly demonstrate that advances in design, reliability, automation, and crashworthiness made this generation of aircraft substantially safer than the previous generation. The 777's safety record is excellent. It first flew in 1994, but it was not until 2008 that the type experienced its first hull loss accident. The Asiana accident in 2013 was the first fatal accident suffered by the worldwide 777 fleet.

Training and procedures, developed over many years with insight gained from accident investigations, academic research, and operational experience helped cabin crew to assist passengers following what was an unanticipated accident. The materials used within the cabin reduced the risk from fire and the crashworthy design of the seats helped to reduce injury severity for many passengers. Plenty went wrong that day in terms of individual and crew performance, but as we discussed in Chapter 3, typically, there was no single problem that caused the disaster. The official ICAO Annex 13 report into the accident concluded the following[1]:

> *The National Transportation Safety Board (NTSB) determined that the probable cause of this accident was the flight crew's mismanagement of the airplane's descent during the visual approach, the PF's unintended deactivation of automatic airspeed control, the flight crew's inadequate monitoring of airspeed, and the flight crew's delayed execution of a go-around after they became aware that the airplane was below acceptable glidepath and airspeed tolerances. Contributing to the accident were (1) the complexities of the autothrottle and autopilot flight director systems that were inadequately described in Boeing's documentation and Asiana's pilot training, which increased the likelihood of mode error; (2) the flight crew's nonstandard communication and coordination regarding the use of the autothrottle and autopilot flight director systems; (3) the PF's inadequate training on the planning and execution of visual approaches; (4) the PM/ instructor pilot's inadequate supervision of the PF; and (5) flight crew fatigue, which likely degraded their performance.*

In simple terms, while the actions of the crew fell short of what may have been required on the day, there was also an acknowledgment that they were the consequence of a training system and the design of the aircraft and its documentation. ICAO Annex 13 investigation reports do not seek to allocate blame or liability, but rather to learn from serious incidents and accidents to prevent recurrence through safety recommendations. In doing so, they try to look beyond the individual actions on a day to the barriers or risk mitigations that could or should have been in place to stop a hazard ever getting past the last line of defense—in this case, the flight crew.

Over the history of commercial aviation, our understanding of the role of the human has developed and with it, our expectations about what is reasonable to expect. In this accident, we would rightly expect the crew to be trained and licensed to operate this aircraft and into this particular airport. We would expect their training to have been validated by a regulatory body in Korea for both international standards, and recommended practices set by the International Civil Aviation Organisation, and to be compatible with those of the destination state—in this case, the United States. We would expect the aircraft to be operated in accordance with standard operating procedures developed by the manufacturer, tailored by the airline and approved by the regulator. We would also expect that the crew had had an appropriate amount of rest prior to the flight and depending on the length of sector, some form of rest during the flight. All these things have developed from experience, much of it hard won through incidents and accidents. None of these barriers or defenses are perfect and rare events and accidents such as Asiana flight 214 still occur.

To better understand the strengths and weaknesses of the human component of air transport, we need to start with understanding a bit more about how humans work.

9.2   The MK1 Human Being

Human beings haven't changed much over the last 50,000 years. While our understanding of the benefits of exercise, education, nutrition, and healthcare have had a positive influence on life expectancy, the way in which the human body works have not changed a great deal. While we have attempted to better understand the "operating system" through the discipline of psychology, this only started to develop in the late 1800s. For much of aviation's early development, there was a clear mismatch between the design of aircraft and those who were to operate and maintain them. As the Australian Transport Safety Bureau's Layman's Introduction to Human Factors in Aircraft Accident and Incident Investigation[2] cites:

> *During the First World War, the British Royal Flying Corps recorded that, of every 100 aviators killed when flying, two met their death at the hands of the enemy, eight died because of mechanical or structural failures of their aircraft, and 90 perished as a result of what the Flying Corps described as their own individual deficiencies. (US Army Technical Manual, 1941)*

It is argued that the birth of what is now known as human factors in aviation came from the realization that aircraft could not be replaced at a fast enough rate. Attention initially focused on the selection of the people with the right skills and attitudes to fly aircraft, but this was not enough to reduce accident rates significantly. Efforts then shifted to reducing variability in performance through pilot training. At the same time, aircraft performance was improving, bringing new problems associated with altitude, vestibular illusions, and so on, as well as from increasingly complex aircraft. As the discipline of human factors developed, so does the need to understand processes such as pilot judgment and decision-making grew.

Aviation psychology focused on understanding, initially at least, how the pilot's mind worked and latterly, other aviation disciplines such as air traffic control and maintenance. The Second World War saw great advances through necessity, ranging from pilot selection to the use of basic flight simulators in training. While research into the effects of spatial disorientation, fatigue, and human information processing was taking place, the focus remained very much on the individual rather than how groups or crews of individuals interacted. The concept of "pilot error" as an explanation of aircraft accidents remained strong until the 1970s when the industry started to focus on why, in many accidents, it was a whole crew that seemed to perform below what may have been expected. A classic example from 1972 involved a Lockheed L-1011 aircraft which crashed in the Everglades after the crew of three became so focused on diagnosing what was ultimately found to be a faulty gear warning light that they failed to notice that the aircraft was descending below its safe altitude. A total of 99 people died in what was a serviceable aircraft through what appeared to be a powerful group fixation on a simple problem.

For all the spectacular performances by elite athletes or discoveries by great scientists, the human being remains flawed or at least inconsistent. Some of these flaws and inconsistencies apply to all humans, some are specific to individuals while others may relate to a particular circumstance. The study of the physiology and psychology of humans has taught us a great deal about how humans work and think. However, translating this knowledge into the design and operation of aircraft is not straightforward. At any one time, the aviation system is filled with people with different personalities, preferences, pressure, and experiences interacting with each other and operating technology designed by similarly diverse people. Add to that the fact that some of those people will be having a bad day, many will be communicating in a different language or over a crackling radio, and it is a miracle that the system works as well as it does!

Humans are excellent problem solvers, but even this strength has a downside. Our ability is based on pattern matching and the application of heuristics—simple mental subroutines. While most of the time, this will lead to a successful outcome, on other occasions, such as when the situation has been misperceived or misunderstood, the consequences may be severe.

9.3   Fatal Confusion on the Flight Deck

In January 1989, a Boeing 737-400 operating as British Midland Flight 92 was climbing out of London Heathrow Airport on a routine flight to Belfast in Northern Ireland. This was a relatively short flight and as it was a Sunday evening, the aircraft was full of 118 passengers returning home. The aircraft experienced a series of compressor stalls resulting in severe vibration as well as smoke and fumes in the cabin and the two pilots immediately started to troubleshoot. Thinking the problem was with the number two (starboard) engine, the crew throttled it back and eventually shut it down. As they did so, the smoke and fumes went away, and the commander asked the cabin crew to prepare for a precautionary diversion to East Midlands Airport (Figure 9-2).

During the diversion, there were minimal indications that the crew had made anything other than the correct decision, but as the aircraft commenced its final approach, vibration increased and the number one engine failed. Efforts to relight the number two engine failed and the aircraft collided with terrain just outside the airport perimeter. The embankment that the aircraft came to rest on was very steep and the occupants experienced severe deceleration forces. Forty-seven people died of their injuries and 74 received serious injuries.

During the investigation by the UK's Air Accidents Investigation Branch, it became clear that the flight crew had in fact shut down the wrong engine and that the accident could have been avoided. However, as we have discovered, no accident is the result of a single action and this one revealed a range of issues that needed to be resolved.

The first related to the aircraft's engines which were a scaled-up version of an earlier design, and which proved to be vulnerable to turbine blade damage. Had the engine

FIGURE 9-2 British Midland Airways Boeing 737-400 G-OBME resting on the embankment of the M1 motorway, short of East Midlands Airport. (Source: Open Government Licence, https://en .wikipedia.org/wiki/Kegworth_air_disaster#/media/File:G-OBME_Aerial_photograph_of_site_(AAIB).jpg)

failure not occurred, there would have been no need for rapid troubleshooting by the crew. The First Officer appeared to be uncertain as to which engine was failing and the CVR records him saying, "It's the le…It's the right one." The Commander appeared to accept this diagnosis and order "it" to be shut down. Faced with an unexpected failure, the investigators concluded that the crew reacted too swiftly and without proper confirmation of the problem. This may not have been helped by the redesigned flight instrumentation which had seen traditional analog "steam" gauges replaced with a hybrid digital display. As shown in Figure 9-3 which is reproduced from the investigation report[3], the widely used white needle is replaced by three yellow digital dots on the outside arc of the instrument. Imagine trying to read such a gauge while experiencing severe vibration, smoke, and fumes and when there is no positive indication other than the position of the "needle" to indicate a vibration problem with a particular engine.

The instrumentation is not the only part of this complex story, the flight crew would likely to have been surprised and even though training for engine failures is routine,

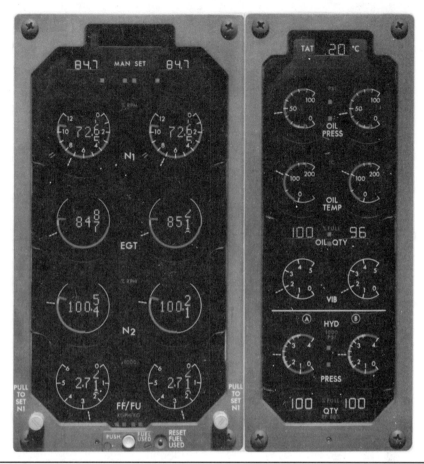

Figure 9-3 Boeing 737-400 instrumentation. (Source: Open Government Licence)

experiencing them for real has become unusual. This is one of the ironies of increasing reliability and on this occasion, the crew must have felt some relief that they appeared to have got the situation under control. With the benefit of hindsight, it is easy to see that the crew could have verified their first decision and avoided the accident, but faced with an apparently improving situation, it is not a natural human tendency to question the situation.

Even acknowledging that the crew made a rushed decision where they had not followed the airline's checklist, the accident could still have been prevented. Subsequent human factors training has reinforced the importance of crew reviewing and evaluating their decisions. In this case, both the cabin crew and the flight deck instrumentation held critical information. In the case of the cabin crew, at least one of them remembered an announcement from the flight deck about a "fire in the right engine" when they had clearly seen sparks coming from the left one. Reflecting on why they had not questioned the flight deck crew, the flight attendant has said that culturally, that would never have been acceptable, and they had reassured themselves that it probably just a slip of the tongue and that the pilots would have known what they were doing.

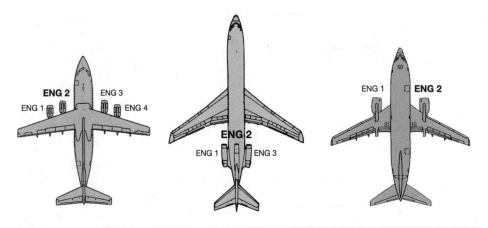

FIGURE 9-4 Relative position of engines on three aircraft types. Note that the "number 2" engine is on the port wing, center, or starboard wing depending on configuration. (Source: Author)

The role of the cabin crew as the "eyes and ears" of the flight deck has been highlighted following a number of high-profile events, especially since the demise of the flight engineer. In the case of United Airlines Flight 232 where a DC10 aircraft suffered a catastrophic failure of its center engine resulting in damage to all three hydraulic systems (see Chapter 11), the cabin crew were able to identify damage to the aircraft control surfaces and communicated it to the crew. However, without a common technical language, they found it hard to communicate what the "back wing" meant.

Aviation is awash with complex terminology and acronyms. The British Midland Flight 92 accident provides a good example. The engine that failed was on the left-hand side of the aircraft, but we all know that many people confuse left from right and for passengers who may have heard the PA that said the problem was with the right engine, how many would know that the position is based on a plan view of the aircraft with the nose facing forwards? To avoid this confusion, airlines chose to refer to engines by number starting with 1 on the far left in plan view—which sounds straightforward.

Imagine then the confusion for flight attendants of a particular airline with a mixed fleet of aircraft that included the BAe146, Boeing 727, and 737 aircraft seen in Figure 9-4. Unlike the pilots, cabin crew could fly on any type and could even find themselves on all three aircraft on one day. During that time the number 2 engine could be found on the inside of the port wing of the BAe146, the center (tail) of the 727, and the starboard wing of the 737! While it may not be terminology they would use very often, in an emergency, it is easy to imagine the confusion.

9.4   The Birth of Modern Aviation Human Factors

In 1969, after completing his psychology degree at the University College London, the British writer and ex-pilot David Beaty published his first nonfiction book, *The Human Factor in Aircraft Accidents*.[i] Beaty was fascinated by what lay behind the possible causes of aviation accidents labeled as "pilot error." During the 1970s, several fatal

---

[i] David Beaty is probably better known for his later book, published in 1991, *The Naked Pilot—The Human Factor in Aircraft Accidents*.

accidents started to highlight the importance of crew performance including the L1011 Everglades crash mentioned above and the loss of a British European Airways Trident aircraft after takeoff from Heathrow Airport, where the captain mistakenly retracted the leading-edge devices causing the aircraft to enter a deep stall. In response to those and other serious accidents, the Flight Safety Foundation (FSF) and the International Air Transport Association (IATA) convened conferences in 1974 and 1975 in Virginia and Turkey to address the human causes of commercial aviation accidents.

The concern was growing within the awareness of aviation safety professionals but was not necessarily high on the consciousness of the public, but that would soon change. Just 2 years after the IATA conference, two Boeing 747 aircraft collided in foggy conditions on the island of Tenerife, Spain, causing the worst loss of life in any single accident in the history of commercial aviation. That terrible record stands to this day. The collision was a result of coordination breakdown, the *false expectancy* of a takeoff clearance in the mind of one of the captains, and the lack of confidence of crewmembers in questioning the captain's decision to take off without a clearance. A total of 583 people lost their lives in that tragic accident.

One year later, in 1978, one more accident occurred that would be the final straw that broke the camel's back. A Douglas DC-8 operating as United Airlines Flight 173 crashed near Portland, Oregon, after running out of fuel, killing 10 occupants. The accident resulted from the captain focusing too heavily on preparing the cabin for an emergency landing due to a gear malfunction, while neglecting both the fuel state and the increasing concerns of the other flight crewmembers who were rightfully worried about running out of fuel.

The string of accidents all occurred due to poor communication and coordination by the humans operating the flights so, building on the momentum of the safety conferences in Virginia and Turkey, NASA decided to host a series of conferences in 1979 out of which the CRM concept was officially born. At that time the term stood for *Cockpit Resource Management* and was narrowly focused on flight deck crewmembers, often comprised of two pilots and a flight engineer. In hindsight we consider the CRM principle that was born in 1979 as the first generation of CRM, since much was to happen over the next few decades that would shape the evolution of the modern CRM—*Crew Resource Management*—of today.

9.5 Evolution of Crew Resource Management Principles

Crew Resource Management has evolved considerably since then. Many experts claim that we are now living in the sixth generation of CRM. The *first generation of CRM* in 1979 focused on changing individual behavior, primarily that of the captain, so that input would be incorporated from other flight deck crewmembers when making decisions. Many of the captains of that era were born in the 1920s and were veteran combat aviators from World War II. The first generation of CRM has since been humorously referred to as "charm school" in that by trying to change the behavior of captains it was often perceived as trying to turn gruffness into charm.

Around 1980 and 1981 two airlines led the CRM movement globally, United Airlines and KLM Airlines. Both programs stressed management and personality styles. As one might imagine, many captains felt personally insulted by such initiatives and were very defensive when told that they were exhibiting accident-prone behavior and therefore had to change. Can you imagine a captain of that era, who may have shot down enemy aircraft and been labeled a hero, suddenly told that he was a problem?

As airline accidents with CRM components continued to happen, such as the very dramatic crash of a Boeing 737 operated as Air Florida flight 90 during a winter storm in Washington DC in 1982, the industry and government continued to cooperate to shape and evolve CRM. In 1984, heavily influenced by the leadership of NASA psychologist Dr. John Lauber, CRM was defined in such a way that was accepted by stakeholders as "the effective utilization of all available resources—hardware, software, and *liveware*—to achieve safe, efficient flight operation." Liveware, in this context, alluded to humans.

As a result of these efforts, around 1984 the second generation of CRM took shape. Instead of focusing on changing individual behaviors, CRM now went deeper in an attempt to change attitudes and focus more on decision-making as a group. This generation also recognized that CRM should involve more than just flight deck crewmembers and that others, such as flight attendants, often possessed key information that should be communicated to the flight deck to prevent accidents. Special emphasis was placed on briefing strategies and the development of realistic simulator training profiles known as Line-Oriented Flight Training (LOFT). When combined with realistic situational scenarios, such simulators host LOFT sessions that are grueling but extremely important for pilots to learn both technical and CRM skills, and how both skills need to work together to assure a safe outcome to the flight.

By 1985 only four air carriers in the United States had full CRM programs: United, Continental, Pan Am, and People's Express. American and the U.S. Air Force Military Airlift Command soon introduced CRM programs, and the U.S. Navy and Marine Corps were on the verge of starting CRM programs. In 1989, a very serious accident happened that provided irrefutable proof that CRM principles worked. A DC-10 operating as United Airlines Flight 232 experienced the uncontained failure of its #2 engine, resulting in a loss of normal flight controls. The captain, Al Haynes (an ex-US Marine Corps pilot and flying instructor) very ably coordinated flight deck and cabin resources to perform a controlled crash of the aircraft at Sioux City, Iowa. The resulting crash killed 111, but there were 185 survivors who likely would not have survived at all had it not been for the CRM prowess of the crew.

In the early 1990s the third generation of CRM took hold, which deepened the notion that CRM extended beyond the flight deck door. That generation saw the start of joint training for flight deck and cabin crewmembers, such as for emergency evacuations, placed emphasis on the role of organizational culture, and also started exploring how flight deck automation was increasingly a key component of communication and coordination protocols for pilots. Some vocal pilots voiced concern that CRM was becoming too diluted by extending it past the flight deck with all the increased emphasis on using external resources. Around 1992 CRM also saw itself being exported to the medical community to address similar group dynamics events associated with medical error in both routine and emergency care at hospitals.

Around the mid-1990s the fourth generation of CRM was introduced, which promoted the FAA's voluntary Advanced Qualification Program (AQP) as a means for "custom tailoring" CRM to the specific needs of each airline and stressed the use of Line-Oriented Evaluations (LOEs). This generation also fostered the pairing of crew behaviors to checklists and advocated the integration of CRM training directly into technical training versus as a stand-alone initiative. In 1998 the FAA made CRM training mandatory for U.S. Airlines through FAR 121.404. It should be noted that this was 19 years after NASA's effort to start what became CRM, which shows the not-uncommon delay in implementing new regulations that promote safety.

Around 1999 the fifth generation of CRM took hold, which reframed the safety effort under the umbrella of "error management," modified the initiatives to be more readily accepted by non-Western national cultures, and placed even more emphasis on automation and, specifically, automation monitoring. Three lines of defense were promoted against error: the avoidance of error in the first place, the trapping of errors that occur so that they are limited in the damage they create, and the mitigation of consequences when the errors cannot be trapped. This generation also promoted the use of incident data in addition to accident data. This generation was influenced by the recently retired NTSB Chair Robert Sumwalt and two co-authors who studied the ASRS database and in 1997 published "What ASRS Data Tell about Inadequate Flight Crew Monitoring." In the paper they state, "One pilot must monitor automated flight systems 100% of the time" and mentioned that "Monitoring is the lifeblood of safe flight operations."

At the start of the 21st century, the sixth generation of CRM was formed, which introduced the Threat and Error Management (TEM) framework as a formalized approach for identifying sources of threats and preventing them from impacting safety at the earliest possible time. The origins of TEM can be traced back to a research partnership between the University of Texas Human Factors Research Project headed by Professor Bob Helmreich and Delta Airlines (the subsequent LOSA initiative is discussed in Chapter 3). Sophisticated and elaborate TEM models were, introduced for intervention and rely heavily on human factors knowledge for improving safety in aviation.

Threats can be any condition that makes a task more complicated, such as rain during ramp operations or fatigue during overnight maintenance. They can be external or internal. External threats are outside the aviation professional's control and could include weather, a late gate change, or not having the correct tool for a job. Internal threats are something that is within the worker's control, such as stress, time pressure, or loss of situational awareness. If the threats are not managed properly, they can impact safety margins and cause errors, which are mistakes that are made when threats are mismanaged. Errors come in the form of noncompliance, procedural, communication, proficiency, or operational decisions.

The TEM framework recognizes the relationship between threats and errors. In fact, many airlines are moving away from CRM training and putting more emphasis on TEM instead. They feel that CRM is too broad and open to too much interpretation. If aviation professionals can identify the threats and manage them, then they can directly mitigate human errors. Safety procedures are in place to resist some risks from having harmful outcomes, such as inspections and operational checklists, but some errors do not have a buffer. However, we as workers also have the opportunity to resolve the error before it leads to a negative impact.

To assess the TEM aspects of a situation, aviation processionals should:

- Identify threats, errors, and error outcomes.
- Identify "Resolve and Resist" strategies and countermeasures already in place.
- Recognize human factors aspects that affect behavior choices and decision making.
- Recommend solutions for changes that lead to a higher level of safety awareness.

TEM builds upon the various skills that underpin crew resource management and provides a structured approach to developing individual and team countermeasures to deal with a particular situation or threat. Former Head of Human Factors for ICAO, Captain Daniel Maurino describes such countermeasures in three categories[4]:

1. Planning countermeasures: essential for managing anticipated and unexpected threats

2. Execution countermeasures: essential for error detection and error response

3. Review countermeasures: essential for managing the changing conditions of a flight

Detailed examples of both individual and team countermeasures are provided in Table 9-1.

Planning Countermeasures		
SOP BRIEFING	The required briefing was interactive and operationally thorough	- *Concise, not rushed and meets requirements* - *Bottom lines were established*
PLANS STATED	Operational plans and decisions were communicated and acknowledged	- *Shared understanding about plans—"Everybody on the same page"*
WORKLOAD ASSIGNMENT	Roles and responsibilities were defined for normal and nonnormal situations	- *Workload assignments were communicated and acknowledged*
CONTINGENCY MANAGEMENT	Crew members developed effective strategies to manage threats to safety	- *Threats and their consequences were anticipated* - *Used all available resources to manage threats*
Execution Countermeasures		
MONITOR / CROSS-CHECK	Crew members actively monitored and cross-checked systems and other crew members	- *Aircraft position, settings, and crew actions were verified*
WORKLOAD MANAGEMENT	Operational tasks were prioritized and properly managed to handle primary flight duties	- *Avoided task fixation* - *Did not allow work overload*
AUTOMATION MANAGEMENT	Automation was properly managed to balance situational and/or workload requirements	- *Automation setup was briefed to other members* - *Effective recovery techniques from automation anomalies*
Review Countermeasures		
EVALUATION / MODIFICATION OF PLANS	Existing plans were reviewed and modified when necessary	- *Crew decisions and actions were openly analyzed to make sure the existing plan was the best plan*
INQUIRY	Crew members asked questions to investigate and/or clarify current plans of action	- *Crew members not afraid to express a lack of knowledge—"Nothing taken for granted" attitude*
ASSERTIVENESS	Crew members stated critical information and/or solutions with appropriate persistence	- *Crew members spoke up without hesitation*

TABLE 9-1 Examples of TEM Countermeasures (Source: Maurino, 2005)

9.6   CRM Beyond the Flight Deck

The principles that lie beneath CRM on the flight deck have benefit across the aviation system and beyond. Many airlines found that the role of cabin crew as the "eyes and ears" of the flight crew was enhanced through joint CRM training, especially as automation led to the reduction of flight deck crew from three to two. Efforts to try and improve communication through the flight deck door were potentially hampered by additional security measures following the 9/11 attacks in the United States, but this only made effective CRM more important, especially in dealing with in-flight events such as disruptive passengers and medical emergencies.

The concept has also been adopted in air traffic control where Eurocontrol—the pan-European civil-military organization supporting European aviation—developed the Team Resource Management (TRM) concept from the mid-1990s. With a training syllabus that was created specifically for the air traffic management community, TRM was also the way that human factors training was introduced to many Air Navigation Service Providers (ANSPs).

Key topics now include:

- Teamwork
- Leadership
- Team Roles
- Situational Awareness
- Decision-Making
- Communication

The effectiveness of this training is assessed through ICAO's ATCO Competency Framework (see Doc 10056 Manual of Air Traffic Controller Competency-based Training and Assessment, 2017) which considers both "Technical" and "NonTechnical" skills. Other examples of translating CRM include Maintenance Resource Management training which was introduced to Continental Airlines, US Airways and United Airlines from the mid-1990s and even single-pilot CRM for general aviation pilots where their definition of "crew" has widened to include air traffic control, maintenance personnel, instructors, and so on.

Outside of aviation, CRM has been translated for use in railroads, shipping, firefighting, and even healthcare. While many of the principles are common, successful implementation is based on acknowledging the differences and ensuring that those receiving the training understand the relevance and opportunities for implementation.

9.7   The Human within a System

Aviation is no different to other transport modes in placing a human being within a complex system of interactions. These are frequently described using a simple but effective model developed by Elwyn Edwards in 1972 and subsequently modified by Frank Hawkins.[5] Writing for the British Airline Pilots Association in a paper called "Outlook for Safety. Man, and Machine: Systems for Safety," Edwards offered a helpful model to explain the interactions and complexities that surrounded major accidents.

S = Software
H = Hardware
L = Liveware
E = Environment

FIGURE 9-5   The SHELL Model. (Redrawn from Hawkins, 1993)

This is known as the SHELL model (see Figure 9-5) as it covers the following components:

- **Software**—these are nonphysical aspects of a system that govern how it operates, such as the rules and regulations, operating procedures, habits, customs, and practices. These are likened to the software that controls a computer system and may be documented in things like checklists, maps and charts or standard operating procedures.

- **Hardware**—these are the physical elements of a system, which in aviation may include aircraft, airports, ground vehicles, air traffic control systems, and their constituent components.

- **Environment**—this encapsulates both the physical environment (which in aviation may cover meteorology, terrain, and the physical environment around an airport, as well as the actual workplaces such as the aircraft flight deck, cabin and hold, maintenance hangar, and air traffic control tower) and the wider context in which aviation operates such as the organizational, political, regulatory, and economic situation.

- **Liveware**—this describes the human element of the aviation system, covering the human performance aspects of pilots, cabin crew, air traffic controllers, maintenance engineers, managers, passengers, and so on. The model features "Liveware" twice to signify the interactions that take place between people.

While each of the four components covers many topics that are relevant to human performance, it is the interfaces between them that is often the source of problems. When using the SHELL model as a tool, it is useful to examine the interactions and potential mismatches. For example, the interaction between a pilot (liveware) and their aircraft (hardware) may be mismatched if equipment is poorly designed or badly located, or if warnings fail to alert the crew as intended. Similarly, of the interaction between the two elements of liveware is dependent on a range of factors including interpersonal skills, teamwork, cultural factors, and personality.

In simple terms, human factors problems can be found in any of the component areas (which are supplemented by "culture" in a subsequent revision of the model) or in the interfaces between them. Accident investigators must work very hard to understand these problems, partly because evidence can be absent or lost and partly because they need to avoid their own biases in terms of what has happened. As Professor Sidney Dekker[6] notes, the investigators job is "…to understand why it made sense at the time" and even with a surviving witness, their recollection may not be accurate.

How we design systems will suit some operators and users more than others. This is partly because of the natural diversity between people and sometimes because the designer has not envisaged something being used the way it is. Imagine having the technical ability to design or program a complex piece of avionics and somehow having insight as to how pilots from a range of cultures and operating in different weather conditions, emergencies or states of fatigue may operate it and you will start to understand why human factors is so important.

The challenge is to try and design a flexible and resilient system that maximizes the use of human's greatest strengths and which compensates for physiological and neural diversity as well as performance variability. If that sounds complex, then it also needs to be able to demonstrate regulatory compliance, be acceptable to the traveling public, affordable, scalable, and able to adapt to new developments!

Chapters 11 and 12 will take us deeper into how humans can be both problems and solutions when it comes to commercial aviation safety. However, before then, we must also remember that aviation safety is not just about the prevention of aircraft accidents. We also must protect employees and passengers within what can be an incredibly inhospitable environment.

9.8   Protecting the Human—Workplace Health and Safety

Around the world, workplace health and safety are variously regulated. Employees are protected by rule about the hours they can work, the substances they are exposed to, the safety systems that must be in place to support them and so on. Some of these protections apply to aviation whereas others do not. Those involved in aviation safety management often must span the two worlds with the added complexity of variety around the world. Working hours restrictions which apply to ground handlers in Europe, for example (known as the EU Working Time Directive), do not apply in the United States.

While aircraft accidents attract the greatest attention, aviation actually suffers chronic losses through minor workplace accidents, which lead to a loss of productivity. Flight attendants, for example, routinely suffer from burns and scalds while working in confined galleys, sometime during turbulence. They also suffer back injuries from helping passengers with heavy baggage and, for some at least, hours of working in high-heeled shoes. Ground handlers can suffer from the effects of extreme weather, be exposed to injuries from dealing with awkward and heavy loads, often in confined spaces.

The National Institute of Occupational Safety and Health suggests that controlling exposure to hazards in the workplace is vital to protecting workers (and passengers). It suggests a simple hierarchy of controls that range in effectiveness.

At the highest level removing or eliminating the hazard is likely to be the most effective, followed by substitution, engineering controls, administrative controls, and finally personal protective equipment as described in Figure 9-6.

Next time you travel, have a look at the potential hazards and consider the ways in which they are addressed. Often the least effective are the most visible—workers in

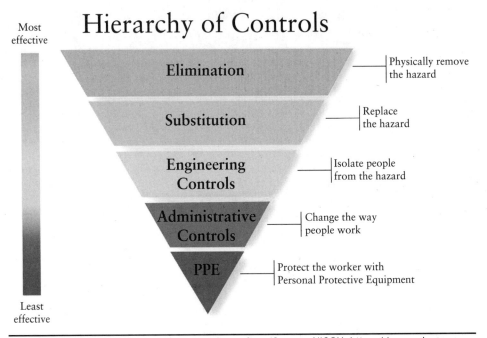

Hierarchy of Controls

Most effective

Elimination — Physically remove the hazard

Substitution — Replace the hazard

Engineering Controls — Isolate people from the hazard

Administrative Controls — Change the way people work

PPE — Protect the worker with Personal Protective Equipment

Least effective

Figure 9-6   Hierarchy of controls for managing safety. (Source: NIOSH, https://www.cdc.gov /niosh/topics/hierarchy/default.html)

hi-visibility clothing wearing ear defenders are a good example. At this level, we can expose them to greater hazards—for example, ear defenders may protect an airport worker from the noise of a jet aircraft, but also mean that they fail to detect a hazard such as a car that is moving toward them.

The more effective controls need management actions, especially during the design process. In some contexts, it may involve the use of machines or autonomy, but this needs to be compatible with the intended use. There are lots of great examples around the airport apron of both highly effective simple designs and overly complex ones which become a nuisance and ignored. For example, ground handling staff at British Airways came up with a simple length of flexible hose that stands above the coupling between baggage carts as a way of eliminating the temptation to step between them. The cost was minimal, but the effect has been highly successful. In contrast caterers are often seen to fail to use the extendable handrail that is intended to stop them falling between the gap between their van and the aircraft because of how awkward it is to move and how nervous they are about causing damage to the aircraft when they move it into place.

Ultimately there is often a trade-off that is invisible until an accident or incident occurs. A good example is the uniform worn by cabin crew for a role which is primarily concerned with safety, but which is often perceived as service. When a Singapore Airlines Boeing 747 crashed on takeoff from Taipei in 2000, the cabin crew's role rapidly shifted to managing the evacuation of a burning aircraft. Seen in Figure 9-7, the iconic "Singapore Girl" uniform includes a long straight sarong kebaya and open-toed sandals, neither of which were especially suitable for this role, not least as many cabin crew lost their sandals during the impact and then had to run through sharp wreckage to safety.

FIGURE 9-7   Singapore Airlines cabin crew uniform. (Source: https://www.singaporeair.com/en _UK/sg/flying-withus/our-story/singapore-girl/)

Since the accident, the uniform has been redesigned to include special sandals to be used on takeoff and landing. However, for many airlines, the focus remains on creating a particular visual image rather than ensuring the health and safety of employees. If the latter were true then we might expect cabin crew to look more like the people found in the pit lane at a car race!

Similarly, the handing of baggage may look very different if we were to concentrate on removing workers from the hostile environment that is the airport ramp. Many aircraft types and airlines prefer the use of people working in confined areas to optimize the loading of baggage rather than using containers. This is a situation that is exacerbated by the pressure to accept passengers closer to the departure time or to ensure rapid turnarounds. The role of the safety manager is to ensure that they meet their duty of care to protect employees and passengers. Failure to do so may result in prosecution and a large fine. In 2021, British Airways was fined £1.8M after one of its workers was crushed by a baggage trolley. Even though the employee was walking in an area that should have only been used by vehicles, the occupational health and safety regulator, the HSE, brought a prosecution based on the fact that it had become custom and practice for employees to use this a short cut and the airline had failed to identify this as a hazard or take measures to change behavior. Following the accident, physical barriers were put in place (engineering controls) and stricter monitoring of behavior (administrative controls) was introduced.

A final added complexity is where the jurisdiction of an occupational health and safety regulator ends and where another begins. In other words, when do rules designed for aviation activity such as pilot duty times overlap or contradict OHS rules such as the European Working Time Directive or OSHA vs the FAA in the United States? This became even more complex during the COVID-19 Pandemic where public health rules pertaining to quarantining, vaccinations, and the use of masks either changed rapidly or added complexity to an already complicated aviation system. In some countries, flight crew found themselves being taken directly to isolation hotels between flights, where they were confined to their rooms. Putting the psychological effects to one side, personal strategies to deal with the effects of time zone changes and disrupted sleep is also a challenge.

9.9 Conclusion

The human component of commercial aviation safety covers a wide range of topics from physiology to psychology. Major aircraft accidents often include a human contribution, but as Chapters 10 and 11 will explore, so does every successful flight. In managing the safety of passengers, crew, and workers, we need to carefully consider the relative strengths and weaknesses of human beings and what measures we can take to deliver optimal performance.

Key Terms

Crew Resource Management

Hierarchy of Controls

Human Error

Pilot Error

SHELL Model

Threat and Error Management

Topics for Discussion

1. What does the term "human factors" mean?
2. Why is the term "pilot error" often misleading?
3. What are "heuristics" and what effect may it have on human performance?
4. What were the main findings of the NTSB investigation into the Asiana 214 accident and how can these be applied to commercial aviation safety?
5. Why may have the crew of British Midland Flight 92 been confused as to which engine had failed?
6. What changes have occurred that may prevent an accident similar to Flight 92 in modern aircraft?
7. Describe the six generations of CRM.
8. What kinds of countermeasures can be applied as part of TEM?
9. What are the six key elements in Team Resource Management for ATCO?
10. What is the hierarchy of controls?

References

1. National Transportation Safety Board. (2014). *Aircraft Accident Report: Descent Below Visual Glidepath and Impact with Seawall Asiana Airlines Flight 214 Boeing 777-200ER, HL7742 San Francisco, California July 6, 2013*, NTSB/AAR-14/01 PB2014-105984, Washington DC.

2. Adams, D. (2006). *A Layman's Introduction to Human Factors in Aircraft Accident and Incident Investigation*. Canberra, Australia: ATSB.

3. Air Accident Investigation Branch. (1990). *Report on the Accident to Boeing 737-400, G-OBME, Near Kegworth, Leicestershire on 8 January 1989*. London: HMSO.

4. Maurino, D. (2005). Threat and Error Management (TEM) Paper presented to the Canadian Aviation Safety Seminar, Vancouver, Canada, 18-20th April. https://skybrary.aero/sites/default/files/bookshelf/515.pdf

5. Hawkins, F. H. (1993). *Human Factors in Flight*. Aldershot: Ashgate.

6. Dekker, S. (2002). *Field Guide to Understanding Human Error*. Aldershot: Ashgate.

CHAPTER **10**

Humans as the Challenge

Introduction

Commercial aviation, like other safety-critical industries such as nuclear power or petro-chemical, is considered a high-consequence environment because even relatively small errors or deviations can lead to disastrous consequences. Some refer to those types of businesses as being conducted by high-reliability organizations, high-risk industries, or high-consequence operations, where all terms are expressions of the same sentiment.

When considering how even small errors can trigger large problems, it becomes clear why commercial aviation operations require special safeguards to protect against potentially terrible outcomes. It also becomes evident why an entire chapter of this book is dedicated to the challenge of dealing with human fallibility. It is an ever-present condition that follows us as a shadow through life, rearing its ugly head all too often to remind us that as humans we are all susceptible to error.

This chapter explores humans as a major challenge for commercial aviation safety. A high degree of technical knowledge does not guarantee a successful outcome to a complex, technical situation. Whenever people are part of the equation, we rely on technology, procedures, communication skills, teamwork, experience, and extensive training to produce an acceptable outcome. Yet, in spite of these defenses, sometimes it all goes wrong—terribly wrong.

10.1   Human-Centered Design

The term "human-centered design" has become a popular way of describing the ideal combination of technical systems and their human operators. In simple terms, it is a way of describing how the benefits of the human can best be exploited by designing a complementary technical system. However, many aspects of aviation do not appear to have been designed this way, which can lead to errors, workarounds, and suboptimal performance.

Lessons have tended to be learned the hard way—sometimes known as "tombstone safety." The design of aircraft has changed many times over the years in an attempt to reduce either the likelihood of making an error and should it occur to minimize its severity. A good example is the altimeter, which is a critical flight instrument, but which could easily be misread when pilots needed to interpret the position of three needles on an analog "clockface" gauge. Modern altimeters generally feature a numerical representation of the aircraft's altitude and consequently, the error rate has dropped significantly.

FIGURE 10-1   Three-pointer altimeter. (Source: Wikipedia Commons)

In Figure 10-1, one hand represents thousands of feet, one tens of thousands of feet, and the other is hundreds of feet, but which one and how rapidly can you calculate it? While those with experience may be able to calculate the answer swiftly, the accident record shows that there was plenty of opportunity for error.

Attitudes toward design have also changed from a world where errors were considered to be due to a lack of skill or professionalism on the part of the individual to one where the risk that someone may make an error can often be reduced. A great deal of effort has been focused on improving the human–machine interface on the flight deck and, more recently, on ensuring that aircraft are designed for maintainability—in other words, so that maintainers are able to properly access components and that the likelihood of misassembly or injury are reduced.

User-centered design is not restricted to the physical environment. It may also include procedures and processes including standard phraseology for communications. The main challenge is understanding the needs of a potentially wide variety of users. While a test pilot may be capable of flying a poorly designed aircraft, real-world operations will involve an airline's entire population of pilots—including the poorest performing one. To better understand how to design for safety, it is important to look at the different ways in which humans can deviate from expected performance.

10.2   Philosophy of Human Error

The ingenuity of humans is beautifully demonstrated by the rapid progress made from the first powered flight by the Wright Brothers at Kitty Hawk in 1903 to Neil Armstrong setting foot on the moon in 1969. Humans can operate at the limits of their physiology and solve complex problems beyond the processing power of supercomputers. However, the downside of being human is the tendency for all of us to make errors or to commit violations. Errors include actions which fail to meet their intended objective.

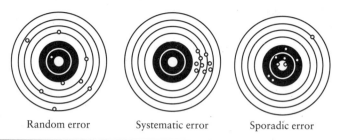

Random error   Systematic error   Sporadic error

FIGURE 10-2   Three types of error as shots on a rifle target. (Source: Reason, 1990)

One way of describing them is whether the error was one of commission, omission, or substitution. An error of omission is where an action is not done, whereas an error of commission is one where something was done but should not have been. Errors of substitution occur when an action must take place, but the wrong action is taken.

Professor James Reason's seminal book *Human Error* published in 1990,[1] provides deep insight into different types of error and the various causes that lie behind them. A range of classifications are discussed including using a simple analogy of the way rifle shots are distributed across a target (Figure 10-2).

Random error may be observed where there is no clear pattern to the shots and many factors may influence such variability. While some of the shots are close to the bullseye, others are at the extreme of the target, the error may also be considered to exhibit large variation. This is likely because of a lack of skill in the shooter—the more accurate shots being a function of the amount of scatter rather than any level of marksmanship. Training in how to use the sight may help this novice to reduce the level of error. An aviation example may be the lack accuracy that a new novice shows in attempting to fly straight and level or to aim for a particular spot on the runway during landing.

In contrast, a systematic error is often caused by only one or two factors. On a target, this may look like a nicely grouped set of shots that are all offset from the bullseye, perhaps because of a misaligned sight on the gun. In an aviation environment, the low level of variability suggests the individual or system can reproduce their actions accurately, but that there is a common source of error such as a misunderstanding from training or miscalibration.

Finally, sporadic error occurs when, despite a generally good performance, an isolated error occurs. Such errors tend to be the most dangerous as they are so difficult to predict. The underlying cause may be easily identifiable (in this case, perhaps the shooter sneezed) or may be an unusual or even novel occurrence. An apparent lack of a trend may also be a function of relatively few data points and further data may be required before the cause is revealed.

Reason also classifies errors into three types: mistakes, slips, and lapses.

A mistake may be considered to be an intended action which proceeds as planned, but yet still fails to achieve the intended outcome. Where the intention was not appropriate, a slip or lapse may be considered to have occurred. A slip is an error where the right intention is carried out incorrectly whereas a lapse is a failure to carry out an action. Reason[2] also discusses the classification of error from the different levels at which people perform their actions. The three levels are called skill-based, rule-based, and knowledge- based actions.

At the skill-based level, we carry out familiar tasks in a largely automated way. At the rule-based level, we need to account for a particular set of circumstances and modify our automated response. The problems tend to be familiar and hence are dealt with using a prepackaged solution. At the knowledge-based level, we are unfamiliar with the problem and have no prepackaged solution. We therefore try and think the problem through on the spot, albeit in the knowledge that we are out of our comfort zone.

At the other end of the spectrum from errors are violations. Reason describes the difference as follows:

1. *Errors are unintended. Violations are deliberate. The deviation from procedures is intended, though the occasional bad consequences are not.*

2. *Whereas errors arise primarily from information problems (i.e., forgetting, inattention, incomplete knowledge, etc.), violations are more generally associated with motivational problems (i.e., low morale, poor supervisory example, perceived lack of concern, the failure to reward compliance and sanction noncompliance, etc.).*

3. *Errors can be explained by what goes on in the mind of an individual, but violations occur in a regulated social context.*

4. *Errors can be reduced by improving the quality and the delivery of necessary information within the workplace. Violations require motivational and organizational remedies.*

Not all violations are committed by individuals who are intending to cause a bad outcome. Some violations arise because there appears to be no other way to get the job done or because the individual is trying to deliver a good outcome for their employer (e.g., a speeding driver trying to catch up with their schedule)—a so-called violation for organizational gain. The kind of violation where an individual sets out to be deliberately reckless or to commit sabotage, is actually pretty rare.

The range of errors and violations mean that different strategies need to be engaged to deal with them. While some may be addressed through training, others will benefit from harder defenses or incentives. Demanding "professionalism" is simply not enough and neither is expecting that all problems can be engineered out.

10.3 Who Is the Ace?

What does the modern aviation professional look like? There is no real answer to this question—aviation is a global industry populated with a wide diversity of people and functions. On every front, old stereotypes have been challenged and although there are still examples where certain demographics are over or under-represented, the industry needs to be able to function safely across an incredibly diverse community.

The variety of roles and working environments will attract different people to the industry. Certain barriers exist for people joining the sector, such as the very high cost of initial pilot training, which is often funded by the individual rather than an employer. Similarly, different roles will attract a range of personality types, for example, cabin crew need a strong service ethic to work with passengers and air traffic controllers require the mental dexterity to control multiple aircraft at different speeds and trajectories.

Certain roles require both careful use of complex technology and the ability to deal with novel or unusual situations when technology does not behave the way in which it was expected. Many aviation roles require a rapid change from one function to another—for example, a cabin crew member who may be focused on serving meals at one moment to dealing with an in-flight medical emergency the next. Some pilots now describe their role as hours of boredom punctuated by moments of terror as the tedium of a long cruise phase on autopilot is followed by a late runway change during the busy approach phase.

Most aviation roles require teamwork, even in the case of a single pilot operation. This teamwork can range from the specialists within a design team to the communication between an air traffic controller and a pilot. The accident record shows that where information is not shared accurately between members of a team, misperceptions can develop, or errors can go unchecked. Aviation relies on multinational, multidisciplinary teams that are often transient in nature as an aircraft passes through multiple air traffic zones or airports. The ability to work under such conditions is just one of the skills required of a modern aviation professional.

On the flight deck, modern commercial airline pilots will operate as a two-person crew supported by a range of technologies. These are designed to try and reduce the likelihood of error within regulatory requirements for safety performance. The regulatory requirements are designed around ensuring a certain level of reliability and redundancy and it is having successfully demonstrated that the crew complement has reduced over time. For the two pilots that remain, each must be able to play different roles on any particular flight as pilot flying (PF) or pilot monitoring (PM). In addition, the pilot in command must always be able to make timely and well-reasoned decisions either as PF or PM. If it were possible to answer the question "who is the ace?" in this context, it is arguably the pilot who is best able to operate in different modes depending on the situation and who uses the best available resources and information to maintain situational awareness in order to make sound decisions.

10.4   Understanding the Human Factor

The people who operate and support the global aviation system are crucial to its safety performance; the resourcefulness and skills of crewmembers, aircraft designers, air traffic controllers, and maintainers help prevent countless accidents each day. However, despite the excellent safety record, many studies attribute human error as a significant factor in most commercial aviation accidents.

The UK Health and Safety Executive describes human factors as the environmental, organizational and job factors, and human and individual characteristics which influence behavior. The Federal Aviation Administration goes on to suggest that "The human element is the most flexible, adaptable and valuable part of the aviation system, but it is also the most vulnerable to influences which can adversely affect its performance."

Commonly cited studies argue that between 70% and 80% of all aviation accidents are attributable to human error somewhere in the chain of causation. Contemporary safety efforts have therefore been heavily focused on trying to understand the human decision-making process and how it can be improved. Many studies have tried to develop a better understanding of how humans react to operational situations and interact with new technology and improvements in aviation safety systems. How human beings are trained and managed affects their attitudes, which in turn affects their performance of critical tasks. In turn, their performance affects the efficiency and, therefore, the costs and safety of the operation.

It is important to understand how people can be managed to yield the highest levels of error-free judgment and performance in critical situations, while at the same time providing them with a satisfactory work environment. A review of cockpit voice recordings from accidents clearly indicates that distractions must be minimized, and strict compliance with the sterile cockpit rule must be maintained during the critical phases of flight (taxi, takeoff, approach, and landing). The sterile cockpit rule, in its different forms, is a ban on nonpertinent conversation among members of a flight crew during critical phases of flight to minimize distraction.

While emphasis often focuses on the pilots, they are by no means the only ones who should be discussed in the sterile cockpit rule philosophy. The same holds for all human factors–related initiatives. Pilots are, however, often the last link in the safety value chain and are usually in a position to identify and correct errors that could otherwise result in accidents and incidents. One problem is poor human decision-making. Essentially, three reasons explain why people make poor decisions:

1. They have incomplete information.

2. They use inaccurate or irrelevant information.

3. They process the information poorly.

Psychologists have traditionally explained the limited information processing capabilities of humans with Miller's law. George Miller was a cognitive psychologist at Princeton University when in 1956 he published what would be known as his law. Miller's "Magic number" concept states that one can make 7 (plus or minus 2) absolute judgments, or differentiations, along a single dimension. For example, we can differentiate between seven plus or minus two sound intensities and seven plus or minus two shades of the color red. This concept has been applied to the number of mental chunks of information an average human can hold in working memory. Working memory is improved when, for example, a pilot uses more than one sensory dimension, such as visual and auditory, to sense and process information because the information is processed differently in the brain.

Although the reality is much more complicated than has been summarized here with regard to Miller's law, the sentiment still holds. Modern research has shown that accidents are more likely to occur during high workload, task saturation periods, and when there is an overload of one or more of an individual's processing channels. As a consequence, and to reduce the workload during critical task saturation situations, new pilots, for example, are taught a task-shedding strategy to focus on the most important task on the flight deck: flying the aircraft. The Federal Aviation Administration (FAA) spells out the importance of A-N-C which means Aviate, Navigate and Communicate as the order in which all pilots should prioritize their activities. This is further reinforced with the three P's—Prepare, Plan, and Practice—maxims which work for pilots of all experience levels (Figure 10-3).

Postaccident investigations usually uncover the details of what happened. With mechanical failures, accident data analysis often leads logically to why the accident occurred. Determining the precise reason for human errors is much more difficult. Without an understanding of human behavior elements in the operation of a system, preventive or corrective actions are impossible. Understanding human factors is especially important to systems where humans interact regularly with sophisticated machinery and in industries where human error–induced accidents can have catastrophic consequences.

General Aviation
Joint Steering Committee
Safety Enhancement Topic

FAA
Aviation Safety

Fly the Aircraft First

NTSB accident data suggests that pilots who are distracted by less essential tasks can lose control of their aircraft and crash. In light of this pilots are reminded to maintain aircraft control at all times. This may mean a delay in responding to ATC communications and passenger requests, or not responding at all unless positive aircraft control can be maintained throughout. In other words,

Fly the Aircraft First!

It's as Easy as A-N-C

From the earliest days of flight training, pilots are taught an important set of priorities that should follow them through their entire flying career: Aviate, Navigate, and Communicate.

The top priority — *always* — is to aviate. That means fly the airplane by using the flight controls and flight instruments to direct the airplane's attitude, airspeed, and altitude.

Rounding out those top priorities are figuring out where you are and where you're going (Navigate), and, as appropriate, talking to ATC or someone outside the airplane (Communicate). However, it doesn't matter if we're navigating and communicating perfectly if we lose control of the aircraft and crash. A-N-C seems simple to follow, but it's easy to forget when you get busy or distracted in the cockpit.

A famous example of a failure to aviate is the December 1972 crash of Flight 401, an Eastern Airlines Lockheed L-1011. The entire crew was

single-mindedly focused on the malfunction of a landing gear position indicator light. No one was left to keep the plane in the air, as it headed towards a shallow descent into the Florida Everglades. Four professional aviators — any one of whom could have detected the descent — were so focused on a non-critical task that they failed to detect and arrest the descent, resulting in 99 fatalities. They did not follow established aviation priorities — they *failed to fly the aircraft first.*

Disconnect from Distractions

As we can see from the Eastern Airlines example, distractions can be deadly in an emergency situation and can rob your focus from more critical items or tasks.

Continued on Next Page

www.FAASafety.gov
AFS-920 18-07

Produced by *FAA Safety Briefing* | Download at 1.usa.gov/SPANS

FIGURE **10-3** Extract from FAA safety briefing. (Source: FAA, 2018, https://www.faa.gov/news/safety_briefing/2018/media/SE_Topic_18-07.pdf)

Human factors principles are increasingly incorporated into design standards, especially for complex elements such as flight deck design and design for maintainability in aircraft maintenance. However, such principles are sometimes not prioritized in the same way as technical decisions in aircraft designs, regulations, and operations which are primarily based on "hard" sciences, such as aerodynamics, propulsion, and structures.

Human capabilities do not often lend themselves readily to consistent, precise measurements. Human factors research often requires much more time to understand complex qualitative data or to capture a representative sample of a highly variable population. Data on human performance and reliability can be regarded by some technical experts as "soft" and receive less attention in aviation system designs, testing, and certification than "harder" engineering applications.

Aviation human factors is a multidisciplinary science that attempts to optimize the interaction between people, machines, methods, and procedures that interface with one another within an environment in a defined system to achieve a set of systems goals. Aviation human factors encompass fields of study that include, but are not limited to, engineering, psychology, physiology, anthropometry, biomechanics, biology, sociology, and certain fields of medicine. Human factors science concentrates on studying the capabilities and limitations of the human in a system with the intent of using this knowledge to design systems that reduce the mismatch between what is required of the human and what the human is capable of doing. If this mismatch is minimized, errors that could lead to accidents will be minimized, and human performance will be maximized.

Figure 10-4 is representative of a modern flight deck—in this case the Airbus A220 twin jet which was originally developed as the Bombardier C-Series and which entered service in 2016. The importance and influence of human factors are seen in many aspects of design including the positioning of the primary and secondary displays, external visibility through the windshield—both in the air and while maneuvering on the ground, the shape and functionality of switches, levers, and the sidestick controller, the comfort

Figure 10-4 The flight deck of a modern Airbus A220 aircraft. (Source: Wikipedia Commons, https://en.wikipedia.org/wiki/Airbus_A220#/media/File:AirBaltic_Bombardier_CS300_mainenance_(33221388195).jpg)

and adjustability of the crew seats and so on. In the real flight deck, these are supplemented by the logic and content of aural warnings, the logic in which safety critical information is prioritized and presented on the various displays, and the interaction that takes place between pilots, cabin crew, air traffic controllers, and ground crew.

10.5 Safety Myths

The safety of commercial aviation is often described in terms of negative performance—aircraft written off or lives lost in accidents, for example. When accidents do occur, the media is often swift to link the event to previous ones to infer a trend or at least a lesson "not learned." It is much harder, by definition, to identify the accidents that didn't happen—the normal flights that land without incident or accident. Consequently, events are often characterized in terms of the last thing to go wrong or at least an action that appeared to lead to a series of consequence. Simplistic descriptions of accidents as a chain of events which, if only one of the links had been broken, could have been prevented have not helped our understanding of the role of human factors.

As discussed in Chapter 9, terms such as pilot error or human error are often used synonymously, even though the former is merely a subset of the latter. Complex accidents that may be the result of many interacting factors are often reduced to single causes by journalists in pursuit of a headline. To do so is not only demeaning to the individual involved, but also misses the myriad opportunities that exist to prevent recurrence. Accidents rarely, if ever, occur because of a single factor and even the concept of an accident chain (which at least acknowledges that multiple factors are involved) can oversimplify what causes an accident within a complex system.

We know that all humans make errors and that although with appropriate selection, training, and experience, we can change the types and frequency of errors people are likely to make, we can never eradicate them completely. This is something that has been increasingly factored into the design of commercial aviation through regulatory requirements, especially in terms of error tolerance and multiple redundant systems. In simple terms, as we have learned more about human factors, so too, technical, and procedural interventions that either reduce the likelihood of occurrence or reduce/mitigate the consequence of them if they do, have increased.

10.6 What about "Pilot Error?"

It is almost impossible to open a book on flight safety or CRM that does not promptly refer to the high percentage of accidents that are classified as being due to pilot error. Although the numbers vary depending on which source is quoted, experts and records typically refer to large percentages of aircraft accidents as being caused by pilot error.

In 1999 the Data Acquisition and Analysis Working Group of the Flight Safety Foundation analyzed 287 fatal approach-and-landing accidents involving jet and turboprop aircraft between 1980 and 1996. The study observed that CRM breakdowns were present as contributory factors in nearly half of the accidents. The same study looked at 76 approach-and-landing accidents between 1984 and 1997 and found that CRM failure was the third most frequent causal factor (63%) and the second most frequent circumstantial factor (58%).[3]

In another example, in 2001 a group of researchers from the Department of Emergency Medicine at Johns Hopkins University School of Medicine analyzed

329 major airline crashes, 1627 commuter/air taxi crashes, and 27,935 general aviation crashes for the years 1983 through 1996. They found that pilot error was a probable cause for 38% of the major airline crashes, 74% of the commuter air taxi crashes, and 85% of the general aviation crashes (Johns Hopkins, 2001).

Although some of the depicted statistics refer to human error, versus pilot error, the two terms are often (incorrectly) used synonymously when discussing it in the aviation context. Of course, experienced investigators are quick to point out that using the term human error opens up the possibility of analyzing actions and omissions that took place outside of the flight deck. Such actions or omissions may have occurred during the design process or during maintenance and helped induce the active errors by the flight crew that immediately preceded the accident.

Society often neglects to consider that humans, outside of the failed system, have also had a hand in all the automation that surrounds us. It is incorrect to ask, "Is it a human problem or an automation design problem" since a human had to design the automation. Similarly, it is incorrect to ask, "Was it human error or mechanical failure" if the mechanical failure was due to an inspector failing to catch an existing fracture in a component. Both the pilot and inspector are human, and human error makes no distinction as to the profession being performed by a person.

There are times, however, when pilot error is identified clearly as a critical factor in an accident. Our own inherent biases mean that when a bad outcome is preceded by errors or violations by the flight crew, it is tempting to assign cause to them and not look at situational, contributing, or exacerbating factors such as training or procedures. This is especially the case when the flight crew was properly trained and procedures existed, but mistakes were still made. For example, the fatal accident that occurred at Tenerife Los Rodeos Airport in 1977 where a KLM B747 collided with a Pan Am B747 with the loss of 583 lives is often characterized as the ultimate pilot error accident. Little is made of the diversion that caused the aircraft to be in an unfamiliar airport, the time pressure the pilots may have felt to ensure passengers did not miss their cruise ship or the pressure they felt not to exceed their duty hour limitations. Indeed, the poor location of the airport in an area that was prone to hill fog, the lack of a ground movement radar system, and the congestion that led to the aircraft backtracking down the runway are also rarely mentioned.

Thorough examination of the event, should lead us to ask the question, how did a highly experienced crew which included a well-regarded training captain end up in such a deadly situation? The accident was not simply a bad decision by the aircraft commander.

Historically, accidents were often conveniently explained by deeming the cause to be "pilot error" and taking scant further action than to remind other pilots not to make the same mistake. The public seemed relieved to know that a single "bad apple" had created the problem, and after a period of shock and grieving, the aviation world would return to business as usual. In a sense, we treated the symptoms of a disease while permitting the underlying cause to go unchecked.

Such a process may possibly have originated because many of the pilots involved in accidents did not survive to defend their reputation, and some companies found it convenient to excuse any possible mismanagement leading to the accident by assigning cause exclusively to the dead pilots. Also, the science of accident investigation has evolved from the point where a cause was identified to a contemporary understanding of situational, causal, and contributory factors. Lastly, the general public is usually not educated in the complexities of aviation and seeks simple answers instead of complicated accident causes.

The first few days following an aircraft accident where media interest is at its greatest can be summarized as an amorphous and emotional mass of conflicting eyewitness reports, misdirected reporters' questions, a tendency to believe initial causal theories because they sound good, and the temptation to oversimplify what happened. To wade through all that chaos, accident investigators must exercise patience and care, focusing on minutiae and yet keeping hold of the big picture.

Is calling "pilot error" a myth meant to remove individual accountability from pilots in terms of error management? Absolutely not, it is merely to point out that such characterization is often an oversimplification and as such, risks those best placed to prevent a recurrence from missing the opportunities to do so.

10.7 Cognitive Error

In the past, human factors' analyses of aviation accidents and incidents have relied on "loss of situational awareness (SA)" or similar generalized descriptions as a simplistic label. "Loss of SA" and "breakdown in SA" are commonly used umbrella terms that mask the multitude of cognitive breakdowns that caused the ultimate loss of SA. The use of "loss of SA" as a causal finding for an accident, although technically correct, proves insufficient for exposing where the breakdown in SA occurred. Therefore, aviation training specialists have been unable to develop effective measures for teaching how to prevent the erosion of SA past a certain superficial level.

Critics argue that investigators need, "… access to the psychological life beneath an error … more guidance beyond a cursory analysis … if one is to truly understand the genesis of human error."[4] This section of the chapter examines some of the underlying cognitive processes that can lead to loss of SA.

"Cognition" is a term that refers to the mental processes generated in the brain and related to thinking. Cognition is defined by the APA Dictionary of Psychology as "all forms of knowing and awareness, such as perceiving, conceiving, remembering, reasoning, judging, imagining, and problem-solving."[5] As such, human cognition lies at the heart of every action taken prior to and during an aircraft's flight.

It proves daunting to consider that every aspect of a flight is controlled by a series of loosely networked, complex, error-prone supercomputers that are subject to a single point of failure—our brains! No user manual has ever been written to teach humans how to effectively use our brains. In contrast, we would not dare fly an aircraft or operate other complex equipment without first studying an operator's manual or receiving training, but that is essentially how we approach the use of our brains every day. Accidents can start forming way back during the design of an aircraft or system while still in the brain of an engineer due to errors in thinking or knowledge gaps. New errors or more errors can then be introduced in the brain of an assembly line worker who manufactures the aircraft and those mistakes, like the ones of an engineer, may not surface until later, during the life of the aircraft (Figure 10-5).

The brains of numerous support personnel serve as external crewmembers for any given flight, and their errors directly impact the success of the flight. The brains of training department instructors and chief pilots create procedural guidance and teach pilots how to follow such guidance. The brains of government regulators provide oversight of operations. At the end of the long safety value chain are the brains of the pilots.

Figure 10-5   The "crew" of modern transport aircraft goes way beyond the flight deck as these C-17 military aircraft of the Royal Australian Air Force, US Air Force, and Royal Air Force show. (Source: Authors, https://commons.wikimedia.org/wiki/File:RAF_RAAF_USAF_C-17s_2007.jpg)

We must recognize that causal errors in commercial aviation accidents almost always have some link to cognitive processes. Scholars of human factors are quick to point out that benefits reaped from automatic cognitive functions generally come at the expense of control over the processes. Automatic mental modeling and SA construction often rely on deep-seated biases that taint our actions and decisions without us even knowing about it but that occur to speed up our cognitive processes. Linked to these biases are heuristics, or "thinking shortcuts" that generally work very well, in that they help speed up our thought processes but can also misdirect our perception of the outside world. In the case of verbal and nonverbal communication, our brain has been trained since childhood to "fill in the gaps" when information is missing.

10.8   Situational Awareness (SA)

As previously noted, the term "pilot error" is generally misleading. In simple terms, errors are often the symptom of deeper issues or at least the consequence of several contributory and exacerbating factors. Humans are only as good as the skills they have to respond to the situations they perceive. In simple terms, slips and lapses generally occur when individuals misremember or forget the way to do something, whereas mistakes come when an action is taken intentionally but it is an incorrect or inappropriate action. Lots of factors may influence the likelihood of such errors occurring and one of the most significant is a loss of what is known as "situational awareness" (SA).

However, to understand how SA is lost, we must first try to understand what this largely intangible state means. Perhaps the simplest definition of SA is "a mental model of one's environment." A more detailed definition of SA is "the ability to perceive factors that affect us and to understand how those factors impact us now and in the future."

Figure 10-6 In 2013, the crew of a British Airways B747 lost situational awareness while taxiing around Johannesburg, South Africa, and collided with a building. The incident was widely reported on social media by passengers including this photo from Harriet Tolputt. (Source: twitter.com and cited https://news.sky.com/story/british-airways-plane-crash-in-johannesburg-10423664)

But such a definition is incomplete since SA is more than perception. Others have defined SA in-flight operations as "the realistic understanding of all factors which affect the safety and effectiveness of a flight," except that such a definition implies that SA is realistic, whereas in reality, we know there can be good SA and poor SA. A person also can have a continuously varying level of SA over a period of time. Still others have defined SA as "the understanding operators have of the system and its environment at any one time." While this reminds us that SA is important to other roles within commercial aviation such as ATC or maintenance, the problem with that definition is that SA is more than just understanding. As we can see, it is a complex topic that deserves further exploration (Figure 10-6).

Interest in SA research grew through the 1980s and accelerated through the 1990s. The surge in research was due to, in great measure, the increased use of automation and technology and the new type of errors that such innovations engendered. A significant moment occurred in 1995 when what would later be termed the Situation Awareness Error Taxonomy was created by a human factors researcher, Dr. Mica Endsley. She researched SA exhaustively and argued that SA is developed, sustained, and destroyed along three distinct levels:

- **Level 1: perceiving** critical factors in the environment
- **Level 2: understanding** what those factors mean, particularly when integrated together with your goals
- **Level 3: projection**—understanding what will happen with the system in the near future

In simple terms, the three levels of SA can be characterized as perception, understanding, and projection. As such, the most appropriate definition of SA for this book is,

"The perception of the elements in the environment within a volume of time and space, the comprehension of their meaning, and the projection of their status in the near future" (p. 36[6]).

The three levels form a hierarchy of SA where people must attain a lower level prior to moving onto higher levels of SA in order to retain good SA. In other words, the accuracy of SA at higher levels is dependent on the accuracy of SA at the lower levels. For example, a flight attendant directing passengers in an emergency evacuation on a taxiway who does not perceive the presence of fire outside of an emergency exit (poor Level 1 SA) cannot be expected to understand the risk of opening the exit (poor Level 2 SA), and therefore may inappropriately proceed to open the exit and direct passengers to exit toward the fire (poor Level 3 SA). This would be considered poor overall SA, potentially leading to death or injury of passengers.

By contrast, the same flight attendant who correctly perceives the presence of fire outside of an emergency exit (good Level 1 SA) will be more likely to then understand the risk of opening the exit (good Level 2 SA), more likely to direct passengers to an alternate exit away from the fire (good Level 3 SA), and therefore be considered to have good overall SA, leading to a safe evacuation.

Based on the previous explanation, we can see that SA can be challenging to build and easy to lose. Imagine an aviation maintenance technician who is taking over the task of replacing a leaky oil pump from a colleague during a shift change. The handoff must be carefully performed in order for the incoming technician to correctly perceive what tasks have already been accomplished (Level 1 SA), to then understand how the completed tasks fit into the overall process for replacing the pump (Level 2 SA), and lastly determine how they will proceed to complete the task (Level 3 SA). An error in the first level makes it hard to correctly establish Level 2 SA. Or, the technician may correctly build Level 1 and incorrectly build Level 2 SA, making it difficult to determine the proper remaining actions (Level 3 SA).

Even if the first two SA levels are correctly built, the technician may incorrectly determine the remaining actions and therefore end up with poor SA and an improperly replaced oil pump.

Why would it be difficult to establish SA at each level? Human factors scientists have stressed the need to investigate the psychological roots of each level of SA rather than simply stating superficial descriptions of how poor SA contributed to an event. It is not uncommon for investigators to simply describe the causes of an event as "loss of SA" without probing deeper. If we are to prevent future accidents, we cannot simply tell aviation professionals, "don't lose SA in the future!" Indeed, one of the insidious problems of losing SA is the likelihood that the individual may not even realize they have done so!

Thinking about different error types can help us explain how SA breaks down. Continuing with our previous example of the technician replacing the oil pump, bad practice may lead to them not consulting a task card or maintenance manual and then omitting an item in the oil pump change procedure (Level 1 SA). Or perhaps an instructor in a training course improperly taught the incoming technician that several clamps were acceptable for connecting the pump fittings to the engine gearbox, whereas only one special type of clamp can be used for the oil scavenge line (Level 2 SA)? Or, perhaps a false expectancy that a more experienced colleague was going to check the work caused complacency leading to a rushed job to complete the replacement and get the aircraft back in service (Level 3 SA)?

It thus proves necessary to study the underlying factors that make loss of situation awareness so insidious to flight safety. One key factor that can both build and destroy SA, as previously mentioned, is expectancies. In the context of human psychology, an expectancy is an attitude or mental model that guides operators to relevant situation cues, thus determining how a person approaches a situation to increase the efficiency of his or her situation assessment.[7]

A false cognitive expectancy is the unreasonable anticipation of an event or condition that does not occur as envisioned. The determination of reason is a subjective assessment by the researcher based on experience and training. The incorrect anticipatory mindset may be the result of conscious or unconscious cognitive processes.[8]

The reason why expectancies function as a crucial aspect of cognition is to increase the efficiency of actions. Expectations about information can impact the speed and accuracy of the perception of information. Continued experience in an environment helps people to develop expectations about future events that predispose them to perceive the information accordingly. In fact, the constant fusion of cognitive expectancies to actual perceived situations to build SA, at the risk of suffering expectancy violations, has been shown as one of the most critical processes responsible for situation awareness creation and diminution. In other words, incorrect expectancies, together with a very long list of other mental biases, can both help create and destroy SA (Cortés, 2011[8]).

There are numerous other factors that affect SA, such as poor communication. This may occur when standard phraseology is not followed, or where what is heard is what is expected rather than what is said. Other important factors impacting SA are fatigue or stress. For example, if an air traffic controller is fatigued or stressed, they may be less likely to perform a visual sweep of a runway ahead of an early morning takeoff or may find themselves less likely to identify a readback error.

In short, there are many actions that need to happen to ensure good SA. Underlying factors can cause a loss of SA, and such a loss is often associated with negative outcomes, such as accidents. Researchers continue studying the ways in which SA breaks down and how aviation professionals can maximize the accuracy with which they build each of the three levels of SA. Professionals in commercial aviation must take full advantage of all the available sources around them, which include other team members, air traffic controllers, dispatchers, and flight attendants, in order to stay in the loop as to developments and thus keep their SA high, while postponing noncritical tasks or delegating tasks in order to prevent workload from getting too demanding. Technology also can help, such as through the use of Head Up Displays (HUDs) on aircraft such as the Boeing 787 and Airbus A350 and in terms of the type and prioritization of warnings presented to the crew, especially during critical stages of flight. Air Traffic Control systems also increasingly benefit from human-centered design to enhance SA ranging from the design and color of displays to audible and flashing conflict alerts if aircraft are given crossing tracks where there is a risk of a breakdown of separation.

As our understanding of SA improves, challenges remain, and new ones develop. Increasing traffic levels can mean that both flight crew and ATCOs need to work even harder to maintain the comprehension of other aircraft's intentions. For example, some fear that the "loss of party line" effect that comes from moving voice communications to controller-pilot datalink communications (CPDLC) may negatively impact on pilots' SA. Others fear that as some air traffic control facilities move from physical towers to digital facilities, which may even control multiple remote aerodromes, ATCOs may

FIGURE 10-7   Cranfield Airport Digital Air Traffic Control Centre supplied by SAAB. (Source: Author)

suffer from a reduced level of SA. As a counterpoint, in practice, it is also possible that with good design, these fears may be unfounded, and SA can actually be enhanced.

The Cranfield University Airport digital air traffic control center was opened in December 2018 and became the United Kingdom's first operational "remote tower." Using a set of 14 high-resolution screens (see Figure 10-7) with the option to overlay information such as taxiway and runway edges and to track aircraft, the situational awareness of air traffic controllers now exceeds that of the conventional tower it replaced.

10.9   Human Performance

Much has been made of the discussion of what constitutes "human factors" and in dispelling some of the myths around human error. However, there is growing interest in the topic of human performance—both in terms of the limitations of the human and in what can be done to maximize performance.

The ICAO Manual on Human Performance for Regulators,[9] which was published in 2021, describes human performance as follows: *human performance (HP) refers to how people perform their tasks. HP represents the human contribution to system performance.* Continuing improvements in elite sport are largely related to better understanding of human performance rather than the evolution of the human being. Training techniques, nutrition, hydration, rest, and recovery techniques as well as improvements in the ergonomics of clothing and equipment have all helped athletes to improve their performance.

In aviation, the focus on human performance has predominantly been on the accuracy component of human performance (whereas elite sport would also focus on speed). If a task is not performed "accurately" then it may either lead to an error or at least a less efficient outcome. For example, a maintenance task may lead to rework upon inspection or could lead to an in-flight shut down of an engine. Mishandling of the latter by a fatigued or overloaded crew could lead to a serious incident or accident. Careful attention should therefore be given to all the factors that may have influenced the people involved. In other words, consideration must be given not only to the human error (failure to perform as required) but also to why the error occurred.

ICAO describes five principles that illustrate how human performance is influenced by different factors:

1. People's performance is shaped by their capabilities and limitations.
2. People interpret situations differently and perform in ways that make sense to them.
3. People adapt to meet the demands of a complex and dynamic work environment.
4. People assess risks and make trade-offs.
5. People's performance is influenced by working with other people, technology, and the environment.

These principles acknowledge the variability that exists between individuals based on personal factors and the influence of external factors including the operating environment and the technology and people that they interact with.

To better understand human performance, it is worth considering each principle in more detail:

1. *People's performance is shaped by their capabilities and limitations.*

   Capabilities and limitations are dependent on a wide range of factors including individual physiology, fitness, aging, education, training, currency, recency, personality, and mood. While some of these factors can be assessed during recruitment, initial or recurrent training, performance can also vary over time. Where athletes can improve their performance through training and with coaching, it is also possible to overtrain. For pilots, airlines need to strike a balance between the practice and recency that comes with regular line flying and courses or simulator sessions it may need to provide to prepare crews for emergency situations or new procedures.

   Systems such as Flight Operations Quality Assurance (FOQA) help to provide a level of assurance of performance to supplement the observations of senior crew or Training Captains. Where airlines have managed to develop a good reporting culture, individuals are able to self-report concerns about physical or mental health, fatigue, and so on. However, there remains a concern among some crew that this can also lead to suspension from flying and therefore a loss of earnings.

2. *People interpret situations differently and perform in ways that make sense to them.*

   As the earlier discussions on cognitive performance and situational awareness demonstrate, it is possible, or indeed likely, that people will interpret situations differently. Interpretation depends on so many factors, not least the experiences and biases that an individual may bring. Whereas training has a vital role to play, especially in supporting the application of standard operation procedures, there will always be situations that are novel and unusual. People can be fooled by the information and stimuli they experience, may be influenced by stressors including the response of those around them and the time they think they have available to deal with a situation. When things go wrong, investigators learn most by trying to see the situation through the eyes of those involved and interpret why the response made sense at the time. From a positive perspective, differences in interpretation can also help by identifying alternatives before committing to a decision. Careful management of such input is an important element of CRM/TRM.

3. *People adapt to meet the demands of a complex and dynamic work environment.*

Human beings are amazing. Their ability to adapt to new or changing situations is generally excellent, but not infallible. Every day the commercial aviation depends on the ability of people across many roles dealing with challenges ranging from technical defects to severe weather and from new air traffic procedures to disruptive passengers. Adaptations can be the way in which novel problems are resolved, but can also become problematic, especially if they become what are referred to as "routine violations." If the only way to get the job done is to adapt a procedure or use a piece of equipment in a way that differs to the designer's intent, then these can lead to an unintended erosion of safety. Once again, the challenge for the aviation safety professional is to find a balance between harnessing this great human strength and not overdoing it to the point of breaking a safety system.

4. *People assess risks and make trade-offs.*

If risk is an expression of a hazard and its probability, people are recurrently having to make rapid decisions about the risks that they perceive and what trade-offs may be associated with it. A simple example could be a flight crew offered an ATC short cut as they approach an airport. While there may be an advantage to the passengers in terms of arrival time or indeed the company in terms of fuel burn, there is also a risk associated with a need to reprogram the Flight Management Computer or prepare the cabin for landing. With experience and practice, individuals can enhance their risk assessment skills, but by supplementing this with standardized tools or approaches, it is possible to improve performance.

5. *People's performance is influenced by working with other people, technology, and the environment.*

Aviation is a complex sociotechnical system where people from myriad backgrounds and cultures interact with complex technology in a highly dynamic operating environment. Human performance is influenced by all these things, sometimes in a way that is hard to predict. Turbulence may make one person sick and leave another unaffected; changes in time zone may make it impossible for one person to sleep for others make little or no difference; an assertive personality may lead to some becoming withdrawn and others to be similarly assertive. These things all highlight the need for high-quality in-service feedback so that those who design equipment or procedures can accommodate such variability.

10.10   Tools to Enhance Human Performance

The FAA describes Aeronautical Decision-Making (ADM) as a systematic approach to the mental process used by aircraft pilots to consistently determine the best course of action in response to a given set of circumstances (FAA Advisory Circular 60-22). Various mnemonics have been produced to help with the process.

For example, the FAA Pilot's Handbook of Aeronautical Knowledge recommends the DECIDE mnemonic in Figure 10-8.

British Airways emphasizes the importance of assessing the time available for using the formal decision-making tool using the "T" of its T-DODAR mnemonic (Figure 10-9).

10.11   Fitness for Duty

Human performance is so important that the FAA has crafted a regulation dealing solely with the topic of *fitness for duty*. For many people, if they wake up not feeling well one morning, there is still a good chance they could survive the workday without putting anyone in danger. However, for aviation professionals, even the smallest disruption of balance in one's body could severely compromise the safety of hundreds of people. Recent accidents highlighted the need for future awareness campaigns and areas of research to improve fitness for duty.

The FAA explains fitness for duty in FAR 117.5 as being "physiologically and mentally prepared and capable of performing assigned duties at the highest degree of safety." Unfortunately, there are grey lines in the meaning of the FAA definition as it is hard to standardize and quantify what physiological and mental fitness is for each professional.

The DECIDE Mnemonic of Aeronautical Decision-Making	
Detect	Detect the fact that change has occurred
Estimate	Estimate the need to counter or react to the change
Choose	Choose a desirable outcome (in terms of success) for the flight
Identify	Identify actions which could successfully control the change
Do	Take the necessary action
Evaluate	Evaluate the effect(s) of his/her action countering the change

FIGURE 10-8   The FAA DECIDE mnemonic. (Source: www.faa.gov)

The T-DODAR Mnemonic of Aeronautical Decision-Making	
Time	Determine the time available for decision-making and proceed accordingly
Diagnose	Develop an understanding of the problem
Options	Consider what courses of action are available
Decide	If multiple options are available, choose the best one
Assign	Allocate who needs to do what to execute the decision
Review	Constant review of the situation allows adaptation to new information

FIGURE 10-9   The T-DODAR mnemonic. (Source: British Airways)

Although the presence of stress in one's life has clear effects on performance, and alcohol can obviously severely impact performance, the *fatigue* element is less obvious and requires elaboration. Although it can be hard to pinpoint, fatigue is classified by the following:

1. Weariness from mental or bodily exertion

2. Decreased capacity or complete inability of an organism, an organ, or a part to function normally because of excessive stimulation or prolonged exertion

Fatigue can decrease short-term memory capacity, impairs neurobehavioral performance, leads to more errors of commission and omission, and increases attentional failures. A study of medical residents looked at the impact of fatigue and found that the risk of making a fatigue-related mistake that harmed a patient soared by an incredible 700% when medical residents reported working five marathon shifts in a single month.[10] Think about how these same findings could cause breakdowns in safety for aviation professionals during long shifts. Some studies have shown sleep deprivation in pilots causing performance reductions equivalent to having a blood-alcohol level of 0.08%.[11] In other words, being fatigued can deteriorate our performance as much as if we were too drunk to drive a car!

There are three types of fatigue: transient, cumulative, and circadian. Transient fatigue is brought on by sleep deprivation or extended hours awake. The second type, cumulative fatigue, is repeated mild sleep restriction or tiredness due to being awake for extended hours spanning a number of days. Lastly, circadian fatigue is the reduced performance during night-time hours, particularly between 2 am and 6 am for those who are used to being awake during daytime and asleep at night. Contributing factors to all types of fatigue include personal sleep needs, sleeping opportunities, physical conditioning, diet, age, alcohol, stress, smoking, sleep disorders, mental distress, sleep apnea, and even heart disease.

It is imperative to recognize the symptoms of fatigue because it is associated with accidents, reduced safety margins, and reduced operational efficiencies. In 1993 the NTSB started including fatigue as a probable cause of some accidents, starting with the report on the uncontrolled collision with terrain by an American International Airways Flight 808, a Douglas DC-8, carrying cargo to Guantanamo Bay, Cuba. Another notable accident occurred at Little Rock, Arkansas, in 1999 when an MD-82 aircraft overran the runway on landing with 11 fatalities.

Fatigue is, if anything, underreported in accidents as evidence can be hard to acquire. However, simulator studies demonstrate the effects of fatigue on human performance on cognitive and motor skills occur in similar ways to increasing blood alcohol concentration.

Fatigue is a significant challenge for aviation because of long shifts, circadian disruptions, and antisocial work hours. Flight crew may experience ultra-long-haul flights such as London to Perth (20 hours) and cross multiple time zones. While "heavy crewing" allows for a rest period during the flight, sleep quality may be poor. Maintenance work often takes place at night when aircraft schedules allow, but when circadian lows can lead to low arousal and poor decision-making.

Attempts to ensure staff have sufficient rest are limited by conditions that are still not adequately addressed, such as the following:

- Reporting for duty while fatigued due to personal factors such as a new-born baby, relationship issues, or health concerns such as sleep apnea.

- Commuting from home to the location where a trip will commence may require significant time awake and induce stress prior to commencing professional duties.

- A crewmember may be provided an adequate rest period prior to a flight but may be unable to obtain quality rest due to ambient noise, such as when there is a party or a loud television in an adjoining hotel room, construction noise, or if the crewmember misuses allotted rest time for personal activities that are not conducive to rest, such as sightseeing or working a second job.
- Crossing time zones and not being able to be on a consistent sleep schedule.

Taking the above into consideration, it is perhaps fortunate that there have not been more aviation accidents and incidents as a result of fatigue. This may be a feature of the difficulty of evidencing fatigue as a causal factor in inducing an error. While investigators are increasingly using evidence from the 7 or 10 days before to build up a picture of the likelihood of impaired performance as a result of fatigue, this can be very difficult. Each person reacts differently to sleep disruptions, but there is still plenty of empirical evidence to support the view that fatigue not only has a negative impact on human performance but is also difficult to self-diagnose. A fatigued person can suffer from poor risk perception and sudden loss of consciousness through microsleeps.

The professional life of many people in aviation can best be described as shift work, which is a schedule that falls outside the traditional 9 am to 5 pm workday and which can include evening, night, morning, or rotating periods. Negative side effects of shift work include acute/chronic fatigue, sleep disturbance, disruption of circadian rhythm, impaired performance, cardiovascular problems, and family/social disruption. In turn, this can lead crew members to become dependent on medication for sleep or to manage circadian dysrhythmia, much of which is unregulated.

Fatigue can affect a wide variety of roles in aviation beyond the aircraft's crew. When operating while fatigued, air traffic controllers may commit an error, such as miscalculating required separation between aircraft or in transmitting information such as the altimeter pressure setting. Potential errors in maintenance may include incorrect assembly of parts, missed steps in task cards, and incomplete inspections.

Fatigue can never be completely eliminated, and as new threats emerge such as from increasing flight times, longer commutes, and "always on" culture attached to mobile personal devices, a risk-based approach is required. In 2010, the president of the United States signed Public Law 111-216, the Airline Safety and Federal Aviation Administration Extension Act of 2010, which focuses on improving aviation safety. Section 212(b) of the Act requires each air carrier conducting operations under Title 14 of the Code of Federal Regulations part 121 to develop, implement, and maintain Fatigue Risk Management Plans (FRMP). Such plans consist of an air carrier's management strategy, including policies and procedures that reduce the risks of flight crewmember fatigue and improve flight crewmember alertness. Further information on FRMP is contained in FAA Advisory Circular 120-103A, entitled "Fatigue Risk Management Systems for Aviation Safety."

These guidelines provide a source of reference for managers directly responsible for mitigating fatigue, detailed step-by-step guidelines on how to build, implement, and maintain an FRMP, and describe the elements of an FRMP that comply with industry good practice. A typical FRMP structure consists of nine elements, all of which the carriers must address in their own plan:

1. Senior Level Management Commitment to Reducing Fatigue and Improving Flight Crew Member Alertness
2. Scope and Fatigue Management Policies and Procedures
3. Current Flight Time and Duty Period Limitations

4. Rest Scheme Consistent with Limitations

5. Fatigue Reporting Policy

6. Education and Awareness Training Program

7. Fatigue Incident Reporting Process

8. System for Monitoring Flight Crew Fatigue

9. FRMP Evaluation Program

By using this system, fatigue becomes a part of the safety management system in the same way other aspects of health, environment, productivity, and safety are managed. However, most carriers focus on managing flight crew fatigue and to a lesser extent maintenance engineer fatigue. In contrast, few would consider the decisions being made at the corporate headquarters by fatigued managers.

10.12   Communication Issues

The term "communication" usually includes all facets of information transfer including verbal and nonverbal communications. It is an essential part of teamwork, and language clarity is central to the communication process. Different types of communication, and verbal communication in particular, remain one of the weakest links in the modern aviation system. More than 70% of the reports to the Aviation Safety Reporting System involve some type of oral communication problem related to the operation of an aircraft.

Technologies, such as airport surface lights or CPDLC, have been available for years to circumvent some of the problems inherent in ATC associated with verbal information transfer. Sometimes, however, solutions bring unintended negative consequences or at least concern that they will. For example, one potential problem with ATC data link communication is the loss of the "party line" effect (hearing the instructions to other pilots), which removes an important source of information for building pilot SA about the ATC environment. However, the so-called party line is also a source of errors for pilots who act on instructions provided to other aircraft or who misunderstand instructions that differ from what they anticipated by listening to the party line. When CPDLC was introduced across the Pacific in the 1990s, the loss of party line was far less significant than anticipated as many crews admitted not listening to the low-quality Long Wave radio transmissions except when they wanted to speak to ATC. Switching ATC communication from hearing to visual, as is the case with reading data link communiqués, also can increase pilot workload under some conditions, especially in congested airspace or when messages are delayed in transmission.

Miscommunication between aircrews and air traffic controllers has been long recognized as a leading type of human error. It has also been an area rich in potential for interventions. Examples are the restricted or contrived lexicon (e.g., the phrase "say again" hails from military communications, where it was mandated to avoid confusing the words repeat and retreat); the phonetic alphabet ("alpha," "bravo," etc.); and stylized pronunciations (e.g., "niner" to prevent confusion of the spoken words nine and five).

Adequate communication requires that the recipient receive, understand, and can act on the information gained. For example, radio communication is one of the few areas of aviation in which complete redundancy is not incorporated. Consequently, particular care is required to ensure that the recipient receives and fully understands a radio communication. There is more to communication than the use of clear, simple, and concise language. For instance, intelligent compliance with directions and instructions requires knowledge of why these are necessary in the first place.

Trust and confidence are essential ingredients of good communication. For instance, experience has shown that the discovery of hazards through incident or hazard reporting is only effective if the person communicating the information is confident that no retribution will follow her or his reporting of a mistake.

The horrific ground collision between two Boeing 747 aircraft in Tenerife in 1977 introduced in Chapter 9 resulted in the greatest loss of life in an aviation accident and featured a fundamental communication error as part of the accident sequence. Today, controllers restrict the word cleared to two circumstances—cleared to take off and cleared to land— although other uses of the word are not prohibited. In the past, a pilot might have been *cleared* to start engines, *cleared* to push back, or *cleared* to cross a runway. The recommendation would have the controller say, "Cross runway 27," and "Pushback approved," reserving the word *cleared* for its most flight-critical use.

The need for linguistic intervention never ends, as trouble can appear in unlikely places. For example, pilots reading back altimeter settings often abbreviate by omitting the first digit from the number of inches of barometric pressure. For example, 29.97 (inches of mercury) is read back "niner niner seven." Since barometric settings are given in millibars in many parts of the world, varying above and below the sea level pressure standard value of 1013, the readback "niner niner seven" might be interpreted reasonably but inaccurately as 997 millibars. The obvious corrective-action strategy would be to require full readback of all four digits.

A long-range intervention and contribution to safety could be to accept the more common (in aviation) English system of measurement, eliminating meters, kilometers, and millibars once and for all, despite the recommendation of Annex 5 of ICAO to adopt SI units across aviation. Whether English or metric forms should both be used in aviation, of course, is arguable and raises sensitive cultural issues. At this time, the English system clearly prevails, since English is the ICAO-mandated international language of aviation, as stressed in a decree by ICAO regarding language proficiency on January 1, 2008.

10.13 Humans and Automation

Throughout the existence of powered flight, engineers have looked for ways to reduce the workload for flight crew and increase the reliability of the aircraft. The first autopilot system was developed by Sperry as early as 1912 to allow the aircraft to fly straight and level along a compass course thereby reducing the pilot's workload. Colloquially known as "George," the autopilot has evolved along with a vast number of onboard systems to allow aircraft to fly further and faster, to provide incredibly accurate navigation as well as the ability to detect other aircraft and rising terrain. Modern aviation has become incredibly reliant on automation and although the huge improvements in the accident rate demonstrate the tangible effect on safety, this is evidenced through a lack of accidents rather than any hard proof. Sadly, it is only the failures associated with automation that seem to get public attention.

A case in point is the Airbus A320 aircraft which first flew in 1987 as the world's first digital fly-by-wire airliner. Using an envelope-protection system and computer-controlled flight control surfaces, the aircraft was revolutionary. The design philosophy aimed to protect the pilot from operating outside of a safe performance envelope for any particular phase of flight.

Unfortunately, the aircraft suffered a high-profile accident in 1988 while conducting a demonstration flight at Mulhouse-Habsheim Airport. While attempting a slow speed pass of the runway, the crew found themselves lower than anticipated and attempted to climb using maximum TOGA thrust (the setting for takeoff go-around). The aircraft recognized it was close to stalling and did not respond to the pilot's elevator input, which would have accelerated the onset of the stall. The aircraft collided with trees and three passengers lost their lives. In simple terms, the aircraft behaved as intended, but this was different to traditional aircraft such as the Boeing 737 which would have allowed the pilot to climb even if it led to a stall.

Since that accident, the Airbus envelope protection has doubtless saved countless lives in situations where flight crew may have inadvertently taken the aircraft beyond its safe performance. Indeed, even in the famous case of the "Miracle on the Hudson," exceptional flying by Captain Chesley Sullenberger was complemented by the "alpha-protection mode" which provided maximum performance for the weight and performance of the aircraft without the risk of stalling the aircraft.

Advances in automation are not always matched with an understanding of how it works. This can be especially problematic when a system malfunctions or when crew find themselves in an unusual situation. In 1992 an Airbus A320 operating as Air Inter flight 148 crashed into terrain near Strasbourg, France, with the loss of 87 lives when pilots misunderstood the descent mode that was selected in the autoflight system and descended at 3300 ft/min rather than at a flight path angle of 3.3 degrees. Two years later, an Airbus A300 operating as China Airlines flight 140 stalled on approach to Nagoya, Japan, when the automation confused the pilots, and an unintentional go-around was initiated during approach to landing.

According to Michael Feary, a research psychologist for NASA, part of the problem was that the information provided in the flight computer displays was written in "engineer-speak" versus "pilot-speak." How automation combines human error is a subject of interest. You could write an entire book on the topic. In fact, many very good books have already been written. As a minimum, we should become familiar with three terms: automation surprise, mode confusion, and GIGO (garbage in, garbage out).

Automation surprise is when a person has a mental model of the expected performance of technology and then encounters something different. In some cases, we anticipate a response to the last selected mode but do not notice that the automation has reverted to a submode, and thus, behaves differently than expected. Designers strive to make automation intuitive; however, operators may encounter functioning that they have not been taught and which is not announced by the automation control heads or associated displays. A study that surveyed 1268 pilots revealed that 61% of them still were surprised by the flight deck automation from time to time.[12]

A specific example of automation surprise is described as mode confusion where the pilots expect their actions to have a different effect than it does. Mode confusion can happen when we do not completely grasp the inner workings of automation or if we fail to monitor the automation as it operates, much as we would monitor a colleague to stay in the loop as to what is happening. In the case of the Air Inter accident at Strasbourg, the interface design was poor and a missing decimal point on the flight control unit was all that would have alerted the crew to their error unless they spotted the aircraft's rapid descent on the primary flight display.

10.13.1 The Startle Effect

If there was a downside for aviation safety from the very high levels of reliability that have become commonplace, it is that problems can develop suddenly and those with little experience of dealing with such events are unable to comprehend and correctly process the situation. The so-called "startle effect" has been discussed in the context of several major accidents including the 2009 AF447 crash in the Atlantic Ocean described in Chapter 1. The startle effect, also known as amygdala or limbic hijack, is a physical and mental response to sudden intense and unexpected stimulus. The physical response is also known as "fight or flight" and according to Skybrary[13] it may elicit the following changes in physiological activity:

- Cardiovascular System: Heart rate increases, blood pressure rises, and coronary arteries dilate to increase the blood supply to brain, limbs, and muscles.

- Respiratory System: Depth and rate of breathing increases providing more oxygen to the body.

- Endocrine System: Liver releases additional sugar for energy. Adrenal glands release adrenalin.

- Muscular System: Muscles tense in readiness for immediate action.

- Excretory System: Sweat production increases.

- Nervous System: Brain activity changes, reactions become less reasoned and more instinctive.

In the case of flight AF447, the investigators concluded that as the autopilot and autothrottle disengaged unexpectedly, the PF "made rapid and high-amplitude control inputs" which led to the aircraft pitching up 11 degrees in 10 seconds. The investigators believed "the excessive nature of the PF's inputs can be explained by the startle effect and the emotional shock at the autopilot disconnection."

In certain circumstances, the effect can lead to extraordinary performance, but this is never guaranteed and is very difficult to train for. While simulators can deliver very high levels of fidelity, the ethics of creating an extreme situation to elicit such a response is questionable. The FAA issued an advisory in 2017 to address the need for Upset Prevention and Recovery Training. It acknowledges that limitations of flight simulation training devices can present a risk of negative training transfer—in other words, simulators may not behave in the same way as the real aircraft when operating outside its normal flight envelope.

In the past, pilots coming from the military or general aviation would likely have greater exposure to problems in the air or unusual flying such as aerobatics. A real engine failure is relatively common in these environments, whereas in commercial airlines, it is rarely experienced outside of the simulator. It is an example of where high engineering reliability can have unintended consequences for the flight crew. This presents a challenge for the training community and those who manufacture flight simulation training devices.

10.14   Conclusion

The accident record reminds us that complex systems such as aviation can be brought down through the actions or inactions of individuals. However, detailed analysis of human performance demonstrates that most of our fallibilities are well known and, to a large degree, predictable. By building systems that minimize the likelihood of an erroneous act and reinforcing them with barriers that limit the consequences, we can deliver a higher level of aviation safety. The various challenges discussed in this chapter represent a brief introduction to a huge body of knowledge that continues to grow. Those involved in aviation safety must take the time to become familiar with the topic of human performance as merely "being human" does not make you an expert!

Key Terms

Aeronautical Decision-Making

Automation

Communications

Crew Resource Management

Human-Centered Design

Human Factors

Human Performance

Situation Awareness

Threat and Error Management

Topics for Discussion

1. What is the purpose of Human-Centered Design and where might it be applied in commercial aviation safety?
2. What distinguishes a slip from a lapse or a mistake?
3. What is the primary difference between a mistake and a violation?
4. What is human error and how does it differ from pilot error?
5. What are the three levels of situational awareness?
6. ICAO describes five principles that influence human performance. What are they?
7. What does the mnemonic T-DODAR stand for and what is its purpose?
8. What is the startle effect and how might it threaten flight safety?
9. What communication errors occurred in the 1977 Tenerife disaster and how were they addressed for the future?
10. How did man and machine work together to achieve a good outcome in the "Miracle on the Hudson"?

References

1. Reason, J. (1990). *Human Error*. Cambridge: Cambridge University Press.
2. Maurino, D. E., Reason, J., Johnston, N., & Lee, R. B. (1995). *Beyond Aviation Human Factors: Safety in High Technology Systems*. Aldershot: Ashgate.

3. Khatwa, R., & Helmreich, R. (1998–1999). Analysis of critical factors during approach and landing in accidents and normal flights. *Flight Safety Foundation Flight Safety Digest, 17–18*(11–12, 1–2), 1–257.

4. Wiegmann, D. A., & Shappell, S. A. (2003). *A Human Error Approach to Aviation Accident Analysis. The Human Factors Analysis and Classification System.* Burlington, VT: Ashgate, p. 150.

5. VandenBos, G. R. (Ed.). (2007). *APA Dictionary of Psychology.* Washington, DC: American Psychological Association, p. 187.

6. Endsley, M. R. (1995). Towards a theory of situation awareness in dynamic systems. *Human Factors, 37*, 32–64.

7. Strauch, B. (2004). *Investigating Human Error: Incidents, Accidents, and Complex Systems.* Burlington, VT: Ashgate.

8. Cortés, A. (2011). A theory of false cognitive expectancies in airline pilots. Unpublished doctoral dissertation. Northcentral University, Scottsdale, AZ.

9. ICAO. (2021). *Manual on Human Performance for Regulators.* Doc 10151. Montreal.

10. Hallinan, J. T. (2009). *Why We Make Mistakes.* New York, NY: Broadway Books.

11. Paul, M. A., & Miller, J. C. (2007). *Fighter Pilot Cognitive Effectiveness during Exercise Wolf Safari.* DRDC Technical Report No. 2007-20. Toronto, CA: Defense Research and Development Canada.

12. Bureau of Air Safety Investigation. (1999). Advanced-technology aircraft safety survey report. *Flight Safety Foundation Flight Safety Digest, 18*(6–8), 137–216.

13. Skybrary. (2022). *Startle Effect.* https://www.skybrary.aero/articles/startle-effect

CHAPTER 11

Humans as the Solution

Introduction

There is an apparent paradox in aviation where the amazing level of reliability that has been achieved through many years of hard work contrasts with the very high profile that any aviation incident or accident seems to get in the news media. When a major disaster strikes, it would be hard to comprehend from the often-sensationalist media coverage that on the vast majority of flights everything happens normally and the aircraft lands safely at its destination. According to the International Air Transport Association (IATA), in 2021, there was 1 accident in every 0.99 million commercial airline flights. If we look just at IATA member airlines, the rate was 1 accident in every 2.27 million flights. These dramatic statistics are a great reminder that aviation does an incredibly good job of achieving safe operations. The reasons for this are many, but as the SHELL model in Chapter 9 demonstrates, it is the human (or "liveware") that sits at the heart of the system, both in terms of failures and successes. This chapter focuses on the latter.

11.1 Professionalism in Aviation Exemplified

On November 4, 2010, a Qantas Airbus A380 aircraft took off from Singapore for Sydney with 469 passengers on board. At the time, the A380 was the largest and most advanced passenger aircraft in the world. Four minutes into the flight and passing 7,400 feet, the crew heard two "BOOMS!" and the aircraft shuddered. Immediately the master warning system activated, and the flight deck was filled with the repetitive "bing, bing, bing, bing…" that accompanies it.

The commander of the aircraft, Captain Richard de Crespigny, needed to stabilize the aircraft and diagnose the problem. As mentioned in Chapter 10, all pilots are taught a simple mantra: Aviate—Navigate—Communicate. In other words, manage the workload by focusing on flying the aircraft first, deciding where to go next, and telling people about it last. In an aircraft like the A380, this meant using the autopilot to hold the altitude (which reduces the stress on the engines and airframe) and then starting to work through what they knew about the cause and effects of the explosion.

Even though it is the largest commercial aircraft in the skies, the A380 is designed to be operated by two pilots. This is achieved through an advanced technology flight deck with a flight management system (FMS) and electronic centralized aircraft monitoring (ECAM) system at the heart. What used to be a bewildering set of dozens of dials, gauges, and switches have been reduced to fewer multifunction displays supported by a logic which aims to present the flight crew with the most relevant information

Figure 11-1   Damage to number 2 engine of the Airbus A380. (Source: ATSB—Public Domain, https://www.atsb.gov.au/publications/investigation_reports/2010/aair/ao-2010-089)

in order of priority. In other words, the automation is designed to help the flight crew make the best choices to achieve the right outcome (Figure 11-1).

At this stage, it was not clear to the crew of QF32 that the aircraft had suffered damage to all four engines and shrapnel from the uncontained failure of the number two engine had ripped through the wings and fuselage causing a fuel leak and damage to the network of wiring connected to the ECAM system. As the commander confirmed that he had control of the aircraft, he asked the First Officer to start to run through the procedures being presented on the ECAM system display. What none of the crew could have anticipated at that point was just how many ECAM messages they were about to be presented with or that it was only by choosing to question the logic of some of those messages that they would achieve what Carey Edwards, a notable human factors expert and member of the Royal Aeronautical Society's Human Factors Group, described as "…one of the finest examples of airmanship in the history of aviation."

As the aircraft circled off the coast of Singapore, the crew continued to evaluate the condition of the aircraft in order to prepare for an emergency landing. The aircraft had departed Singapore with 105 tons of fuel for the 4000 mile sector to Sydney, but as the crew monitored the aircraft they realized that 9 tons of fuel had already leaked out of at least 10 holes in the aircraft's fuel tanks. On the one hand, such a leak presented a fire risk and on the other, it reduced the endurance of the aircraft and therefore the time the crew had to manage the situation and attempt a safe landing. The ECAM alerted the crew "FUEL: WINGS NOT BALANCED" and instructed them to transfer fuel to the lighter wing. It was at this point the commander called out "STOP!" He recognized that transferring fuel this way would move fuel away from the two least damaged engines and potentially lose it overboard.

QF32 was an unusual flight for many reasons beyond the explosion. Although the A380 is designed to be safely operated by two pilots, on the sort of long-haul sectors that Qantas operates there is often a third "relief" pilot—in this case known as a Second Officer.

This allows one pilot to take a rest in flight, often in a dedicated bunk area within the flight deck. On this occasion, the three primary flight crew—Captain Richard de Crespigny, First Officer Matt Hicks, and Second Officer Mark Johnson—were accompanied by two more—a "check captain in training," Harry Wubben, and "supervisory check captain," Dave Evans. Harry was conducting a line check for Captain de Crespigny and Dave was checking the checker! It is probable that this flight had more A380 experience onboard than anything outside the original test program. This could have been good or bad. Who was the most senior pilot onboard? Did Dave outrank Harry and Richard? Thankfully the answer was straightforward—Richard was the pilot-in-command (PIC) on this flight and as soon as the explosion occurred, the check was suspended, and all the crew pivoted their role to support the PIC.

Captain Dave Evans remains modest about their performance as a crew, often saying that "we just did what we were trained to do," but the actions of the crew cannot be underestimated. Faced with a novel problem, the details of which were not fully apparent to the crew at the time, they were able to apply their experience and training to work as a team to achieve a successful outcome. That meant learning how the aircraft was performing, which systems were unavailable or damaged, and even deliberately entering false data into the Airbus laptop to get at least approximate answers to questions that the software was not written to answer! It is hard to imagine any level of automation that could deal with such situations so effectively.

Of course, with all the challenges presented to the flight deck crew, it is easy to forget that the emergency was also being experienced by 440 passengers and 24 cabin crew. Managing the cabin was Michael Von Reth, a highly experienced cabin crew member who spent many years as a member of the Asia Pacific Cabin Safety Working Group and as a safety investigator for the Flight Attendants Association of Australia and for Qantas. He immediately recognized the seriousness of the situation and he needed to switch to a different role.

There were no emergency procedures to cover the specifics of the situation, but Michael knew from the events that he had studied over the years what a crucial role the cabin crew could play as the "eyes and ears" of the flight deck. He relayed crucial information to the pilots as to the visible damage and managed the welfare of those in the cabin by broadcasting clearly the role of the cabin crew and asking for cooperation and compliance. He knew that he was dealing with passengers who would be frightened and who in some cases, may not speak English. He was also acutely aware that while cabin crew members train for their primary role as a safety professional, this wouldn't stop them from feeling anxious.

His role became providing information to the flight deck, clear direction for the cabin crew, and giving reassurance to the passengers. His performance on the day resulted in him being presented the Flight Safety Foundation's "Professionalism Award in Flight Safety"—the first cabin crew member to receive the award in the FSF's 64-year history. Leadership, especially through a crisis is critically important and yet so difficult to define precisely. It is crafted through selection, training, experience, personality, and culture.

If all the above was not enough to demonstrate the positive effect of professionalism in trying circumstances, the crew still needed to land the aircraft and get everyone safely off. With the level of damage that the aircraft had suffered, this was not going to be easy. For starters, only two of the eight hydraulic pumps were working; the left wing was electrically dead; the autobrakes were showing as failed; and some antiskid modules were inoperative. In addition, the outer ailerons had failed, reducing the aircraft's roll capability, and the fuel imbalance meant they would be landing 40 tons overweight.

There were plenty of more problems besides and readers are encouraged to study the captain's full account in his book *QF32*.[1]

Against the odds, the crew pulled off a successful landing in Singapore, but even then, there was the potential for defeat to be snatched from the jaws of victory. The landing needed to be fast to retain control of the damaged aircraft (166 knots compared to the calculated 145 knots) and the crew needed to brake hard knowing engine and aerodynamic braking were compromised. As the aircraft decelerated, fuel poured out of the holes in the wing, and the brakes heated up beyond 900°C. With jet fuel having a flash point of +38°C and an auto-ignition temperature of 220°C, the potential for a severe fire was huge. Once again, the crew needed to comprehend the unique situation they found themselves in and dynamically assess the risk. An evacuation would likely cause injuries, especially for those sliding from the upper deck and as people ran through fuel or foam. To make matters worse, the number one engine (outboard from the exploded engine) would not shut down despite attempts from the flight deck and would eventually require the airport fire service to "drown it" using water and foam.

Eventually, passengers and crew were disembarked by stairs and taken to the terminal with the last passenger leaving the aircraft one hour and 52 minutes after it had stopped. From the initial explosion to this point had been three hours and 39 minutes. The crew were exhausted and emotional but continued to think about the welfare of their passengers. Captain de Crespigny insisted on spending 50 minutes briefing passengers and answering their questions about what had just happened. He even went as far as to give everyone his mobile number with the personal guarantee: "If you don't agree with how Qantas is treating you or if you think Qantas doesn't care, then call me on this number" (Figure 11-2).

FIGURE 11-2   Captain Richard de Crespigny visiting Professor Graham Braithwaite at Cranfield University to speak of his experience to the Aircraft Accident Investigation class of 2014. (Source: Author, Graham Braithwaite)

The QF32 incident is an incredible story of human resilience, adaptability, leadership, and teamwork. While perhaps not as famous as the "Miracle on the Hudson," the achievements of the crew of QF32 are arguably superior. The damage to the aircraft was beyond what could have been reasonably anticipated by the designers. The complexity of the scenario they faced far exceeded what could have been plausible in a simulator session. Captain de Crespigny puts the outcome down to teamwork saying, "I don't think great teamwork is an accident. It's the result of knowledge, training, and experience. As the leader of the QF32 team that day, I believe my approach was never that of a hero, but that of an experienced leader who absorbs and gains from the wisdom of those who know more than I do and who are willing to share and work as a team."

11.2   Leadership and Followership for Safety

Leadership can be explained in many ways. In the context of safety, leadership can be defined as "the process by which an individual influences the behavior and attitudes of others toward a common goal." The role of such influence in commercial aviation has a direct bearing on the safety of an operation but is not always addressed when discussing accident prevention. Followership is as important as leadership and describes the ability to take direction and to deliver on what is expected by a leader.

The cumulative effect of such influences, when taken across the entire employee group, fosters a certain safety culture that sets the expectations for behaviors and attitudes about risk. There are many leaders in commercial aviation. Some are more obvious than others on board the aircraft (such as PIC or the cabin manager), whereas others are in leadership roles at local tactical levels such as at airport managers, or at the more strategic "head office" end. Where the effect of the PIC's leadership may have an immediate impact on the safety of a flight operation, others may have a less direct influence.

A straightforward way to consider how aspects of leadership and followership may influence safety and efficiency in flight is through the concept of the CRM pyramid model as shown in Figure 11-3. The shape is designed to illustrate that the higher elements of the model are built on broader foundations or prerequisites. Safety and efficiency—balanced at the top of the model—are the consequence of sound *aeronautical decision-making* (ADM) and which in turn relies on a *shared situational awareness* (SSA). ADM refers to how pilots use mental processes to consistently determine how best to respond to a given set of circumstances. ADM is only as good as the SSA of the team members who are involved in making the decision. SSA refers to the common perceptions and comprehension of an environment and how they impact the future, as held by two or more people. The accuracy of SSA creates the reality and the common mental model of the team members, and therefore, directs their decision making. The SSA is a product of how well a team coordinates its actions, which depends intimately on the verbal and nonverbal communication of members, which explains why the *coordination* component underlies SSA and why the *communication* component underlies coordination in the model.

The foundation of the model is *leadership and followership*, since the effectiveness of the working relationship between leaders and followers directly impacts the quality of communication that is used to coordinate actions that build SSA for peak ADM.

To further understand the importance of leadership, it is important to address the common confusion over the difference between management and leadership. Management is the process of "planning, organizing, directing, and controlling"

Safety & Efficiency

ADM

SSA

Coordination

Communication

Leadership ⇔ Followership

FIGURE 11-3   The CRM pyramid model. (Source: Authors)

behavior so as to accomplish a given workload by dividing up tasks.[2] Most managerial tasks to be completed by flight crews are carefully described in each organization's standard operating procedures (SOPs). This reduces the managerial workload of an aircraft commander or cabin manager, but does not address the pressing need for leadership, which is about setting the example for others to follow and consists of designing a system of incentives and disincentives for encouraging the behavior of followers. Leadership revolves primarily around the concept of creating a positive work atmosphere, so the crew effectively manages resources and complies with procedures.[3]

An aircraft commander, for example, must perform both managerial and leadership functions as part of their job and both aspects have a direct impact on safety. This may be characterized as an ability to "direct and coordinate the activities of other team members, assess team performance, assign tasks, develop team knowledge, skills, and abilities, motivate team members, plan and organize, and establish a positive atmosphere."[4] Again, it is hard to envisage how such a blend can ever be replaced by fully autonomous systems.

11.3   Safety Leadership

Management textbooks are filled with descriptions of what makes for a great leader, but what about safety leadership? On the one hand, it is hard to separate safety leadership from leadership in a safety-critical business such as aviation, but on the other, it is worth focusing on those elements of leadership that prioritize and promote safety.

Tom Krause[5] described seven key safety leadership characteristics and associated behaviors which in turn influence safety culture:

1. **Credibility**—What leaders say is consistent with what they do.

2. **Action orientation**—Leaders act to address unsafe conditions.

3. **Vision**—Leaders paint a picture for safety excellence within the organization.

4. **Accountability**—Leaders ensure employees take accountability for safety-critical activities.

5. **Communication**—The way leaders communicate about safety creates and maintains the safety culture of the organization.

6. **Collaboration**—Leaders who encourage active employee participation in resolving safety issues promote employee ownership of those issues.

7. **Feedback and recognition**—Recognition that is soon, certain, and positive encourages safe behavior.

Those *leaders* may include the traditional business leaders such as the Board Chair or Chief Executive Officer (CEO) but will also include those in specific areas of the business, such as an Airport Station Manager or Fleet Manager, as well as on the aircraft itself.

A famous example from outside aviation is Paul O'Neill who became the CEO of Alcoa in 1987. Alcoa is a major producer of aluminum and is involved in all stages from mining and smelting to fabricating and recycling. In his first speech to shareholders, O'Neill surprised the audience by saying, "I want to talk to you about worker safety." He continued, "Every year, numerous Alcoa workers are injured so badly that they miss a day of work. … I intend to make Alcoa the safest company in America. I intend to go for zero injuries."

His audience was confused. There had been no mention of profits or the usual business buzzwords. When someone asked a question about inventories in the aerospace division and another asked about capital ratios, O'Neill replied, "I'm not certain you heard me. … If you want to understand how Alcoa is doing, you need to look at our workplace safety figures." Profits, he said, didn't matter as much as safety.

Some investors were spooked and almost stampeded out to the lobby to call their clients to sell their stock. One famously recalls saying, "The board put a crazy hippie in charge and he's going to kill the company." However, within a year, Alcoa's profits hit a record high and by the time O'Neill retired in 2000 to become US Treasury Secretary, the company's annual net income was five times larger than before he arrived. Its market capitalization had risen by $27 billion, all while Alcoa became one of the safest companies in the world.

O'Neill's leadership provides an excellent case study for safety professionals and beyond.[6] His realization was that by demonstrating care for Alcoa's employees, he could create a culture where they cared about what they did, and ultimately this would translate into superior business performance. There is a clear read-across for aviation, but it is also reassuring to know that a company like Alcoa supplies aluminum products to the aerospace sector for the construction of aircraft such as the Boeing 737. Safety is the sum of the entire supply chain and not just the end-user.

Leadership can also be demonstrated by an organization. A good example is that shown by CHC Helicopters who when headquartered in Vancouver, Canada, brought people together from across the company for a Safety and Quality Summit. The idea was to share best practice and to promote the importance of safety and quality to their business which involved operating helicopters in inhospitable environments such as in support of offshore oil and gas exploration (Figure 11-4).

Following the initial success of the event, attendees were asked how it could be improved. Their answer was to invite their customers—in this case the companies who

FIGURE 11-4 The Annual CHC Safety and Quality Summit website. (Source: http://www.chcheli .com/SafetySummit)

were using CHC to transport their workers. It became obvious that there was a huge amount of knowledge to share and that they shared a mutual goal in ensuring the safety of workers. The next step was to invite their competitors—other helicopter operators such as Bristow, Cougar, and PHI who would compete for the same contracts, but who also shared common safety values. The event has become an annual fixture with many hundreds of attendees. The focus is on workshops where attendees develop new skills and exchange their experiences. The atmosphere is open and honest—all the companies involved face common safety challenges, but there is something amazing about seeing rival CEOs stand shoulder to shoulder to ensure that no one ever needs to tell the family of an employee or customer that their loved one will not be coming home.

11.4   Enhancing Human Performance

There are many ways in which human performance can be enhanced or optimized. The challenge is to recognize that our workforce is both physically and mentally diverse, and that we are aiming for a sweet spot of making it straightforward for someone to perform well, while not taking away the satisfaction they may derive from their job. We are also trying to enhance performance for the majority of time when operations are relatively normal, while ensuring, so far as is reasonably practicable, that individuals are able to perform effectively in the rarer cases when things go wrong.

11.4.1   Recruitment and Selection

Safety-critical roles within aviation require essential criteria to be met during recruitment, especially in terms of physical and increasingly also mental health. Pilots need to be able to function under conditions of physiological and psychological stress which include the effects of altitude, time zone change, vestibular illusions, and so on. They must be able to complete complex mathematical calculations accurately and at speed, work with a frequently changing group of people and make sound decisions. Air traffic controllers need to cope with a variety of hazards while demonstrating excellent communications skills. Cabin crew need to cope with the effects of travel and a high workload while being ready to pivot to their safety role at little notice. They also need to deal with passengers who can often be quite challenging!

Various tests are used to evaluate new recruits, the most challenging of which is the pilot medical. For those who wish to fly as an airline transport pilot, the criteria are even more demanding. The FAA First Class Airman Medical Certificate is issued based on medical examination of eyes; ear, nose, throat, and equilibrium; mental; neurologic; cardiovascular and general medical condition. These are described in the US Federal Aviation Regulations (CFR part 67 Medical Standards and Certification). Certain conditions such as angina pectoris, epilepsy, coronary heart disease, and diabetes mellitus requiring hypoglycemic medication will result in disqualification, as will mental health conditions such as bipolar disorder, psychosis, and personality disorders which have been serious enough to have repeatedly manifested itself by overt acts. A previous history of substance abuse or dependence will also lead to disqualification.

Following initial screening, regular medical checks are designed to ensure that changes in health are caught early. As pilots age, some of the checks become more frequent or in-depth. For example, FAA Part 67 requires an Electro-cardiogram (ECG) at age 35 and then annually from age 40. Air traffic controllers have less stringent medical requirements whereas maintenance engineers have little or nothing. Cabin crew tend to be required to pass an initial medical assessment, but only usually require follow-up examinations every 5 years, or following a prolonged period of illness, or if the employer has concern about the continued fitness of an individual. Some airlines require their cabin crew to pass an annual assessment of their physical ability to do the job such as opening doors or overwing exits, pushing galley carts, and so on.

The murder-suicide that led to the loss of Germanwings Flight 9525 in 2015 with the loss of 150 lives shocked the global aviation industry. The copilot, Andreas Lubitz, locked the cockpit door after the captain had left to use the toilet and put the aircraft into a dive while accelerating. Attempts to break the cockpit door down failed. Investigators soon discovered that Lubitz had previously been treated for suicidal tendencies and had been declared "unfit to work" by his medical examiner. However, he had not declared this to his employer and instead had reported for duty. While no one will ever know his exact motive or whether he was experiencing some form of psychosis, the event brought into sharp focus the relatively light-touch approach that was being taken toward mental health beyond initial recruitment. It also highlighted the danger of a single crew member on the flight deck in the post-9/11 world of locked flight deck doors.

Consequently, the European Aviation Safety Agency (EASA) developed new safety rules on mental fitness of flight crew and in 2018 introduced safety measures including pilot support programs, alcohol testing of pilots and cabin crew, and psychological assessment of pilots before they start their employment.

11.4.2 Competency-Based Training

Since 1948, ICAO Member States have agreed to adopt common standards for pilot qualifications as contained in Annex 1: Personnel Licensing. The 1970s and 1980s saw improvements to the way in which pilots were trained through the adoption of Crew Resource Management and the use of Line Oriented Flight Training (LOFT) which aimed to provide the opportunity to put CRM skills into practice in transferable in-flight scenarios in a simulator. As routine safety data collection and analysis developed in the mid-1990s, the opportunity to tailor training to address specific risks arose. In parallel, there was a recognition that experience could not be judged solely by hours flown and that the quality of a pilot's training and experience were both important and measurable.

Competency-based training and assessment (CBTA) has been widely adopted by ICAO and its Member States for pilots and air traffic controllers. According to ICAO, competencies refer to the dimension of human performance that are used to reliably predict successful performance on the job. Competencies are manifested and observed through behaviors that utilize the relevant knowledge, skills, and attitudes to carry out activities or tasks under specified conditions. The nine pilot competencies are:

1. Application of Knowledge
2. Application of Procedures and Compliance with Regulations
3. Communication
4. Aeroplane Flight Path Management, automation
5. Aeroplane Flight Path Management, manual control
6. Leadership and Teamwork
7. Problem Solving and Decision Making
8. Situation Awareness and Management of Information
9. Workload Management

In addition, there are five instructor/evaluator competencies:

1. Pilot competencies (see above)
2. Management of the Learning Environment
3. Instruction
4. Interaction with the trainee
5. Assessment and Evaluation

An example of competency and observable behaviors is presented in Figure 11-5.

As a methodology, CBTA focuses on outcomes—in other words, the performance of the individual in key areas which are defined as competencies. These areas may include specific tasks and observed behaviors, although it is the total performance that is critical.

According to IATA, evidence-based training (EBT) aims to "...identify, develop and evaluate the competencies required to operate safely, effectively and efficiently in a commercial air transport environment" while addressing the most relevant threats according to evidence collected in accidents, incidents, flight operations, and training. EBT and CBTA have allowed training organizations and operators to focus their training resources on areas of greatest need, for example, loss of control in flight (LOC-I) and upset prevention and recovery training (UPRT).

Competency Description	Observable behaviors
Application of Knowledge Demonstrates knowledge and understanding of relevant information, operating instructions, aircraft systems and the operating environment	OB 0.1 Demonstrates practical and applicable knowledge of limitations and systems and their interaction OB 0.2 Demonstrates required knowledge of published operating instructions OB 0.3 Demonstrates knowledge of the physical environment, the air traffic environment including routings, weather, airports and the operational infrastructure OB 0.4 Demonstrates appropriate knowledge of applicable legislation OB 0.5 Knows where to source required information OB 0.6 Demonstrates a positive interest in acquiring knowledge OB 0.7 Is able to apply knowledge effectively

FIGURE 11-5   Example of a pilot competency. (Source: IATA)

11.4.3 Practice Makes Perfect

Synthetic training devices such as flight deck simulators have transformed the way in which flight crew are trained. From basic procedural trainers to full flight simulators (FFS) with six degrees of motion, such devices have created a safe and largely cost-effective training solution. This means that flight crew and their trainers can practice safety-critical maneuvers such as engine failure on takeoff without the risk of losing an aircraft. The level of fidelity also means that training transfer is extremely high—in other words, experience demonstrates that skills learned in a simulator transfer with a high degree of effectiveness into real situations on the flight deck.

Simulators have their limitations, especially in terms of an aircraft that is flying outside its normal operating envelope. The software and models that power a simulator will always find it difficult, if not impossible to reproduce the handling characteristics of a damaged aircraft. They also cannot reproduce the effect on the crew of knowing that they are all that stand between success and catastrophe. At this point, the human can demonstrate ingenuity, flexibility, and a level of determination that cannot be seen in a simulator.

A classic example involved the crew of United Airlines Flight 232, a DC-10 aircraft that suffered an uncontained failure in its center (tail) engine that severed all three hydraulic systems. On paper, this should have been impossible, but for Captain Al Haynes, First Officer William Records, and Flight Engineer Dudley Dvorak, this was exactly what they experienced on July 19, 1989. Without hydraulics, everything was difficult as all the principal controls, plus flaps, brakes, and landing gear all depended on them. As the crew struggled to keep control of the aircraft, a DC10 flight instructor who had been traveling as a passenger, Captain Denny Fitch joined the crew on the flight deck to help.

No simulator session could have reproduced what they were experiencing and even after the disaster, no one was able to do as good a job as the crew of Flight 232. Instead, the crew drew upon all their training, experience and survival instinct to use differential thrust on the two functioning engines (on either wing) to turn the aircraft. Flight Instructor Fitch knelt behind the throttles to maneuver them. Although the aircraft was slow to react, the crew found they could alter the pitch of the aircraft and make unstable turns to the right (Figure 11-6).

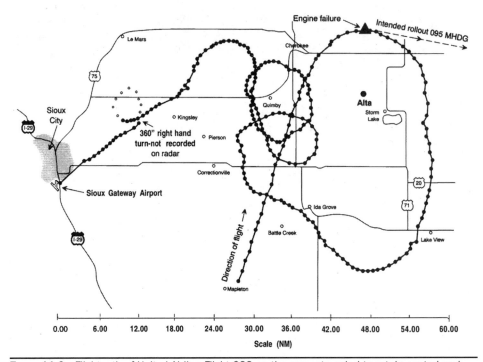

FIGURE 11-6   Flightpath of United Airline Flight 232 as the crew struggled to retain control and divert to Sioux City Airport. (Source: Wikipedia Commons, https://en.wikipedia.org/wiki/United_Airlines_Flight_232#/media/File:UA232map.png)

The aircraft diverted to Sioux City Airport, Iowa to prepare for an emergency landing. On board, the flight crew were not optimistic. Asked by one of the cabin crew about whether they should evacuate the aircraft after landing, the captain said, "…if we have to evacuate, you'll get the command signal to evacuate, but I think—really have my doubts you'll see us… standing up, Honey" before pausing and adding "Good luck, Sweetheart."

Controlling the final approach was all but impossible. Just before touchdown, the aircraft dipped a wing and cartwheeled along the runway. For the emergency responders and news crews that were looking on in horror, it seemed unlikely that anyone could survive. While a total of 112 people died that day, the miracle was that 184 people survived; some of them were able to walk from the wreckage. In theory at least, this was an unsurvivable accident. The manufacturer, regulator and operator considered the total loss of hydraulic-powered flight controls so remote that there was no procedure to counter such a situation (NTSB, 1990[7]). On the day, the flight crew thought differently, doing everything they could to try and land the stricken aircraft.

The four crew on the flight deck all survived the accident even though the remains of the cockpit were not found until 35 minutes after the crash. All of them went back to flying duties including Captain Al Haynes who went on to give over 2000 talks about the accident, never once accepting the title of "hero." In its final report, the NTSB stated that it viewed "…the interaction of the pilots, including the check airman, during the emergency as indicative of the value of cockpit resource management training."

The Board also went on to state that "…under the circumstances the UAL flight crew performance was highly commendable and greatly exceeded reasonable expectations."

As a case study for the benefits of CRM, United Airlines Flight 232 remains one of the best, but the performance of the flight crew was not the only good news story that day. The emergency response plan for the local authority worked well, not least because only 2 years before, a full-scale disaster drill had been hosted at Sioux City Airport involving airport and county emergency responders. Simulation, whether high-fidelity or tabletop is a valuable tool for ensuring preparedness for challenging or infrequent situations such as aircraft emergencies.

11.5   Information Sharing

Many accidents have demonstrated what happens when information is not adequately shared, either between aircraft crew, or between others in the aviation system. Human factors training including Crew Resource Management has helped to provide an understanding of the importance of information sharing as well as a range of tools and techniques to facilitate it.

11.5.1   Tactical Information Sharing

In some airlines, this is about empowering junior crew members to speak up, and if necessary, supporting this with SOPs that require cross-checking. The Australian airline Qantas introduced a simple but powerful phrase, "Captain you must listen" which was only to be used as a last resort and which required the aircraft commander to listen to the information that was being presented. They also recognized that over time, one of their national cultural traits—that of a relatively flat hierarchy and open communications was gradually being eroded. Junior crewmembers who in the past would routinely speak up without fear of repercussions were starting to become more inhibited. This prompted a reinforcement of best practice through procedure, in this case, a specific requirement for junior crewmembers to monitor others and provide feedback. Rather than getting in trouble for pointing out a captain's error, a crewmember was more likely to be reprimanded when the captain realized their own error and realized their actions had *not* been questioned!

Two less well-known examples of highly effective information sharing occurred during the high-profile accidents that involved British Airways Flight 38 at Heathrow in 2008 and US Airways Flight 1549 in New York in 2009. On both occasions, the air traffic controller provided rapid information—in the former example to following traffic and in the latter example to the flight crew of the stricken A320 (Figure 11-7).

At London Heathrow, which sees around 650 arrivals each day and an aircraft landing every 45 seconds at peak times, air traffic controller Greg Kemp was responsible for giving landing clearances to inbound aircraft. At around 1240, he noticed something wasn't right about a Boeing 777 on final approach where the pilot was clearly trying to keep the nose up while sinking. When the crew radioed "Mayday, Mayday," there was only a few seconds before it hit the grass short of the runway. Hitting the "crash" button to alert the emergency services, Greg read out the details of what happened, where, and the relevant rendezvous point, just as he had trained. He then went on to contact the aircraft that were lined up behind the crashed aircraft to make sure they abandoned their landing approach, coordinated with the emergency services, and ensured that aircraft were either switched to the other runway or told to find an alternative airport to land.

FIGURE 11-7 British Airways Boeing 777 crash-landed at Heathrow in 2008. (Source: Wikipedia Commons, https://commons.wikimedia.org/wiki/File:BA38_Crash.jpg)

The audio recording of the controller's actions is an incredible example of calmly reacting to a frightening and unexpected situation. This is, at least in part, down to selection and training.

In the "Miracle on the Hudson," where US Airways Flight 1549 suffered a double engine failure after hitting a flock of geese after takeoff, the story of Captain Chesley Sullenberger's amazing flying is well documented. However, the role of the LaGuardia air traffic controller Patrick Harten in providing options to the crew of what was known as "Cactus 1549," diverting other aircraft, and alerting potential airports is less well known. The audio and transcript are freely available on the internet and are worth a listen.

Bear in mind that we all now know the details of the situation and the outcome, but at the time, the controller would not have known the exact position of the aircraft or whether the options for diversion were feasible. When he offers a return to LaGuardia, he is told "unable" and that they may end up in the Hudson. He replies, "Okay what do you need to land?" and is asked, "what's over to our right in New Jersey … maybe Teterboro?" The controller immediately coordinates with Teterboro tower and offers it to Cactus 1549 but gets the reply "We're gonna be in the Hudson." It is hard to comprehend what the controller must have thought at this point, but he still goes on to provide options. "Cactus 1549 radar contact is lost. You also got Newark airport off your right at 2 o'clock in about 7 miles." By this stage the aircraft was about 300 feet over the Hudson River with no thrust. Harten later told a Congressional Panel that he felt sick when Sullenberger said "we're gonna be in the Hudson." "People do not survive landings on the Hudson River, and I thought it was his own death sentence," he said. "I believed at that moment I was going to be the last person to talk to anyone on that plane alive."

While calm actions of the flight deck and cabin crew will always be at the heart of the Miracle of the Hudson story, the air traffic controller's contribution should not be underestimated. In slightly different circumstances, it could have made all the difference.

11.5.2 Strategic Information Sharing

Tom Palmer, Senior Vice President Services for Rolls-Royce Civil Aerospace, observed, "Hundreds of millions of people fly every year and the industry's safety record is superior, better regulated and better implemented than any other means of transportation on the planet. This has not just come about because of legislation but because of the aviation industry's lifelong commitment to sharing data and implementing change, based upon lessons learned."

Sharing of the lessons from serious incidents and accidents has been standard since ICAO first published Annex 13: Aircraft Accident Investigation in 1951. However, more recently, sharing of safety data has become more widespread. The collection and analysis of routine flight data through Flight Operations Quality Assurance (FOQA) or Flight Data Monitoring (FDM) programs transformed individual airlines' understanding of how crews were performing. Subsequent pooling of this data in a de-identified form through schemes such as IATA's Flight Data eXchange (FDX) has allowed benchmarking of performance between operators of the same type of aircraft or those operating in the same regions. A further extension has been to include safety and security incident reports from flight, cabin, and ground operations through IATA's Incident Data eXchange (IDX) program.

Shared safety data creates the opportunity for deep learning through trend analysis. Large datasets allow earlier detection of emerging issues and benchmarking of performance. The principle of human-centered design mentioned in Chapter 10 creates the opportunity to ensure that the human is best positioned to draw on their strengths and complement the technology. This sounds simple, but in practice, it can be very hard for, say, the designer or manufacturer to understand how the end-user will operate their equipment or the full breadth of situations they may face. It is in such circumstances that data shared from company reporting systems or formal investigations plays such an important role.

11.6 Seeing the Big Picture

When any aviation incident or accident hits the headlines, the immediate aftermath is often filled with speculation and opinion as to what went wrong. Traditional media outlets face the challenge to be "first with the news" whereas increased use of social media allows anyone with an account to share images, information, and opinion as well as, in some cases, deliberate misinformation. So-called "Monday Morning Quarterbacks" often look at a selection of information without consideration of its quality or completeness. The apparent 20:20 hindsight may create interesting stories, but thankfully is not the way aviation learns about safety.

The detailed and thorough analysis of incidents and accidents (see Chapter 3) may take longer than a newspaper or television report to complete but allows for consideration of all the facts. Increasingly, this can mean going deep into the organizational, cultural, and political factors, some of which were in place years or even decades before an accident. From an investigation level, this is important in developing a complete understanding of causal and contributory factors. From a human performance perspective, it is also important to help us determine where remediation efforts are likely to have the greatest effect or at least gain widest adoption.

For the safety professional, deciding how to spend the finite safety dollar is more than a financial decision. It also requires consideration of the implications of doing not or not doing something and what unintended consequences may arise. Take the example of putting the seatbelt sign on in flight when experiencing turbulence. Numerous in-flight events have occurred when passengers and cabin crew have been seriously injured due to severe turbulence. In many of those cases, being seated with a seatbelt fastened low and tight would have prevented injuries, but is it therefore reasonable to expect both passengers and crew to remain in their seats with a seatbelt on for their entire journey? For most flights, the answer must be no, otherwise how would anyone go to the restroom or receive a meal or drink in flight? The answer lies somewhere in between based on an assessment of the overall risk (which may inform an airline's policy) and a dynamic assessment of conditions on a particular flight based on weather forecasts, weather radar, and reports from preceding aircraft. Even then, certain things can be done to proactively manage risks from turbulence such as in the design of cabin interiors or coffee pots through to suspending service of hot drinks. While the nature of risk is that not every decision will prove to have been correct, a human strength is to take all these factors into consideration and apply them to a specific setting.

However, in a system as complex as commercial aviation, effective decision-making will always rely on timely access to high-quality information. This comes from many sources and therefore may vary in quality. When something has gone wrong and damage, injury, or death has occurred then getting hold of information can be difficult, especially if an individual or indeed the organization fears the consequences. This is one of the reasons that ICAO Annex 13: Aircraft Accident Investigation has always been based on a "not for blame" approach to investigation and has recommended that States legislate protections for certain data (such as interviews and cockpit voice recorder transcripts). This ensures that investigators have greater access to information that may help them to understand why people made decisions or performed the way they did.

More recently, operators, maintainers and air navigation service providers have started to adopt a "just culture" approach. Just culture is cited as one of the five elements of safety culture by Professor James Reason who described it as "… an atmosphere of trust in which people are encouraged, even rewarded, for providing essential safety-related information, but in which they are also clear about where the line must be drawn between acceptable and unacceptable behavior." The approach has been adopted by organizations such as Eurocontrol, the European Organization for the Safety of Air Navigation, who use the following definition: "Just Culture is a culture in which front-line operators and others are not punished for actions, omissions or decisions taken by them which are commensurate with their experience and training, but where gross negligence, willful violations, and destructive acts are not tolerated."

The notion of just culture is built on a sense of fairness where the subtleties of human errors and violations are acknowledged. This can be hard for some to appreciate in the emotion that follows an event, but when done well, promotes both openness and accountability. Getting to the correct answer requires patient and dispassionate

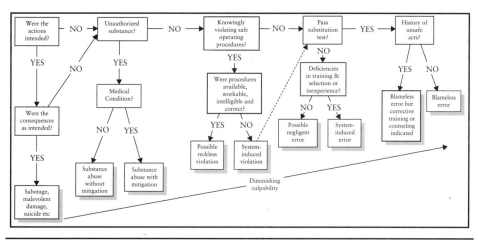

FIGURE 11-8   Reason's Decision Tree for assessing culpability. (Source: Reason, 1997[8])

collection and analysis of the facts, along with a sense of humility and compassion in trying to understand what factors influence the actions or decisions of others.

To aid decision-making as to the culpability or responsibility of an individual's action, Reason developed a decision tree as shown in Figure 11-8. There have been many different interpretations of this since to address a broader range of circumstances and sectors. For example, the British military flowchart differentiates between five types of violation: routine, exceptional, situational, unspecified, and optimizing.

However, no decision aid is infallible, so users need to be careful in their application. The effects of a poor decision on the safety culture of an organization may be severe, especially as anyone feeling like they have been treated unjustly is likely to share their dissatisfaction widely. Far better to be overly lenient and ensure that people remain confident to report than to risk punishing someone inappropriately. In organizations that have successfully implemented a just culture approach, the level of trust among employees has grown. This in turn supports a more open reporting culture and therefore better information from which risks can be assessed and decisions made.

11.7   Conclusion

In the last two chapters, we have seen two extremes—the human as the challenge and as the solution to commercial aviation safety. It is often said that "human error is the downside of having a brain." It is a good reminder that although humans remain fallible, there are plenty of ways in which we can, and have improved human performance and benefited from the ingenuity, resilience, and adaptability that so far only a human being can bring to the aviation system. When things do go wrong, it is important to ensure we fully understand the reasons that lie behind them and to also ask whether the system should have been able to cope with something as predictable as human error. Just as no aircraft accident has a single cause, no safety-critical system should ever fail because of a single person. It is very clear that the human element will always play a key role in commercial aviation safety.

Key Terms

Communication

Competence Based Training and Assessment

Evidenced Based Training

Followership

Human Performance

Just Culture

Professionalism

Safety Leadership

Topics for Discussion

1. In the Qantas Flight 32 incident, what factors influenced the success of the team?
2. What role did the Cabin Manager play in the Qantas Flight 32 event?
3. What is the difference between leadership, management, and followership?
4. What role do leadership and followership play in the CRM pyramid?
5. According to Tom Krause, what are the seven key safety leadership characteristics and associated behaviors?
6. Why did Paul O'Neill frighten investors during his first conference with them?
7. What principle did Alcoa adopt and what was its effect?
8. What seven key areas are covered by a class 1 pilot medical examination?
9. What is Competency Based Training and Assessment?
10. What is the difference between "not for blame" and "just culture"?

References

1. De Crespigny, R C. (2012). *QF32*. Sydney, NSW: Macmillan Press.
2. Wagner, J. A., III, & Hollenbeck, J. R. (2005). *Organizational Behavior: Securing Competitive Advantage*. Mason, OH: Thomson South-Western.
3. Lumpé, M. (2008). *Leadership and Organization in the Aviation Industry*. Burlington, VT: Ashgate.
4. Northouse, P. G. (2007). *Leadership: Theory and Practice*, 4th ed. Thousand Oaks, CA: Sage.
5. Krause, T. R. (2005). *Leading with Safety*. Hoboken, NJ: John Wiley & Sons, Inc.
6. Duhigg, C. (2012). *The Power of Habit: Why We Do What We Do, and How to Change*. New York: Random House.
7. NTSB. (1990). Aircraft Accident Report: United Airlines Flight 232, McDonnell Douglas DC-10-10, Sioux Gateway Airport, Sioux City, Iowa, July 19, 1989. PB90-910406 NTSB/AAR-90/06.
8. Reason, J. (1997). *Managing the Risks of Organisational Accidents*. Aldershot: Ashgate.

CHAPTER 12

Safety Culture

Introduction

One of the most promising ways of improving safety performance in commercial aviation is through the effective management of culture. Safety culture is often referred to as the shared beliefs, attitudes, values, and behaviors that relate to safety within an organization. A considerable body of research has recognized the significant influence of culture when explaining the professional, organizational, and ethnic influences on human performance, and in particular the effect of culture on decision-making.

Contemporary decision-making theories emphasize the need to understand human cognitive processes contextually, that is, from within their environment. The dynamic, highly technical, and procedural environment of commercial aviation has a powerful influence on human behavior. As humans are highly attuned to their social environment, they have a natural tendency to conform to social protocols, and this phenomenon is accentuated by what are sometimes referred to as "cultural norms."

Although links between cultural norms and human behavior are well established within academic and safety management circles, there is some ambiguity about what these links mean and how they can be practically managed. What safety science is becoming increasingly aware of, and as we will see in this chapter, is how dangerous an organization with a poor or dysfunctional safety culture can be.

12.1 What Is Culture?

12.1.1 Defining Culture

The *Oxford English Dictionary* describes culture as the "attitudes and behavior characteristic of a particular social group." Although there are numerous definitions of culture, safety researchers, such as Professor James Reason, established that the most common revolve around the anthropological concept of "attitudes and values." Reason described culture as a "construct," that is, an idea or a theory that is put together from people's opinions, rather than built up from empirical evidence. Culture is something that is inferred from statements and behaviors of individuals related to group behavior. As we have discussed in Part I of this book, traditional approaches to safety and risk management have heavily relied on statistics, perhaps because this quantitative approach better aligned with engineering or "hard science." It is not difficult to see that the introduction of amorphous ideas such as culture into a traditional sphere of engineering was not an easy transition. Culture is not easy to define and by its very nature is subjective.

A typical study of culture would involve surveys and assessments of prevalent attitudes and opinions to influence or manage improvements in workforce behaviors. The Dutch social scientist Geert Hofstede suggested that to understand their context, cultural influences should be described by their source; from national, organizational, or a further subgroup, such as gender, occupation, or region. In summary, culture needs to be described from within the landscape of its origin to make any sense. We need to know where people are coming from, ethnically, socially, and professionally, to even begin to understand why they think and act in a particular manner.

12.1.2   National Culture

The majority of research into the influence of national culture has its roots in anthropology. In his early work,[1] Hofstede initially proposed four dimensions of national culture: Uncertainty Avoidance, Power-Distance, Individualism, and Masculinity.[i] Utilizing Hofstede's criteria, the British academic Professor Graham Braithwaite[2] assessed the positive elements of Australian national culture with those of other nationalities as possible lead indicators to influencing safety performance. Linking national cultural traits to levels of safety performance has attracted considerable interest from within safety science and shows a significant level of correlation.

Using Hofstede's cultural characteristics, safety researchers[3] identified a higher chance of accidents with some national cultures that displayed high Uncertainty Avoidance (UA); a high UA culture would typically display a trait of avoidance of the anxiety caused by uncertain, risky, and ambiguous situations. More recent research has identified a significant relationship between national cultural characteristics and safety culture suggesting that there is a correlation between high levels of UA and low safety performance.

The basic theory is that high UA promotes a culture that inhibits positive safety-related behaviors, such as *inter alia* open reporting, operational flexibility, and non-process-driven decision-making. A high UA culture would emphasize the importance of adherence to rules and process above a safety culture that promoted the flexible and active engagement of risk. To use the language of Charles Perrow within the language of his Normal Accident theory, loosely coupled systems can produce higher safety performance than their tightly coupled counterparts.

Another study by Li, Harris, and Chen,[4] based on Hofstede's criteria, determined that countries with high Power-Distance (PD) characteristics made different types of errors, from those with low PD cultures. Significantly, they noted the increased frequency of contributory errors instigated at the higher end of the organization. In other words, the more hierarchical organizations, which displayed rule-governed behavior, tended to produce errors originating within the management ranks rather than within the operational community. Relatively high levels of compliance or obedience within high PD cultures allowed these errors to filter through to the operational environment without being re-assessed and challenged.

---

[i] Uncertainty Avoidance relates to how process driven an organization is: the concept is closely linked to tightly coupled organizations. Power-Distance is how important hierarchy is to members of an organization. Individualism is the extent to which individual need exceeds those of the collective within an organization. Masculinity rates the importance of masculine power traits as an influencer within organizations. Later work by Hofstede expanded this to five with Long Term versus Short Term Goal Orientation and then later, six with Indulgence versus Restraint.

Countries such as Australia or the United States, which tend to demonstrate lower PD values, do tend to have above average levels of safety performance. However, it would be overly simplistic to relate safety performance directly to Western cultural traits. There are many other factors at play in influencing safety performance, not least the fact that many Western cultures also have more resources to allocate to equipment and training. The positive elements of Western culture traits have been emphasized in accident reviews in contrast to non-Western cultural traits which have sometimes been highlighted as the direct cause of a breakdown in CRM principles.

Most of the cultural research work in commercial aviation originates in Western societies and institutions and, therefore, inevitably promote the benefits of individualistic traits such as assertiveness above compliance.[5] Assertiveness is thought to be an effective approach in developing a lower PD cockpit environment, but it can also promote competition or even confrontation. Collectivist cultures, typically found in Eastern nations, promote communication, cooperation, and interdependence. The subtle cultural differences illustrated by the difference in body language in Figure 12-1 not only allow us to recognize differences between cultures but provide a means to understand the dynamics of our own cultural perspectives. Future research in safety culture and global CRM training programs must include more diverse perspectives to understand a more global approach to safety culture.

An ideal safety culture would benefit from the positive elements of both low PD (open communication between the crew) and collectivism (recognition of the interdependent nature of effective teamwork).[6] To deduce that national cultural boundaries define safety performance is inaccurate, it is a far more complex pattern.

FIGURE 12-1 Collectivist cultures promote communication, cooperation, and interdependence. (Source: Dreamstime)

Cultural influences can be multidimensional or transnational; the strength and type of cultural influence is not necessarily reflected in national boundaries. To get to a level of granularity where we can begin to understand and manage safety culture, we must look more closely at the organizational level.

12.1.3 Organizational Culture

There are two main approaches to the study of organizational theory described as the normative and the anthropological. The normative is the approach of the more pragmatic or functionalist researcher. This approach identifies the numerous systems, structures, policies, and procedures identified within an organization. While this may seem to be a sensible or even intuitive approach, this area of research has struggled to show how the identification of visible organizational structures is utilized to improve safety performance. The normative approach has overly focused on *what* is going on but gives little or no indication of *why* things are happening.

The alternative anthropological approach (sometimes referred to as the interpretative approach) focused on the more intangible characteristics of organizations. However, the interpretative approach has yet to provide any scientific validation that metrics concerning the underlying beliefs, attitudes, and values can be directly attributed to safety performance or accidents. Both the normative and the anthropological approaches have perhaps been too siloed within their own methodology. To understand the complexities of organizational culture, a multifaceted approach would seem appropriate.

Sometimes referred to as "the father" of the study of organizational culture, Edgar Schein's theory (Figure 12-2) brings together elements of normative anthropological research into a singular model. The model consists of three basic tenets: observable artifacts, espoused values, and underlying assumptions. Observable artifacts are effectively the look and feel of an organization. They are the way an organization would "dress" itself up. It's what an impartial outsider would see and hear were they to visit a corporate headquarters or a company training establishment. It is the objects, images, language, or sounds that organizations project. These overt indicators can give an impression of how an organization may function and what some of the key drivers

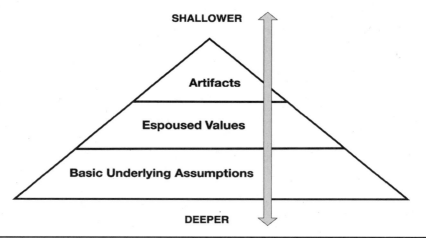

FIGURE 12-2 Schein's model emphasizes the importance of collective assumptions. (Source: Authors)

are, but they can also provide very misleading impressions. It has become something of a cliché that an airline or aircraft manufacturer would display some image or form of words that in effect proclaim that safety comes first. This most famous of espoused values is clearly not always true. The real nature of espoused values of an organization can be explored through various forms of surveys and assessments; they can give an explanation of the nature of more visible artifacts by placing them into some form of context. Espoused values in essence are what a company wants the world to believe it is all about. They are the self-proclaimed principles sometimes depicted in a company constitution or corporate philosophy. However, although they may explain the reasoning behind company artifacts, such as language, process, or rules, they are not necessarily an indication of the prevailing cultural motives. Schein goes further than quantifying espoused values of an organization as a means of understanding culture.

The most influential element of his model is that of the underlying assumptions of an organization. To Schein, these are the essence of its culture. In relation to safety, these assumptions relate to the core reason an organization's attitudes and values collectively see itself as fundamentally safe or dangerously vulnerable. Therefore, even a basic grasp of accident causation, as we described in Part I is so important to our broader grasp of safety and safety culture. These underlying assumptions are a function of the collective learning process of an organization's personnel, related through various forms of storytelling. Rather than being the output of the group the assumptions are embedded into organizational psyche, typically shaped by significantly negative or traumatic experiences.

To understand the underlying assumptions, Schein recommended research techniques focus questions and self-analysis by group members, designed to challenge what has evolved to become unquestioned principles within an organization. However, he warns against the use of oversimplified models of culture and recommends a deeper methodological approach adopted from anthropology. If we look at the image of office workers "fist bumping" in Figure 12-3, how much can we really infer from such a superficial snapshot of apparently friendly and supportive behavior? We we might want to engage in some level of conversation, perhaps even over a period of time before even trying to understand the real dynamics of this exchange. Schein is saying there is no quick and easy assessment method that can characterize organizational culture, investing time into understanding the often-unique nuances of an organization is the key to comprehension of culture.

12.1.4   Defining Safety Culture

The influence of organizational culture and policies on the effectiveness of safety and risk controls is acknowledged and defined within the opening pages of ICAO's SMM, "Culture is characterized by the beliefs, values, biases and their resultant behavior that are shared among members of a society, group or organization."[7] In its fourth edition, the document describes four eras of safety management based on contemporary knowledge: from the early 1900s to the late 1960s—the technical era; from the early 1970s to the mid-1990s—the human factors era; and latterly, from the mid-1990s to the early 21st century—the organizational era; and now up to the present day, the systems era. It is within these latter two eras of safety management that safety culture holds particular relevance. Safety researchers have established that organizational influence filters down to operational decision-making not through formal rules and procedures but through the established cultural norms of the organization.[8]

FIGURE 12-3 Artifacts of organizational culture can include objects, language, and process. (Source: Dreamstime)

The International Nuclear Safety Advisory Group defines safety culture as "that assembly of characteristics and attitudes in organizations and individuals, which establishes that, as an overriding priority, safety issues receive the attention warranted by their significance."[9] The interest in safety culture from the nuclear power industry was undoubtably prompted by the explosion of the Chernobyl Nuclear Reactor in 1986. The potential scale of this disaster should not be underestimated. The explosion was eventually contained but at certain points during the catastrophe the Russian and Ukrainian nuclear authorities feared the chain reaction would leave much of eastern and central Europe uninhabitable for many years.[10] The subtle but deep impact on our collective thinking when events of this magnitude occur question our confidence in new technology and its ability to deliver consistently safe performance. It also forces us to question our confidence in ourselves as a collective to consistently deliver safe performance, and for that reason, the study of safety culture is so important.

12.2 Concepts of Safety Culture

12.2.1 Describing Safety Culture

Despite a lack of agreement in accurately defining safety culture, its popularity as a conceptual mechanism to understand safety-related behavior within organizations has steadily increased. In a similar way that organizational theorists have approached organizational culture, with normative and anthropological approaches, the study of safety culture has originated from two different perspectives: the functionalist and the interpretative. The functionalist approach sees safety culture as a component of the larger organization. An organization's safety culture is seen as a feature, it is something that the organization "has." This approach is favored by managers and practitioners as it implies something that can be measured and therefore managed and controlled. This simplifies a lot of problems for managers as it assumes an objective assessment can be made of safety culture that is separate from their own activities and can be ascribed to the generic workforce rather than their own performance.

The interpretative approach creates far more conceptual problems for those favoring metrics and apparent certainty. An interpretative approach sees culture as a metaphor for the organization itself. The culture *is* the organization, and its cultural traits define its essential nature.[11] The organization has certain traits that determine organizational behaviors. Accepting that safety culture determines the nature of an organization, undermines the mechanisms of functionalism that manage and control it. The implication ensures all levels of the organization are in effect influenced (or even controlled) by the nature of its safety culture. This is perhaps an uncomfortable thought for those who wish to directly control their environment and those people within their organization.

A simple description comes from the American Institute of Chemical Engineers which describes safety culture as "how the organization behaves when no one is watching." This insightful description acknowledges that safety culture is less about the espoused values that might be on display to an auditor and more about what influences employees to make the "right decision" in a novel situation or one for which there is no standard operating practice.

While there are many writers on the subject that have emphasized that wherever different elements of safety culture exist it is difficult to explain or model how they interact. This was perhaps, best explained by Albert Bandura and his concept of reciprocal determinism. Bandura suggests that people's thoughts and actions are not wholly determined by their environment nor are they entirely self-determined. They are in effect a product of both. This approach has in many ways been reflected in more recent research by Professor Lisa Feldman-Barret.[12] Feldman-Barret is a professor of psychology at Northeastern University and holds positions at Harvard and Massachusetts General Hospital. She suggests that there are significant areas of the brain which have no obvious predetermined role. The brain's success is largely due to its plasticity and therefore our ability to adapt and thrive. It allows us to adapt to new environments and learn from experiences. Feldman-Barret suggests that certain parts of the brain have designed-in redundancy that is specifically set aside for us to absorb knowledge and behavioral patterns from our surrounding culture. In effect, we are physically designed to be influenced by the cultural environment in which we exist. Culture is powerful stuff.

12.2.2 Modeling Safety Culture

There have been a number of attempts to model safety culture, but as we have discussed, as a construct of people's thoughts and opinions, it does not lend itself easily to being modeled. This approach is by no means unusual in science, but it would be fair to say some areas of science lend themselves more conveniently to deconstruction. Culture's complexity and contextual nature open the door to criticism that any model that tries to represent the total function of safety culture as being overly simplistic.

One of the early attempts to model the elusive concept of safety culture was by E. Scott Geller in 1996.[13] Geller proposed a triad model of safety culture comprising of the environment, the person, and behavior. However, the descriptive model lacked an explanation of how the three elements interacted which undermined its utility.

Cooper[14] saw safety culture as a subset of organizational culture. Like Geller, he opines that based on theories of organizational culture, safety culture comprises three interacting elements: psychological, behavioral, and organizational. Cooper suggests an outline mechanism of interaction based on Albert Bandura's model of reciprocal determinism but like Geller does quite explain the process of interaction between the three components.

12.2.3 Reason's Model of Safety Culture

In his classic book *Managing the Risks of Organizational Accidents*, Reason's 1997[15] approach influenced many subsequent models of safety culture in its attempt to align the competing conceptual approaches of whether safety culture is a property or an element of an organization. Reason identified five key elements of safety culture, which build on each other and interact through an organization's lifecycle (Figure 12-4).

Reason's now widely adopted model held its foundation in the establishment of a just culture. In a world increasingly dependent on the use of mechanisms in litigation and criminal law to mitigate safety fears, safety-critical industries such as commercial aviation have become concerned with the impact on the open reporting of safety events or even safety concerns among front-line staff. The concept of a just culture was further developed in work by the American academic David Marx. Marx used his combined experience as a systems engineer and his studies of law to formulate the concept of just culture within healthcare. Marx wanted to create an equitable balance between profession accountability and open reporting. This was not without its challenges as we shall discuss in Chapter 16 where we discuss criminalization within commercial aviation. In line with the differing levels of authenticity described in Schein's model of organizational culture, just culture is not achieved by labeling nor is it a policy imposed from

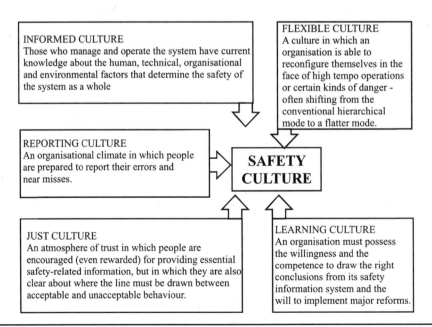

FIGURE 12-4 Reason's model of safety culture. (Source: Adapted from Reason's 1997 Model)

senior management, but the level of genuine belief at the operational level that safety improvement and not blame is the result of reporting safety concerns.

The motivation to create a just culture is not simply to introduce equity into the workplace, or to create an environment of psychological safety, but to create an atmosphere of trust specifically to promote safety reporting. In complex safety-critical occupations such as commercial aviation, there is a need to develop safety intelligence which is the lifeblood of safety management systems. There are three fundamentals that can impede effective safety reporting: fear, futility, and functionality. Fear of reprisal from management has been noted in many safety-critical industries including commercial aviation.[16]

In a study of one European Airline with an established just culture and best practice Crew Resource Management training schedule, the researchers found that of the 1751 airline's crew, half had remained silent over a safety-critical issue. Status was the greatest determinant of silence: highest for First Officers and pursers followed by main crew and the lowest for Captains.[17] Without any feedback mechanism within the reporting process, reporting is reduced through a notion of futility in the process. Effective reporting cultures need to demonstrate the value of safety intelligence. Reason suggested that even simple rewards can dramatically improve not just the amount of reporting but its quality. Finally, there is the issue of functionality. If a reporting system is overly complex or unreliable, it will deter users. There is only a finite amount of time that reporters will have the motivation and requisite recollection of events to provide a suitable quality of safety report. In order to be effective, a safety reporting system needs to be intuitive and accessible.

It is only through the establishment of reliable reporting channels within the organization can it be considered an informed culture. Of course, that in itself is not guaranteed without the information being effectively distributed and communicated. The organization has to then be able to adapt and flex as new safety intelligence is analyzed. Through multiple iterations, the organization can now learn and monitor the safety impact of any adaptations through open reporting. Reason's model implies that an organization's safety culture follows a path of maturity and this concept has been developed by subsequent theorists.

12.2.4 Measuring Safety Culture

One of the earliest attempts to measure safety culture was in 1996 by the American sociologist Dr. Ron Westrum.[18] Westrum proposed three scales of safety culture maturity, as illustrated in Figure 12-5. The scale suggests that each increased level of improving safety culture is detected by an improved level of the handling of safety-related data.

The importance of open reporting systems should not be understated; without it, an organization leaves its risk or safety management system blind to threats. However, the way safety data is handled is not the only significant function of the mechanics of safety culture. The lead indicators of reporting and data handling do not necessarily represent the internal process of how safety processes are managed.

This methodology of measuring culture was adapted and expanded by Parker, Laurie, and Hudson in 2006[19], also shown in Figure 12-5, who added James Reason's suggested further categories of reactive and proactive grades. The model grades the most ineffective of safety cultures as pathological, that is in denial of reality through to generative where threats to safety are actively sought out and managed. Although the refined models added more breadth in their similar five-scale models of continual improvement, the model's output misses the opportunity to conceptually describe

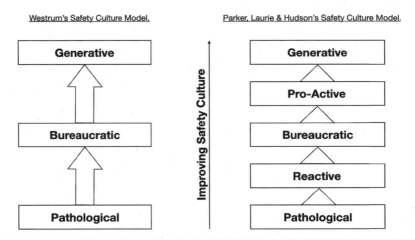

Westrum's Safety Culture Model.

Parker, Laurie & Hudson's Safety Culture Model.

FIGURE 12-5   Models of safety culture maturity. (Source: Authors)

internal technical, managerial, and operational processes that can combine to cause accidents; these models suggest simple, linear causation rather than complex interactive causation.

A further and perhaps more fundamental criticism of this approach to modeling safety culture is the attempt to encapsulate the whole organization. Reason's model not only identified a process of maturity but also described different attributes of safety culture. Numerous studies have highlighted much of the literature describing safety culture consistently tends to describe the whole when in fact it relates to dimensions, elements, or a specific subculture of an organization. Rather than attempting to encapsulate one singular definition of safety culture, more recent modeling of safety culture has attempted to incorporate aspects of contextualization to the area or industry of study. There would also appear to be a general recognition that models of safety culture need to evolve to promote further utility and management of safety performance.

This component approach to modeling safety culture was adopted and developed as part of a doctoral thesis at Cranfield University in the United Kingdom.[20] Figure 12-6 illustrates nine attributes of safety culture. WBL or willful blindness is a legal term which in effect places an obligation of due diligence on those in positions of responsibility for safety, a subject explored by the British consultant and academic Margaret Heffernan in her excellent book, *Wilful Blindness: Why We Ignore the Obvious*.[21] This attribute measures the authentic commitment of senior management to safety.

Another attribute LRP is a measure of the perceived effectiveness of corporate policy on safety performance. EAC assesses how compatible employee attitudes are with safety policy and the CAP attribute identifies the emergence of incongruent subcultures within the larger organization.

The ALS attribute is a measure of operational management effectiveness. The importance of the role of middle management is sometimes overlooked, but the "bottle-necking" of critical safety information is invariably determined by the style and therefore effectiveness at this level.

ISC measures the success of any efforts to improve safety culture and MXR the level by which excessive risk encouraged by senior management. SBR indicates the level of deviation from standard operating procedures and XRD measures if excessive risk is challenged at operational level.

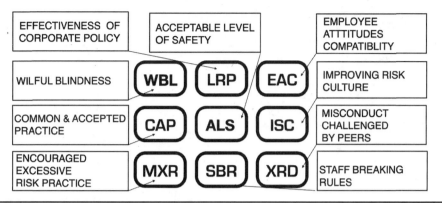

Figure 12-6 Suggested attributes of safety culture. (Source: Authors)

By displaying the values of each individual component as well as the overall patterns of interaction within the organization, the model provides a visual metric of both the sum and the parts of an organization's safety culture. The monitoring and measurement of safety culture are now promoted across commercial aviation as an essential element of safety management process. It also aligns well with the systems approach to safety that we shall explore in Part IV. It has also achieved greater significance as a metric of organizational due diligence and has increasingly been recognized as a key factor in organizational failure.

12.3   The Power of Safety Culture

12.3.1   Trust

Trust is a crucial aspect of safety management, yet it is rarely discussed in academic literature or media commentary. Trust is an intangible element of commercial aviation which allows the system to produce not only high levels of safety performance but also to maintain its viability as a business. On March 10, 2019, the passengers of Ethiopian Flight 302 trusted their pilots, Captain Yared Getachew and First Officer Ahmed Nur Mohammod Nur, to be competent, well trained, and to follow the correct procedures. In turn, these professional pilots trusted their airline to have appropriately trained them and to have presented them with a safe, well-maintained, and serviceable aircraft. The passengers, the operating crew, and the airline based all these assumptions on their trust in the aircraft manufacturer, Boeing. They trusted Boeing to have designed and produced a safe aircraft and to have communicated all appropriate operating information to airlines and their employees. They also trusted the highly respected regulatory bodies, such as the FAA, to have ensured their trust was not unfounded (Figure 12-7). Sadly that trust was misplaced, when design flaws in the Boeing 737 MAX resulted in the destruction of the aircraft and the loss of all 157 passengers and crew.

The strength of this contract of trust within the system of commercial aviation should not be underestimated; the trust in this particular aircraft type had endured, despite the fact that another Boeing 737 MAX had crashed in Indonesia less than five months prior to the date of their flight, killing all 189 passengers and crew. Everyone listened and took note of the then Boeing CEO, Dennis Muilenburg's words, that despite the Lion Air Flight 610 crash the 737 MAX was a safe aircraft and all appropriate technical

Figure 12-7   The Ethiopian 737-MAX ET-AVJ that was destroyed on March 10, 2019, killing all 157 passengers and crew. (Source: Authors)

information had been passed to the operators.[22] Trust in Boeing persisted when it was stated by a US Congressman that the accidents were something to do with the pilots being trained outside of the United States,[ii] despite official reports to the contrary.[23] Everyone trusted Boeing to have done the right thing, even when their own engineers and test pilots had suspected, and internally discussed that the Lion Air crash was a result of their poorly designed MCAS[iii] control and sensor system. Everyone assumed because Boeing was a such highly respected and trusted institution that they would find the technical problem and solve it. But the problem wasn't just technical, and the ultimate cost for Boeing wasn't just financial.

The core issue was the slow deterioration of its organizational safety culture over many years and the cost was far greater than the significant drop in its share price or the billions in criminal fines[24] and ongoing litigation—Boeing lost people's trust and without it, the whole system of commercial aviation safety fails.

12.3.2   A Culture of Concealment

Boeing's commercial success in marketing the Boeing 737 MAX to potential customers was based on its claim that existing 737 operators did not need further (and expensive) simulator training. It had effectively placed a bet with Southwest Air that its pilots

---

ii In what is arguably the most ill-informed and unsubstantiated comment on the whole Boeing MCAS issue, Congressman Sam Graves stated in a Congressional Hearing in May 2019 that "pilots trained in the US would have successfully been able to control this situation."

iii MCAS stands for Manoeuvring Characteristic Augmentation System, which is designed to enhance the pitch stability of the 737 MAX to make it feel like other 737 aircraft in its class. The system in operation at the time of both the Lion Air and Ethiopian Airline crashes had significant design flaws.

would have no more need than the two-hour computer-based training mandated by the FAA to become competent on the MAX variant and could seamlessly operate both types. By January 2019 Boeing had delivered 350 of these new aircraft with a further 4661 on order. The stakes were high and pressure was mounting to deliver these aircraft.

The House Committee on Transport and Infrastructure carried out an extensive report on how Boeing had operated as a company in the run-up to the two fatal accidents. The committee extract over 100 pages of internal communication which gave some background to the attitudes which had formed within the organization. When some 737 MAX operators, including Lion Air, requested simulator time to become more familiar with operating the new aircraft, a Boeing employee wrote a text in June 2017, "Now friggin Lion Air might need a sim to fly the MAX, and maybe because of their own stupidity. I'm scrambling trying to figure out how to unscrew this now! idiots."[25] This text was one of many examples uncovered by the House Committee who described Boeing of hosting a "culture of concealment."

The extent and seriousness of the deceit that existed within the company shocked many, but perhaps most acutely those who worked within the aerospace industry. Many had operated under the assumptions that organizations like Boeing would never allow such things to happen; the revelations questioned some core beliefs in the level of integrity of commercial aviation safety. Boeing's safety culture had degraded to a level where it was not only concealing the existence of the MCAS system from 737 Max pilots but also failing to disclose that the AOA (Angle of Attack)[iv] Disagree Alert was inoperable on the vast majority of the 737 MAX fleet. The company was not only concealing information about its previous process and design failures, but according to the House Committee had been knowingly mass-producing and selling unsafe aircraft.

12.3.3   Why Is Safety Culture Important?

Looking back at examples like the Boeing MACS story, it is apparent how insidious poor safety culture can be in reducing safety performance. We shall review the Boeing MCAS story from a systems perspective in Chapter 13 and from a regulatory perspective in Chapter 16. It is crucial that we fully understand how these events unfolded as the challenge to safety science is to take these lessons and project forward and prevent re-occurrence. Traditional safety management has been heavily dependent on statistical data. However, with dwindling accident data to rely on especially within an ultrasafe system like commercial aviation, this approach can only give a limited indication of safety performance. The other major source of information has been through individual deconstruction of aviation accidents. These two approaches, which had underpinned safety management now provide limited predictive capability.[26]

More sophisticated safety management models have developed to attempt to address the issues that have arisen from traditional approach to monitoring and managing aviation safety. These traditional retrospective methods of measuring safety performance have less relevance in this increasingly complex system where accident causation is difficult to conceive let alone, explain. More recent analysis techniques have increased emphasis on predictive techniques. These new techniques can quantify risk types, model exposure, and then reprioritize these risks. As discussed throughout Part IV of this book, there is growing consensus by safety researchers to take a top-down

---

iv The angle of attack (AOA) is the angle between the oncoming air or relative wind and a reference line on the airplane or wing. It is one of the most important sensors on the aircraft as the AOA determines the ability of the aircraft to generate adequate lift.

or systems-based approach. In the words of Professor Nancy Leveson, "Safety is a system property, not a component property, and must be controlled at the system level rather than the component level."[27] Safety culture is seen by many safety specialists as a key area that helps us understand from a total system perspective and develop more predictive capabilities in safety management.

12.3.4   Does Poor Safety Culture Cause Accidents?

The Confederation of British Industry refers to safety culture as "… the way we do things around here." However, despite the various definitions, many accident investigation bodies such as the NTSB are reticent about describing "safety culture" as a "probable cause" of accidents. "Investigators should be particularly cautious about attempting to assess safety culture after an organization has experienced an accident or incident."[28]

This unease about attributing safety culture as an accident cause stems from the lack of any consensus as to what actually defines it. Without a clear definition of safety culture, there is the danger of taking a subjective perspective on any sequence of events and attributing blame that suits individual or collective agendas. The only commercial aviation accident investigation body, we are aware of, which even approaches a formal postaccident assessment of safety culture attributes, is the Transport Safety Board (TSB) of Canada. TSB Canada, the country's multimodal accident investigation body, has developed a common framework for its staff to investigate organizational and management factors. The stated aim of TSB Canada policy is to address systemic issues that shape human performance.[29]

The causal significance of organizational factors in commercial aviation safety analysis found formal recognition through the dissenting comments of John Lauber of the NTSB.[v] Commenting on the loss of Continental Express Flight 2574 near Eagle Lake, Texas, in 1991, Lauber stated that in his opinion the probable cause of the accident should read, "The failure of Continental Express management to establish a corporate culture which encouraged and enforced adherence to approved maintenance and quality assurance procedures" (NTSB 1992:54). Since its emergence as a concept, debate has continued, discussing how best to employ safety culture as a means of improving safety performance. Safety culture is now recognized as such in a number of major accidents and analyses of systems failures including the King's Cross Underground fire in London in 1987 and the Piper Alpha oil platform explosion in the North Sea a year later.

The developing significance of safety culture is apparent in more recent accident investigations. The primary recommendation of then-Queen's Counsel, Charles Haddon-Cave, in his extensive Inquiry into the loss of Nimrod XV230 in 2006 in Afghanistan was for the military to develop a "New Safety Culture." Following the loss of Air France 447, an Airbus A330 in 2009, one of the first reactions by the operator, Air France, was to commission an independent review board to assess and report on its own safety culture.[vi]

---

[v] The National Transportation Safety Board (NTSB) is an independent US federal government agency charged with determining the probable cause of transportation accidents and promoting transportation safety, and assisting victims of transportation accidents and their families.
[vi] The Independent Safety Review Team (ISRT) was established in December 2009. It consisted of a number of global aviation safety experts and produced 35 recommendations for Air France. The document containing the recommendations has not been released to the public. Source: http://news.aviation-safety.net/2011/02/09/air-france-acts-on-independent-safety-review-team-recommendations. Accessed July 1, 2017.

Similarly, the Presidential report of the National Commission in 2011 into the explosion of BP's Deepwater Horizon in 2010 made numerous references to the inconsistency and ultimate failure of the company's safety culture.[vii] It is apparent that several investigative reports and numerous academic papers suggest that a dysfunctional, weak, or absent safety culture is a significant feature of many serious accidents. Safety culture has been used in some of the above-mentioned investigations as shorthand for a whole plethora of organizational failings rather than being singled out as the scientific cause of the accident.

Scientific explanation of cause is not the responsibility of legal institutions. As we shall discuss in more depth in Chapter 16, when considering legal liability following an accident, it is important to consider that accident causation is as much an issue of societal perception of culpability as it is an *ex-poste* technical explanation of events. Accident causation and therefore any subsequent criminal liability are as much a function of presiding social opinions on the nature of causation as the vagaries and technicalities of legal mechanism. There are often very different perspectives between lawyers and accident investigators. The professional obligation of a prosecuting or plaintiff lawyer was clearly stated by Robert Lawson KC in an address at the Royal Aeronautical Society in London in 2015, which is to establish liability or blame in order to pursue criminal charges or to recover civil damages. Their professional obligation is to pursue that goal and adapt an interpretation of events in a way that best suits their client's interests. If that interest is to persuade the court that poor or weak safety culture "caused" an accident, then that becomes the narrative.

Despite its utility from a legal perspective as an explanatory concept of organizational failure, from the scientific perspective of causation, discussed in Part I, it is difficult to establish safety culture as either a necessary or sufficient precondition to an accident. What is, perhaps, a more realistic position is to state that the absence of an effective safety culture inhibits or even removes organizational learning. This absence of learning is a critical flaw in any organization as described by the renowned organizational theorist Peter Senge. In his best-selling book *The Fifth Discipline*, he states, "Learning disabilities are tragic in children, but they are fatal in organizations. Because of them, few organizations live even half as long as a person—most die before they reach the age of forty."

12.3.5 Regulating Safety Culture

The common theme supporting Peter Senge's assertion is that without organizational learning (an effective safety culture) an organization lacks the ability to identify threats, adapt, and ultimately react in sufficient time to avoid catastrophe. However, within commercial aviation, regulatory efforts to encourage the utilization of safety culture assessment and management have been sporadic. Programs of safety culture assessment have been developed by some regulatory agencies such as the Swiss Federal Office of Commercial Aviation (FOCA) and the Australian Transport Safety Bureau (ATSB). Accident investigation agencies, such as the NTSB, have mentioned safety culture as a causal factor within their formal reports for some time but to date only the Canadian TSB practically utilizes safety culture as part of their investigative process. Eurocontrol is the organization responsible for Air Traffic Control Management across mainland Europe.

[vii] BPs own investigation report, Deepwater Horizon Accident Investigation Report, September 8, 2010, does not mention the term "safety culture." The report focuses on the technical and human failings of the BP and Transocean rig teams. The National Commissions report to the President mentions "safety culture" 25 times.

Since 2012 it has introduced a safety culture assessment following Regulation EU390/2013 where safety culture is formally identified as a key performance indicator.

The overall pattern is that the concept of safety culture does not sit well with regulatory or process-driven programs. This leaves something of a dilemma for safety managers. If there is no regulatory mandate and little consensus as to what a positive or effective safety culture should look like, then there is little motivation for commercial aviation operators to invest in assessment techniques. But the establishment of a universal regulatory defined standard of safety culture has more fundamental obstacles to overcome.

Several academics have challenged the idea that safety culture can actually be measured and therefore managed or regulated in any conventional sense. The Swiss academic Gudela Grote suggests the utilization of safety culture to achieve broad consensus across an organization through "loose coupling," which is key to enabling organizational adaptability in the face of uncertain outcomes. She suggests that the "abstract concept" of safety culture itself is an inappropriate regulatory objective and focus should instead be directed at rule adaptation and interpretation by the workforce by which the quality of safety culture can then be inferred. In other words, allowing operatives the latitude to maintain high standards of operation under a broad guideline approach is preferable to being overly specific and creating a "one-size-fits-all" solution. When a generic program of improving health and safety culture was introduced to the Norwegian petroleum industry, variable interpretations from the regulator, management, and the workforce resulted in significant disparities in the program's output. One of the fundamental problems with safety culture has been that there are many opinions of what good looks like.

12.3.6 Managing Safety Culture

If managers do not understand or are not actively engaged with their organization's safety culture, it is more likely than not that their organizational culture is managing them. One of the difficulties of managing culture is that from within an organization it is almost impossible to separate ourselves from its influences and effects. Just as Albert Bandura stated, our relationship with our cultural environment is reciprocal and deterministic: just as the individual influences their surrounding culture, the culture in turn influences the individual.

From the study of organizational culture, several accepted principles have emerged. Possibly the most widely accepted is that culture is driven from the top of an organization. This is derived from the observation that senior management in high-performing safety cultures demonstrates a genuine belief in the importance of safety. The importance of authentic safety leadership at the very top of an organization cannot be understated. IATA[viii] has recently emphasized the importance of this principle by enacting a safety leadership charter for senior airline executives. The charter has a number of guiding principles:

1. Fostering safety awareness with employees, the leadership team, and the board

2. Integrating safety into business strategies, processes, and performance measures

---

[viii] The International Air Transport Association, IATA is a trade federation of 290 airlines across 120 nations. It acts as a forum for the industry over issues of safety, technical innovation and tariffs.

3. Creating the internal capacity to proactively manage safety and collectively achieve organizational safety goals

4. Developing an atmosphere of trust, where employees are encouraged and confident to report safety-related information

5. Establishing a working environment in which clear expectations of acceptable and unacceptable behaviors are communicated and understood

6. Building an environment where all employees feel responsibility for safety.

These principles are aimed at giving airline executives a basic framework with which they can tailor their individual safety management strategies. The principles align with many of the theoretical approaches to safety culture which we have discussed in this chapter such as trust and a strong emphasis on an authentic just culture rather than making policy proclamations.

The correlation between an authentic commitment to safety culture from senior management and an organization's safety performance may suggest how safety culture should be managed. Accepted wisdom suggests that an effective safety culture starts at the top and all the positive safety messages are allowed to trickle down. However, research carried out by Dr. Damon Centola of the University of Pennsylvania using his theory of complex contagion suggests that by identifying the causal effects of social networks, the emergence of new social norms can be controlled.[30] Centola, who consults on risk culture within the banking sector, emphasizes the importance of personality within smaller subcultures. Managing at this level of granularity within organizations is key to promoting and introducing positive behaviors. In summary, although the perception of authentic commitment from the top of an organization is a necessary attribute of positive safety culture, it is not sufficient. Peer-to-peer training initiatives using well-trained and highly motivated personnel is an effective strategy to alter behavioral values and norms.

Another key principle of safety culture management is in the ownership of risk. The deliberate imposition of cognizant risk ownership has been extensively discussed by academics such as Rhona Flin and Laura Fruhen.[31] In an ultrasafe environment such as commercial aviation, with complex causal relationships that are obscured by technical complexity, organizations, and the individuals within them, forget to be afraid. Several studies concerning what is known as "chronic unease" of senior and safety managers have concluded this is a desirable state of mind to maintain a suitable level of concern to effectively manage safety-critical operations. Utilizing safety culture as a means of deliberately managing anxiety in your organization may seem paradoxical to the principles of trust and justness but it serves as an example of how the management of safety culture can require some counterintuitive strategies.

In the world of safety culture, nothing should be quite so terrifying as complacency.

Key Terms

Artifacts of Culture

Basic Underlying Assumptions

Just Culture

Power-Distance

Process Driven

Psychological Safety

Uncertainty Avoidance

Topics for Discussion

1. How would you define culture?
2. Explain two of Hofstede's cultural characteristics.
3. What are the most influential characteristics of organizational culture according to Edgar Schein?
4. In James Reason's model of safety culture, how do the different elements interrelate?
5. What are the three biggest influencers of an effective reporting culture?
6. Why is "trust" an important element of safety?
7. How is safety culture linked to Charles Perrow's concepts of tight and loose coupling?

References

1. Hofstede, G. (2001). *Culture's Consequences: Comparing Values, Behaviours and Organizations Across Nations*. Sage Publishing.
2. Braithwaite, G. R. (2017). *Attitude or Latitude?: Australian Aviation Safety*. Routledge.
3. Soeters, J. L., & Boer, P. C. (2000). Culture and flight safety in military aviation. *The International Journal of Aviation Psychology, 10*(2), 111–133.
4. Li, W. C., Harris, D., & Chen, A. (2007). Eastern minds in western cockpits: Meta-analysis of human factors in mishaps from three nations. *Aviation, Space, and Environmental Medicine, 78*(4), 420–425.
5. Hudson, P. (2007). Implementing a safety culture in a major multi-national. *Safety Science, 45*, 697–722.
6. Hofstede, G., Hofstede, G. J., & Minkov, M. (2005). *Cultures and Organizations: Software of the Mind*. Vol. 2. New York: McGraw-Hill.
7. ICAO (International Civil Aviation Organisation). (2018). *Safety Management Manual (SMM)*. 4th ed. Doc 9859 AN/474. ICAO.
8. Orasanu, J., & Connolly, T. (1993). The reinvention of decision making. In: G. A. Klein, J. Orasanu, R. Calderwood, & C. E. Zsambok (eds), *Decision Making in Action: Models and Methods*. American Psychological Association. Washington, DC: Ablex Publishing.
9. "Safety Culture"—A report by the International Nuclear Safety Advisory Group (Safety Series No.75-INSAG-4).
10. IAEA. (1986). *Summary Report on the Post-Accident Review Meeting on the Chernobyl Accident*. INSAG Series No. 1.
11. Edwards, J. R., Davey, J., & Armstrong, K. (2013). Returning to the roots of culture: A review and re-conceptualisation of safety culture. *Safety Science, 55*, 70–80.
12. Barrett, L. F. (2017). *How Emotions Are Made: The Secret Life of the Brain*. Pan Macmillan.
13. Geller, E. S. (1996). *The Psychology of Safety: How to Improve Behaviors and Attitudes on the Job*. Chilton Book Company.
14. Cooper, M. D. (2000). Towards a model of safety culture. *Safety Science, 36*, 111–136.
15. Reason, J. T. (1997). *Managing the Risks of Organizational Accidents*. Vol. 6. Aldershot: Ashgate.

16. Townsend, A. S. (2016). *Safety Can't be Measured: An Evidence-Based Approach to Improving Risk Reduction.* Routledge.

17. Bienefeld, N., & Grote, G. (2012). The silence that may kill, when aircrew members don't speak up and why. *Aviation Psychology and Applied Human Factors*, 2(1), 1–10, Hogrefe Publishing.

18. Westrum, R. (1996). Human factors experts beginning to focus on organizational factors in safety. *ICAO Journal*, 51(8), 6–8.

19. Parker, D., Lawrie, M., & Hudson, P. (2006). A framework for understanding the development of organisational safety culture. *Safety Science*, 44(6), 551–562.

20. Lawrenson, A. J. (2017). Safety culture: A legal standard for commercial aviation. Doctoral dissertation. Cranfield University, United Kingdom.

21. Heffernan, M. (2011). *Wilful Blindness: Why We Ignore the Obvious.* Simon & Schuster.

22. Bloomberg UK. https://www.bloomberg.com/news/features/2021-11-16/are-boeing-planes-unsafe-pilots-blamed-for-corporate-errors-in-max-737-crash. Accessed October 2022.

23. BBC News. Pilots were not to blame for 737 crash. https://www.bbc.co.uk/news/business-48528383

24. Department of Justice. https://www.justice.gov/opa/pr/boeing-charged-737-max-fraud-conspiracy-and-agrees-pay-over-25-billion. Accessed October 2022.

25. The House Committee on Transport & Infrastructure, Final Committee Report on The Design, Development & Certification of the Boeing 737 Max., at p. 156. https://www.govinfo.gov/app/details/CHRG-116hhrg40697/CHRG-116hhrg40697/context. Accessed October 2022.

26. Lofquist, E. A. (2010). The art of measuring nothing: The paradox of measuring safety in a changing civil aviation industry using traditional safety metrics. *Safety Science*, 48(10), 1520–1529.

27. Leveson, N., Dulac, N., Marais, K., & Carroll, J. (2009). Moving beyond normal accidents and high reliability organizations: A systems approach to safety in complex systems. *Organization Studies*, 30, 227, Sage.

28. Czech, B. A., Groff, L., & Strauch, B. (2014). Safety cultures and accident investigation: Lessons learned from a National Transportation Safety Board Forum. Adelaide, Australia, October 13-16, 2014.

29. Morley, J., & Stuart, J. (2014). Improving investigation of organizational & management factors forum. *Journal of ISASI*, April-June. http://www.isasi.org/Documents/ForumMagazines/Forum_Spring_2014_Web.pdf. Accessed August 12, 2022.

30. Centola, D. (2018). *How Behavior Spreads: The Science of Complex Contagions.* Vol. 3. Princeton, NJ: Princeton University Press.

31. Flin, R., & Fruhen, L. (2015). Managing safety: Ambiguous information and chronic unease. *Journal of Contingencies and Crisis Management*, 23(2), 84–89.

PART IV

Organizational and System Safety

CHAPTER **13**

Introduction to System Theory and Practice

"Every system is perfectly designed to get the results it consistently achieves."
(Promoted by Dr. Don Berwick, then John Nance, BSS Kansas City, October 21, 2008)

Introduction

In most accident analyses, a sequence, or chain of events, is found that resulted in the accident. This sequence of events is then analyzed and when one of the "links in the chain" is identified as a factor, changes are implemented to prevent that *particular* factor from causing another accident. This factor, or factors, is commonly identified as a *root cause* of the accident, as described in Part I. After the root cause events are discovered and, if possible, mitigated, the industry tends to "move on" and the problem is considered "solved." Would mitigating the one or two factors identified in the chain of events be enough to prevent another accident or have only prevented an accident that might occur with almost the same precise set of circumstances? The root cause most often found involves the operator (in aviation, often a pilot) who did not perform some action correctly or at the right time. It is very convenient to blame the pilot, particularly a dead pilot, as they cannot defend themselves. Plus, blaming a pilot allows the rest of the industry, the airline, the manufacturer, the regulators, and others to avoid responsibility and avoid having to spend any money by making changes. But have we learned enough from the accident to prevent another one? Are there deeper systemic problems that remain unresolved? Are we learning enough from accidents?

13.1 The Boeing MAX Accidents[1]

The aftermath of these accidents[i] resulted in one of the most public exposes of aircraft certification in history. Interestingly, many of the problems that were eventually identified have occurred several times. Not the specific problem these flights encountered but rather the more general problem of changes over time, decisions that seemed good

---

[i] Lion Air flight 610 crashed on October 29, 2018, after a faulty angle of attack sensor made the flight control computer "think" that the aircraft AoA was too high, which resulted in the flight control computer attempting to input nose-down trim to reduce the AoA, quickly overwhelming the pilots. On March 10, 2019, Ethiopian Airlines flight 302 crashed after the aircraft's AoA sent incorrect information to the flight control computers in a similar manner to the Lion Air 610 accident.

at that time they were made, regulatory oversight, funding limitations, and political pressures. These are accidents where everyone was doing what they thought they were supposed to do, and all decisions made were well within industry standard practice. So how did it all end so badly?

The B-737 series of aircraft has been one of the most important sources of revenue for Boeing since its introduction in 1968. With a relatively simple design, predictable flying qualities, high reliability, and the ability to operate into fairly small airports, it has been very popular with airlines. In 1984 Airbus launched the A320 series, which was designed to be the primary competitor to the B-737 and, at the time, the McDonnel Douglas DC-9 and MD-80 series. Initially Boeing had a natural advantage over Airbus as those airlines that had already been operating the B-737 generally preferred not to change to another type as the cost of introducing a new model into a fleet is quite high. Additionally, the B-737 was a "proven" model, as it had been successfully operating for a number of years. To overcome this, Airbus aggressively marketed the A320, and it came to be a serious competitor with sales approaching the B-737 by the 1990s and surpassing Boeing by 2019.[2] A large part of the competition was improvements on the design. Both the A320 and B-737 were stretched to be larger, retrofitted with larger engines and newer technology that improved fuel efficiency, decreased the noise footprint, and included newer safety features. For Airbus this meant expanding the A320 "family" to include the A321, A318, and A319, each with its unique capabilities. Boeing expanded the B-737 family from the initial B-737-100 and -200 to include the B-737-300, 400, 500, and on to the -900.

In the ongoing battle for market share, in December 2010 Airbus introduced an upgraded A320 NEO, or *new engine option* to the Airbus family, which was both larger and had much more fuel-efficient engines. In response, Boeing had to create something that was comparable—either a brand-new design or a previous design that could be adapted. The latter choice is preferable simply because it is more appealing to airline customers and substantially less expensive to design and certificate. Having a new version of the same type of aircraft requires little or no training for pilots who were flying previous versions of that model, and the costs for training other personnel, dispatchers, cabin crew, aircraft mechanics and ramp is also much less. Additionally, the airline does not have to maintain a stock of different parts as it is, for the most part, the same aircraft as one they were operating previously. From the perspective of Boeing, modifying an existing design would also be substantially faster than starting with a "clean sheet" design as it is a much simpler process and requires much less to obtain regulatory approval. Faster development was important as if a new type was not available for a few years after Airbus launched their new version it would cause a substantial loss of market share.

Still, adding a new variant to an existing airplane type is not as simple as just having it look the same and what it is named. In addition to several other requirements, it is important that, from the perspective of the pilot, the airplane closely handles the same way as previous versions and that the aircraft systems are very similar. Some minor variation is allowed by the regulations but not much. Adding bigger engines to an airplane can change the way the airplane handles significantly, particularly when those engines are installed such that when thrust is changed it changes the aircraft pitch. Underwing engines will pitch the aircraft up when thrust is added and reduce the pitch when thrust is reduced. The much larger engines installed on the Airbus A320 NEO as well as the B-737 MAX LEAP (Leading Edge Aviation Propulsion) engines can create

FIGURE 13-1 Boeing B-737-204A[iv] featuring the long "stovepipe" engines used on the B-737-100 and -200. (Source: Pixabay, Kitmasterbloke)

a lot of pitch force and, potentially, at certain speeds and conditions, make the aircraft not as aerodynamically stable[ii] as is required for certification.[3] For Airbus this was not a problem as the aircraft is fully fly-by-wire (FBW) so the fly-by-wire system could easily absorb the changes in thrust and the pilot will not feel any difference between the NEO and previous versions of the A320. For Boeing, though, under certain circumstances, this might be problem as the aircraft uses conventional controls.[iii]

The problem of insufficient natural stability is not a new problem for the B-737. In 1984 Boeing introduced the Boeing 737-300, 400, and 500 series aircraft, which were equipped with the larger CFM-56-3 engines. The larger diameter of these engines resulted in them being centered a bit lower beneath the aircraft than the earlier models. In addition, these newer engines produced significantly more thrust. These factors, plus the longer fuselage in these new models, resulted in enough of a difference from the earlier B-737-100 and 200 models, that there was a need to improve the natural stability (see Figures 13-1 through 13-3). Boeing added a *speed trim system* (STS) which

---

[ii] To meet certification requirements aircraft must meet pitch stability requirements as defined by the regulations (see 14 CFR § 25.173—Static longitudinal stability).

[iii] "Conventional controls" refer to flight controls that do not use computer enhancement. These can be as simple as having a cable directly connected between the pilot controls and the control surfaces, such as the elevators, or, as in the case of the B-737 where the control forces would be too large for a human to move (even with the use of pulleys and levers) the pilot controls cables that are connected to hydraulic actuators, which, in turn move the control surface itself. This does necessitate the use of some sort of artificial feel mechanism so the pilot will not move the controls too much at high speeds, but those systems do not change the basic architecture. With FBW controls it is possible to have a computer moderate the movement. In the case of the Airbus (and Boeing FBW aircraft such as the B-777 and B-787), the controls command a rate of movement of the aircraft. For example, at lower speeds pulling back on the controls commands a pitch rate, and higher speeds a g-rate. The computer moves the surfaces as necessary to accomplish the commanded rate, thus, if, as in the case of the NEO, the aircraft might be pitching up sharply with more thrust, the system would automatically add downward elevator to offset this and the pilot would not notice the difference.

[iv] Boeing appends a unique number for each customer, so, in this case, the designation -204A refers to a -200 series aircraft that was originally built for Britannia Airways. See https://www.avcodes.co.uk/boeing.asp

Figure 13-2   Boeing 737-529. Notice the larger engines moved forward of the wing. (Source: Pixabay, Aleksandr Markin).

Figure 13-3   Boeing 737-400. The bottom of the engines is "flattened" to provide sufficient ground clearance to accommodate the larger diameter while preserving the landing gear length. (Source: Pixabay, Aero Icarus).

FIGURE 13-4   Boeing B-737-8 MAX. (Source: Pixabay, Anna Zvereva)

automatically moved the trimmable horizontal stabilizer (THS)[v] under certain conditions, essentially making the aircraft feel more stable to the pilots. Pilots became accustomed to the trim commonly moving without their input when the autopilot was off.

The B-737 MAX had an additional twist. Not only were the new engines even more powerful, but the new LEAP engines were much larger than the engines on the earlier models (see Figures 13-4 and 13-5). To accommodate the engines Boeing had two choices. One would have been to make longer the landing gear, but this would be a substantial change to the airplane. The other was to move the engines forward and higher to have enough ground clearance. Boeing chose the latter, but this resulted in the engine nacelles themselves developing lift at high angles of attack. As the lift from the nacelles was forward of the wing this would result in the aircraft control forces getting too light at high angles of attack. To mitigate this problem, Boeing added a new system that is called the Maneuvering Characteristics Augmentation System (MCAS) to manage this risk. After the design was developed and in the testing stage it was discovered that the MCAS authority, as originally designed, was inadequate to meet the requirement. The MCAS was modified to add additional THS authority.

The MCAS inadvertently created an unsafe condition under certain circumstances. The THS has an enormous amount of authority and can easily exceed the elevator control authority (which is what pilots primarily use) in specific scenarios. However, because the system was not meant to be more than a mitigation for handling qualities it was deemed not a critical flight item, and Boeing designed MCAS so it only

---

[v] The THS, commonly called "stabilizer trim" is a system whereby the entire stabilizer is moved rather than just a trim tab, as is the case on many smaller and non-jet powered aircraft.

Figure 13-5   Boeing B-737-8 MAX. Note the large engines moved forward and upward. (Source: Pixabay, Hawkeye UK—Lest We Forget, edited by the authors)

needed information from one of the two angle of attack sensors. It was when the system received incorrect information indicating that the angle of attack was extremely high that the MCAS reacted with an extreme degree of THS movement. Combined with some other situational aspects, this resulted in two crashes.[4]

It is important to note that the aircraft met all certification requirements. The crews were fully qualified. MCAS worked exactly as designed. MCAS was designed in accordance with all the rules and procedures in place at the time. The pilots performed well within the performance range that one would expect. The accidents occurred despite this.

13.2   Systems, Accidents, and Hazards

How can an accident occur when the system works as designed? The key is in the word "as designed." A design is an abstraction and as such cannot "fail." Why is this important? First, some definitions.

A *system* is "a set of things (referred to as system components[vi]) that act together as a whole to achieve some common goal, objective, or end."[5] In engineering, accidents occur when a set of circumstances create a hazardous state which is not mitigated. If there is no hazard, there cannot be an accident. If our goal is to understand and prevent future accidents, then we must find a way to analyze the hazards that can lead to an accident. What is a hazard?

A *hazard* is a system state that occurs within a certain context. For example, a mountain is not a hazard by itself, it only becomes a hazard if we fly too close to it. The MCAS on the Boeing 737 was not a hazard, nor was how it functioned. It only became a hazard

---

[vi] A component in this context can be a physical part, a combination of parts (like an aircraft), a person, an organization, or a computer.

when specific scenario that included environmental conditions became true. People who design systems are *engineers*.

From an engineering perspective, a hazard is something that the engineer can do something about. Put another way, an engineer has no control over the mountain or a thunderstorm, but we can design either technology or policy and procedures to prevent flying close to, or into, the mountain or thunderstorm. This prevents us from creating a hazardous scenario that might lead to an accident. The key to prevent accidents is to identify hazards and design our system so that we avoid hazardous conditions.

13.2.1 Models

Although this chapter's topic is system theory, let us briefly revisit the topic of accident models. We use models as they are the best way to understand complex things. A model, by definition, is a simplification of reality but it still contains the features we believe are the most important.[6] As vital as that is, before using a particular model to make decisions, it is equally important to also understand what the model might be missing. The most common trap is to continue simplifying the model to make it easier to understand, but in doing so there is a real risk of oversimplifying the model to the extent it is no longer useful.[7]

Several of the most popular models in use today to identify hazards are popular in a large part because they are simple to understand. Intuitively, and based on a simple model of linear causality, accidents are the result of a linear chain of *events*, each one creating the conditions for the next to happen. An *event* could be a component failure, or the action a person or computer made. The concept of linear causality infers that the subsequent event could only happen if the previous event created the conditions for it. It is *linear* because we can draw a straight line from one event to the next. To analyze a particular event, we can then simply look at a few factors that might affect it. These can be influences from a preceding event, organizational aspects (for example, see Reason's *latent conditions* in Chapter 3),[8] as well as environmental factors that might come into play. Each event in the chain can then be analyzed independently. This includes both component failures and how a person might react. This approach, known as "reductionism,' first described by Rene Descartes[9] can work well for relatively simple systems, and forms the basis for what is known as *reliability theory*.

13.2.2 Reliability Theory

Reliability theory is the underlying theoretical basis for the analysis of the rate in which systems fail, and most current hazard analysis models are based on it.[10] The basic premise of reliability theory is that accidents are caused by failures, therefore, to prevent accidents all that needs to be done is to ensure that each component is reliable. Hazard analysis for a system can then be deduced by analyzing each component separately. By examining the reliability of each of the components and assigning a probability of failure to each, the total reliability of the entire system can be calculated without regard to how each component might interact when put together. For example, if there are three redundant components, each with a probability of failure of 10^{-2} then the total probability of all of them failing at the same time would be 10^{-6}. There is no accounting for a common-cause failure scenario in this calculation.

Very complex structures can be a challenge to analyze due to the minute variations in forces and stresses. An example might be the landing gear on a large aircraft and how

it interacts with the wing structure to which it is attached (Figure 13-6). Calculating exactly how any particular metal component might fail requires knowing so many details that exact calculations are not possible with current computing power. In a structure this can be managed by just making the parts a bit stronger than they need to be, a term engineers call "overdesigning." This works because each component that cannot be quantified can be just made stronger, hence more statistically reliable. An individual metal part will, on average, have similar characteristics to a duplicate part, so we can use statistics to quantify the probability of failure. The more computing power available the less overdesign is needed. This is because computers are good at calculating a lot of things at the same time, so we can more finely adjust our parameters as more possible interactions can be quantified. But what happens when working with a system that has parts that do not react in a simple way?

In many designs it is anticipated that if a component fails the human (pilot in the case of an aircraft) will be able to intervene to prevent an accident. For example, if part of the flight control system is not working the pilot should have the skill to fly the aircraft without it. It is in this manner that the concept of reliability theory can be expanded beyond hardware and electromechanical components to include humans and software. In other words, humans and software are considered just another component in the system design and ostensibly a probability can then be estimated to their ability to capture a problem before an accident results. As almost any problem can be potentially mitigated if a human happens to identify and fix it in time, the design becomes founded on the expectation that humans to do the "right thing" to prevent an accident. Typical values assigned to humans are in the nature of 10^{-4}, meaning that the human will only "fail" to prevent an accident once during every 10,000 hours.

The application of reliability theory to humans poses some interesting problems and paradoxes. If humans do make errors does adding more humans increase the probability of an error? Do two people react the same exact way to the exact same circumstances? Even computer software will behave differently depending on the scenario and other factors. It is for this reason that applying a reliability-based model cannot provide reliable answers for a hazard analysis when considering humans or software. Specifically, because the behavior is contextually based so does not behave the same way each time, whereas, for example, a metal part will consistently behave the same way to the same stresses, or an electrical motor will reliably behave the same way for a given amount of applied current.

When reliability theory is applied to humans the implication is that the behavior of people is reliable or consistent. The assumption is that, given a set of conditions, people will, on average, perform the same way. To further enhance reliability, it is common to create detailed procedural steps for people to comply with. If a person does not follow the steps they are viewed as noncompliant, and if an accident occurs the person is held responsible. The higher the reliability requirements the more adherence to procedures is demanded.

Human actions are generally judged by the outcome of those actions and there is a tendency not to view the context in which those actions or decisions were made.[12] When a person does not manage the situation, it is viewed as a human failure, rather than considering that the combination of events that are occurring at that time are beyond what a person could be expected to cope with. Reliability theory thus necessarily reframes the problem from looking to understand why a certain action made sense to the pilot (or other person) at the time, to viewing the lapse as a "failure." Lack of procedural compliance that results in an accident is viewed as a failure as well. This problem is even more acute when considering the implementation of computer software into the system.

The hazard analysis methods based on reliability theory were developed prior to the advent of computer technology, which is now ubiquitous throughout aviation systems, including air traffic control systems, dispatcher and maintenance tools, flight training systems, avionics and automation technology on the flight deck, and automatic monitoring of all these systems.[12] At the time those methods were developed, the reliability of hardware, structures, and electromechanical devices was limited and computers were still several decades in the future. Modern aircraft components are now much more reliable than previous generations, but these new designs are now infused with computer software.[13] The assumptions underlying reliability theory are no longer valid for modern aircraft designs that are software-intensive and more complex than previous generations of aircraft. The same is true for air traffic control systems and other implementations of computer technology. Although the implementation of software has led to improvements in safety, it has also created new vulnerabilities. Hazards in modern aircraft have thus shifted from being a consequence of random hardware failures and pilot error to flaws in the design itself.[14]

Newer systems have become increasingly software dependent, and the way people interact with the software has become more critical. Unfortunately, it is very difficult or, oftentimes, impossible using reliability theory-based methods, to identify problems in the software design or the interface between software and humans until the design has reached the point where it has been built and is being tested. This is because there is no way to quantify the reliability of a software or procedure design without violating statistical assumptions.[vii] Once the system is built then problems are often identified during testing that did not become apparent earlier. If this occurs the manufacturer either must start from scratch or find a way to mitigate the problem with a modification or procedural steps or both. Often this results in secondary problems as the modification can increase the complexity of the design in ways that were not anticipated. Some problems remain unidentified due to the limits of the testing program and can appear unexpectedly during service. This occurred most publicly in the case of Boeing's 737 MAX.

Despite these problems, the models that are based on reliability remain quite popular as it is easy for decision makers to grasp and the assignment of probability values (even when obtaining those values violates statistical assumptions) provides a tangible way to justify those decisions later, even if an accident occurs. In other words, it provides a way for managers to avoid responsibility.

The problem is *complexity*, and to address that, a new approach is needed.

13.3   Systems Theory

The whole is greater than the sum of its parts. Quote often paraphrased from Aristotle.

What is meant by the term *complexity* when referring to a system? A complex system is one in which the behavior cannot be inferred by the behavior of its individual components, but rather is an *emergent* property of the way those components interact with each other[16]. Although we are here discussing human-engineered systems, an analogy to the field of biology is useful. The most obvious for us is to consider a living person. We cannot infer how a person will react to a situation by examining the properties of that person's cells, or even their brain. The person will behave based on not only the properties of their individual parts, but also to the environment around them plus their own experiences, knowledge, intelligence, and other factors, such as reaction time, age, fatigue state, and so on.

In a similar manner, a computer will respond based on its programming, information it is receiving, and the types of controls it has been provided to take actions and other aspects. A large organization will behave based on a number of factors, including everything from the organizational structure of the company, the type of business, the safety culture, the environment and external and internal pressures and more. Even with that, without changing anything else except a few key people in various positions, how the organization behaves can become entirely different. As might now be apparent, any system that incorporates people in decision-making roles (which is any human role in a system as it is always expected that they make some decisions) or software control, will become complex.

How can we model complex systems? As we have seen, the use of just reliability is not enough. To address the limitations in understanding complex systems a new approach was needed. System theory was developed to understand how complex

---

[vii] All statistical tests rely on a set of assumptions about the data being true. These assumptions vary, and some tests require more than others, for example, for the *t*-test the assumptions include *independence, normality, homogeneity of variances,* and *random sampling*. Software and human behavior can violate all of these in various ways.

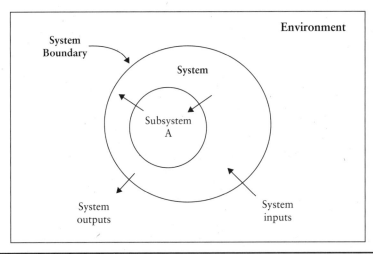

FIGURE 13-7 A simple system. (Source: Leveson and Thomas, 2018, p. 17[5])

systems functioned. System theory, which is founded on the same mathematical basis as other engineering principles, is the idea that many of the characteristics and behaviors of a system emerge as a consequence of the interactions between the components in that system.[15] Figure 13-7 is a pictorial representation of a simple system model.

As described by Leveson,[12] system behavior falls into three general types. The first are those in which it is possible to decompose the system into subcomponents, analyze them independently, and that doing so does not distort the system behavior. They are not subject to feedback loops or nonlinear interactions, and the behavior of each component is the same whether examined individually as it is when part of the whole. These are relatively simple systems that can be analyzed statistically.

The second type of system is complex, but still regular and random enough that the assumptions underlying the use of statistics are not violated. These systems can be studied under the field of statistical mechanics. The earlier example of a complex landing gear design would fall into this category. The third type of system exhibits what is termed as *organized complexity*. These systems are both too complex to analyze easily, but at the same time they are too organized for statistical analysis as the events within them are not random. Human behavior is not random but controlled. The same is true for software, it does not react randomly to the environment, but rather as it was designed to react. The interaction between these components can result in a new system behavior that might not be possible to foresee when the components are analyzed independently. System theory was developed for this third type of system.

13.3.1 The Third Kind of System

In the complex systems developed in modern engineering there are electromechanical components, coupled with human and software controllers. In general, these complex systems can be presented as a hierarchy, with each level more complex as it moves higher. At the lower levels there are simple mechanical systems that are not subject to emergent properties (Figure 13-8).

Consider a simple bicycle which uses different components such as a frame, wheels, handlebars, a seat, and pedals. At this level there are no emergent properties.

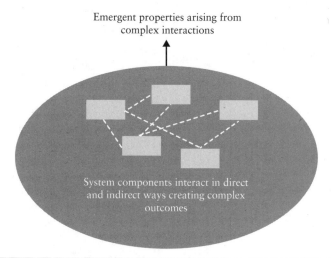

Emergent properties arising from
complex interactions

System components interact in direct
and indirect ways creating complex
outcomes

FIGURE 13-8   Emergent properties. (Source: Authors, adapted from Leveson and Thomas, 2018[5])

Things may change, such as weather may rust the metal, or cause the rubber in the tires to decompose, but this will happen in a very predictable way that can be statistically analyzed. It is only when we add a human that the behavior becomes unpredictable as the more basic components are going to react based on how the human rider is manipulating them. Of course, they, and the human, will also interact with the environment, and that interaction will vary depending on how the rider reacts.

Where the rider directs the bicycle will directly affect the rate of wear on the components and not in a way that can be predicted without a lot of other information. What makes this interesting, though, is that if we view the combination of the rider and the bicycle components as a system, then the behavior of that system will be dependent on how it works together. Consider that the type of bicycle, a road bike, or a mountain bike, will likely play a role in where it is ridden. The physical fitness of the rider, the previous experiences and training, all will go into how that bike is used. In this way, the actual system behavior *emerges* from the combination of these aspects, plus factors outside the control of the rider, such as the weather or terrain.

How the bicycle will behave is constrained by how the rider is choosing to control it, or put another way, the rider places a constraint on the degree of freedom that the bike components have[12] (Figure 13-9).

System theory considers communication and control (see Figure 13-10). The control is the application of the constraints on the hierarchy below, limiting the behavior within the bounds that the higher level deems to be meaningful. To control something, the controller requires feedback of the state of the component being controlled, and that requires some sort of communication.[12]

Control requires having a goal, the ability to affect the system state, a process (or mental) model of the system itself, and the ability to "see" the condition of the current state of the system. This can be drawn in the form of a control loop, as can be seen in Figure 13-10.

A slightly more refined diagram is shown in Figure 13-11. Here the controller is using actuators to control the process, and receiving feedback from the process, allowing the

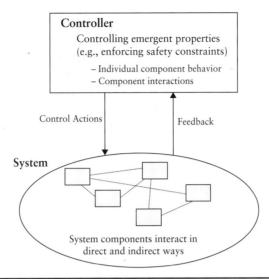

FIGURE 13-9   Control of a system. (Source: Leveson and Thomas, 2018, p. 11[5])

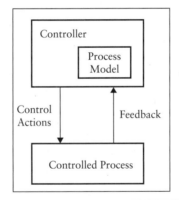

FIGURE 13-10   Simple control loop. (Source: Leveson, 2012, p. 88[13])

controller to modulate its control actions as necessary via a *control algorithm*.[viii] Changes to the environmental conditions or needs of the controller can be met. This controller can be, in turn, controlled by another controller.

Consider a person who is trying to remotely control the temperature of a building. There is a temperature sensor in the building which provides a reading of the temperature to the person in the remote location. The person has a controller that can adjust a furnace to provide more or less heat to the building as required. The *control* is through that controller, and the *feedback* is via the temperature sensor and reading. As can be imagined, if the controller is not responsive the temperature will not be controllable.

[viii] An *algorithm* is a process or set of rules to be followed by a controller (usually computer software) to complete a particular task. For example, a human or computer follows an algorithm if given a set of step-by-step instructions to solve a mathematics equation.

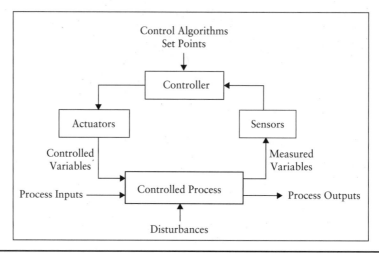

Figure 13-11   A control algorithm. (Source: Leveson, 2012, p. 6[13])

Similarly, if there is a problem with the sensor or the temperature reading indication the building temperature also will not be controllable.

Now we add in a computer to help us. The computer senses the temperature and has a controller to adjust the temperature via a programmed algorithm. Still, the computer controller has the same limitations as the person did, and now the person is one step removed from the process making monitoring more difficult. Even more challenging for the person is sorting out how to know if the system is working correctly. On the other hand, if the system is very reliable the person might start relaxing their attention to that building temperature and start focusing on things that person deemed to be a better use of their time. If some situation is encountered that the computer software was not designed to manage the temperature might exceed the desired limits while the person is attending to another task. The person gets blamed for this even though the computer's programming created the situation. System theory provides a way for us to model this type of scenario.

In a simple way, these examples show how a system can be controlled. It also provides a way for us to ensure we consider all of the parts of the system. Those mechanical components still need to be reliable but for the human or software controller the focus is not on reliable behavior itself, but rather on what controls and feedback the controller has and how well it understands how to operate those controls. For a person this requires a mental model of the process and how the controls and feedback work. A computer has a process model, which is similar but more limited in that a computer cannot work outside that process model.

13.3.2   Humans in a System

A human not only has a mental model of the system they are trying to control, but also understands the system and the reason the system is in place, or stated differently, the person can understand the *meaning* behind what they are trying to accomplish even if they do not perform each task precisely the same way each time. A simple example of this might be a situation of actors in a live performance, where the actors, the physical components such as the set and props, the production team, director, and audience are all parts of the system. A play is highly choreographed and rehearsed but sometimes things do not go as planned. An actor might forget their lines, or a prop might not work. Part of the stage may physically get stuck when it is supposed to be moved between scenes. Remember the performance is live, so unlike a recorded performance there is no way to fix the problem, instead the actors must find a way to modify the performance so that the scenes fit together to still bring the play to the intended conclusion. The actors understand how the play is supposed to end, which in this case, is the *system goal*. The play reaches a successful conclusion even though the actors did not precisely follow the script.

By contrast, although a computer will perform a given task precisely the same way each time, the computer is also generally constrained to that performance. Some might argue here that computers can be programmed to deviate from a rigid path. This is true to an extent. During the design of the computer requirements the designer can, and should, consider various scenarios that might occur and program those into the software design. The limitation here is that the design is still limited to the imagination of what scenarios need to be managed and is unable to deviate beyond what it has been designed to do. A person has no such limitations and can perform as necessary, even when a scenario occurs that nobody thought was possible during the design phase. If the system is not doing what is desired the person can find other ways to reach the desired system goal, while the computer is limited to only its designated functions. It is important to point out here that in most current software design the scenarios that are considered are generally considered based on the believed probability of occurrence, so those scenarios deemed very low probability are almost always left out of the design. This, again, is a limitation of reliability theory.

The role of the pilot is to manage those scenarios that were not envisaged by the system designers. Unfortunately, this often means that the automation hands the aircraft to the pilot only when it cannot manage the situation. The pilot may now have to intervene when the consequences of any misstep can be catastrophic—and they are expected to quickly correctly diagnose and respond to the unusual situation under extreme stress with limited information.[17] Unfortunately, if the pilot is not able to "save the day" the pilot is usually still blamed for the accident in this type of scenario.

13.3.3   Control Structures

In system theory the system is modeled as a hierarchical control structure. What this means is that starting at the most basic level are the physical components of the system, which are at the bottom of the structure. In our bicycle analogy this would be the bicycle itself. In the building example, we are controlling the furnace. The next item over that is what is controlling the physical system. That can be a person, it can be computer software or even an organization. Regardless of what type of controller we are using, that controller has the capability to make decisions that will affect the physical system. That controller may be, in turn, controlled by other, higher-level controllers. As we move up higher the controllers generally have more authority and a broader scope of control but less ability to control the system at a granular level.

An example can be seen in Figure 13-12. As can be seen, the aircraft automation is directly controlling the physical aspects of the aircraft, and the flight crew is, in turn, controlling the aircraft automation. Above the flight crew, air traffic control is controlling the flight crew's actions. Notice that the flight crew can choose to directly control the aircraft, bypassing the automation, but doing so changes the cognitive role of the flight

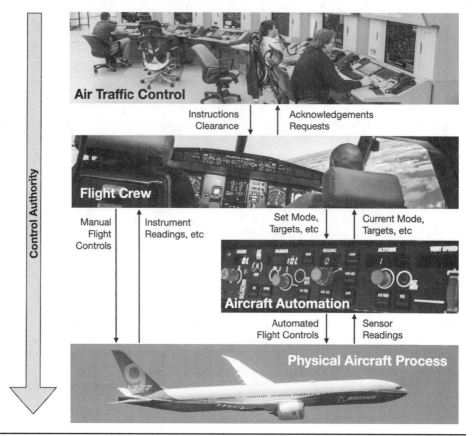

Figure 13-12 Generic aircraft control structure. (Source: Adapted from Leveson and Thomas, 2018, p. 24[5])

crew as they have to devote attention to the direct control of the aircraft. It should also be noted that air traffic control's control over the flight crew is limited to constraining certain actions, and the flight crew can choose to deviate from the controller's instructions. This is not uncommon. Consider the situation with a military officer. The soldiers under the officer's command are duty bound to follow the officer's orders but do have the ability, and are often expected, to deviate under certain extreme circumstances.

What becomes important in analysis is understanding why the controller was not, or might not be, able to control the system for the desired outcome. It could be due to a problem with the physical system, but it could be that the controller had a problem with the control. Perhaps the feedback was not correct, or perhaps the command did not get through. Considering again the officer commanding the soldier. If the soldier did not do what was commanded the problem could be that the command was issued but did not reach the soldier. It also might be that the soldier did execute the command, but the officer was unaware. It also is possible that the soldier could not execute the command due to a problem in the physical system.

As the reader might surmise in this example, using system theoretic methods it is possible to examine the entire sociotechnical system, including all aspects of the

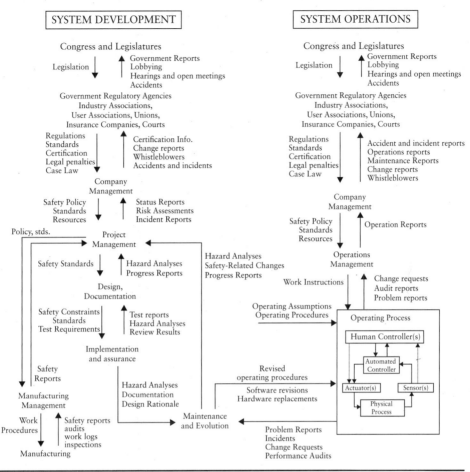

FIGURE 13-13 Sociotechnical control structure. (Source: Leveson, 2012, p. 82[13]).

system's physical components such as software and computers, but also human controls such as management and government regulators (see Figure 13-13).

The original intent of system theory was to understand and analyze complex systems. In the late 1990s through the early 2000s, Dr. Nancy Leveson of MIT expanded system theory to apply it to creating an accident causality model and through that, new methods of causal and hazard analysis. The next sections review those developments.

13.4 STAMP

It can be said that the greatest insights often appear obvious and simple in retrospect. Throughout history, the most interesting concepts emerge out of the boundary of two disciplines. Leveson's genius was applying the concepts of system theory to accident models. System theory holds that the behavior of the system is constrained through the system design. The behavior of a hardware or electromechanical component is constrained by its physical design. The behavior of a software component is constrained by the design of its software—the coded algorithm process model, combined with the physical design of its circuitry and the controls it has been provided, coupled with the information that it is given. The behavior of how these components interact with other components and the environment is constrained by the controller above it. Humans are also constrained in the system by the training they receive, the controls they are provided, the information they are provided and the boundaries of the policies and procedures they have been given but humans have the unique ability to move outside of those boundaries and constraints to ensure the system goals are met.

13.4.1   Safety Analysis Using System Theory

When a system is analyzed, the boundaries of the system, the upper and lower bounds, are defined by the person doing the analysis. There is, for example, no need to look at how the atoms bond together to create a specific material unless that information helps understand the system (for example, the system that is being analyzed might be one that is creating a component that has certain strength requirements). Similarly, although a problem might be possible to trace through the highest level of a government the ability of the analyst to effect change at those levels may render analysis at those higher levels moot.

As previously discussed, the behavior of the entire system emerges from the inter-action of the components[ix] in the system. Through constraining the behavior of those components, it is thus possible to ensure that the entire system behaves in a safe manner. In this regard, recall that systems theory frames safety as an emergent property that arises from the way the system components interact with the environment through controlling and constraining their behavior so they cannot do something unsafe. It follows that an accident results from not properly constraining the behavior of the components in the system, be they physical, software, or human. The purpose of the accident investigation then becomes an investigation into why the controls that were designed to constrain the system to prevent the accident did not do so. Prospectively, if we can identify the hazard scenarios that can lead to accidents, we can design our system to constrain the behavior to prevent those hazards from becoming accidents. Based on systems theory, accidents result when there are flaws in the control structure itself. This is, then, a system theoretic accident model.

Leveson[13] used system theory to create System Theoretic Accident Models and Processes (STAMP), and then, based on STAMP, further developed methods that provide a way for engineers or accident investigators to analyze systems. These are then the processes. Leveson created two processes that are based on STAMP, the first a post hoc accident analysis method, Causal Analysis using System Theory (CAST), the second is a prospective system analysis method that can be utilized to identify hazards called System Theoretic Process Analysis (STPA). The STPA process results in an analysis of the controls that need to be in place to prevent a given hazard. The analysis then leads to the creation of a set of requirements, which can be used by engineers and researchers to analyze the design, policies, training, and procedures that can lead to hazards. STPA is a formal hazard analysis methodology that has been found to identify many more scenarios that could lead to a hazard than other methods (mostly based on reliability theory), which in turn can be used to generate controls to mitigate those hazards.

[ix] Please see footnote v for further description.

13.5   The B-737 MAX Series Aircraft

Returning to the B-737 MAX that was discussed at the beginning of this chapter, what went wrong? Simply put, Boeing and the FAA used methods based on reliability theory. Probability values were assigned to the failure modes and assumptions on human response were based treating the humans in the system as a component that has an expected compliance rate. Reliability theory is dependent on probability, but computers and humans simply do not fit into that paradigm. Of course, it often happens that we get lucky. Most of the time humans do what we expect them to do, as do our computers. Unfortunately, sometimes we run into a scenario that was not anticipated by the system designers. When that happens all the probability numbers become moot. In fact, those numbers already were moot before they began—we cannot assign probability values to humans or computers.

From a system theory perspective, the examination of what went wrong on the B-737 MAX would be through understanding why the controls in place did not constrain the system. Starting at the bottom there are at least two paths we can take, although we would also study the interactions between them. The first is looking at the controls from the manufacturing side. The second is understanding the controls up through the pilot. The interrelationship between them is a result of assumptions made about the design from both sides.

A full analysis would be quite extensive and is outside the scope of this book. The analysis would look to understand why the controls in place were not adequate to allow for a design that contributed to unsafe actions. This would start at the basic design of the aircraft itself and what the MCAS was designed to do and what controls was it given. Each control action would be analyzed to see if it could be unsafe. Would there be scenarios where not trimming could lead to an unsafe condition? Would there be scenarios where trimming could be unsafe? What if the trimming was done too soon, or too late, or in the wrong order? What if the trimming was done for too long or too short a period of time? Scenarios would be created for each of these scenarios and then mitigated in the design itself. The same would be true of the controls over the design team, which would include management, regulatory oversight, and testing requirements. In each case, the object would be to identify scenarios that could create an unsafe control action.

The same would apply to the actions of the pilots, the pilot training management, and so on. In doing so it would be possible to identify many aspects that could be unsafe before any accident occurred.

If STAMP's CAST were to be applied when examining these accidents, the entire sociotechnical system must be considered. When examining an event that has already occurred, what is important is understanding how the scenario in which the accident occurred was not captured in the hazard analysis.

The hazard that resulted in the accidents related to the B-737 MAX was the MCAS computer rapidly and repeatedly trimming the THS nose-down when the angle of attack was not high. This scenario was not contemplated in the design and, therefore, the pilots did not receive any related training. If it had been considered, then the design would have reduced or eliminated this hazard (the MCAS has subsequently been redesigned to mitigate this hazard). The pilot "error" of not correctly responding to the MCAS commands would not have occurred—an example of the safety engineering principle that human error is a symptom of a poor design.

In the aftermath of the accidents, it was found that the vast majority Boeing engineers sincerely *believed* that the aircraft was safe to fly. There were flaws in the communication and control between Boeing and the FAA, largely due to slow changes over time. Many of the changes at the FAA were due to increasing budget constraints over several decades.[18] In the end, those involved were predominantly trying their best to make a safe airplane within the system constraints that they were given. The accidents resulted from the way the entire sociotechnical system interacted together.

13.6   Question: Why Was This Not a Problem Before?

The honest answer is that in many respects it has. The history of the Douglas DC-10 cargo doors, which included a close relationship between the FAA and the manufacturer, discounting problems based on probability and other similar aspects closely mirrors many of the problems identified with the design of the MCAS, even though the specific accident factors were quite different. Still, the reader might be surprised that there have not been more accidents considering these problems.

There has been one aspect that has served as a bit of a buffer to the application of reliability theory to system design. Very experienced engineers and regulators have often been able to "see through" the reliability methods and identify potential problems. This is literally the product of deep experience. After designing aircraft or being involved with aircraft certification for many decades a person has seen so many scenarios that they will often recognize those areas that the reliability theory-based methods fall short. One of the identified problem areas in the Boeing 737 MAX saga was that senior staffing levels were reduced under the errant belief that the engineering analysis methods such as Failure Modes and Effects Analysis (FMEA) would adequately address the safety aspects. In addition, the regulator ceded more control to Boeing. These were somewhat gradual changes over time which acted as a ticking time bomb.

Similarly, extremely experienced, highly trained pilots can also serve as a buffer to a poor design. It is important to differentiate here the meaning of experience in this context. There are many pilots today who might be considered extremely experienced in terms of flight hours but who have never experienced any sort of equipment failure of any sort, or, if they did, it was only a relatively simple one that was well predicted in the policies and procedures, so well trained for. In times past pilots were more likely encountering unique and novel scenarios that nobody thought possible, and due to much less reliable components, most pilots that had reached a certain level of experience had encountered several such events. In addition, the stories of other pilot's experiences were widely shared. The pilots learned a high degree of adaptability as opposed to just procedural knowledge and were able to apply that knowledge. Unfortunately, today, for a variety of reasons, neither of these is true. Whether or not those that design the training are aware of it, current training is also founded on reliability theory-based methods. These methods are simply not enough to provide pilots with the capability to manage many of the complex problems that were not anticipated in the design.

It must be stated clearly that even with experience and training it is far from certain that a pilot could manage the most challenging scenarios that might present themselves, but at least they would have a chance of doing so. Further, the aircraft designers still *expect* pilots to manage those scenarios that they could not anticipate in the design. This assumption is not being met. The combination of the assumptions in the design, the unexpected software interactions, with the reduced adaptability of the pilots resulted in two horrific accidents.

It must again be emphasized that this story is not unique to Boeing. Many other accidents are similar in that they have involved assumptions in the design as to both pilot behavior, pilot training, maintenance procedures, problems in oversight, and many more problems in the controls that have been designed to prevent accidents. Flawed assumptions have resulted in accidents for Airbus, Douglas, Lockheed, and many other manufactures. In fact, it is unlikely that any manufacturer has been immune to similar problems. These issues are surprisingly similar across a variety of industries from nuclear power, military systems to the petrochemical, automobiles, healthcare, and all of aerospace. As stated by Leveson, the same accident appears to be occurring over and over with just a few details changed.[17] The reason is that we are not learning enough from accidents. Most accident reports result in superficial findings that are simply not enough to change the entire sociotechnical system in the manner necessary to actually prevent future accidents. A system method, such as STAMP, is necessary to identify problems that might create an accident as we capture the information that is needed to prevent a repeat of the decision making that resulted in an accident.

> STPA is a relatively new hazard-analysis technique based on an extended model of accident causation. It assumes that accidents can occur from not just component failures but also the unsafe interactions of system components, none of which may have failed. STPA, developed by professors at the Massachusetts Institute of Technology, is being adopted by a growing number of industries as well as numerous Boeing organizations.[19]

Today STAMP is being applied by aerospace manufacturers, the military, and airlines around the world, as well as promoted by many government and safety agencies. The number of organizations using it is expanding rapidly as it is the only way to manage and identify problems in complex systems. STPA is being used as a proactive way to identify problems and accident potential before they happen. Some examples of applications of these methods are included in Chapter 15.

Key Terms

Causal Analysis Using System Theory (CAST)

Control Structure

Emergent Property

Hazard

Models

Reliability Theory

System

Systems Theory

Systems Theoretic Accident Model and Processes (STAMP)

System Theoretic Process Analysis (STPA)

Topics for Discussion

1. What is reliability theory and how is it used today?
2. What is a hazard?
3. What is a system?

4. Why does reliability theory lead to problems in our designs?
5. What is system theory?
6. What is an emergent property?
7. Describe STAMP.
8. Describe the difference between CAST and STPA.
9. Describe a type of problem that STAMP-based methods could manage that methods based on reliability theory could not

References

1. Malmquist, S., & Rapoport, R. (2020). *Grounded: How to Solve the Aviation Crisis*. Chicago, IL: Lexographic Press.
2. Slotnick, D. (2019). Boeing's 737 officially lost the title of world's most popular airplane. Airbus' competitor just passed it in sales. *Business Insider*. https://www.businessinsider.com/airbus-beats-worlds-most-popular-plane-a320-737-2019-11
3. FAA. (2022). 14 CFR § 25.173—Static longitudinal stability. https://www.law.cornell.edu/cfr/text/14/25.173
4. https://www.faa.gov/foia/electronic_reading_room/boeing_reading_room/
5. Leveson, N. G., & Thomas, J. P. (2018). *STPA Handbook*, p. 169. http://psas.scripts.mit.edu/home/get_file.php?name=STPA_handbook.pdf
6. Weisberg, M. (2012). *Simulation and Similarity: Using Models to Understand the World*. New York, NY: Oxford University Press.
7. Box, G. E. (1976). Science and statistics. *Journal of the American Statistical Association*, 71(356), 791–799.
8. Reason, J. (2016). *Managing the Risks of Organizational Accidents*. Brookfield, VT: Routledge.
9. Descartes, R. (1988). *Discourse on Method and Related Writings*. Clarke, D. M. (trans.). New York, NY: Penguin Books. ISBN 9780521358125.
10. Goodall, R., Dixon, R., & Dwyer, V. M. (2006). Operational reliability calculations for critical systems. *IFAC Proceedings Volumes*, 39(13), 771–776.
11. National Transportation Safety Board. *Crash During Landing*. DCA97MA055. https://www.ntsb.gov/investigations/AccidentReports/Reports/AAR0002.pdf
12. Leveson, N. G. (2009). Applying systems thinking to analyze and learn from events. *Safety Science*, 49(1), 55–64.
13. Leveson, N. G. (2012). *Engineering a Safer World: Systems Thinking Applied to Safety*. Cambridge, MA: The MIT Press.
14. Abbott, K., McKenney, D., & Railsback, P. (2013). *Operational Use of Flight Path Management Systems*. Washington, DC: Performance-based operations Aviation Rulemaking Committee.
15. Leveson, N. G. (1995). *Safeware: System Safety and Computers*. Boston, MA: Addison-Wesley.
16. Bertalanffy, L. V. (1969). *General System Theory: Foundations, Development, Applications*. New York, NY: George Brazillar, Inc.
17. Leveson, N. (2022). *An Introduction to System Safety Engineering*. Cambridge, MA: MIT Press.
18. DeFazio, P., & Larsen, R. (2020). Final Committee Report: The Design, Development and Certification of the Boeing 737 MAX. https://transportation.house.gov/imo/media/doc/2020.09.15%20FINAL%20737%20MAX%20Report%20for%20Public%20Release.pdf
19. Boeing. (2022). Dutch Roll. https://www.boeing.com/resources/boeingdotcom/features/innovation-quarterly/2022/04/dutch-roll-in-IQ-2022-Q2.pdf

CHAPTER 14

Safety Management Systems and Safety Data

Introduction

Safety Management Systems (SMS), as the name implies, is a system to manage safety. In general terms, the approach is similar to the management of any other system, and the methods that are incorporated into a SMS are similar to the approaches used to manage a large organization. The key difference is that safety itself is managed rather than the manufacture of the product or goals of the organization. As such, SMS is a standardized approach to controlling hazards across an entire organization through the promotion and sharing of safety data and best practices. SMS requires crafting policy, methodically calculating and controlling risk, measuring how everything is working, and communicating it all to employees. A commercial airline, like any other major business, has to integrate and manage a complex array of tasks to deliver products and services to the customer. Safety must be built into every aspect of the organization, from the overall structure to being an integral part of every employee's job responsibility.

The airline industry has always placed great emphasis on safety and has moved aggressively to identify and control problems that cause accidents. Airlines have learned, sometimes the hard way, that active management of risk is an absolute requirement for a healthy company. Even though the management of safety is not a new concept, it was not until the 1990s that safety professionals started seriously asking the rather profound question of whether they were managing safety at peak efficiency. This introspective question may have been prompted by recognition of the safety professional's role in an organization, by increased public awareness that advances in scientific and business principles can be applied to management for making any operation more efficient, and that safety is a state that requires management.

It was the convergence of those lines of thinking, combined with the traveling public's increasing intolerance of inefficiency, which led to the birth of a standardized approach to minimizing risk, which today we call SMS. Although safety management as a practice has existed since the birth of aviation, in one way or another, SMS is a formal and standardized framework of best practices for running a safety program based around sophisticated business and scientific processes. Those processes are not just used to create safety, but also used to monitor the health and effectiveness of a safety program so that it operates at peak performance. A key component of the architecture mirrors the business world's use of performance indicators. In SMS, these are called safety performance indicators. It should be noted here that these are based on reliability theory. The implications of this will be discussed later in the chapter.

An ideal airline SMS would have unlimited resources to manage the safety demands of the business. However, since commercial aviation is in the business of making profit and any investment in safety requires people, time, and money, limited resources force airlines to choose wisely about how to allocate their funds. Safety efficiency, therefore, is not just about making risk management more streamlined and effective, but also about saving money by operating a safety program to be as lean as possible without seeing significant safety issues arise.

Another helpful way to understand the history of how Aviation Safety Management Systems have improved overall safety performance is to consider Figure 14-1. Safety has now moved from the *Reactive Stage,* to the *Proactive Stage,* to the *Predictive Stage* of identifying potential/future safety problems.

Safety implemented correctly can improve cost-effectiveness, augment predictability, ensure reliability, and protect the brand name. Unfortunately, the perception of a dichotomy between safety and profit being in conflict is common. When no accidents have occurred, there is a tendency to lower the guard and start diverting resources into other aspects of the organization. If cutting a corner without an accident occurred a few times, then people start believing that it is not a problem. If cutting that corner made the operation more efficient there is a bias toward repeating the seemingly harmless change. The problem here is that, as discussed in Chapter 13, hazards are contextual, and so it is all too easy to repeat an action many times without any accident even though the same action can become hazardous through changing a few details.

An organization's corporate culture must emphasize safety jointly with profit, controlling risk as a core value to help the bottom line, not to hinder profit. The best approach is a balanced allocation of resources where safety management is a core business function, closely intertwined with business objectives and not in competition with the profit aspects of the business. Safety and profit should be perceived as inseparable

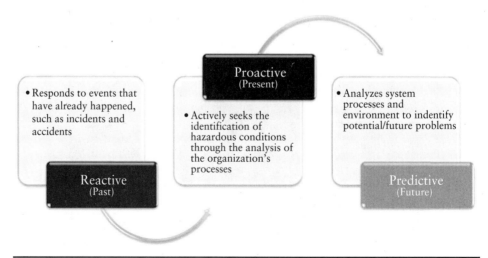

FIGURE 14-1 Reactive, proactive and predictive safety. (Source: Federal Aviation Administration—FAA)[1]

Figure 14-2 The ultimate objective of SMS is maximizing safety efficiently. (Source: Authors)

since an unsafe commercial operation will soon cease to be used by customers. In Figure 14-2, going around storms may add time (affecting profit) but improve safety.

An airline manager might prefer to leave the fastening of seat belts to the passenger discretion, or at the very least, not tightly enforce the requirement. This can appear to be a reasonable decision on face value, but passengers are not fully aware of the risks involved and many passengers are injured each year because of unexpected turbulence. Despite this, passengers are generally only advised, but not required, to have seatbelts fastened when the seat belt sign is not illuminated. Passengers need some leeway in order to be able to get up and use the lavatory or stretch on long flights, but many treat air transport more like travel on trains which have much more predictable motion and so do not require the installation of seatbelts. How can an airline balance these aspects without disenfranchising passengers? This is one example of how enforcement of safety standards is vital to prevent personal injury.

Fortunately, today there are outstanding online resources to consult in order to learn the details of SMS. The most important references are provided for your use:

- The *ICAO Safety Management Manual*, 4th ed. (Doc 9859—2018)[2]

  https://www.icao.int/safety/safetymanagement/pages/guidancematerial.aspx

- SKYBRARY—The aviation website for Europe and EASA containing useful information

> https://skybrary.aero/articles/safety-management-system#:~:text=The%20
> objective%20of%20a%20Safety,related%20to%20safety%20of%20operations.

- Federal Aviation Administration—Safety Management Systems Website

  https://www.faa.gov/about/initiatives/sms

In many chapters of this book, the reader has already been introduced to aviation SMS concepts. This chapter will bring together many of the previously discussed ideas as we formally introduce and explain the framework of SMS. We will discuss the history of SMS principles, the fundamental concepts, the tools and techniques used by aviation safety professionals, and the future of SMS in commercial aviation safety.

14.1   Evolution of SMS

It is clear that modern safety principles have significantly evolved in the years following World War II. Following the war, commercial aviation saw significant growth and so did the enabling technology with the introduction of the B-47 Stratojet, the jet engine, and swept back wing configuration. This *technical era* of aviation safety concentrated on the improvement of technical factors during the decades of the 1950s and 1960s.

In the 1970s, the Japanese gained world market share in the electronics and automobile industry using the "Total Quality Management" (TQM) concepts of quality experts, W. Edwards Deming and Joseph Juran, among others. This movement in Japan focused on improving the quality of manufactured products through teamwork and employee involvement. In the aviation world, the focus shifted to the *human era* by using such tools as Crew Resource Management (CRM) and Line-Oriented Flight Training (LOFT). As outlined by ICAO in the Safety Management Manual, the period from mid-1970s to the mid-1990s has been called the golden era of aviation human factors.

In the early 1990s, aviation safety thinking shifted its focus to the *organizational era*, which included technical, human, and organizational factors. As previously discussed, Dr. James Reason's *Swiss Cheese Model* was based on the concept of the "organizational accident," which is a breakdown of organizational processes resulting in an accident. Although the concept of organizational accidents was not new, Reason's prominence endorsing the concept and putting it into a framework that was easy to visualize, contributed to a rapid acceptance of the concepts. Reason's publications, culminating eventually in 1997 a book titled *Organizational Accidents* occurred at a time when the aviation industry was looking for ways to cope with accidents that were occurring due to factors that were not based on simple component failures.

Using quality-based employee involvement and cultural empowerment concepts employed by the Japanese and adopted by the International Organization for Standardization (ISO), aviation SMS theory has evolved into a worldwide movement which pays great safety dividends. It should not surprise the reader that current SMS is closely tied to business efficiency concepts. As the reader might surmise, although a large improvement over simpler approaches, the Swiss Cheese Model is still based on reliability theory, and the focus is entirely on human behavior while not capturing many of the nuances involved when humans, software, and technical systems all interact in complex ways.

14.2 ICAO Annex 19: Consolidation of SMS Standards

In view of the increasing complexity and tighter interdependence among various aviation sectors, ICAO decided to consolidate all SMS standards into a new annex which allows a higher degree of integration of safety management functions. Annex 19 was the first, new ICAO Annex in over 30 years and was developed in two phases. The first phase involved the organizing of existing safety management–related content, modifications to improve the language for clarity, and modifications to ensure standardization and harmonization of SMS information across the ICAO Annexes. It was adopted by the ICAO Council on February 25, 2013, and became applicable in November 2013. A second phase resulted in Amendment 1 to Annex 19. It raises the components of SMS to the "Standard" level and addresses the protection of safety data and information. The Amendment to Annex 19 was effective in July 2016, becoming fully applicable in November 2019. In the *ICAO Safety Management Manual*, Doc 9859, Fourth Edition, published in 2018, ICAO also began promoting a broader view for SMS to include considering the entire aviation system. Despite this, the analysis methods outlined in the document reflect methods that date from the mid-1990s and before.

14.3 Structure of SMS: Four Components (Pillars of SMS)

In order to simplify and organize the best practices of managing safety, the majority of SMS regulation follows the structure outlined by ICAO, although the reader should be aware that some aviation organizations have different organizing plans as approved by the regulator for the country in which they operate. Regardless of SMS structures, the philosophies that guide the systems are all similar. SMS is an organized approach to controlling risk from unintentional damage or harm, including the necessary organizational structures, accountabilities, policies, and procedures.

The SMS framework consists of 4 components (sometimes referred to as *pillars*), and 12 elements within those components. ICAO provides guidance on these components—which are generally included in the SMS policy documents by the ICAO member countries—as they are keys to implementing SMS. The four components of SMS are as follows:

1. *Safety Policy*. The focus here is on creating an environment for effective safety management which lays the foundation for the organization's safety culture. This is dependent on establishing a safety policy and having senior management commitment to safety and continuous safety improvement. Safety policy defines the methods, processes, and organizational structure needed to meet safety goals. This can be remembered via the expression *document it in writing*, referring to the need to have key aspects of a given SMS readily available to employees as a written reference.

2. *Safety Risk Management (SRM)*. This involves the identification of hazards, risk assessment, and risk mitigation. SRM determines the need for, and adequacy of, new or revised risk controls based on the assessment of acceptable risk. This can be remembered via the expression *hunt the hazards, then assess and mitigate the risks*, referring to the fundamental need of having employees aggressively seek safety problems and having a means of effecting change to produce better conditions.

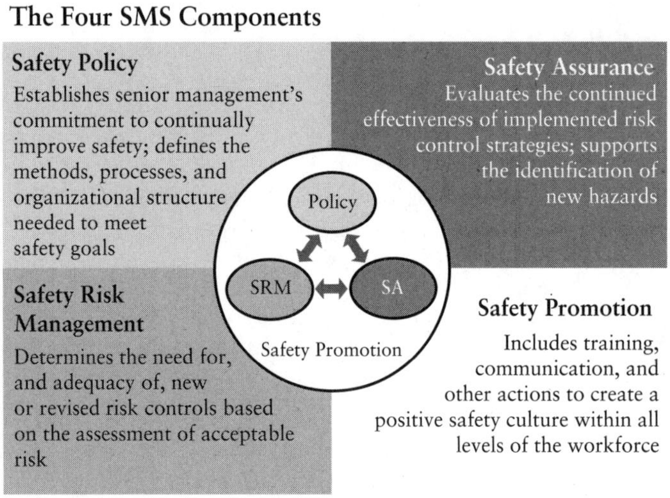

The Four SMS Components

Safety Policy

Establishes senior management's commitment to continually improve safety; defines the methods, processes, and organizational structure needed to meet safety goals

Safety Assurance

Evaluates the continued effectiveness of implemented risk control strategies; supports the identification of new hazards

Safety Risk Management

Determines the need for, and adequacy of, new or revised risk controls based on the assessment of acceptable risk

Safety Promotion

Includes training, communication, and other actions to create a positive safety culture within all levels of the workforce

FIGURE 14-3   The organizing scheme for SMS. (Source: FAA)[3]

3. *Safety Assurance (SA)*. Evaluates the continued effectiveness of implemented risk control strategies, and supports the identification of new hazards. Is the SMS working as it was expected and required to do? This also includes the management of change and continuous improvement of the SMS. This can be remembered via the expression *measure and improve it*, referring to the need to use scientific principles to determine the effectiveness of the actual management change processes being used.

4. *Safety Promotion*. Includes training, communication, and other actions to create a positive safety culture within all levels of the workforce. "Senior management provides the leadership to promote the safety culture throughout an organization."[1] This can be remembered via the expression *learn it and share it*, referring to the need to continuously learn about the SMS, about new and old hazards, and disseminating such life-saving information to all employees who can affect safety.

FAA's approach to organizing SMS by placing concepts into four components is illustrated in Figure 14-3 . Note how the promotion component is shown to impact all three other components.

14.3.1   Component 1: Safety Policy

Safety policy is the written commitment for safety within an organization. The policy should include a commitment for continuous improvement, promotion, and regulatory compliance. To accomplish this the policy should outline and provide the resources necessary for the SMS, clarify that safety is the primary responsibility of all managers and ensure that the policy is understood and implemented throughout the organization.

There are five key components of safety policy as outlined by ICAO (2018). These are:

1. Management commitment and responsibility
2. Safety accountabilities

3. Appointment of key safety personnel

4. Coordination of emergency response planning

5. SMS documentation

It is the role of management to establish the safety objectives in an organization. This should be demonstrated through the senior management commitment. Safety policy encompasses the leadership of the *Accountable Executive (AE)* to provide clear safety objectives, methods, processes, and the organizational structure needed to meet safety goals. As stated by Professor Nancy Leveson of MIT, "this must be genuine, not just a matter of sloganeering".[4] If the safety program is to be effective, the safety department must be part of the mainstream operational or engineering management structure. Too often the safety department is placed under the quality assurance structure, which creates an expectation that safety is only an "after-the fact or auditing activity…safety must be intimately integrated into the design and decision-making activities."[i] Placing the safety program under quality assurance can result in the safety department having less influence on operational decisions that can be subject to organizational pressures to put profit over safety.

Safety promotion refers to the nonpunitive, positive safety culture (also known as a "Just Culture") which empowers all employees to have a role in safety. This is accomplished through comprehensive SMS training, communication, and awareness techniques. It is appropriate that this element is placed first on the list since it sets the stage for the other elements to exist and reach full potential. The starting point for SMS Policy is identifying the AE, who is a single, identifiable person having final responsibility for the organization's SMS, including decisions regarding operations and everything that supports operations. For purposes of this chapter, the reader can assume that the AE of a large airline will be the Chief Executive Officer (CEO), and of a small commercial aviation operation, the AE may be the owner.

That AE must then oversee how his or her "organization defines safety policy in accordance with international and national requirements and will sign the overall policy. It should reflect the organization's commitment to safety, should include a clear statement about the provision of the necessary resources for the implementation of the safety policy, and should be communicated throughout the organization." Identifying precisely who that AE is becomes critical because senior management is in control of the safe and efficient operations of the aviation organization. It is through management that employees are hired, trained, and dismissed when necessary. The safety policy should be developed by senior management and clearly communicated to all employees. It is extremely important that leadership be strong, demonstrable, and visible if the airline's SMS program is to succeed. At the same time, if the hierarchy is too strict it can result in limited communication depending on the interests of managers as the information moves up the chain of command. This can result in a misperception of the risk in the organization.[2]

Safety policy should also make clear that safety is the responsibility of all members of management, not just the safety department. The policy should clearly lay out who the safety manager is and how the company's SMS shall be administered so that responsibility is widely distributed for safety. As an example, this can take shape by

---

[i] See reference (4), p. 434 for further reading.

using a *Safety Action Group (SAG)* and *Safety Review Board (SRB)* to assist management officials in accomplishing safety responsibilities.

Under normal circumstances, the safety manager communicates to the AE through either the SAG or the SRB. The SRB is a very high-level strategic direction group chaired by the AE, which includes senior line functional managers. Once strategic direction has been developed by the SRB, the members of the SAG are tasked with implementing the decisions at a tactical level. In this typical SMS organization, the Safety Manager serves as the secretary of the SAG to record its actions and keep score of corrective actions taken.

Safety policy should also ensure that an *Emergency Response Plan (ERP)* allows for a smooth transition from normal everyday processes to emergency operations and then back to normal conditions once the emergency has been managed. Airlines prepare an ERP document in writing to describe what actions should be taken following an accident and who is responsible for taking each action. An annual emergency response drill is usually required to ensure that the plan is adequate and that designated personnel are appropriately trained.

Lastly, careful documentation is an essential element of an SMS organization. Each AE should ensure that the safety manager has an SMS implementation plan that provides objectives, processes, and procedures and communicates who is accountable for what aspect of the plan. There also should be a manual that explains the whole SMS so that any employee can understand the operation.

14.3.2   Component 2: Safety Risk Management

The heart of any SMS organization is the interrelationship between safety risk management and safety assurance. Basically, the safety risk management process provides for the identification of hazards and assessment of risk. Once it is assessed, risk should be controlled (mitigated) to a level *as low as reasonably practicable (ALARP)*. At this point, the safety assurance process takes over to assure that the risk controls are effective using system safety and quality management concepts of monitoring and measurement.

Before a risk can be managed, a potential hazardous condition must be identified. It is easy to confuse the concepts of hazard or risk. In Chapter 13 we provided the engineering definition of a hazard, which is a system state that occurs within a certain context that can result in an accident, so the same event can be a hazard in one condition and benign in another. An autopilot that suddenly just disconnects in cruise is usually not a hazard, but one that disconnects just prior to touchdown on an Autoland with very low visibility can create a hazard. ICAO defines it in a similar, but a bit differently, stating that a *hazard* is "a condition or activity with the *potential* to cause death, injuries to personnel, damage to equipment or structures, loss of material, harm to the environment or reduction of the ability to perform a prescribed function."[1] An example of a hazard under the ICAO framework would be a leak in the hydraulic line of an aircraft braking system, but because it is not a risk in and of itself, the leak does not let us know how it impacts safety.

As discussed in Chapter 2, risk is defined as the assessment, expressed in terms of predicted probability and severity, of the consequences of a hazard taking as reference the worst credible effect. Thus, risk is expressed in a formula as: $R = P \times S$, where

- R is the expected risk (loss) per unit of time or activity.

- P is the probability that a given hazard may materialize (likelihood of a loss event per unit of time or activity).

- *S* is the severity of consequences (effects) that the materialized hazard can generate (loss per event).

  So, returning to the previous example:

- A leak in the hydraulic line of an aircraft braking system when the aircraft moving is a hazard. It is important to keep in mind that hazards are always conditional. A leak in the in the hydraulic line if the aircraft is in storage may be a nuisance, but in most cases is not a hazard.

- The potential for running off the runway because the pilot might not be able to stop the aircraft on landing using brakes is one of the consequences of the hazard. This might also be considered a scenario that is a consequence of the hazard.

- The assessment of the consequences of literally running off the runway expressed in terms of *probability (P)* and *severity (S)* is the *safety risk (R)*.

Stated another way for clarification, in the equation $R = P \times S$:

- *P* is the probability that the aircraft would run off the runway given that its stopping capability is impaired.

- *S* is the severity of the above event which could vary from no aircraft damage or injuries (near miss) to total destruction, or both (depending on variables such as the aircraft speed and runway remaining when the braking fault is noticed, availability of other means of stopping, etc.).[ii]

Although this is a popular and common way to define risk, it is problematic. Probability, also often called likelihood, is generally not knowable and not possible to calculate. It can only be estimated using very subjective methods that are subject to human bias. One problem with this approach is that if an accident or incident has not occurred the probability is often adjusted downwards. The actual probability has not changed, but the perception of it has. Similarly, the severity measure can be subjective. Problems with probability will be discussed further later in this chapter. Severity largely depends on who is making the evaluation and how they are choosing to approach this metric. If, for example, fatalities were used, the result might be that an aircraft with 100 passengers has twice the risk as one with 50 passengers, even though nothing else has changed. What is severe to a passenger might not be severe to a company shareholder, and one person may identify an event as more severe than another one. This will be discussed further, along with some approaches to actually identify and manage risk, in Chapter 16.

Notwithstanding the problematic nature of determining actual risk, an understanding of the concept will make understanding the importance of the elements in the second component, safety risk management, easier. Analyzing a large volume of safety data from flight operations and determining risks are only one part of the equation. The information we gather is an integral component in creating, implementing, and validating the effectiveness of risk mitigation in an SMS. For this reason, the second

---

[ii] The astute reader will recognize that this approach is based on reliability theory, as described in Chapter 13.

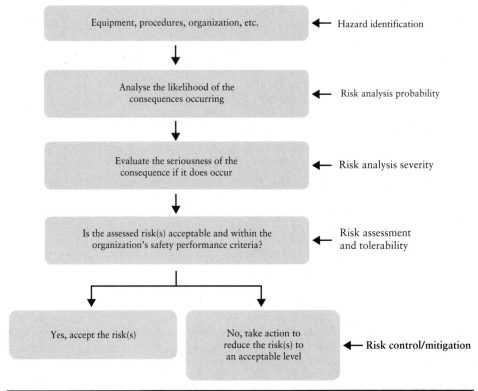

FIGURE 14-4   Hazard identification and risk management process. (Source: ICAO, 2018[2])

component of SMS incorporates both hazard identification and risk assessment mitigation. ICAO's process is outlined in Figure 14-4.

Safety risk management begins with a clear understanding of an organization's functional systems, which are analyzed by experienced operational and technical personnel to detect the presence of hazards. Under SMS, each organization should create and maintain a formal process that ensures that operational hazards are identified. Such a process should be based on a combination of reactive, proactive, and predictive methods of safety data collection. How hazards are identified will depend on the resources, culture, and complexity of the organization. The identification of external safety data will be discussed in Section 14.6 of this chapter.

After the safety hazards have been identified, they must be analyzed to determine their consequences to the organization. This step entails creating and maintaining a formal process that ensures analysis, assessment, and control of the safety risks in operations. The conventional method to analyze risk is to break it into two components, the probability of occurrence and the severity of the event should it occur, as previously described. Typically, one tool often used is called a "risk tolerability matrix," which visually displays a risk to determine if it is unacceptable or needs further mitigation to reach the ALARP or acceptable region.

LOSA, FOQA, and reporting systems were described in Chapter 2. In addition to the government reporting systems described, SMS also includes internal organizational safety reporting programs. These programs generally include anonymous reporting as well as self-identified reports. In order to encourage more reporting, American Airlines began a program that has since been adopted broadly into the US system called the Aviation Safety Action Program (ASAP).

ASAP is a program that includes participation of the operator (airline), FAA, and, if applicable, labor union. Each ASAP program is unique to the operator and must be approved by all parties. When a report is received it goes to a group called the Event Review Committee (ERC). The ERC then determines whether to accept the report, and that decision will be made based on whether the report is timely, if the content of the report is sole source (meaning that the report is the only way that the event would be known), and if there is criteria that would automatically exclude it, such as criminal activity. If it is "sole source," then the report is incorporated into the SMS for informational purposes. If it is not "sole source" the ERC makes the determination based on whether the event involved intentional noncompliance with regulators or procedures. If it did then it will generally be referred for enforcement action, but if not the ERC will determine if additional training is needed or to just accept the report with no further action. Regardless, if the report is accepted the reporter is given immunity to enforcement action by both the organization (company) and the FAA. The goal of the program is to encourage more safety reporting, and thus identify more potential hazards. Similar programs have been developed for air traffic control (ATSAP), dispatchers (DSAP), and maintenance (MSAP).

ICAO illustrates this risk tolerability as an inverted triangle as shown in Figure 14-5. Note the coding of the risk matrix.

- Dark gray—unacceptable risk (intolerable)
- White—acceptable with mitigation (tolerable)—i.e., Approaching ALARP
- Light gray—acceptable region

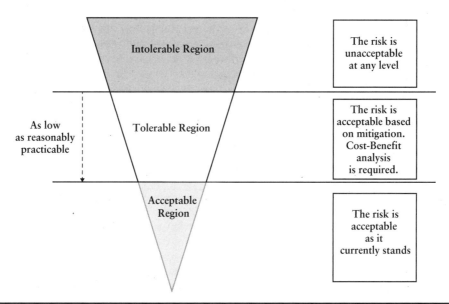

FIGURE 14-5 Safety risk management—tolerability matrix. (Source: Derived from ICAO Safety Management Manual[2])

To conduct a safety risk assessment, it is very helpful to plot it visually to clearly see the relationship between severity of consequences and likelihood of occurrence. The classic chart was initially developed by the US Department of Defense and is contained in the MIL-STD-882, which has been adopted in various forms across the aviation industry. An example of this chart used by FAA was provided in Figure 2-3, Chapter 2.

This Safety Management System processes that use such a risk matrix require the development of criteria that explain each classification of severity and likelihood appropriate for the organization's type of operations and their operational scenarios. For example, one organization operating smaller and less complex equipment may depict severity in terms of the dollar value of potential damage that is quite different from another organization utilizing very expensive equipment.

Such risk assessments are often a combination of professional experience and expert judgment, including input from data collection tools. As previously discussed, there are many problems with the risk matrix. A simple one is that the scales are not necessarily measured in a consistent way. Is "remote" twice as likely as "improbable" or just 20% more? Is the scale just an ordinal scale, where the hierarchy is just one is greater than another, or is it an interval or ratio scale[iii]? These factors make a difference and are almost never defined on the table.

Control strategies, techniques, and protective equipment impact the risk equation ($R = P \times S$) in the following ways:

- *Design and engineering* will aim to reduce/eliminate the hazard and works on the probability (P) side of the risk equation, which subsequently reduces/eliminates the severity (S).

- *Safety devices*, which can be active or passive, reduce risk in a similar way to *design and engineering* but are considered less effective. They can be used to reduce severity and/or probability. Active safety devices require human action and are not as effective as passive devices. Examples include seating heights and windshields that provide increased visibility for pilots or collision avoidance systems. Passive devices require no human action to activate. They do not perform active monitoring but are always present. Examples include guardrails and safety nets.

- *Warning devices* increase awareness of the hazards, and hence, reduce the probability of being impacted by hazards. Should the warning be ignored, the full extent of severity will be felt.

- *Procedures and training* are mainly intended to reduce probability (P) but can also reduce severity(ies) as in emergency evacuation drills.

- *Personal protective equipment* (PPE) serves to reduce the severity (S) from a materialized hazard. PPE should be considered the last line of defense, as it requires human input to be worn correctly, and even if worn correctly, it can be uncomfortable and cumbersome to wear.

[iii] An *ordinal* scale is just listing items in a particular order. Stating that *severe* is worse than *moderate* does not provide any information for *how much* different they are. An *interval* scale provides some numerical basis, but no information still not way to quantify how much more. For example, the Celsius scale is measured using an interval scale. 10 degrees Celsius is not twice as hot at 5 degrees, as the zero-point is not a true zero (negative numbers are possible). A *ratio* scale provides a true measure. If we start at zero we know that 50 is half as many as 100. Unfortunately, this is virtually never defined in the risk matrix.

To wrap up some of the most important concepts about safety risk management, let us use the example of a voluntary reporting system for aviation maintenance technicians performing routine maintenance to an Airbus A320. An employee reported that the shift change process does not ensure that a work task has proper continuity between technicians. A proper SMS will institutionalize a routine method for assessing the risk associated with such a hazard, would implement controls for the risk, and then would provide a means to measure the control in the future to assess adequacy.

14.3.3 Component 3: Safety Assurance

Once the risk assessment and controls are in place, and "ALARP" has been declared, the SMS process shifts to the safety assurance side of the ledger to provide feedback on how the safety risk management process is performing. By leveraging efficiencies gained in SMS data and combining cross-functional trends into a centralized safety database, safety professionals can analyze what is happening over time better and target their mitigations strategically. If performing well, the safety assurance process provides positive reinforcement that risks are properly managed. However, new hazards may inadvertently be introduced into an SMS whenever a change occurs. The safety information system provides the "change management" source of information about the state of safety for the SMS 1.

Safety assurance is the validation that the safety performance is meeting the goals set by the organization. Safety assurance consists of those processes and activities that are used to assess the SMS. There are several tools that are used for safety assurance, and, it turns out, these are very similar to the approaches used for quality assurance of a product. In this case, safety is the product that we are producing. Internal audits are often used to assess the effectiveness of the SMS and identify areas that might need improvement. Safety risk controls and mitigation techniques need to be examined and audited to ensure they are working as designed.

14.3.3.1 Safety Performance Indicators

It is tricky to measure how well SMS is working because health does not necessarily correlate to a reduction in the number of accidents. As a result, a key component of the SMS model that was adopted from the business realm is the use of performance indicators to assess the effectiveness of safety programs.

Under SMS these are called *safety performance indicators (SPIs). ICAO Annex 19* defines an SPI as "a data based safety parameter used for monitoring and assessing safety performance." It measures whether or not a system is operating in accordance with the goals of the safety program as opposed to simply meeting regulatory requirements. Using SPIs represent a shift from traditional data collection and analysis methods to the development of mechanisms that continuously monitor safety risks, detect emerging safety risks, and determine any necessary corrective actions. Safety performance monitoring is a vital aspect of the safety assurance component. SPIs are the basis for data collection. The SPIs should encompass a wide variety of indicators.[1]

Unfortunately, the *ICAO Safety Management Manual* only provides the basic definition and guidance on the use of SPIs with a few generic examples. Consequently, individual organizations must develop their own meaningful measurements. Traits that determine a well-crafted SPI can be referenced from the ICAO manual and are not included here. The challenge of creating SPIs lies in narrowing the focus of a specific measurement. Organizations should focus on selecting indicators that can impact future operations from the unwanted events that they are trying to eliminate.

Each SPI should be created for a different purpose, and the more critical the issue, the narrower the focus should be. Every organization has the flexibility to create its own SPIs and should do so according to its needs and operations, although some SPIs may be predetermined and required by a regulatory authority as a mandatory measurement. Flexibility is important because there are no predefined SPIs for specific aeronautical services. This characteristic enables organizations to develop the cornerstone for their performance-based oversight. Since ICAO's verbiage lacks advice on creating SPIs, ICAO and IATA are currently developing material to assist operators in this task.

To provide an SPI example, Figure 14-6 shows a commercial jet operating in the vicinity of thunderstorms, which of course could produce inflight turbulence. Such operations are common in commercial aviation and an associated SPIs chosen by the airline may depend on the culture and operational specifics for the organization. For example, one air carrier may choose to create an SPI that tracks the very few passenger injuries that occur due to turbulence, while another one may use an SPI that measures a wider number of incidents by tracking how many flight attendant injuries result from encounters with inflight turbulence. Another operator may prefer to create an even wider SPI that measures the number of unforecast turbulence encounters as reported by pilots, whereas another airline may prefer an SPI that measures an even larger population of events, such as moderate and severe turbulence encounters as measured through a flight data monitoring (FDM) program (see Chapter 15).

Creating SPIs paves the way for setting safety alert levels. Organizations should select *statistical process control (SPC)* limits of the maximum number of safety events that

Figure 14-6 A variety of SPIs could be created to manage the common problem of inflight turbulence. (Source: Wikimedia Commons)

can occur for them to still be operating within the acceptable range of safety. Statistical analysis of data from SPIs enables organizations to pick a quantifiable threshold for the alert zone. When the alert levels are reached, then there should be further investigations, or a causal analysis combined with a plan that uses SMS tools to bring performance back to an acceptable safety level.

Although alert levels are somewhat subjective, they can be very useful to help determine between significant and insignificant trends. As the aviation industry shifts to using SPIs, organizations will benefit from the conclusions they can derive from the metrics. This data will provide meaningful information that cannot be manipulated and that truly reflects operational trends. Additionally, the flexibility of customizing SPIs allows each organization to focus on its own operational context and challenges.

Commercial aviation can use numerous data acquisition approaches for monitoring and measuring safety performance as part of an SMS, which include *Continuous monitoring, Internal audits, Internal evaluation audits, External audits, Investigations, employee reporting systems, Analysis of data,* and *System assessments.*

14.3.3.2   *Audits and Inspections*
The mention of audits and inspections bears special discussion, because as noted above, a primary element of safety assurance is the need to continuously improve a company's SMS. Rather than maintain the safety status quo, the SMS process seeks continuous improvement using tools such as internal evaluations and independent audits to obtain timely feedback information. Each SMS should create and maintain a formal process for detecting the reasons for substandard performance of the SMS and also to determine how that substandard performance impacts operations, in hopes of either eliminating the reasons altogether or at least making them less damaging. Continuous improvement can be greatly enhanced by a strong safety culture and committed management officials. Proactive evaluation of facilities, equipment and procedures, and vigilance are a must for SMS effectiveness.

Internal audits should be conducted by people or departments that are independent of the functions that are being audited and provide feedback to the AE. ICAO outlines the following aspects that should be audited:

1. Regulatory compliance
2. Policies, processes, and procedural compliance
3. Effectiveness of risk controls
4. Effectiveness of SMS

If internal audits are not possible then external audits should be used to ensure that the people who are doing the audits are not influenced by the management they are auditing. Audits follow a fairly standard process as there are three basic aspects. The first is to determine if the required process or policy exists. This is accomplished by reviewing the documentation and determining if it meets the requirements. The second phase is to determine if the process or policy is actually being followed. Is it used in operations? Do users understand the process or procedure? Finally, it is necessary to assess whether the process or procedures are resulting in the outcome that is desired. The results of the audits become one of the inputs to the SRM. Identified noncompliance and lack of effectiveness are used to modify the SMS to make it more effective.

Continuous improvement of the SMS is a never-ending goal and several overall processes can be used toward that end:

- *Preventative/corrective action* assignments must be tracked and managed. This is an active process requiring routine follow-up and continuous assessment.

- *Top management review* should be regularly conducted by the SMS Accountable Executive and other safety boards and committees at all levels of management. This review should include the inputs and outputs of SRM and the lessons learned from the safety assurance process.

- *Comprehensive audits.* These are extensive, detailed safety inspections/surveys (site conditions and programs) of the entire organization (an air carrier), a facility within an organization (an air carrier hub or maintenance facility), or an operation within a facility (fuel handling at the ramp). These audits are usually conducted on an annual basis by a team of safety professionals from within and outside the organization, facility, or site. They are usually led by the company safety director or the divisional or site safety manager. This type of audit results in a formal written report of findings and recommendations and usually requires a response within an agreed time frame.

- *Self-audits.* These are informal daily, weekly, or monthly audits that resemble inspections and form the crux of the internal inspection program (IEP). The frequency of checks depends on the operation being assessed. For example, certain areas of airports have to be checked several times per day, whereas some operations in a hangar or on the ramp may only require one inspection each day. These audits are conducted by on-site line management at a given facility and usually result in a checklist of items that either are in acceptable condition or need fixing. No reports are normally generated as part of these audits. One of the functions of the Director of Safety is to help set up and provide expertise to local internal inspection teams.

- *Status audits.* These audits are intended to determine the status of compliance at the time of observation. These are usually subject-oriented audits and are conducted in areas of high risk and/or areas of known or perceived concern.

Evaluating a welding operation is an example of a status audit. These audits may generate reports if the findings are serious enough to require management's attention.

14.3.3.3   Change Management

Change management is a critical aspect of safety assurance. A formal process for managing change is necessary for an efficient airline SMS program. Every airline should create and maintain a formal process that identifies changes that are occurring within the company that may impact previously developed processes. The formal change management system should consider how critical the system is to the airline, and whether the change introduces new, unforeseen hazards or risks.

There are numerous instances from safety history that demonstrate how changes can create new hazards into operations. An example is the implementation of the incredible precision available in modern flight deck avionics instrumentation. Although precision is usually quite helpful to promoting safety, it is important to be aware of the assumptions in the current safety system design and make sure that any changes do not invalidate those assumptions. In 2011 Gol Airlines and an Embraer jet had a midair collision over Brazil. At the time the airspace over that region was not tightly controlled, using methods based mostly on time. The lack of precise altimetry provided a bit of extra assurance that aircraft would not collide if the timing was off. In this case it is likely that had the altimeter or navigation systems been less precise the accident may not have occurred. A mature risk mitigation process which is ostensibly under control may become unsafe when faced with unexpected changes to the system.

Similarly, new operational procedures or system changes need to be verified by the safety risk management process to see whether they have an impact on other processes. Changes can occur for a number of reasons. Growth, or contraction of the size of the organization can both create challenges. A growing company will have shortages in resources of various kinds. Challenges hiring more people can create stress on the remaining employees, potentially resulting in fatigue as they are having to work longer hours or contend with fewer days free of duty. Rapidly expanding companies have to acquire more aircraft, which may mean operating more flight hours on the existing fleet. Maintenance under such conditions can be a challenge as the aircraft spend less time on the ground. The expansion can lead to a business cutting corners trying to keep up with the growth. A contracting company can also be resource limited. Companies normally shrink due to financial hardships, which can also lead management to cut corners in an attempt to remain profitable.

Business improvements can also lead to unexpected challenges. In the mid-2000s, airlines adopted tablets for pilots to use for their charts and manuals. Tablets are lighter, can be updated easily and cost less than printing and shipping updated charts and manuals. The weight savings can be measured in lower fuel costs, and prior to tablet implementation there were pilots who missed work due to injuries maneuvering heavy chart and manual cases around the tight quarters on the flight deck, where chart cases needed to be moved around the seats and placed or lifted while the pilot twisted to reach the handles. On balance, it is hard not to see the tablet as an improvement, yet several problems have identified themselves.

As indicated, a key aspect of change management is being able to measure what effect we have had with actions taken to sustain or enhance safety. To determine whether change is occurring, and to detect whether it is happening to the benefit or detriment of risk management, safety performance monitoring, and measurement is a critical part of an efficient SMS safety assurance process. After the hazards are identified and the risks are assessed, properly mitigated, and controlled, operational performance must be monitored and measured in order to ensure optimum effectiveness of the SMS. This is an internal process of the SMS team, focusing on data and information regarding the performance of the safety system. Finally, the last element of Safety Assurance is the continuous improvement of the SMS operation. Quality feedback monitoring processes are strongly recommended by both ICAO and the FAA to enhance the entire operation of the system.

14.3.4   Component 4: Safety Promotion

An integral part of a strong SMS culture is safety promotion. If the attitude of employees is that the program is only for compliance with regulations, then regard for safety and the SMS will not permeate through all aspects of daily operations. For this reason, the final component of SMS calls for the organization to continuously promote safety as a core value of the organization.

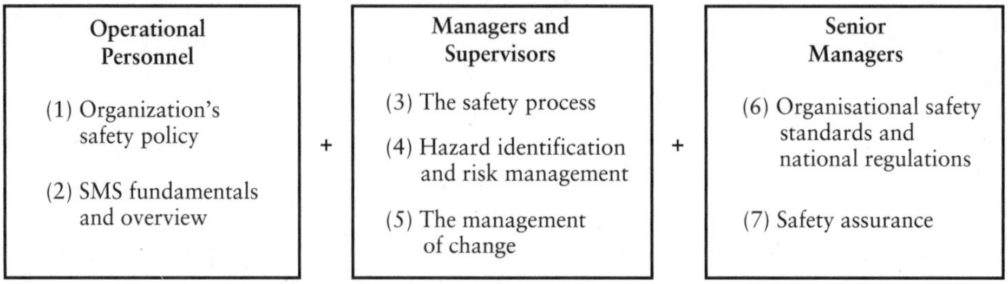

Operational Personnel		Managers and Supervisors		Senior Managers
(1) Organization's safety policy (2) SMS fundamentals and overview	+	(3) The safety process (4) Hazard identification and risk management (5) The management of change	+	(6) Organisational safety standards and national regulations (7) Safety assurance

FIGURE 14-7   Safety training and education fundamentals. (Source: Derived from ICAO)[2]

The building blocks of a good program are depicted in Figure 14-7 and training programs should be tailored to an employee's job function in the organization. The left-most block in the figure describes how operational personnel should understand the organization's safety policy and have at least a functional understanding of how SMS operates. Added to that, the center block in the figure portrays how managers and supervisors must also have a more in-depth understanding of the safety processes used by their organization to include hazards identification and change management to address the risk associated with such hazards. Lastly, the right-most block of the figure depicts how senior managers must go beyond SMS fundamentals to ensure that organizational standards and regulations are observed.

14.3.4.1   Safety Training and Education

Within the safety promotion component of SMS, there needs to be a formal and recurrent process for training employees in safety. This requires creating and maintaining a *safety training program* that ensures personnel are trained and competent to perform SMS duties. The scope of the safety training should be appropriate to each individual's role in the SMS, meaning that the training programs should be customized to the different roles played by employee groups in the organization. An aviation maintenance technician working in a hangar and a customer service agent staffing a check-in counter at the terminal will naturally have different impacts on the SMS of their airline and need to be trained on how their role impacts the safety of the operation. It is important to note that the AE should also receive special SMS training regarding roles and responsibilities, including an understanding of the SMS and its relationship to the organization's overall business strategy.

Safety training is most effective when it is integrated into a company's performance and job practice requirements.

14.3.4.2   Safety Communication

A large part of safety promotion involves the critical need for communication. A robust SMS culture encourages a learning environment where employees communicate openly up and down the management chain without fear of reprisal. Each organization should create and maintain a formal means for safety communication that ensures that all personnel are fully aware of the SMS, conveys information that is critical to safety, explains why particular safety actions are taken, and details why safety procedures are introduced or changed. This guidance means that employees should have easy access to their organization's SMS manual and associated safety process and procedures. Communication processes should also include airports and other providers of service to the airline.

A healthy SMS may also feature safety newsletters, notices, and bulletins, plus the use of Web sites or email to keep employees in the loop regarding safety issues and resolutions. Safety publications are an important component of a safety program. Most airlines publish regular internal safety documents to maintain high individual awareness of safety and risk management. Videos are also used to disseminate safety information to all employees. Communication and education are critical elements of any proactive safety program; there must be an effective mechanism in place to ensure the flow of critical safety information within the company.

In conjunction with the safety risk management process, the communication of pertinent information necessary for reduction of the risk must be provided to the safety user. Depending on the type of risk involved, the user may be a pilot, flight attendant, aviation maintenance technician, or related support person. Recurring material failure causes, such as fuel pump failures, would probably only be distributed to technical services or the maintenance department, while human factors issues would be best communicated to all employee groups for example.

Figure 14-8 shows a pair of aviation maintenance technicians working under the cowling of a Boeing 737-800 power plant. A safety manager viewing such an operation may contemplate what communication processes are used to relay important safety

FIGURE **14-8** Empowering technicians to report hazards without fear of reprisals is a key aspect of a healthy SMS. (Source: Wikimedia Commons)

information to them and whether the corporate culture he or she is helping to create is one that encourages such technicians to freely report the hazards that they encounter.

Communication takes on many forms in an SMS. The airline safety, maintenance, and flight operations organizations are often mistakenly viewed as separate entities with little or no shared mutual interests, when in actuality, the three organizations are closely aligned by complex relationships. While the maintenance department provides virtually all the technological expertise necessary to maintain the aircraft fleet, the pilots and flight attendants are its end-use customers and, therefore, must maintain a user's level of technical knowledge and related safety issues.

Another key link of the relationship is forged by the airworthiness concept. While the captain is responsible for ensuring the final airworthiness and safety of the aircraft, it is the maintenance department that maintains or returns an aircraft to an airworthy condition. A fundamental component of this link is the aircraft maintenance logbook (which can be in paper or electronic form), which serves to document the status, degradation, and restoration of the aircraft between variable levels of serviceability.

The third component of the flight–maintenance–safety relationship is the regulatory-procedural link. Both procedural and technical regulatory issues must be closely coordinated. While regulatory requirements emanate from the FAA, they are often precipitated by NTSB investigative findings, thereby necessitating direct communication between agencies for effective implementation of evolving safety requirements. Because of the highly complex nature of modern aircraft, a means to quickly analyze and disseminate critical FAA and NTSB safety information is required. In addition, conduct of both major and minor technically oriented investigations will require in-depth and thorough coordination between these agencies. This is also true with foreign regulatory agencies; however, the specific processes vary widely between nations.

14.4   How to Implement SMS: A Phased Approach

The classic implementation roadmap is set forth in a four-phase process in the *ICAO Safety Management Manual* and FAA documentation. The phases of implementation are generally arranged into four levels of "maturity" similar to that developed in the proven capability maturity model for software engineering at Carnegie-Mellon University in Pittsburgh, PA. Also note the similarity to Westrum's safety culture maturity levels discussed in Chapter 12. Figure 14-9 is an FAA depiction of this maturity model, which is further described below:

Each level of the SMS maturity model is explained as follows by the FAA:

1. *Level Zero*: Orientation and commitment is not so much a level as a status. It indicates that the Aviation Product/Service Provider has not started formal SMS development or implementation and includes the time period between an Aviation Product/Service Provider's first requests for information from the FAA on SMS implementation and when they commit to implementing an SMS.

2. *Level One*: planning and organization. Level One begins when an Aviation Product/Service Provider's Top Management commits to providing the resources necessary for full implementation of SMS throughout the organization. Two principal activities make up Level One:

   a. *Gap Analysis*: The first step in developing an SMS is for the organization to analyze its existing programs, systems, and activities with respect to the

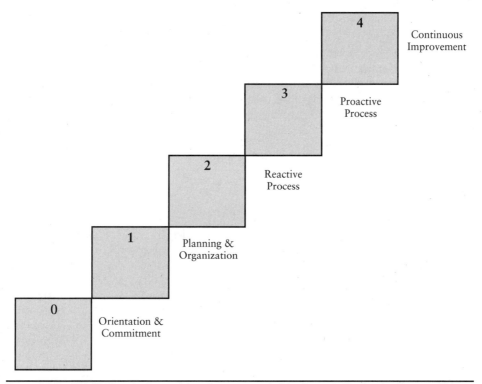

FIGURE 14-9   FAA SMS maturity model and recommended implementation basis. (Source: FAA Advisory Circular AC 120-92B)[5]

SMS functional expectations found in the SMS Framework. This analysis is a process and is called a "gap analysis," the "gaps" being those elements in the SMS Framework that are not already being performed by the Aviation Service Provider.

b. *Implementation Plan:* Once the gap analysis has been performed, an Implementation Plan is prepared. The Implementation Plan is simply a "road map" describing how the Aviation Service Provider intends to close the existing gaps by meeting the objectives and expectations in the SMS Framework.

3. *Level Two*: reactive process, basic risk management. At this level, the Aviation Service Provider develops and implements a basic safety risk management process. Information acquisition, processing, and analysis functions are implemented and a tracking system for risk control and corrective actions are established. At this phase, the Aviation Service Provider develops an awareness of hazards and responds with appropriate systematic application of preventative or corrective actions.

4. *Level Three*: proactive processes, looking ahead. At this level, the activities involved in the safety risk management process involve careful analysis of systems and tasks involved, identification of potential hazards in these functions, and development of risk controls. The risk management process

developed at Level Two is used to analyze, document, and track these activities. At this level it can be said that the organization has a full-up, functioning SMS.

5. *Level Four*: continuous improvement, continued assurance. The final level of SMS maturity is the continuous improvement level. Processes have been in place and their performance and effectiveness have been verified. The complete safety assurance processes, including continuous monitoring, and the remaining features of the other safety risk management and safety assurance processes are functioning. A major objective of a successful SMS is to attain and maintain this continuous improvement status for the life of the organization.[iv]

14.5   Future SMS Challenges

Today, the transformation of commercial aviation safety into the SMS framework is in full swing on a global basis. Using quality principles as a successful guide, SMS will continue to employ new methods with empowerment of employees and positive safety culture techniques to obtain safety levels never before achieved in commercial aviation.

This is not to say that there are no challenges ahead as SMS is integrated into daily operations. Stolzer and Goglia[6] talk about some of these difficulties in their book *Safety Management Systems in Aviation*, 2nd ed. A strong safety record is not necessarily an accurate predictor of future safety performance. Currently, the literature available about SMS leans more toward developing and implementing programs, and there is somewhat of a gap of information regarding how to measure its effectiveness. For example, early metrics for measuring effectiveness were more reactive and measured accidents and incidents after they occurred. That method did not lend way to preventing future problems from occurring. Strong evaluation programs, then, should be developed to focus on both strengths and weaknesses of the SMS program in an attempt to improve the system. Methods that are based on the evaluation of, and violations of, assumptions are more likely to identify the potential for risk. This will be discussed further in Chapter 16.

Currently, SMS effectiveness centers on the collection and aggregation of information from sources such as ASAP, FOQA, FAA surveillance, internal evaluation programs, and accident investigations. However, there is currently research assessing two new quantitative methods, both adopted from the business world, which could be used for future evaluation of SMS programs.

- The first quantitative method is data envelopment analysis (DEA), a powerful technique to evaluate and benchmark organizations. DEA relies on linear programming, which is frequently used in operations research to evaluate the effectiveness of certain operational segments of an organization. It does this by using a math algorithm to consider all the inputs and resources used by each segment of an organization and comparing the relationships of the variables involved. Stolzer and Goglia explain that DEA is particularly useful when comparing "best practices." The benefits of DEA would be that multiple inputs and outputs could be compared, and peer measurement could be ascertained.

[iv] www.faa.gov[1]

- The second quantitative method of SMS effectiveness is Input–Output (IO) which analyzes the interdependencies between the various branches of an economy. Stolzer and Goglia state in their book that IO demonstrates how parts of a system are affected by any change in the system. Thus, the IO process would be very useful in the safety assurance phase of change management.

Researchers are currently exploring these two methods for use on future SMS effectiveness studies. With the employment of these new DEA and IO tools, SMS effectiveness will potentially be more scientific and less equivocal in the future. Still, the problems previously stated with the subjectivity of metrics used need to be carefully evaluated.

First, to be successful, senior leadership must have a strong, visible commitment to the four SMS components, or the four "Pillars" upon which SMS is based. Executives must also adopt a more democratic style of management while maintaining that not all ideas will be accepted and implemented despite being appreciated and considered. Companies should ease employees into this system using the phased implementation process, thus allowing the organizational culture to adapt and evolve to these new concepts. Adopting this enlightened mindset as soon as possible will facilitate successful implementation and smooth SMS execution in the years ahead. There is a natural advantage here to smaller operators. Due to fewer numbers of people and smaller fleets, these carriers can be nimbler in responding to problems and needs. This, however, is counterbalanced by the fact that a major reason that commercial aviation has been so safe has been extreme conservatism in adopting changes. An ideal culture balances these two aspects.

14.6   Safety Data

Comprehensive hazard analysis requires using all available resources to identify potential hazards. Although an organization can collect a significant amount of information internally, the fact is that data rates are often too small to identify trends. In Section 14.3.2, outside data sources were listed as a potential source to identify hazards for this reason.

In passenger transportation, *safety factors* (some of which are *causal factors*) are events that are associated with or influence fatality rates. A *safety indicator* is a measurable safety factor. The *probability* of death or injury as a result of traveling in a given transportation mode, when quantified, is the primary benchmark of passenger safety. Car accident rates also are commonly used as safety indicators since most passenger fatalities occur as a result of terrestrial vehicle accidents and not aviation ones. If one were to look at risk as the probability of death, then the risk due to traveling can be established from fatality rates. However, commercial aviation fatality rates are poor indicators to show change in short-term risk since a large jet accident can potentially result in the deaths of a few hundred people. Hence, a single accident can greatly influence fatality rates. Therefore, the trending of fatality rates requires data from extended time periods (5 years or more). As an alternative process, accident rates can be used instead of fatality rates as indicators of safety levels. The number of fatalities, even for a specific type of accident, varies greatly with each crash. However, while the number of accidents may have a narrower range of annual variance than the number of fatalities, it still presents challenges for analyses. While the way that accidents are defined may mean that there are more numbers of accidents than fatalities, modern commercial aviation thankfully presents few accidents making for difficult statistical analyses. Also, the number of accidents in a given country, geographic area, or type of operation can vary significantly from one year to the next, thus resulting in accident rates also being poor indicators of short-term estimates of risk trends. This problem is exacerbated if an airline is attempting to identify trends using only their own data.

Before using accident statistics for safety analyses, it is necessary to understand the denominators of *exposure* used to calculate transportation accident or fatality rates. *Exposure data* can be thought of as an indication of the amount of opportunity an event has to occur. Cycles, distance, and time for passengers and vehicles are the principal exposure types. They are used in the denominator of rates, such as fatalities per passenger departure or electrical system failures per flight-hour. The choice of type of exposure data will affect how rates are compared between and within transportation modes. We must be careful to only compare rates that have the same units of exposure. For example, it would be highly confusing to compare fatalities per departure in one year to fatalities per flight hour in another year.

It is clear that the probability of an accident is significantly higher during takeoff or landing given the increased risk in such phases of flight over others such as cruise. Since the cruise phase is usually the longest part of any commercial flight, then most travel time occurs under a lower average risk. Such variation is not present during surface transportations, such as during the daily commute to one's office, although specific locations may pose more driving hazards than others. Aviation accidents can vary greatly in severity, ranging from a flight-attendant back injury from pushing carts during in-flight service or handling overhead luggage to several hundred deaths due to an aircraft crash. To render a more equitable perspective on aviation risk, some databases categorize accidents as fatal and nonfatal. However, even when using such a process, a single death on a ramp during an aircraft pushback would still be classified as an aviation accident even though the aircraft involved in the accident is untouched.

To address this issue, the International Civil Aviation Organization (ICAO) classifies accidents as *major, serious, injury, or damage* (see definitions in Chapter 3). Another classification popular with insurance and aircraft manufacturing industries is to refer to accidents as either an (aircraft) *hull loss* or nonhull loss. A hull loss is airplane damage that is beyond economic repair, which implies that very serious damage was incurred.

Accident counts by themselves cannot be reliably used to measure relative safety among different airlines or flight departments because there are just not enough accidents for statistics to be usable. Most airlines will, thankfully, not have a hull loss in any given year, making comparison based on such criteria meaningless. Small numbers of accidents, even if including all four accident categories used by organizations such as the NTSB, may produce a false picture due to highly unpredictable events, such as turbulence encounters.

Furthermore, all other considerations being equal, an airline that has a larger fleet of aircraft could be expected (statistically speaking) to have a larger number of accidents than an airline with fewer planes. Similarly, aircraft models that are flown more often would be expected to be involved in more accidents than less frequently used models. For this reason, the *accident rate*, which is the number of accidents divided by some common base variable (e.g., flight hours, departures, and miles flown), is a more valid indicator of relative safety than just accident counts. In this context, time (flight hours) is a popular measure of exposure in many types of risk analysis. Since flight-hour data is needed for economic, operational, and maintenance reasons, airlines keep accurate records of such data which are readily available for safety rate comparisons. Flight-hour data is also used for determining airworthiness criteria. Finally, rate data is used by various governmental and nongovernmental bodies to compare safety performances among airlines and other sectors of the air transportation industry.

Even when comparing rates versus number of accidents, context must be considered in order to perform a fair analysis. For example, weather is often a factor in accidents and some airlines operate primarily in relatively benign weather, such as is often found in the desert southwestern United States or Mediterranean Europe. While others operate in very challenging weather, such as in Canada, Alaska, Scotland, or Scandinavia, where large portions of the year require operating in frozen precipitation, contaminated runways, and low ceilings. Another example is the variation in navigation aids and airport conditions faced in different parts of the world. Airlines providing service within Europe may expect radar control and precision approaches to long paved runways, whereas some airlines in South America and Africa routinely operate in nonradar environments to dirt runways in the jungle. There are also variations between airlines in terms of safety support, financial support for training and operations, and regulatory oversight.

Aviation accidents are rare occurrences, and the risk of death or serious injury by air travel is miniscule. With such small numbers, making statistical inferences from the data to evaluate safety performance of the industry is very difficult. Regulatory agencies and the aviation industry maintain a wide variety of safety-related information. However, one must take into consideration the accuracy and completeness of these data when developing safety trends and recommending control measures. Safety factors other than fatalities or accidents that include the nature and causes of accidents (covered in Chapter 3) should be studied and used for preemptive strikes against safety hazards and to provide timely feedback to safety managers and policymakers.

In summary, no single measurement provides the complete safety picture. Passenger-exposure data are used when passenger risk is to be described. Passenger-miles are used when the influence of vehicle size or vehicle speed on the data is nonexistent. Departure-exposure data accounts for nonuniform risk over a trip. Time is the most widely used exposure measure, and time-based data are often readily available.

14.6.1 Aviation Accident and Safety Statistics

In Chapter 2 of this book we discussed various data reporting systems, both voluntary and mandatory, that are used in aviation. While ensuring compliance or consistency in safety reporting is difficult in general, voluntary systems have added complexity in that they are inconsistent and less reliable because they are subject to reporting bias and erroneous content. Such reports can still be highly valuable, of course. There are several reliable sources of accident data. Two of the most easily accessible accident databases are maintained by Airbus and Boeing, each publishes an annual statistical summary of commercial aviation accidents, the *A Statistical Analysis of Commercial Aviation Accidents* and *Statistical Summary of Commercial Jet Airplane Accidents* respectively. Another source is the Aviation Safety Network of the Flight Safety Foundation, found at http://aviation-safety.net/. Other databases include the International Air Transport Association (IATA), the Australian Air Transport Safety Board (ATSB), United Kingdom's Accident Investigation Branch (AIB), Transport Canada, and in the United States, the NTSB database and the FAA's Aviation Sharing Information Analysis System (ASIAS).

14.6.2   Industry Involvement with Safety Data

Safety professionals within the airframe and engine manufacturer (also called OEMs, or Original Equipment Manufacturer) community will always be found at accident sites participating in the investigation due to their unique, equipment-specific knowledge. They collect data and analyze them with their expertise and their extensive network of fellow experts on given systems back at their company. They recommend improvements in the way aircraft are designed, built, operated, and maintained. The goal for all who participate in accident investigations is to learn from the event, and therefore, prevent future accidents. Such prevention is best accomplished when the equipment experts are allowed to yield their unique insights as part of the investigative process.

While different manufacturers all participate in the investigative process when their respective equipment is involved, organizations are often organized differently from each other and, consequently, may also approach the investigative process differently. For example, accident investigation and safety data analysis are separate units at Airbus and Boeing, while at other manufacturers, they are combined and often linked to customer support in a single organization. The closest thing to a constant among manufacturers' safety departments is their responsibility for accident investigation. When a company's product is involved in an accident, the company has a duty to help find the causes. In the parlance of the NTSB, the company becomes a *party* to the investigation, as discussed in Chapter 4. That means the manufacturer's representatives work alongside NTSB investigators in examining evidence at the accident site. The parties conduct subsequent tests and suggest findings, but the final analysis and published report are produced by the NTSB. Safety departments with manufacturing companies scrutinize incidents as well as accidents. Committees are formed to represent various departments using SMS procedures.

Based on the events reviewed, the committees recommend design changes or revisions in maintenance or operating procedures. Accident reports produce a lot of data, but the lack of frequency of accidents compared to the frequency of incidents means that still more data come from incident reports and other report files compiled by airlines, manufacturers, and government agencies. The FAA, for example, records *Service Difficulty Reports* (SDRs). Manufacturers' safety departments have come to view these reports as a resource to be developed and cultivated. The data come from a variety of sources. Manufacturers' service representatives around the world report regularly on anomalous events both large and small.

Airlines often report issues directly to manufacturers for both prompt, remedial action input and overall tracking of discrepancies for any given part or system. Other data sources, as previously discussed, include civil authorities and international organizations such as the International Air Transport Association, the UK CAA, and US FAA, as well as insurance underwriters and publications. Even with all those collection efforts, safety staff do not claim that a record of every event unfailingly finds its way to the appropriate database. When little or no damage is involved, airlines sometimes make no external reports due to the very high tempo of technical operations and the never-ending list of repairs and inspections that demand constant attention.

While manufacturers of large airframes maintain what probably are the most extensive databases, other manufacturers also are collecting such information. Makers of engines and other components are also in the business of collecting statistical data. The purpose of gathering the data is trend analysis. Computers may be able to discern patterns that a human analyst may miss. All of the manufacturers have taken a proactive stance in the area of safety. Manufacturers can alert carriers to problems they did not know they had. Teams assembled by OEMs use accident records and other data to identify current safety issues. For example, Boeing was one of the leading proponents in stressing the importance of installing ground proximity warning system (GPWS) and the need for proper training in its use. Similarly, to prevent approach and landing accidents, the OEMs have urged that every runway used by commercial transports be equipped with an instrument landing system (ILS) and several training measures have been advocated for pilots and controllers to alleviate the dangers associated with unstable approaches.

The manufacturer's analysis will, of course, be reflective of their interests. For example, Boeing research addressed *crew-caused accidents*. Their analysis of their data and other sources found that flight crew error was the primary cause in close to 70% of commercial jet hull-loss accidents. Other manufacturers have similarly identified crews as the primary cause of accidents. As described in Chapter 13, accidents are much more complex than this but finding the crew as the primary cause certainly diverts attention away from the manufacturer.

Airlines maintain databases and conduct trend analyses of their own. These analyses can benefit from work done by the manufacturers. For example, American Airlines studied a series of Boeing 757 tail strikes. After assessing its own records, the carrier checked with Boeing about the experiences of other operators of the same type of aircraft. Once the problem was identified, American Airlines and Boeing worked to solve it by revising training methods.

The large airframe manufacturers also publish annual reports that are compilations of industry-wide accident data. It should also be noted that, for day-to-day problem-solving, airlines usually have more contact with a manufacturer's accident and incident investigators than with the analysts behind the data. In the discussions that follow, databases from Airbus, Boeing, the Flight Safety Foundation, FAA, and the NTSB will be explored to analyze aviation accident statistics. These databases are chosen as they are quite large. The Boeing data will be discussed first as the emphasis in the Airbus data is a bit different, with some overlap.

14.6.2.1 Boeing's Accident Statistical Summary

According to the Boeing's accident statistical summary, in the over 60 years worldwide history of scheduled commercial jet operations (1959–2020), there have been 2082 accidents resulting in 30,625 onboard fatalities and 1256 external fatalities. Of the 2082 accidents, 638 resulted in fatalities and 1025 resulted in hull losses.

If one were to consider accidents by type of operation, 80% (1666) of the 2082 accidents were passenger operations, 14% (292) were cargo operations, and the remaining 6% (124) occurred during testing, training, demonstration, or ferrying. US and Canadian operations collectively incurred about 29% (607) of the 2082 worldwide accidents, contributing to about 20% (6206) of the 30,625 global onboard fatalities. Over this period, there have been in excess of 854 million cumulative departures and more than 1638 million cumulative flight hours. The industry is now operating 27,012 jet aircraft, although departures dropped in 2020 to 19.2 million departures as a result of the COVID-19 pandemic.

If we examine a plot of all accidents for the worldwide commercial jet fleet for the period 1959–2020, as shown in Figure 14-10, we can see that the rates for all accidents and those involving hull losses appear fairly stable for the past 40 years.

Accident Rates and Onboard Fatalities per One Million Departures

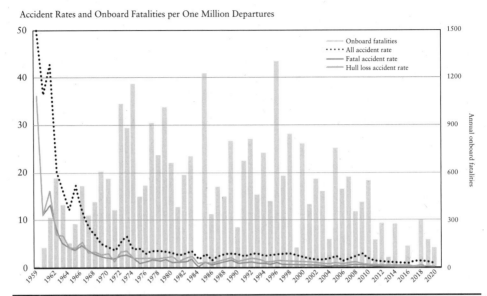

FIGURE 14-10 Accident rates and fatalities by year; all accidents, worldwide commercial jet fleet, 1959–2020. (Source: Boeing 2020 Statistical Summary, September, 2021)[7]

Closer inspection, though, reveals a steady and continuous decrease until reaching an asymptotic state in the past two decades. However, even if this low accident rate remains constant over the next few years, we can expect to see an increase in the actual number of hull-loss accidents each year as the fleet increases in number of departures.

Hull losses and fatal accidents were also analyzed according to the phase of flight in which they occurred (Figure 14-11). The combined final approach and landing phases accounted for 54% of the hull-loss and fatal accidents, followed by the combined phases from loading through initial climb (22%). Cruise, which accounts for about 57% of flight time in a 1.5-hour flight, occasioned only 13% of hull-loss accidents.

In past years and in the previous editions of this book, the summary also considered primary causal factors for commercial operation hull-loss accidents.

Classifications occurred for primary causes as attributed to flight crew, airplane, weather, maintenance, airport/ATC, or miscellaneous/other. However, Boeing has recently stopped reporting this data in adherence with emergence philosophies described earlier in this book about the risks of such monocausal assignments. Finally, fatalities by accident categories were covered for the period 2011–2020 (Figure 14-12). Loss of control in flight accounted for 694 onboard fatalities, 322 were listed as unknown, and CFIT accidents accounted for 229 fatalities, and 36 external fatalities during the period.

These three figures are representative of the whole report and of the even larger use of data and statistics for managing safety. As such, it is evident that their use helps prompt questions and analyses that shed light on different issues in safety and assist with evidence-based decision-making by government officials, airline managers, and manufacturers.

It should be noted that Boeing's accident data exclude turboprop aircraft; Soviet Union and Commonwealth of Independent States accidents; and accidents resulting from sabotage, hijacking, suicide, and military action. Although it is not an exhaustive

Percentage of fatal accidents and onboard fatalities | 2011 through 2020

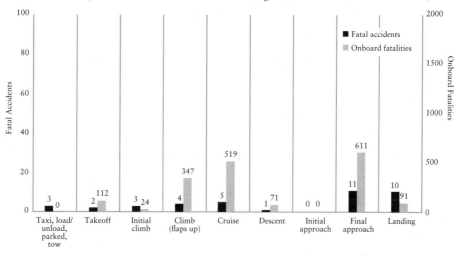

	Taxi, load/ unload, parked, tow	Takeoff	Initial climb	Climb (flaps up)	Cruise	Descent	Initial approach	Final approach	Landing
			13%						54%
Fatal accidents	8%	5%	8%	10%	13%	3%	0%	28%	25%
Onboard fatalities	0%	6%	1%	20%	29%	4%	0%	34%	5%
		8%							40%
Exposure (percentage of flight time estimated for a 1.5-hour flight)		1%	1%	14%	57%	Initial approach fix 11%	Final approach fix 12%	3%	1%

Note: Percentages may not sum to 100% because of numerical rounding.

Distribution of fatal accidents and onboard fatalities | 2011 through 2020

FIGURE 14-11 Accidents and onboard fatalities by phase of flight; hull-loss and/or fatal accidents, worldwide commercial jet fleet, 1992–2015. (Source: Boeing 2020 Statistical Summary, September 2021)[6]

compilation of data, and therefore only offers a partial statistical analysis, the content still proves highly useful for determining safety initiatives.

14.6.2.2   Airbus' Accident Statistical Summary

The Airbus 2022 statistical summary includes all Western-built air transport jets that carry over 40 passengers, and the Sukhoi Superjet from 1958–2021. The Superjet is an advanced fly-by-wire aircraft that includes many technological advances. The Airbus analysis has an emphasis on the influence of technology on safety data, so, unlike the Boeing data, includes an analysis of the technological level of the aircraft associated with accident trends and categories.

Airbus notes that, although the number of flights increased by 4 million globally in 2021 as compared to 2020, that number is still 40% below prepandemic levels. It is worth noting that the accident rate decreased from 2020 to 2021 even though there

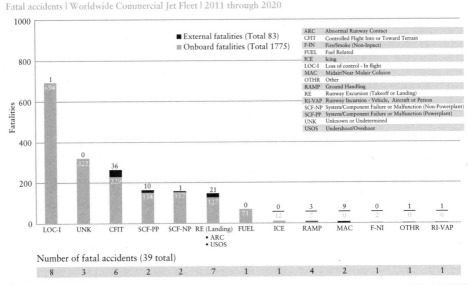

Fatal accidents | Worldwide Commercial Jet Fleet | 2011 through 2020

FIGURE 14-12 Fatalities by accident categories, fatal accidents, worldwide commercial jet fleet, 2011–2020. (Source: Boeing 2020 Statistical Summary, September 2021)[6]

was an increase in the number of flights. Still, with the number of flights decreased significantly from prepandemic levels it is not possible to determine if the reduction in accidents was the result of improvements or other factors. Looking at the accident rates since 1958, it is clear that the number of fatal and hull loss accidents has decreased, as were also reflected in the Boeing data, however, Airbus then examines accident rates in relation to the generation of technology employed in the design.

Airbus classifies commercial aircraft into four generations. First generation characterizes the early commercial jet aircraft designed starting in around 1952. These aircraft have many dials and gauges and somewhat simple autopilot systems, generally with just a few basic functions and no flight management systems. Aircraft in this category would include (but not be limited to) the Boeing-707, 720, the Douglas DC-8 (Figure 14-13), the Convair 880/990, the Comet, and Caravelle.

The second generation appears similar to the first but includes much more sophisticated autopilot and autothrottle systems. Many of these aircraft had the first Autoland capabilities to allow them to land in very low visibility. Aircraft in this category would include (but not be limited to) the first Airbus A300, later models of the Boeing 727, 747-100/-200, DC-10 (see Figure 14-14), L-1011, the Mercure, and Bae146.

The third generation added the first of the electronic flight instrument systems (EFIS), as well as flight management systems (FMS). A classic example is the Airbus A300-600 shown in Figure 14-15, but others include the later B-737 series, the Boeing 717, 757, 767, 747-400/800, and the McDonnel Douglas MD-11.

The third-generation category is quite wide as it accommodates a broad range of aircraft in terms of technology. Airbus classifies these aircraft as being equipped with EFIS and FMS, but some of these aircraft, such as the MD-11 flight deck in Figure 14-16, include features that are not present on many newer designs, such as highly automated aircraft systems. The reason that these are classified as third generation is not due to

FIGURE **14-13**   A Douglas DC-8 flight deck. (Source: NASA)

FIGURE **14-14**   DC-10 flight deck. (Source: Wikicommons, by Piergiuliano Chesi[v])

---

[v] https://commons.wikimedia.org/wiki/File:DC-10_Cockpit.jpg

FIGURE 14-15   Airbus "third-generation" A300 flight deck. (Source: Airbus, 2022)

FIGURE 14-16   MD-11 Flight deck. (Source: Authors)

the sophistication of some of the other systems but due to limiting the definition to the factors that relate to the factors of flight guidance, flight management, and flight control systems.

Airbus classifies fourth-generation aircraft as those with fly-by-wire (FBW) flight control systems. These include the Airbus A320 family, A330, A340, A350, A380, Boeing B-777 and B-787, the Embraer E-Jets, and the Sukhoi Superjet. The flight deck on the Boeing B-777 does not look significantly different than the third generation, as can be seen in Figure 14-17, but under the surface the flight control system uses computer technology to control the aircraft in a manner very similar to Airbus. The FBW Airbus aircraft uses sidesticks as opposed to a conventional control column, as can be seen

FIGURE 14-17   Boeing B-777 flight deck. (Source: Authors)

in Figure 14-18, although it is important to point out that both types of inceptors provide commands to the flight control computers and are not connected to the flight controls even indirectly through a cable-hydraulic system. The addition of FBW enables the designers to constrain the aircraft within safe parameters as determined by the engineers. In the case of Airbus, the FBW system will automatically attempt to prevent the aircraft exceeding extreme bank or pitch angles as well as overspeed and stall

FIGURE 14-18   Airbus A350 flight deck. (Source: Wiki Commons, unedited, by Joao Carlos Medau[vi])

vi https://commons.wikimedia.org/wiki/File:Airbus_A-350_XWB_F-WWYB_cockpit_view.jpg

conditions and will not allow the pilot to intentionally do so. Boeing aircraft will also attempt to prevent the excursions through making exceeding extreme values difficult for the pilot to exceed.

According to Airbus, 75% of the fourth-generation commercial jet aircraft flights were flown by Airbus aircraft. By 2021 there were virtually no first-generation aircraft still flying, only a small number of second-generation aircraft, and quite a few third and fourth generation. By 2021, 54% of the commercial jet fleet operating were fourth-generation jets.

Airbus then examines the relationship between technological advances and accident rates. As depicted in Figure 14-19, the accident rates are significantly lower for second generation as opposed to first, and much lower again for third and again fourth-generation aircraft. This should not be surprising as the increase in technology at each stage mitigated broad spectrums of accident categories. According to Airbus, there was an 86% reduction in CFIT accidents from the second generation to the third, and LOC-I decreased 89% from third to fourth generation. CFIT has been mitigated by Terrain Awareness Warning Systems (TAWS) and the design of the approach and air traffic procedures, plus improvements in air traffic technology. Runway incursion accidents have also been greatly mitigated through improved technology both on aircraft (Runway Awareness Advisory Systems, RAAS, and similar). Loss of Control In-Flight (LOC-I) now remains as the highest category for fatal accidents worldwide.

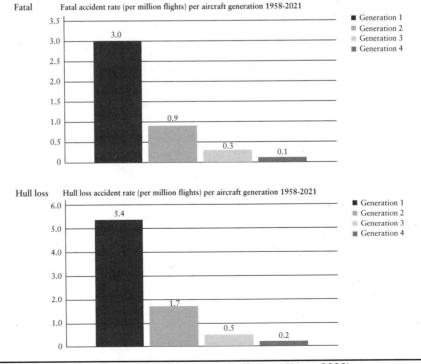

FIGURE 14-19   Accident rates by generation of aircraft. (Source: Airbus, 2022)

14.7 Other Sources of Safety Data

Although flight safety and accident statistics are available from other countries, due to the size of the aviation industry, the United States can provide useful metrics for research, particularly for general aviation. Accident data sets are available for US civil aviation from 1962 through 1981 and 1981 through the present. In addition, NTSB provides a monthly list of aviation accidents. The NTSB also includes a statistical reviews page that covers annual statistical reviews of US civil aviation accidents by calendar year. There are also safety research reports that examine risk and hazards, techniques and methods of accident investigation and the effectiveness of safety countermeasures.

NTSB's annual summary parses the categories by US Title 14 CFR part 121 Air Carriers, part 135 Commuter and On-Demand Carriers and General Aviation. Part 121 is further divided into accidents, accident rates, and severity as well as the "defining event," flight hours, departures, and passenger enplanements, all for a moving 10-year period which lags two to three years behind the current calendar year.

Because of the NTSB's very broad definition of an accident, the United States experiences an average of close to 36 accidents involving scheduled and nonscheduled air service each year. However, serious accidents, those involving fatalities, are much rarer (Figure 14-20). The NTSB's official compilation of accident statistics from 2010 to 2019 for Part 121 scheduled and nonscheduled airline service provides excellent insights into the state of modern commercial aviation safety in the United States.

Interestingly, leading up through 2019, so prepandemic values, the accident rate was increasing slightly, as shown in Figure 14-21. Analyzing the values with the associated spreadsheet there was an average of 0.32 accidents per 100,000 departures and 0.17 accidents per 100,000 flight hours.

Another excellent source from the United States is the FAA's Aviation Safety Information Analysis System (ASIAS) database. This provides not just statistical information but also a searchable record of individual Aviation Safety Reporting System

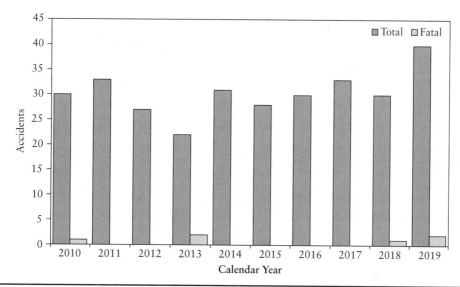

FIGURE 14-20 FAA FAR Part 121 accidents, 2010–2019. (Source: NTSB)

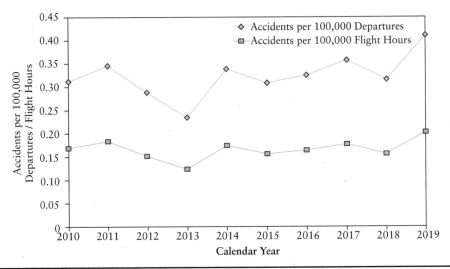

FIGURE **14-21** Accident rates. (Source: NTSB)

(ASRS) reports that can be used to identify potential safety trends. From the asias.faa. gov website the user can access a number of databases that are maintained by the US government. These include:

- FAA Accident and Incident Data Systems
- Air Registry, which includes information on all aircraft registered in the United States
- NASA ASRS database
- Bureau of Transportation Statistics, which includes traffic and capacity statistics on individual air carriers
- FAA New Midair Collision System database
- NTSB Aviation Accident and Incident Data System
- NTSB recommendations with the FAA and responses
- Runway Incursion database
- Preliminary Unmanned Aircraft System accidents and incidents
- World Aircraft Accident Summary, produced on behalf of the British Civil Aviation Authority

Government organizations tasked with the oversight of work-related statistics and injuries also maintain databases. One example is the US Bureau of Labor Statistics (BLS), which tracks workplace illness and injuries. Similarly, the North American Industry Classification System (NAICS) records and reports work-related injuries and illnesses across various industries and categorized by incident, industry, geography, occupation, and other characteristics.

Additionally, there are a number of statistical analyses performed by the Airlines for America (A4A), which is an American trade association and lobbying group that represents the largest airlines and which is based in Washington DC. Their analysis draws on other sources of data, such as those from the National Safety Council, the American Public Transit Association, the Federal Railroad Administration, and the NTSB and provides an excellent source for aviation research.

Every year ICAO publishes a readily accessible and very detailed safety report that depicts accident statistics and concomitant analyses. IATA, a trade association for the world's airlines comprising 268 airlines from 117 countries, represents a different way of measuring the same aspects of safety as those of ICAO. Both approaches show continued improvement and that commercial aviation is overall a very safe mode of transport.

14.8   Caveat Dealing with Safety Statistics

Statistics are often the basis for making a safety case, but one must be careful when presented with them to not just accept them at face value. We will start with a simple example. A cursory internet search will show that most automobile accidents occur close to the driver's residence, generally within 10 km (6 miles), although some variations of the values are found. The reasons generally given for this almost always attribute the primary factor to being that drivers become more complacent due to being comfortable with their surroundings. What is missed here is the obvious: the majority of driving for most people is within close proximity of their home. Based on that, it should not be surprising that most nondriving injuries to people that are not work related are also occurring near a person's home. The exposure to the risk is what makes the difference.

Along with this, an examination of safety by mode is of interest. Flying is often stated to be the safest mode of transportation, but it is worthwhile delving a bit deeper into numbers to understand the potential risk of just accepting these numbers at face value. The problem is how we measure the risk, which, when comparing modes of transportation, is actually an exposure to potential hazards. Exposure can be quantified as exposure per distance (miles or kilometers flown), time or on a per-trip basis. The choice of the metric matters. Here is an example.

According to the US bureau of transportation statistics for US carriers, from 1980 through 2020, on average there have been:

- 0.36 total accidents per 100 million aircraft statute miles[vii]
- 0.31 total accidents for every 100,000 departures
- 0.16 fatal accidents for every 100,000 flight hours

Other problems occur when viewing trends. Random variation, with no other factors, can result in some normal variation. Over time these values will average out to the mean, in statistics this is called *regression to the mean*. This variation can present a problem if a person does not understand this phenomenon. The intentions may be entirely good, or not, but either way, the presentation of the data may result in the appearance of a trend when there is none. Other factors need to be examined to ensure that any changes are due to real factors and not just random variation.

This leads to a discussion of statistical assumptions. Many types of statistical analysis are possible. Most methods are used to find out what the probability of a certain

---

[vii] 1 statute mile is equal to 1.61 kilometer.

outcome or finding is as compared to random variation. We want to make decisions based on whether a factor was real. This is called *inferential* statistics. In order to be able to make predictions on a set of data, certain things need to be verified and assumed to be true about the data set. Although the assumptions needed vary a bit based on the statistical test used, most of the assumptions are to ensure that the data is a good sample. If we are trying to measure something and our measurements are wrong that will obviously result in a useless answer. Although the explanation of statistical assumptions is outside the scope of this book, the reader should always consider this factor prior to assuming the accuracy of inferential statistics that may be presented.

Another problem comes from what may appear to be a simple analysis of data. The human mind is often not able to determine probability and risk. This can be a major problem with the use of a risk matrix, as described earlier in this chapter. Probabilities might not be what they seem. Take for example the problem of unstable approaches. Regulatory agencies and safety organizations have identified an unstable approach as a primary factor in landing accidents. In fact, it is known that 66% of landing accidents list an unstable approach as a causal factor, but is that a fair measurement against other risk factors?

Another problem can occur when dividing data sets. Assume that we were to examine the safety records of airlines regarding installation of a Terrain Awareness Warning System (TAWS). Not all airlines have installed the system. On examination of the data we are surprised to find that when we examine the airlines that have installed the system they are having more incidents involving the terrain than those that have not installed it! Does this mean that the system is adding risk? In actuality this is something called Simpson's Paradox. Data can be skewed when we separate it into groups, depending on what those differences might be. In our example it is likely that the companies that installed the TAWS were also operating into regions where there was higher risk of terrain. This can be calculated using Bayes Theorem, as previously described.

Finally, a word about the use of averages. Sometimes a practitioner may make a decision based on the belief that the average value is adequate and will provide protection. There is a fallacy called "the flaw of averages," and the simplest way to understand this is an analogy. Suppose a river has an average depth of 3 feet. Is it safe to walk across? Although the average may well be 3 feet, a river likely has some very shallow regions and then some that are quite deep. The fact that they average out to 3 feet will not provide protection from drowning!

14.9   Conclusion

Despite aviation being one of the safest modes of transportation in the world, accidents still happen. To counter the accidents that do occur, over the past several decades, our industry has adopted new mindsets and practices to make the aviation environment a safer place. One important way we have done this is by requiring the adoption of SMS both on the international and domestic level.

Although safety management as a practice has existed since the birth of aviation, SMS is a formal and standardized framework for managing a safety program based on accepted business and scientific processes. Although it may appear that most safety efforts should be concentrated on the flight deck, this is only a result of the belief that human error, or more commonly, pilot error, is the primary cause of accidents. The reality is that SMS starts deep in the back offices of air service providers where executives and safety professionals, often teamed with union safety representatives, work to adopt the tenets of SMS. The same occurs in aviation venues ranging from airports, maintenance offices, air traffic control towers, and all other aviation organizations. Each of those parts in the commercial aviation safety framework must have a robust and continuously improving SMS as their lifeblood (Figure 14-22).

The four components of SMS, as developed by ICAO and FAA, and adopted by many throughout the world, provide a systematic approach for achieving new levels of safety that were originally unattainable. Simply stated, *Safety Policy* means *document it in writing*, referring to the need to have key aspects of a given SMS readily available to employees as a written reference. *Safety Risk Management* is the means to *hunt the hazards, then assess and mitigate the risks*, referring to the fundamental need of having employees aggressively seek safety problems and having a means of effecting change to produce better conditions.

Safety Assurance is the way to *measure and improve it*, referring to the need to use scientific principles to determine the value of the actual management processes being used. Lastly, but affecting all other components, *Safety Promotion* is the *learning and sharing* of everything in all the components.

Our measurements of safety more often are founded on statistics. In his autobiography, Mark Twain references the following quote: "there are three kinds of lies: lies, damned lies and statistics." This saying goes on to show how people often use numbers to strengthen their arguments at the risk of misrepresenting meaning if not used carefully. In aviation, statistics are used to measure safety and to provide evidence to justify safety recommendations. The range of safety issues in aviation includes minor sprains and injuries all the way to major catastrophic accidents. Therefore, we cannot rely solely on one calculation to accurately paint the picture of the safety landscape. Instead, we must put everything together to get a holistic view of the safety scene. It is also important to scrutinize those numbers and understand the basis for them.

Figure 14-22 Pilots are but one of the many links in the commercial aviation safety value chain. (Source: Wikimedia Commons)

Although the road to SMS has been long and sometimes difficult, for students of SMS the future is very bright indeed. The next few years will bring increased international collaboration and employment opportunities. Aviation safety professionals will find that SMS is a journey of continuous improvement, not a final destination.

Key Terms

Accident Rate

Accountable Executive

Air Traffic Safety Action Program (ATSAP)

Analysis of Data

As Low As Reasonably Practicable (ALARP)

Aviation Safety Action Program (ASAP)

Aviation Safety Information Analysis and Sharing (ASIAS)

Bureau of Labor Statistics (BLS)

Continuous Monitoring

Crew-Caused Accidents

Design and Engineering

Emergency Response Plan (ERP)

Employee Reporting System

Event Review Committee (ERC)

Exposure Data

External Audits

Gap Analysis

Hazard

Hull Loss

Human Era

ICAO Annex 19

Implementation Plan

Internal Audits

Internal Evaluation Audits

Investigations

Major, Serious, Injury, or Damage Accident

North American Industry Classification System (NAICS)

Organizational Era

Personal Protective Equipment

Probability

Probability

Procedures and Training

Risk

Safety Action Group (SAG)

Safety Assurance

Safety Devices

Safety Factors

Safety Indicator

Safety Management Systems (SMS)

Safety Performance Indicator (SPI)

Safety Policy

Safety Promotion

Safety Review Board (SRB)

Safety Risk Management

Safety Training Program

Service Difficulty Reports

Severity

System Assessment

Technical Era

The Four Components of SMS

Warning Devices

Topics for Discussion

1. Discuss the evolution of SMS principles.
2. List and discuss the salient features of the Four Components of SMS.
3. Explain how safety assurance relates to safety risk management.
4. What is the concept of ALARP and why is it significant?
5. Compare and contrast the differences between hazards and risks.
6. Provide an example of why change management is such a critical aspect of SMS.
7. Discuss which SMS component you think is the most important. Back up your answer with a specific example.
8. Select a sample and very specific hazard for a commercial aviation operator.
9. Assess the risk presented by the hazard and recommend several controls.
10. Why is communication important to SMS?
11. Describe the steps of the SMS maturity model used for SMS implementation.
12. Discuss the future of Safety Management Systems in aviation for today's aviation safety professionals.
13. Discuss some of the issues associated with analyzing and comparing commercial aviation accident statistics
14. Why is it important to open and sustain a dialogue about safety statistics among all players in the aviation industry?
15. How does trend analysis contribute to analyzing safety?

References

1. FAA (n.d.). *Safety Management System.* https://www.faa.gov/about/initiatives/sms/explained/basis
2. ICAO (2018). *ICAO Safety Management Manual.* Doc 9859 https://www.icao.int/safety/safetymanagement/pages/guidancematerial.aspx
3. FAA (n.d.). *Safety Management System.* https://www.faa.gov/about/initiatives/sms/explained/components
4. Leveson, N. G. (2012). *Engineering a Safer World: Systems Thinking Applied to Safety (Engineering Systems).* The MIT Press. Cambridge, MA.
5. FAA (2015). AC 120-92B - Safety Management Systems for Aviation Service Providers. https://www.faa.gov/documentLibrary/media/Advisory_Circular/AC_120-92B.pdf
6. Stolzer, A. J., & Goglia, J. J. (2016). Safety management systems in aviation. Routledge.
7. Boeing (2021). *Boeing 2020 Statistical Summary, September, 2021.* https://www.boeing.com/resources/boeingdotcom/company/about_bca/pdf/statsum.pdf

CHAPTER 15

Proactive System Safety

Introduction

The goal of safety management systems is ultimately to identify hazards and mitigate them before they lead to an accident. Chapter 14 reviewed safety management systems. This chapter will explore some of the ways that the industry is proactively mitigating hazards and outline some new possible approaches to mitigation.

There have been several approaches to mitigating hazards, and, as outlined in Chapter 13, there is a divide based on whether the practitioner is applying methods based on *reliability theory* or *system theory*. Further, the social sciences approach focuses on a host of human factors versus those that specialize in engineering and design. As those that are applying reliability theory conclude that the human is the weakest (least reliable) component in a system, so it should not be surprising that they then turn to those in the social sciences—human factors and cognitive psychology—to develop ways to mitigate the "flaws" in that weakest element. This, of course, increases the demand for those in social sciences, and, as the social science field does not consider either reliability or system theory, a self-reinforcing feedback loop has become prominent between those that base their hazard analysis on reliability theory and those on the social sciences, each adding credence to the other and pushing out the competing system theorists. Unfortunately, the effort to engineer humans out of a system becomes dominant and so eventually the human factors researcher will have little impact on the design of new aircraft.

Social science practitioners delving deeply into understanding human cognition and behavior are actually critical in the effort to improve safety in the system theory model as well. In system theory the human rather being a component is the ultimate controller of the system (aircraft in the case of a pilot) and has a vital role in enforcing constraints on the aircraft systems and any software so that the entire aircraft stays within the boundaries of a safe state. In order to accomplish this, it is vital that human have the requisite skills, controls, and feedback to make this possible. Any research that can improve the capabilities of the pilot (in this case) is vitally important in improving their ability to make a flight safer.

As can be seen, regardless of the theoretical basis then, the role of understanding the human factor is critical. In the case of reliability theory, the goal becomes ensuring humans are trained and conditioned to behave in a reliable way and *proceduralization* is emphasized. In the case of system theory, the goal is to enable humans to have the capabilities both physically and cognitively to have enough control over the system to maintain it in a safe state. Some of the most interesting work in this regard is in the field of cognitive systems engineering.

15.1 Managing Ultrasafe Systems

One of the outputs of commercial aviation being an ultrasafe system is the distorted view sometimes offered by statistics. Accident rates are so low that one accident can entirely skew the apparent safety trends. With a lot of actual accident data, it is not difficult to identify areas that need to be improved. This was done in areas such as mechanical failures. Here the fix was improving component reliability and the introduction of engineering methods such as Failure Modes Effects and Criticality Analysis (FMECA), where the reliability of components could be analyzed, and improvements made. Another example might be what is termed Controlled Flight into Terrain (CFIT), where changes in procedure design and the addition of tools such as Enhanced Ground Proximity Warning Systems (EGPWS), plus warning systems in air traffic control software, have resulted in a significant decrease in this category in the past few decades. Mitigating these types of accidents required information to identify the problems. Today's accidents are more often the result of a complex combination of factors, many seemingly unrelated. Making improvements on a system where there is scant data to rely upon has become one of the most vexing challenges of our time.

Amalberti,[1] stating that safety is founded on the elimination of both technical failures and human error, outlines several factors involved in the ultrasafe systems of today. The first looks at high reliability—as the reliability of systems has increased, the focus has turned more and more to the elimination of human error. Unfortunately, automation has focused on putting people in a safe electronic bubble to prevent error, and it has been the implementation of this sort of protective feature that has contributed to accidents today. Amalberti points out that in less safe systems it is very possible for a safety manager to see tangible results, and thus receive credit for making safety improvements. In ultrasafe systems, any improvement is difficult or impossible to detect. As the accidents that do occur are a result of a combination of factors rather than a serious error or mechanical breakdown, traditional safety reporting systems become less relevant. With the low rates and lack of ability to see longer term trends, the short-term results become favored over tangible improvements.

Lofquist[2] similarly argues that measuring safety improvements when there are almost no accidents is problematic. As stated by Reason, we measure safety by the absence of bad events. We know when safety is absent but are much less certain about when it is present, aside from just knowing that it is not absent. There is, therefore, no way to easily identify those times when the safety of a system has degraded, and some approaches that, on the surface seem common sense will not work. These "common sense" approaches can often migrate into safety myths.

In this section we will explore how such systems have been studied from the perspective of the exploration of how successful organizations have achieved these goals; we will also study what facets are contained in resilient organizations.

15.1.1 High Reliability Organizations (HRO)

Le Coze[3] outlined the history of HROs, explaining that the concept originated out of the social sciences, beginning with research by La Porte in the early 1980s on understanding how some organizations can be almost free from errors. This was in response to a premise promoted by Perrow,[4] who argued that accidents in very complex, high-risk systems are inevitable due to being *tightly coupled*, meaning that any change in one component will affect how the other components behave. In Perrow's view, as the accidents in these complex organizations start in small, and seemingly trivial, ways it is not possible to fully prevent them. Perrow argued that any attempt to mitigate the problems that result in accident just adds more complexity, which, in turn, exposes the organization to more accidents. Yet, HROs seem to somehow counter this. The early research groups focused on organizations and systems that they deemed at very high risk yet had a much lower accident rate than might be expected.[5] They researched US aircraft carriers, the US Federal Aviation Administration's Air Traffic Control system (ATC), and Pacific Gas and Electric's Diablo Nuclear Canyon powerplant. Researchers have also looked at activities such as firefighting.

The essential premise of HRO is that, through studying and understanding how an organization in a high-risk domain can become, and maintain, a very low error state, those concepts can be applied to other organizations in other contexts. More recently, researchers in HRO such as Weick and Sutcliffe[6] have promoted the concept that what creates an HRO is a concept of organizational "mindfulness," where managers in an organization are instructed to be mindful of small aspects that may escalate, terming these "weak signals." Weick and Sutcliffe describe five attributes that all of the HROs possess:

1. Preoccupation with failure. The HRO will ensure that they identify and fix errors.

2. Reluctance to simplify findings. The HRO will not stop at the simple explanation but continue to investigate to identify the true root cause.

3. Sensitivity to operations. The HRO monitors the safety and security barriers that they have put in place to make sure that they are working as expected.

4. Commitment to resilience. The HROs ensure that they have the capability to recover from unexpected problems.

5. Deference to expertise. Rather than just following the chain of command, an HRO will ensure that problems are delegated to those with actual expertise. This can include identifying and using subject matter experts rather than limiting to safety professionals.

The critique of the HRO concept follows several themes.[7] The first is that the definition of an HRO can be generally stated as an organization that could, potentially, have many thousands of accidents, but somehow averts them.[8] By this definition, most organizations are HROs. The second is that the types of organizations chosen for the foundation do not meet Perrow's definition of a complex organization. In the cases of aircraft carriers (which actually have a fairly large number of accidents), ATC and the nuclear powerplant in question, the operations are carefully planned, decoupled, and organized. Although they are risky, the nature of the risk is quite certain and so, manageable. Yet another critique is that advocating concepts such as just paying attention to weak signals is too vague to be of any practical use.

Marais et al point out that a more promising path forward is the systems approach, as advocated by Woods, Hollnagel, Leveson, and Rasmussen. This approach will be reviewed in the following sections.

15.1.2 Adaptability

Accidents almost always occur after changes have been made.[9] Even in a simple system this has been demonstrated, such as the fact that most aircraft engine failures occur during or immediately following a change in power or a change in mode that the software might be in. The same is certainly true for a larger system, where changes can have unexpected consequences. A new type of aircraft or route for an airline will result in new problems that have not been managed before. Changes to the operating environment can similarly impact the way the system operates, resulting in pushing the operators and the equipment in ways that can result in accidents, particularly in combination. As an example, suppose an airline begins operations at a new airport in the mountains when they had previously only operated into low elevation fields. The engines will produce less thrust due to the thinner air. When this is coupled with high terrain and less predictable mountain weather it can create combinations that are much more challenging. If this is not planned or anticipated, it can create significant risk.

Similarly, pressures from competition and budget constraints can have a devastating impact, as we saw with the Boeing MAX. Woods describes the concept of a brittle system, one which can sustain almost no change at all, to a system that can, if necessary, go beyond the boundaries that were anticipated and be able to manage new and unexpected problems.[10] The ideal system can easily, and gracefully, move beyond the initial design concept to some degree, without resulting in an accident. It is resilient.

15.1.3   Resilience Engineering

Resilience is, generally, the ability of something to be able to exceed its normal boundaries and then recover to a safe state. A resilient organization is one that is able to effectively adapt to surprise.[11] This is different than the one that is characterized as robust. Robustness has been defined by Woods as the ability to manage *known* failure modes. Resilience, by contrast, involves the ability to manage unanticipated problems, or *unknown unknowns*.

Woods describes four concepts or views of resilience. The first is the ability to rebound after a disturbance. Why are some organizations, systems, or individuals able to recover from traumatic events more easily than others? The second is robustness, or the ability to expand to meet more and different types of problems. The third is the ability to extend the system's capacity to adapt to surprise, also termed graceful extensibility, and the fourth views resilience as a sustained adaptability.[12] But how can we design a system to be resilient?

The general historical approach, as described in other sections, has been to reduce the variance as much as possible. Reducing variance has continued to be a popular approach, and one that is based on the concepts contained in reliability theory. The concept that reducing the variability of components is the best approach to reducing accidents, and that humans are a component in a system that is subject to errors is a common one when viewed from the perspective of reliability. In that model safety and reliability are conflated, as is commonly found in flawed analysis.

As outlined in Chapter 13, there are several models of accident causality. Although, as stated by Box that all models are wrong, what is important is that the model is useful.[13] Models that view humans as components have had limited success in predicting accidents in complex systems. As much as the simpler model based on reliability may be appealing, it simply is not up to the task when a complex system is under consideration. Although it is certainly true that humans are part of a larger system, their role is not as a component but as *controllers* in the system. This moves the view that humans are the *problem* to a view that humans are a *solution* to improving the safety of any system. Blair and Helms describe what they term the *capabilities view* that views technology as a tool to enable people to fulfill the mission.[14] The other is what is termed the *cybernetics view* where people are considered just subsystems, that is, part of a larger system.

But what is important in a safety critical system? Human resilience is critical, but in a larger system one human is not enough. The overall design of the system must be able to adapt as conditions change. Sustaining the ability to be adaptable is then a critical feature for a safety-critical system. Can a theoretical basis be created that can be used to predict the existence of such a system? This is the premise of Wood's theory of *graceful extensibility.*

15.1.3.1 *Graceful Extensibility*

As characterized by Woods, factors that create sustained adaptability have been difficult to determine. The cases of interest are those that are able to continual adapt to changing conditions, as opposed to those complex systems that are unable to do so. Woods formulated a set of 10 theorems to capture the concepts of what forms of control were necessary to create sustained adaptability. These are then divided into subsets as outlined below, but to understand them it is first necessary to define the meaning behind some of the terms.

The assumptions as a foundation to the theory are that there are finite resources, meaning that whether it is a human or software component, the ability to change and adapt to changing requirements is limited. The second one is that there is continuous change. In the theory, the term "unit" is a single part of a system, such as a person, computer, or even a physical component. The basic concept is that everything has a limit on how much it can be pushed beyond the normal expected design requirements. We can push a metal component quite far before it breaks, and a person can also push their endurance, physically or mentally, up to a certain point. After that, things will break down. For example, a person can continue for a long time despite being extremely tired and missing sleep, but at some point they will simply not be able to perform. By extension, an entire organization can be pushed to its limits and manage fairly well for a period of time, but there is a point where it just cannot keep up with any more changes, and if that happens it can collapse completely. The following from Woods are the axioms of the theory of graceful extensibility and are self-explanatory:

Subset A: Managing risk of saturation

- S1: The adaptive capacity of any unit at any scale is finite; therefore, all units have bounds on their range of adaptive behavior, or capacity for maneuver.

- S2: Events will occur outside the bounds and will challenge the adaptive capacity of any unit; therefore, surprise continues to occur and demands response; otherwise the unit is brittle and subject to collapse in performance.

- S3: All units risk saturation of their adaptive capacity; therefore, units require some means to modify or extend their adaptive capacity to manage the risk of saturation when demands threaten to exhaust their base range of adaptive behavior.

Subset B: Networks of adaptive units

- S4: No single unit, regardless of level or scope, can have sufficient range of adaptive behavior to manage the risk of saturation alone; therefore, alignment and coordination are needed across multiple interdependent units in a network.

- S5: Neighboring units in the network can monitor and influence—constrict or extend—the capacity of other units to manage their risk of saturation; therefore, the effective range of any set of units depends on how neighbors influence others, as the risk of saturation increases.

- S6: As other interdependent units pursue their goals, they modify the pressures experienced by a Unit of Adaptive Behavior (UAB) which changes how that UAB defines and searches for good operating points in a multidimensional trade space.

Subset C: Outmaneuvering constraints

- S7: Performance of any unit as it approaches saturation is different from the performance of that unit when it operates far from saturation; therefore, there are two fundamental forms of adaptive capacity for units to be viable—base and extended, both necessary but interconstrained.

- S8: All adaptive units are local constrained based on their position relative to the world and relative to other units in the network; therefore, there is no best or omniscient location in the network

- S9: There are bounds on the perspective of any unit—the view from any point of observation at any point in time simultaneously reveals and obscures properties of the environment—but this limit is overcome by shifting and contrasting over multiple perspectives.

- S10: There are limits on how well a unit's model of its own and others' adaptive capacity can match actual capability; therefore, miscalibration is the norm and ongoing efforts are required to improve the match and reduce miscalibration (adaptive units, at least those with human participation, are reflective, but miscalibrated).

Although more research is needed, it is theoretically possible that through using these axioms, it might be possible to identify the conditions where a system may be nearing the edge of its capacity to adapt. If so, it would then be possible to modify the system prior to an accident.

15.1.4   Safety II[15]

As stated earlier, the concept that many in the safety sciences promoted involved the reliability theory approach, which views humans as an unreliable component in the system. According to Hollnagel the result is that accident and incident investigations become the focus of safety work, and, in turn the focus is on the ways that humans erode safety through their performance. In the design of work, researchers such as Hollnagel point out that increasing the reliability of machines has served to highlight that human reliability has not improved. The approach to improve human reliability has been to design strict procedures that need to be followed, but this involves sitting behind a desk and *imagining* the exact methods and procedures humans need to follow. This often works quite well, but, as pointed out in Chapter 13, there can be a gap between what is imagined and reality, something practitioners often call *work as imagined* verses *work as done*. Despite this, accident and incident investigation and risk assessments often focus on whether or not the operator followed the procedures as designed with little regard to whether or not those procedures were actually appropriate for the situation that presented itself.

There is ample sources of evidence that can be used when studying accidents and incidents with a view toward the reliability of the components, including the human one. In aviation, flight data monitoring, safety reporting, and accident reporting systems can all be used as part of the analysis. This has been characterized as fly-fix-fly, where problems are found after a flight, those particular problems are addressed, then another flight is accomplished, with the cycle repeating itself.[16]

Also encapsulated here is the concept that the goal of safety is to reduce the number of unwanted outcomes to as low as is practicable. Of course, as pointed out by Hollnagel, this more generally turns out to be "as low as affordable." This approach is classified as Safety-I. According to Hollnagel, in the Safety-I paradigm, the way to make the system safer is by reducing the unwanted outcomes, and that can be accomplished through reducing errors or by preventing the system from going from a normal to an abnormal state through the use of barriers. To become safer, the argument goes, methods must be implemented to ensure that humans consistently behave in an expected way. Although these methods were viable when systems consisted of simple causal relationships, they are no longer adequate for today's complex systems. Hollnagel explains that the foundations of Safety-I were published almost a century ago—and founded on principles developed over the preceding centuries—and in today's complex accidents we no longer can assume that we will find the cause of an accident, or even a plausible explanation. According to Hollnagel, systems are becoming more complex, changes are occurring more rapidly, the principles of how they function are often not entirely known and they are becoming more diverse. These facets make the analysis more difficult using traditional methods.[17]

Safety-I is, by definition, reactive. As an alternative, Hollnagel, based on the concepts of resilience engineering, promotes a new approach, which he categorizes as Safety-II. Hollnagel's alternative is viewing safety as a "dynamic nonevent" where safety is a situation where bad events do *not* occur. It is "dynamic" because it takes continuous effort to prevent a bad outcome. Hollnagel points out that all of the published data involve the events we did not want. We can obtain data on how many people were killed in traffic accidents, but it is much more difficult to find information on how many people were in cars that were *not* in any accident. This leads to a problem known as the regulator paradox, as described by Weinberg and Weinberg where the safer something is the less there is to measure.[18] If a regulator, such as EASA or FAA, is trying to reduce accidents (using reliability theory, as most currently do), the goal then becomes work to reduce variation. However, if all that is being reported are *deviations* then the more the variation is reduced the less information becomes available. This is a current challenge for safety professionals. The number of accidents and incidents in commercial aviation is getting so low that it takes just one accident to create the appearance of a trend in the entire database, going up one year and down the next—but does this really represent a trend? Meanwhile, what is the role of the variability of human performance, and is it always a bad thing? If humans are having to work around and create their own constraints this is an indication that the organizational system is not functioning correctly.

Safety-II is a situation where, rather than ensuring that as little as possible goes wrong, instead we concentrate on ensuring a condition where as much as possible goes right. The concept of Safety-II is that, rather than just limiting safety research to those methods that examine human errors, investigators working on safety should also, and perhaps focus on, the normal work that humans do. The argument is that humans are preventing accidents every day just in the normal course of their actions, and through a better understanding of how they are accomplishing that it might be possible to learn from that and implement new forms of instruction and training. The focus becomes human resilience rather than their faults. Looking at it another way, Hollnagel argues that the Safety-I approach would see the probability of a risk as, for example, 10^{-5} but that means that "there will be at least 99,999 cases of acceptable performance for every case of unacceptable performance." Hollnagel points out that it might be more productive to investigate human performance variability and sort out why those same facets that result in things going right most of the time will, sometimes, result in things going wrong.

Hollnagel argues that performance variability (on the part of humans) is what enables systems to have positive outcomes most of the time. However, if we are to use this then it becomes imperative to identify what goes right. From the perspective of Hollnagel, we already know what these things are, but we just do not notice them as they are just part of normal behavior. Safety-II assumes that people adjust their behavior as required to keep things going right. People also adjust their behavior if they see something going wrong. In either case, it is the performance variability that keeps things moving in the direction we want them to move. According to Hollnagel, people usually mix Safety-I and Safety-II, in that they both avoid things that go wrong and work to ensure that things go right.

Hollnagel explains that "Safety-I assumes that things go well because people simply follow the procedures and work as imagined. Safety-II assumes that things go well because people always make what they consider sensible adjustments to cope with current and future situational demands." So how can safety practitioners identify the habits that lead to things going right? Hollnagel offers several ideas.

The first would be the obvious case of just watching what others do or paying attention to our own actions. Unfortunately, this requires some expertise and observation skills. In many respects, this is what was done with the Line Observation Safety Audit (LOSA) described in Chapters 2 and 14, although LOSA, as designed, is focused on identifying errors and omissions. Hollnagel suggests conducting interviews of employees to learn as much as possible about how the work is actually accomplished and how they work around the challenges. Other methods involve approaches to looking at how problems are solved rather than identifying what problems exist. Hollnagel outlines the methods of *appreciate inquiry, cooperative inquiry, and exnovation.* All of these methods require the practitioner to study an organization in depth, working with the people involved in the operation directly in positive ways. Using Safety-II, Hollnagel states, is a holistic approach that looks at the entire system, as opposed to Safety-I, where the details are the focus.

There are several critiques of the Safety-II concept. The first is that the concept of Safety-I is only true from the perspective of what is generally considered workplace safety. Workplace safety is concerned entirely with the actions of humans and, thus, human error is the focus. Commercial aviation does, of course, have large swaths of practitioners who solely concentrate on the workplace and those actions, but engineering methods include many more considerations that would not fit into the Safety-I framework.[19] The second aspect is that the fact that the lack of an unwanted event, an accident, incident, or loss, is not an indication that the actions were "right." In fact, it could easily be that the particular actions by a person just worked because they did not occur in combination with other factors that would lead to an accident. As outlined by Dekker, most of the time we can avoid an accident even if we are not doing something the correct way, and that experience can result in the wrong conclusions. An example is an aircraft accident that occurred during landing when the pilots were following the procedures as outlined in their manuals. The scenario that can lead to an accident are rare, and a pilot might be able to use a technique that is vulnerable to an accident 10,000 times with no bad outcome, until one day the conditions are just right to make things go wrong. If interviewed or observed prior to the accident, it would almost certainly be reported that the technique (which was following the recommended method) was "doing things right" yet it masked a hidden threat. Another example might be a pilot exceeding recommended "stable approach" parameters and, after a while, coming to believe that the parameters are not important. Each time a parameter is slightly exceeded and that appears successful strengthens the belief that doing so was safe.[20]

While Leveson is concerned with the system safety aspect of Safety-II, a second set of critiques, as outlined by Cooper focuses on the occupational safety and health professional.[21] Cooper argues that the Safety-II (also referred to as the "New-View") concepts are largely untested. This critique is also described by Le Coze; the concepts of resilience engineering derived from the work in Normal Accident Theory and the HRO research, in an effort to improve on the concepts that have been termed "Safety-I." Cooper describes several challenges to the New-View. Cooper (2020), like Leveson, point out that most of what is being advocated by the New-View are already actually being done currently, such as leading and lagging indicator to monitor risk. Cooper points out that both Dekker and Hollnagel promote the concept that people, left to their own initiatives, will do a better job if not centrally controlled, so the management of safety should be left to the workers themselves. This approach, which can be considered a form of anarchy, may be workable in a supermarket, but in a complex organization, would almost certainly increase risk.

Perhaps Cooper's largest critique is that the literature for the New-View contains no practical methods, tools, processes, activities, or any combination that might improve safety. The writings are quite interesting reading, but mostly serve to frame the problem and provide some vague recommendations. Additionally, as also stated by Leveson, the absence of accidents does not mean the system is safe.

15.1.5 Resilience Engineering or HRO?

Le Coze argues that perhaps the way forward is to accept that both HRO and Resilience Engineering offer concepts that may be complementary. The brief history of HRO has already been presented. Le Coze starts by pointing out that the history of resilience engineering is connected with cognitive systems engineering, which is the topic of the next section. The early research in this area was initiated by Jens Rasmussen and James Reason. The emphasis has been on concentrating on ways to improve the expertise of individuals and their ability to cope with the increasingly complex systems, rather than trying to "fix" people so they would have no error. This is closely correlated with the system view of safety, promoting the concept that errors are not the result of practitioners making mistakes but rather systematic factors that make those errors almost inevitable. Le Coze points out that resilience engineering stands on the concepts of adaptive and self-organizing patterns, with the adaptive properties outlined by Woods, and then later Dekker.

Le Coze states that both HRO and resilience engineering have significant similarities. One example is the equivalent "situational awareness" (more on this in the next section) and the HRO term of "having the bubble." Errors are often assumed to be a consequence of not having situational awareness. It should be noted that arguing that an error is a consequence of losing situational awareness is really just replacing one vague term for another, as neither really is possible to easily address without understanding the context in which it occurs. There is no question that many of the concepts of HRO were borrowed by the resilience engineering community and vice versa. This should not be surprising as social science researchers all concentrated on similar problems, even with different approaches. Le Coze argues that both the complementary and diverse components contribute to a better understanding of the problems as well as toward future mitigations.

15.2 Cognitive Systems Engineering

It has been stated that the first publication using the term *cognitive systems engineering* was in a 1982 paper by Woods and Hollnagel.[22] In that paper, they describe what was then a new approach where, instead of just looking at the physical and physiological aspects of system design, the concept is that human cognition becomes interwoven with the entire system thus leading the system to use knowledge about the environment and itself to perform. Today researchers focus on human-centered design of complex systems and how to make such systems more resilient to events that might result in them not accomplishing the system goals. The question remains: just how do we best integrate humans into a complex system and how to best facilitate their needs in the system? This topic has been addressed by a number of researchers, but the overarching view is that humans need to be viewed as an asset that contributes to the safety of the system, and not as a problem that needs to be mitigated.

In the last section the macro view of systems and organizations was explored, but are there opportunities to improve human performance? This topic has been the focus of a large body of research, with a focus on a combination of the design of the systems and interfaces to training and selection of people. Much is involved in understanding human cognition, and through that, being able to identify how we might be able to improve our approach to creating systems and training that enable humans to perform better.

15.2.1 Improving Automation Design

There has been a great deal of work in the area of improving automation design. This work has been in contrast to what has been a push from some, based on reliability theory, where humans are considered a liability and so the focus should be on reducing their ability to create problems. Other researchers have, instead, focused on how we might design systems to support human cognition.

Designing systems so that humans can benefit from them, rather than constrain them, is a critical way forward to make systems safer. There is continual work in this regime, from the work of Dr. Nadine Sarter on understanding how people can misunderstand software modes to the large body of work conducted on the direction and leadership of Dr. Kathy Abbott. Much of this work has been to analyze how people are using and being challenged by current automation design. Much less effort has been spent on how the software might have been designed to better facilitate human performance in the first place.

One of the principal challenges has been to ensure that there is commonality between the platforms from various manufacturers. Following some horrendous accidents that were correlated with differences in instrument layouts, groups such as SAE International created standards in an attempt to improve the situation. Today, though, with the easily adaptable characteristics of computer software, the capability to create divergent designs has facilitated a plethora of different concepts. As a consequence, the problems that were associated with the early designs have now returned. Pilots moving from one type of aircraft to another now face serious challenges that have not been seen in decades.

Although most of the work concerns understanding pilot decision-making and cognitive choices, there is another body of work that examines how future aircraft can be designed to support pilot decision-making. It was from this type of work that systems like Heads Up Displays were introduced, which simply overlay critical flight readings, such as airspeed, altitude, heading, and attitude as well as projected flight path information. One example of proposed designs that would take advantage of computer capability to enhance pilot performance and decision-making is a system proposed by Dr. Mumaw of NASA that can correlate the aircraft maintenance status, runway, and airport capabilities with other factors to rapidly provide pilots with available options.[23] Others, such as Van Baelen, have proposed systems to use haptic feedback to inform pilots if they are approaching a flight limitation that may adversely affect performance.

15.2.2 Training

The application of reliability theory has created a problem in pilot training. As discussed in Chapter 13, if humans are treated as just an unreliable component in a system then the efforts will be to try to make them more reliable and have less authority. In practice, we have seen pilot training that has been continually reduced to what has been deemed they need to know. In this approach, pilots are trained to follow procedures and the belief is that providing them more knowledge and skills could lead them to experiment in unpredictable (and hence, unreliable) ways. Pilot training manuals have been simplified during a time when the complexity of modern aircraft has increased due to the introduction of computerized systems. The proponents of this approach argue that flying is safer than it ever has been. Although this is true, it may be just an outcome of being lucky as the combination of circumstances that result in accidents can be relatively rare (such as microbursts). More likely, though, there are still enough pilots in the system who were trained and flew during earlier times where they learned from challenges, plus, even absent that, humans are highly adaptable and creative—in either case, it is the human adaptive capacity, or resilience, that is actually keeping the commercial aviation safe.

To that end, identifying how to improve human resilience in aviation may provide one of the strongest ways to improve future safety. Currently implemented in programs such as Evidence Based Training (EBT) or Advanced Qualification programs (AQP) is the use of data collected to implement into safety programs. This data is collected in a number of ways as was described in Chapter 15 and will be supplemented in section 16.4, and when a hazard is identified that scenario can then be implemented into the training for other pilots. In this way, training is expanded to not just capture the scenarios that have resulted in accidents and simple mechanical failures, but also challenging situations in which the pilots were able to manage the flight and prevent an accident or incident. Still, there are significant limitations in terms of time and costs that should be recognized. Training generally does not involve the stress that occurs in real-world scenarios, nor do pilots tend to think of it as real. There is evidence that pilots alter their behavior and decision-making during a simulator event in ways that are particular to the event itself.[24]

Moving beyond this, Carroll and Malmquist outline an approach using Tactical Decision Games (TDGs), which could be used to present more complex training scenarios at much lower cost than just using expensive flight simulators. Pilots would work through the problems, and this would be followed by an instructor-facilitated discussion, either individually or in a group, to look at what other options might be available.[25] Another key component to make training more effective is to increase the stress that might occur in a real flight. Through the introduction of training methods that induce stress, the training can become more effective—but this is a challenge as previously stated. In an investigation conducted by Dahlstrom and Nahlinder, it was found mental workload in a simulator was lower than actual flight for unexpected events, so the design of training needs to consider those aspects.[26]

Might it be possible to design training such that people can more rapidly achieve high levels of expertise? Researchers led by Hoffman envisage several approaches that would be used to improve decision-making and analysis. This is a different type of skill than playing a musical instrument, where it is difficult to bypass the hours of practice required to become an expert. In decision making we learn through experience, so providing ways that expose people to more complex scenarios with carefully constructed training can, in some cases, allow an individual to obtain skill sets that would normally take several years to just a few months. This ground-breaking research has the potential to provide significant improvements in aviation safety as the types of problems that pilots are expected to solve become more and more cognitive in nature.[27]

15.3   Certification and Reliability Theory

In 1984, NASA's space shuttle Challenger exploded shortly after launch. In 2006, a Royal Airforce Nimrod MR2 aircraft was lost, and in 2003 the space shuttle Columbia was broke up during re-entry. In these prominent but numerous other accidents, a "safety case" had been built that showed that the requirements for safe flight had been met. What is a *safety case* and how might it work against a safe outcome? In all of these cases, people who were charged with ensuring that safety requirements were met made assumptions that turned out to be invalid. This manifested as waiving or minimizing certain requirements based on the data showing that the particular issue had not created a problem previously or finding some basis that would support that the particular issue was not critical. Some in leadership positions may justify such decisions based on the motive for risk taking as well. In the case of the Challenger, it was found that problems *were* identified, but senior leadership demanded that the engineers *prove* that the problems might create a safety issue. This is the exact opposite of what should occur! Nimrod similarly included factors related to organizational pressures which overwhelmed sound judgment. In the case of Columbia, managers created barriers against opinions that went against their plan, discounting the risks presented. Previous success was used as a justification in all of these examples.

Arguing that a certain factor that should be addressed has been adequately resolved or does not need to be mitigated, when it has not been, is often the result of operational pressure. This is a common theme in several prominent accidents in recent years, including the Boeing 737 MAX. Still, one reoccurring thread has been unrealistic risk assessment. This is an outcome of reliability theory, where past results dominate the statistical risk assessments and there is an overreliance on redundancy. Certification of aircraft and equipment commonly uses a method known as Failure Mode, Effects, and Criticality Analysis or FMECA. This method is predicated on using linear causality and assigning probability, severity, and the ability to detect the problem to determine a total score, which can then be ranked and subject to cost and return on investment analysis. See discussion on risk matrixes in Chapter 15 to better understand the problems with this approach.

There is also a significant risk to the manufacturers. As can be seen from the examples, it is very possible for a company or organization to miss a risk factor in a design due to the component reliability. It may take decades to manifest, or it might be fairly quickly, but the problem is that it is very possible to have reached the point that design changes are almost impossible. At best the company can patch the existing design. This can result in enormously expensive mistakes that might undermine the financial viability of the organization.

15.4   Collecting the Data

Attempting to ensure safety with no knowledge of the system or problems that might be present would not be possible. Control of safety requires the same aspects as control of any other process. It is necessary to have a control path and a feedback path. We have reviewed some approaches to the control aspect, and more will be discussed later in this chapter. Here we will address some of the feedback sources that might be beneficial to safety professionals when assessing the safety of their organization. Clearly, for these to work an organization requires a strong safety culture, as outlined in Chapter 12.

There are several sources of information that might be obtained in addition to the systems previously discussed with accident investigation in Chapter 4. Here we will review the programs discussed briefly in Chapter 2. The most basic of these is the simple safety report. Some sort of safety reporting has been in place within many organizations for decades. Many of these include the choice to submit reports anonymously to encourage people to report items without fear of retribution from management. In general, these reports are "owned" by the organization that created the reporting system and, for legal reasons, generally are not shared. It is then the decision of the management whether to act on the reported problem or not. This can, of course, lead to same types of decisions that resulted in minimizing the risks in the safety case examples discussed in the previous section.

In an effort to encourage more reporting, the NASA in the United States developed a program titled the Aviation Safety Reporting System (ASRS). The program is automatically anonymized, but to encourage reporting the submitter is sent a receipt which they can use to obtain limited immunity from regulatory enforcement action from the FAA. The logic behind this is that the value of the increased reporting that could be available with this immunity outweighs the value of certificate action against a person that might have violated a regulation. The system does not provide immunity for intentional or criminal actions. Additionally, the report cannot, itself, but used as the basis for enforcement action. Similar systems have been created in other parts of the world.

The concepts of organizational and system safety were joined together, initially by American Airlines, to create a program that would encourage even more robust reporting behavior reporting that has come to be known as the Aviation Safety Action Program (ASAP). This program provides limited, but more robust immunity for the reporting of events, in this case affording the reporter protection from both FAA and company disciplinary action as long as the reporter did not intentionally create a substantial risk. This means that a pilot can be protected even after an aircraft accident! The report is shared with the company, the FAA as well as any employee association, such as the pilot's union. All three organizations then coordinate to identify mitigating actions. Similar programs have been developed for ATC (ATSAP), maintenance (MSAP), and dispatchers (DSAP). This program is currently limited to the United States, although some similar programs may exist.

With the advent of the capability to record and track information from various aircraft systems, initially to improve maintenance reliability, additional parameters were added to enable operators to monitor their aircraft operations. This came to be known as Flight Data Monitoring (FDM) and has come to provide a rich source for obtaining data about flight operations. The first programs were implemented by British Airways and TAP Air Portugal. Companies and organizations have been able to use this information to identify problems for a range of issues from crew procedures to ATC and more. As the organization that owns the aircraft also owns the equipment, they have full control as to what to do with it. This, of course, leads to the same problems are previously addressed. Such systems have also been the basis for disciplinary action against employees that were found to have exceeded set values.

FDM can be programmed with "event sets" so that certain exceedances automatically trigger an alert to the company safety department and so can be easily identified. The system works by recording information and parameters via the aircraft data bus onto a solid-state recording system. This can include a removable memory card, or the information can be transmitted wirelessly. FDM programs are now mandatory in many countries.

Although not required in the United States, the Flight Safety Foundation recommended that carriers implement a program to use FDM calling it Flight Operations Quality Assurance (FOQA). From a cultural perspective, the FDM program can result in employee distrust of management. To counter this, FAA regulation does not allow the use of recorded information for enforcement. Today, FOQA programs, where implemented, provide the data of flights without identifying information as to what flight was involved. This precludes the possibility that such data might be used for enforcement action. In order to assure that all data can be collected, FOQA programs include a "gatekeeper" position, a designated employee (usually a member of an employee union) who can access the information as to who was operating a flight and contact them for follow-up without jeopardizing the program.

Checks, evaluation, and training continue to be a source of information as to the quality of the operations. Company check airman and regulatory check airman ride on flights to audit the operation. This can include random "ramp checks" where documents, maintenance history, and pilot qualifications are checked, as well as in-flight "line checks." In addition, pilot training on the line or simulators can provide useful information. Perhaps the largest negative factor is that pilots being observed will modify their behavior and adapt to the expectations of the person providing the evaluation, as previously discussed regarding training in this section.

Another source of information, although less frequent, is the LOSA. With this program, trained observer, often including pilots with the airline being audited, ride along on flights and record their observations in order to identify weaknesses in the operation and recommendations for improvements. Although pilots are still being observed, as the observers have no enforcement power and are not in positions of authority, their impact on how pilots operate is far less than those on the checks or evaluations.

Although these programs are effective, there is still a remaining gap due to people not reporting problems where they do not feel they are in jeopardy. Part of this can be addressed through safety monitoring initiatives, as described in Chapter 2. One proposed solution would be to identify methods to capture what might appear to be routine inquiries. For example, a pilot might encounter a situation and rather than report it just sends an email to someone they believe can answer the question of why it happened that way. Similarly, a maintenance person might contact the company engineering department, or managers of different aircraft fleets might have internal conversations. The problem is that often the people involved do not identify these as a safety issue, even though, the same event in a different circumstance could have resulted in a dangerous situation. As an example, consider the following scenario:

- Airlines are provided certain areas of the airport to operate. They have ramp areas to use for parking aircraft, gates, and often designated maintenance areas or hangars. At times an airline will "own" (generally lease) more gates than they need at certain airports so they can accommodate extra aircraft should they need to—such as weather diversions or charter flights. The airport itself has a number of vehicles, let us suppose that the airport has a number of trucks they use for de-icing aircraft. These trucks take up a lot of room, and at times the airport management parks some of the trucks on the unused space available at an open gate an airline is not using. The airline ramp management complains about this to the airport, and also notifies the airline's properties department. Large airlines have a designated group to negotiate contracts for the various airline's gates. These communications go between these different groups, but no hazard is identified.

- One night a major storm hits the region. Numerous aircraft are diverting. The airline now needs all of their ramp space, but they do not have access due to the airport de-ice trucks parked there. The airport does not have the personnel on hand to move these trucks on short notice. Trying to work around the trucks results in very tight off-gate parking that is not within the normal boundaries, but what choice do they have? The aircraft are towed into and out of tight spaces, but one of the aircraft is accidently pushed into another in the rain and darkness. The trucks parked on the ramp were not a hazard by themselves but became a hazard when the context changed. Could there be a way to capture this type of information and identify it as a hazard prior to the accident? Currently there are no systems in place to do this.

Other sources of data, as discussed in Chapter 14, are the *Service Difficulty Reports* (SDRs) that are filed by airlines and aircraft operators when they encounter a problem with one of their aircraft. These reports include information on problems and malfunctions that have occurred on an aircraft. Additionally, in a program that is similar to the ASAP, airlines subject to US regulatory enforcement can submit a *Voluntary Disclosure Report* (VDR) reporting when they have discovered that they have inadvertently violated a regulation. These reports can be used to identify potential risks to proactively manage safety.

15.5   Industry and Trade Associations

Commercial aviation is incredibly safe, and much credit should be given to the role of trade associations that have involved themselves in rule making, recommendations, and audits. These can be divided into nongovernmental organizations and trade unions. These organizations maintain an active role in ensuring the promotion and safety of aviation.

15.5.1   Nongovernmental Organizations

The oldest organization in the world promoting the safety and expansion of aviation is the Royal Aeronautical Society (RAeS), founded in London in 1866. As seen on the plaque in Figure 15-1, the RAeS is dedicated to the improvement and safety of aviation and maintains a strong association with both government and other industry groups. The RAeS expands its influence through members that hold prominent positions throughout the global aviation industry.

In 1905 there were many different companies working on automobiles. Engineers recognized that there were common design issues and there was a need to develop engineering standards, and that manufacturers preferred to follow accepted practice. Although the original engineers were working on automobiles, their 1916 meeting was joined by the American Society of Aeronautic Engineers, the Society of Tractor Engineers, and engineers from the boating industry. To encompass all these forms of machines, the term "automotive" was chosen to describe any machine that was self-powered, and so became the Society of Automotive Engineers (SAE). More recently the name was changed to SAE International.

FIGURE 15-1   Plaque outside the Royal Aeronautical Society, London. (Source: Author)

Today SAE International provides a platform where industry, government, and research organizations can collaborate to create design standards and recommended practices for industry. These standards often serve as a "boiler plate" for regulatory agencies when writing rules and regulations. Some examples include standards for flight deck instruments, flight controls, and system design for commercial aviation.

Taking up the task of aviation safety, the Flight Safety Foundation (FSF) was founded in the 1940s. This organization is exclusively focused on improving flight safety through risk mitigation and safety investigation. FSF promotes research and safety advocacy and is recognized throughout the world as a leader in aviation safety through the development of standards, providing technical assistance, and promoting safety information as a nonprofit independent organization. Supplementing this work is the International Society of Air Safety Investigators (ISASI). Recognizing that best practices for aircraft accident investigation were needed, the ISASI was established in the 1960s to provide a platform for experienced accident investigators to learn from sharing information on recent investigations and investigation and analysis methods.

Contributing to this work were other organizations, such as the American Institute for Aeronautics and Astronautics (AIAA), the Institute of Electrical and Electronics Engineers (IEEE), the American Society for Testing and Materials (ASTM), the RTCA (formerly known as the Radio Technical Commission for Aeronautics), the European Organization for Civil Aviation Equipment (EUROCAE), and many other smaller organizations around the world.

In 1997, in an effort to reduce the commercial aircraft accident rate by 80% within 10 years, the United States established the Commercial Aviation Safety Team (CAST). This group is an ongoing government/industry partnership that charters groups to research major categories of accidents and identify mitigations for them. The group has been very successful, meeting the first goal and then establishing a new 10-year goal. Government and industry partnerships have proven to be quite effective in gaining the resources as well as the will to make significant changes.

15.5.2 Labor Unions

Quite a few airline pilot unions exist today. The first pilot union was the Air Line Pilot's Association (ALPA), which was initially formed to combat severe safety concerns. Established in 1931 at a time when pilots were being pressured to fly when tired or in bad weather, or face termination, ALPA has come to the largest pilot union in the world with over 65,000 members in the United States and Canada. There are some other, smaller independent unions in the United States, such as the Southwest Airlines Pilots' Association (SWAPA), with more than 8000 members, the Allied Pilots Association affiliated with American Airlines, which has around 15,000 members, the Independent Pilots Association and the International Brotherhood of Teamsters (IBT). In Europe, the collective voice of 40,000 pilots from 33 national pilot associations across the continent are represented by the European Cockpit Association (ECA). At national level, pilots have formed their own unions, such as the British ALPA (BALPA), Vereinigung Cockpit (VC), in Germany, the Syndicat National des Pilotes de Ligne (SNPL) in France, the Spanish Air Line Pilots' Association (SEPLA) and the Japanese ALPA.

The International Federation of Air Line Pilots (IFALPA) is a labor union association founded in 1948, headquartered in Montreal, Canada, and accounts for over 100,000 members in over 100 countries, with members including the US ALPA as well as most of the pilot unions representing pilots from around the world are members. Air safety is a primary responsibility of every airline pilot. As the oldest and largest airline pilots' labor representative, the main role of IFALPA's air safety structure is to provide channels of communication for line pilots to report air safety problems. An additional role is to stimulate safety awareness among individual pilots to enable flight crewmembers to be constructive critics of the airspace system.

Finally, the air safety structure helps investigate airline accidents. Over the years, ALPA's air safety structure has contributed significantly to air safety. Airline pilot unions have trained safety volunteers and led industry projects on aircraft wiring safety, human performance, and nonpunitive corporate cultures, among many other hot topics.

Each individual airline union has its own safety committee. That committee will have a chairperson, who normally reports directly to the elected union leadership, and other members as determined to be needed by that particular pilot group. Larger unions, such as ALPA, also maintain a safety and engineering staff to support broader work. As an example, the national safety structure of the Air Line Pilots Association is composed of five management groups, headed up by Executive Air Safety Chairman, in these specific safety areas:

- Aircraft Design and Operations
- Airport and Ground Environment
- Air Traffic Services
- Human Factors and Training Group
- Accident Analysis

The Accident Investigation Board of the Air Line Pilots Association oversees investigation of accidents involving the air carriers they represent. The board coordinates ALPA's participation in accident or incident investigations conducted by the NTSB or the FAA and ensures that the appropriate ALPA subgroup determines the significant factors in each air carrier accident. ALPA also maintains a worldwide accident/incident hotline for its pilots to call for help if involved in an aircraft critical safety event. If deemed necessary, ALPA will also employ its Critical Incident Response Program (CIRP), which counts on trained counselors who work to mitigate the psychological effect of an accident before harmful stress occurs.

More specific to SMS, since 2008, ALPA has been a leader in supporting SMS principles as set forth in the ICAO Safety Management Manual. The ALPA policy statement is that they support SMS when it is developed and implemented in accordance with the following:

- A documented, clearly defined commitment to the SMS from the CEO—a written SMS policy, signed by the CEO, which recognizes the business benefit of asset preservation and mishap prevention. The policy must show commitment to continuous improvement in the level of safety, management of risk, and creation of a strong safety culture.
- Documented lines of safety accountability.
- Active involvement of the affected employees in a nonpunitive reporting system and a commitment to a "just" safety culture.
- A documented, robust Safety Risk Management (SRM) program. The SRM program requires the participation of labor organization(s) as the representative of their employee groups in both the identification of hazards and in the development of risk mitigation strategies.

- A documented process for collecting and analyzing safety data and implementing corrective action plans.
- A documented method for continuous improvement of the SMS.

Working to maintain the highest levels of airline safety is an enormous challenge. The depth and breadth of ALPA's air safety structure, and the critical role it plays in meeting that challenge, are unique in the air transportation industry.

Pilot unions are just one of several, powerful, and resource-rich labor organizations that can be used as an extension of an airline's SMS. Aviation maintenance technicians often work as part of labor unions, as with the International Association of Machinists and Aerospace Workers (IAMAW), which represent over 600,000 workers in more than 200 industries. The IAMAW's Safety & Health Department has promoted safety in topics such as hazardous material, occupational health, and as proponents of the Ground Operations Safety Action Program. Safety leaders' unions representing cabin crew, such as the Association of Flight Attendants (AFA), have been strong champions of finding better ways of managing crewmember fatigue, investigating aircraft cabin air quality, fighting the spread of communicable diseases, and passenger security screening. As shown in Figure 15-2, cabin crew often have the best perspective of issues affecting passenger safety on a routine basis, such as turbulence and health ailments. In addition, cabin crew are an essential part of keeping the overall flight safe as they are able to monitor a number of aspects that the pilots are not able to see, from unusual odors or sounds to being able to see things through cabin windows.

All of these labor groups provide information and feedback for the SMS. Labor unions are often viewed as antagonistic to the organizations their represented employees work for, but this can translate into candid unfiltered feedback that is not available

FIGURE 15-2 Flight attendant performing a predeparture passenger safety briefing. (Source: Wikimedia Commons)

from other sources, as the union representatives can openly discuss issues without fear of being disciplined or fired due to the nature of their positions and associated labor laws.

It should be noted that absent legal protection the ability of labor unions to facilitate positive change is greatly diminished. If companies have the power to weaken or eliminate those laws the ability of labor unions to improve safety will be compromised.

Whether they are safety leaders in pilot unions, maintenance technician unions, flight attendant unions, or other organized labor groups, individuals who perform safety work for labor unions are often highly motivated to find solutions and can contribute greatly to an airline's SMS.

15.5.3 Industry Trade Associations

As safety is a selling point for the aviation industry, several industry associations also have a robust safety role. These include the International Air Transport Association (IATA), founded in 1945, to promote interairline cooperation on all aspects that affect safety and security. Headquartered in Montreal, IATA represents 290 airlines from 120 nations and has created standards from airport codes to common methods for applying safety principles. IATA safety programs include:

- Corporate safety, which supports SMS and its implementation;
- Operational safety, which includes industry work and programs involving cabin safety, loss of control in-flight (LOC-I), controlled flight into terrain (CFIT), runway safety, midair collision, fatigue management, and more;
- Safety audit programs. IATA conducts a number of different safety audits of their member airlines, such as the IATA Operational Safety Audit (IOSA);
- Hazardous material standards.

As with the pilot associations, there are also airline trade groups in individual countries, such as the US Airlines for America (A4A) and the Regional Airline Association (RAA). Each, to varying degrees, participates in numerous industry safety working groups with the goal to improve aviation safety. Additionally, there are manufacturing industry trade associations.

15.6   Other Government Oversight

Some types of events, although arguably a potential threat to flight safety, are not always managed by aviation regulators. This can vary depending on the criteria set by the individual country but can include issues that can be considered workplace safety or environmental issues. For example, noxious fumes onboard an aircraft that affect the cabin crew might be managed by a government entity such as the US Occupational Health and Safety Administration (OSHA), the United Kingdom's Health and Safety Executive (HSE), or similar. These organizations are tasked with ensuring that employees have a safe working environment.

Similarly, most countries have governmental oversight for environmental protection, which can include rules and regulations pertaining to how to manage a fuel spill, for example. Local authorities also will have regulations regarding crash, fire and rescue equipment, or factors that impact the design of airports and facilities.

Although these regulations may not be aviation specific, they can provide additional oversight and improve aviation safety indirectly.

15.7   The Influence of Oversight

The influence of all of these groups is profound. There are many more than have been covered here, including associations of airline passengers and many individual and smaller groups and companies, all focused on safety. It is hard to match the degree of oversight that all of these combined organizations place on the airline industry.

The scrutiny has affected how the airline industry operates and contributed to strong impetus for very slow change that can be managed and controlled. Everyone from the manufacturers, the airlines, the regulators, and all others involved are under continual pressure to ensure safety, knowing that everything that they do may be monitored, questioned, and audited.

15.8   How to Measure Nothing

As previously discussed, there is very little information available to determine the safety of an ultrasafe system. Lofquist points out that there are three primary phases in a safety management system, which are *proactive, system operation,* and *system outcomes.*

The proactive stage is involved in the initial system design, the introduction, and the development of operational assumptions—the latter of which guide the design itself. There will inevitably be a gap between how a system is expected to perform and how it performs in the actual world. During the development phase, it is, thus, vital that those working on the design work closely with those who would actually be operating the system. The interactive stage entails the measures and requirements needed for real-time operation, where the interactions are affected by changing environment and human variability. The leadership has only limited control of what actually happens in the field, away from their direct oversight, so the system is largely dependent on a sound design and the behaviors of the humans involved, who will be actively attempting to compensate for any shortfalls in the system design. Identifying and measuring these compensations can provide an opportunity to improve the system and mitigate risk. Identification of any deviations between expected and actual performance can serve as an indication of problems. Finally, the reactive phase includes the investigation of accidents and incidents. A comprehensive investigation will enable the organization to identify the problems and, hopefully, prevent a future accident.

Lofquist argues that we need to bridge the gap between the interactive, proactive, and reactive stages if we are to make gains in improving the safety of these ultrasafe systems.

15.9   Myths of Big Data

As computers become involved in more and more aspects of our lives, it is easy to believe that, at some point, all problems will be solvable with computers. There is no doubt that computers have improved our lives in many ways. Some of these may not be obvious. As an example, computers have enabled more precision in designs that make it possible to reduce weight in designs without compromising safety. Computers are also much better than humans and rapidly locating information. A good example can be seen in something as simple as an internet search, where a user can instantly access millions of articles on a particular topic—this would clearly be impossible without the use of computers. This, in turn, allows for much more informed decisions and the ability to rapidly conduct tasks that would at one time been impossible. Because of their ability to process tremendous amounts of data, some believe that computers may help us predict the potential for future events. We shall explore these ideas in Chapter 17.

The idea that we might be able to predict the future with enough data has been a popular concept for decades, being highlighted in science fiction writing from the 1940s, such as Isaac Asimov. The concept here is not a complicated one. The idea is that we are trying to identify precursor events in the data. Precursors are often fairly easy to see in retrospect, as the patterns prior to an event are clear in hindsight. However, without knowing the outcome they tend to just look like part of normal variation. Is it possible that there were differences, but they were just too small to notice with the amount of data available? This would equate to an effect size in statistics. To see a very small effect takes a lot of data, hence, if we can store and process more data it might be possible to find that small effect.

Is this realistic? The problem is that in order to identify a small effect we would first need a model for what we are looking for. Also, using this approach, is it possible that there is never enough data available for the time period where precursors might occur? The idea that we can identify these problems before they occur, commonly referred to as *leading indicators* is based on the assumption that major accidents are a result of a random set of unique events, and so we can use data to identify these unique events using statistics. Attempts to identify the statistical models have, so far, not been effective.

There are indications that the use of Bayesian statistics can improve our capabilities to estimate risk for those processes that are stochastic with known risk factors, but these still require identifying *a priori* information can lead to an accident. One example might be the statistical probabilities for mechanical failure—for example, an aircraft wing must not fail more than once in every 1,000,000,000 hours in flight. However, this does not consider the conditions. Is that probability the same for an aircraft that is intentionally flying into a hurricane for research as it is for an aircraft flying on a clear calm day? Using Bayesian statistics, it can be possible to revise our probability estimates for situations that we know might increase, or decrease, risk.

15.10   Identifying Leading Indicators

As can be seen, most efforts to identify future risk concentrate on trying to identify what we do *not* know. Is there a way to turn this around and use the information we can obtain? Information is available from FOQA/FDM, safety reporting, employee report, and similar sources. It is also important to ensure that we are learning enough from accidents and incidents. Currently, all too often, only surface issues are identified, leaving wide open the potential for future accidents, so a systems view, such as STAMP, is extremely useful. An important additional avenue is that the potential for risk is also available through the identification of assumptions. Prior to taking any action, whether it is walking into a room or flying an aircraft, we make assumptions about the operating environment and what we might expect to occur. For most of us, we *assume* the room is safe to walk into prior to walking in, or we might take certain precautions if we think it is not safe for some reason. Similarly, no airline will dispatch a flight unless they *assume* it will be safe. If these assumptions can be identified, it can be possible to then confirm that they are valid prior to launching the aircraft, so how might that be accomplished?

Even this task requires we make certain assumptions. We start with an assumption, based on system theory, that most accidents are *not* a consequence of a random and unique set of events, but are instead due to the migration of systems or organizations to a state of increasing risk over time as controls are relaxed due to conflicts in goals and reduced perception of risk. There will be, of course, some accidents that are a consequence of random events. It is possible that the low 10^{-9} probability occurs and a wing breaks, but it is extremely unlikely.

Accidents almost always occur when assumptions are not correct.[28] There are two ways this can occur, the first is that they were incorrect at the outset, and the second is that they became incorrect over time. Carefully documenting the assumptions provides a foundation from which it is possible to check to see if the assumptions are being violated. Assumptions may be found to be invalid very early—in fact, one of the roles of flight test is to ensure that the aircraft performs as the models assumed it would. Many other times assumptions can become invalid over time due to changes in expectations, operating environment, wear, material fatigue, or any other factors. Another common violation of assumptions occurs with human performance. If humans are expected to perform in a specific way regardless of the situation (a common design assumption, unfortunately) that assumption is not valid from the outset. People will perform differently depending on a variety of factors, from distractions, variations in the way, and order, of information presented, fatigue, emotional issues, and more.

Castilho outlined an approach where STPA (see Chapter 13) is used to document assumptions. These assumptions are checked against actual performance, and if an assumption is not being met then controls can be designed so that the assumptions match the actual conditions. Actual performance can be checked using the various data collection methods discussed earlier in this chapter. New assumptions are then documented and monitored. This iterative approach is based on actual information that is available rather than attempting to forecast accidents and incidents based on collecting data. The reader is encouraged to research these concepts.[29]

What about those truly random events? Here, again, it is important to know the assumptions that went into the statistical analysis. This is true for any statistical approach, including Bayesian inference. It is far more likely that a statistical assumption was violated than a one in a billion-chance event occurred, so here again, documenting and checking assumptions provides a path forward. If the assumptions have been violated the risk assessment needs to be redone with accurate information.

Finally, vital to any safety organization is a critical monitoring of operational pressure. Every individual is vulnerable to operational pressure, regardless of their status, experience, or expertise. It is often subtle. All people have a strong desire to perform to expectations, particularly higher-ranking people who have been given a position of authority. Additionally, speaking up to authority is virtually always challenging and can be risky. Executives are answerable to the shareholders, and lower-level employees are answerable to their managers. If the reader wonders how a person might allow this sort of pressure to influence them, they should consider whether they were ever in a classroom or work situation where they recognized that a manager or a professor made an error, misspoke, or presented something that was actually incorrect. If so, did they point it out to the person in a position of authority? If so, they are a rare individual as such an action can be risky.

How can operational pressure be monitored? One concept might be similar to the event sets that are used to facilitate FDM, described earlier in this chapter (see Section 15.4). The event sets in this example would be set for the assumptions built into the high-level system and organizational goals, including (but not limited to) the design of the system, which could be an aircraft, the operating environment, policies, procedures, environment, human performance, and other identified areas. If the event set flagged a deviation, then the safety group would need to be empowered to take action to mitigate the risk. The latter would require specific organizational design that would empower a safety department to undertake such task. This in turn requires a strong organizational culture that is designed for safety. For more specific information on the design of a safety culture, the reader is referred to Leveson.[10]

Key Terms

Adaptability

Advanced Qualification Program (AQP)

Aviation Safety Reporting System (ASRS)

Big Data

Cognitive Systems Engineering

Evidence Based Training (EBT)

Failure Modes Effects Causality Analysis (FMECA)

Flight Data Monitoring

Flight Operational Quality Assurance (FOQA)

Gatekeeper

Graceful Extensibility

High Reliability

High Reliability Organization (HRO)

Leading Indicators

Resilience Engineering

Safety-II

Safety Case

Ultrasafe System

Voluntary Disclosure Report (VDR)

Topics for Discussion

1. Describe the concept of reliability.
2. Would you consider all of aviation to be an ultrasafe system? Justify your answer.
3. What are the advantages and disadvantages of the HRO concept in safety?
4. What is adaptability and how does it relate to the theory of graceful extensibility?

5. What are the advantages and disadvantages of EBT and AQP?
6. What is the core concept underlying Safety-II?
7. Explain how the concept of "safety case" may have contributed to accidents.
8. How have unions impacted commercial aviation safety?
9. Which reporting method, from this or the previous chapter, can make the biggest impact on flight safety? Defend your answer.
10. How can documented assumptions be used to identify leading indicators?

References

1. Amalberti, R. (2001). The paradoxes of almost totally safe transportation systems. *Safety Science, 37*(2-3), 109–126.
2. Lofquist, E. A. (2010). The art of measuring nothing: The paradox of measuring safety in a changing civil aviation industry using traditional safety metrics. *Safety Science, 48*(10), 1520–1529.
3. Le Coze, J. C. (2019). Vive la diversité! High reliability organisation (HRO) and resilience engineering (RE). *Safety Science, 117*, 469–478.
4. Perrow (1984). *Normal Accidents; Living with High-Risk Technologies.* New York, NY: Basic Books.
5. Rochlin, G. I., La Porte, T. R., & Roberts, K. H. (1987). The self-designing high-reliability organization: Aircraft carrier flight operations at sea. *Naval War College Review, 40*(4), 76–92.
6. Weick, K. E., & Sutcliffe, K. M. (2003). Hospitals as cultures of entrapment: A re-analysis of the Bristol Royal Infirmary. *California Management Review, 45*(2), 73–84.
7. Marais, K., Dulac, N., & Leveson, N. (2004). Beyond normal accidents and high reliability organizations: The need for an alternative approach to safety in complex systems. In *Engineering Systems Division Symposium.* Cambridge, MA: MIT Press, pp. 1–16.
8. Roberts, K. H. (1990). Some characteristics of one type of high reliability organization. *Organization Science, 1*(2), 160–176.
9. Roberts, K. H. (1990). Managing high reliability organizations. *California Management Review, 32*(4), 101–113.
10. Leveson, N. G. (2011). Engineering a Safer World: Systems Thinking Applied to Safety (Engineering Systems). Cambridge, MA: MIT Press, pp. 18, 39–76. http://www.worldcat.org/isbn/0262016621.
11. Woods, D. D. (2015). Four concepts for resilience and the implications for the future of resilience engineering. *Reliability Engineering & System Safety, 141*, 5–9.
12. Woods, D. D. (2016). Resilience as graceful extensibility to overcome brittleness. In: Florin, M.-V., & Linkov, I. (eds.). *IRGC Resource Guide on Resilience.* Lausanne: EPFL International Risk Governance Center (IRGC), pp. 258–263.
13. Woods, D. D. (2018). The theory of graceful extensibility: Basic rules that govern adaptive systems. *Environment Systems and Decisions, 38*(4), 433–457.
14. Box, G. E. (1976). Science and statistics. *Journal of the American Statistical Association, 71*(356), 791–799.
15. Blair, D., and Helms, N. (2013). The swarm, the cloud, and the importance of getting there first. *Air and Space Power*, pp. 13–37. https://www.airuniversity.af.edu/Portals/10/ASPJ/journals/Volume-27_Issue-4/F-Blair.pdf
16. Hollnagel, E. (2016). *Barriers and Accident Prevention.* London: Routledge.

17. Hollnagel, E. (2018). *Safety–I and Safety–II: The Past and Future of Safety Management.* New York, NY: CRC Press.

18. Goglia, J. (2006). Aviation industry embraces SMS. *AINonline.* https://www.ainonline.com/aviation-news/aviation-international-news/2006-12-21/aviation-industry-embraces-sms

19. Weinberg, G. M., & Weinberg, D. (1979). *On the Design of Stable Systems.* New York, NY: John Wiley & Sons.

20. Leveson, N. (2020). *Safety III: A Systems Approach to Safety and Resilience.* Boston MA: MIT Engineering Systems Lab. Accessed June 15, 2021. http://sunnyday.mit.edu/safety-3.pdf

21. Dekker, S. (2017). *The Field Guide to Understanding "Human Error."* New York, NY: CRC Press.

22. Cooper, M. D. (2022). The Emperor has no clothes: A critique of Safety-II. *Safety Science, 152,* e105047-e105047.

23. Woods, D. D., & Hollnagel, E. (1982, October). A technique to analyze human performance in training simulators. In *Proceedings of the Human Factors Society Annual Meeting, 26*(7), 674–675. Los Angeles, CA: Sage Publications.

24. Mumaw, R. J., Feary, M., Fucke, L., Stewart, M. J., Ritprasert, R., Popovici, A., & Deshmukh, R. V. (2018). Managing complex airplane system failures through a structured assessment of airplane capabilities (No. NASA/TM-2018-219774).

25. Heenan, H. (2022). Do airline pilots consciously alter their behavior during a simulator check. Master's Thesis. Coventry University, Coventry, UK.

26. Carroll, M., & Malmquist, S. (2021). Resilient performance in aviation. In *Advancing Resilient Performance.* Cham, Switzerland: Springer, pp. 85–95.

27. Dahlstrom, N., & Nahlinder, S. (2009). Mental workload in aircraft and simulator during basic civil aviation training. *The International Journal of Aviation Psychology, 19*(4), 309–325.

28. Hoffman, R. R., Ward, P., Feltovich, P. J., DiBello, L., Fiore, S. M., & Andrews, D. H. (2013). *Accelerated Expertise: Training for High Proficiency in a Complex World.* New York, NY: Psychology Press.

29. Dewar, J. A., Builder, C. H., Hix, W. M., & Levin, M. H. (1993). *Assumption Based Planning: A Planning Tool for Very Uncertain Times, RAND, Santa Monica, CA.* MR-114-A. https://apps.dtic.mil/sti/pdfs/ADA282517.pdf

CHAPTER 16

The Role of Government

Introduction

The role played by government entities in aviation safety is neither a glamorous nor a sultry topic of discussion in the commercial aviation community. The discourse is a necessity since regulation and official guidance are critical components of the commercial aviation system. Government agencies promote safe operations and aim to protect everyone involved in operations, whether it be passengers, office workers, ramp personnel, or pilots. However, a sad fact about such efforts is that it only takes one decision made by one person to remove all the protection that the agencies has worked to build. Safety guidance, be it advisory or compulsory in nature, is one of the most important defenses against failure, but only if followed. With a single choice, all the safeguards disappear.

Often government regulations are evolutionary, advisory, and asymmetric. It is unfortunate that many of the safety regulations produced by government agencies are the result of blood priority. By this we mean that, historically, if lives had not been lost in an airplane accident there was very little chance that changes would have been made to existing policies. As we saw with the comment made by the former chairman of the NTSB, Deborah Hersman, in Chapter 2 (Section 2.1.3) of this book, it has traditionally taken the spilling of blood to modify legislation or regulatory change. This process is slowly changing as the industry and regulators increasingly value the concepts of proactive safety, but the process is slow to change.

This chapter explores the role of semigovernment agencies and international, regional, and national aviation authorities such as ICAO, EASA, FAA, and a selection of other national authorities explaining how these organizations affect aviation operations. Each section will talk about the background, organization, and rulemaking authority to give us a better insight into how the rules and regulations come into effect. These organizations are focused primarily on promoting safety before accidents occur, as opposed to the investigation of accidents.

16.1 The International Civil Aviation Organization (ICAO)

16.1.1 ICAO Background

The Origin of ICAO: The International Civil Aviation Organization (ICAO) is a UN specialized agency and serves as the global forum for international civil aviation. Reflecting on the destructive power of aviation that was demonstrated during World War II, the leaders of the United States and Great Britain began studies of probable postwar civil aviation problems with a view to promoting economic development to heal the wounds of war. These studies resulted in US invitations to 55 states to attend an International Civil Aviation Conference in Chicago. One of the goals of this conference was to obtain uniformity in international regulations and standards that was lacking world-wide. Important work was accomplished in the technical field because the Chicago Convention[1] paved the way for a common air navigation system throughout the world. The major accomplishments of the Chicago Convention include the following:

- The Convention on International Civil Aviation provided the basis for a complete modernization of the basic public international law of the air.

- Twelve technical annexes were drafted to cover the technical and operational aspects of international civil aviation such as airworthiness of aircraft, air traffic control, telecommunications, and air navigation services.

- Regions and regional offices were established in specific areas where operating conditional and other relevant parameters were comparable.

- The permanent ICAO organization came into being in April 1947.

The Growth of ICAO—Post–World War II Developments: As aircraft grew in size, speed, and dependability following the war, it was clear that air transportation was quickly becoming a major economic force. Under US and British leadership, civilian and military aircraft aviation operations were blended into a common system of air navigation and communication procedures. Montreal, Canada, was chosen as the new ICAO head-quarters. During this postwar period, several major technical decisions were reached regarding air navigation and communication matters, such as:

- The standard aid for aircraft approaches and landing was the Instrument Landing System (ILS).

- The standard short-range navigation aids were chosen as the VHF Omnidirectional Radio (VOR) supplemented with Distance Measuring Equipment (DME), while standard long-range navigation was chosen as the Long Range Air Navigation (LORAN) radio system.

- VHF voice communication was chosen as the primary means of air-to-ground communication.

- Air traffic corridors were adopted to ensure safe aircraft separation. This concept would evolve into our modern airway system.

- As the speed of jet aircraft increased, more air traffic cooperation was required between states and ICAO regional offices. Under the new ICAO organization, congested air routes were efficiently managed.

In 1944, ICAO set forth aims and objectives to develop "the principles and techniques of international air navigation and to foster the planning and development of international air transport." These aims and objectives, as found in Article 44, were to:

- Ensure the safe and orderly growth of international civil aviation throughout the world.

- Encourage the arts of aircraft design and operation for peaceful purposes.

FIGURE 16-1 ICAO Headquarters in Montreal, Canada. (Source: November 2018, Jerome Cid, Dreamstime)

- Encourage the development of airways, airports, and air navigation facilities for international civil aviation.
- Meet the needs of the peoples of the world for safe, regular, efficient, and economical air transport.
- Prevent economic waste caused by unreasonable competition.
- Ensure that the rights of Contracting States are fully respected and that every Contracting State has a fair opportunity to operate international airlines.
- Avoid discrimination between Contracting States.
- Promote safety of flight in international air navigation.
- Promote the development of all aspects of international civil aeronautics.

Locations: With approximately 900 employees, ICAO has a headquarters, seven regional offices, and one regional suboffice in Beijing, China. Its headquarters are in the beautiful city of Montreal, Canada (Figure 16-1). These regional offices are very important for coordinating international aviation policy and standards, especially in third world and developing nations which do not have a modern aviation infrastructure.

- Headquarters—Montreal, Quebec, Canada
- Asia and Pacific (APAC)—Bangkok, Thailand
  - Suboffice—Beijing, China
- Eastern and Southern African (ESAF)—Nairobi, Kenya

- Europe and North Atlantic (EUR/NAT)—Paris, France
- Middle East (MID)—Cairo, Egypt
- North American, Central American and Caribbean (NACC)—Mexico City, Mexico
- South American (SAM)—Lima, Peru
- Western and Central African (WACAF)—Dakar, Senegal

The vision of ICAO is to achieve sustainable growth for the civil aviation systems of the world. To implement this vision, ICAO has established the following five strategic objectives:

1. **Safety**
   - Enhance global civil aviation safety. This strategic objective is focused primarily on the state's regulatory oversight capabilities. The Global Aviation Safety Plan (GASP) outlines the key activities for the triennium.

2. **Air Navigation Capacity and Efficiency**
   - Increase the capacity and improve the efficiency of the global civil aviation system. Although functionally and organizationally interdependent with safety, this strategic objective is focused primarily on upgrading the air navigation and aerodrome infrastructure and developing new procedures to optimize aviation system performance. The Global Air Navigation Capacity and Efficiency Plan (Global Plan) outlines the key activities for the triennium.

3. **Security and Facilitation**
   - Enhance global civil aviation security and facilitation. This strategic objective reflects the need for ICAO's leadership in aviation security, facilitation, and related border security matters.

4. **Economic Development of Air Transport**
   - Foster the development of a sound and economically viable civil aviation system. This strategic objective reflects the need for ICAO's leadership in harmonizing the air transport framework focused on economic policies and supporting activities.

5. **Environmental Protection**
   - Minimize the adverse environmental effects of civil aviation activities. This strategic objective fosters ICAO's leadership in all aviation-related environmental activities and is consistent with the ICAO and UN system environmental protection policies and practices.

16.1.2   ICAO Organization

ICAO is made up of three governing bodies: an Assembly, a Council, and a Secretariat. The chief officers of ICAO are the President of the Council, currently Salvatore Sciacchitano, and the Secretary General, Juan Carlos Salazar.

The *Assembly* is the first, sovereign governing body of ICAO and meets every 3 years. ICAO's 193 member states and a large number of international organizations are invited to the Assembly which review the complete work of the organization in the economic, legal, and technical fields. Each state is entitled to one vote, and the decisions of the Assembly are decided by a majority of votes cast, except when otherwise provided for in the Convention.

The 41st Triennial Assembly recently concluded in October 2022. See https://www.icao.int/Meetings/a41/Pages/default.aspx

ICAO's 193 member states and a large number of international organizations are invited to the Assembly, which establishes the worldwide policy of the Organization for the upcoming triennium. During Assembly sessions, ICAO's complete work program in the technical, economic, legal, and technical cooperation fields is reviewed in detail. Assembly outcomes are then provided to the other bodies of ICAO and to its member states in order to guide their continuing and future work, as prescribed in Article 49 of the Convention on International Civil Aviation.

Each member state is entitled to one vote on matters before the Assembly, and decisions at these sessions are taken by a majority of the votes cast—except where otherwise provided for in the Convention.

The *Council* is the permanent, second governing body of ICAO, which is elected by the Assembly and is composed of 36 Contracting States. The Assembly elects the council member states using three criteria:

- States of chief importance in air transport
- States which make the largest contribution to the provision of facilities for civil air navigation
- States whose designation will ensure that all major areas of the world are represented

As its primary governing body, the Council gives continuing direction to the work of ICAO. The primary focus is to adopt and incorporate ICAO Standards and Recommended Practices (SARPs). To develop SARPs, the Council is assisted by the Air Navigation Commission in technical matters, the Air Transport Committee in economic matters, and the Committee on Unlawful Interference in aviation security matters. The primary work of the Air Navigation Commission is to advise the Council on aviation navigation issues using international experts with appropriate qualifications and experience in this area. It is important to note that these experts are expected to serve worldwide aviation interests and function independently, and not as representatives of their member states.

The *Secretariat* is the third governing body of ICAO and is headed by the Secretary General. It is divided into five main divisions or bureaus:

- Air Navigation Bureau
- Air Transport Bureau
- Legal Affairs and External Relations Bureau
- Technical Cooperation Bureau
- Bureau of Administration and Services

Of these five bureaus, the Air Navigation Bureau (ANB) is by far the most important for the purposes of this text. The Standards (Rulemaking) process of ICAO are updated by ANB through updating ICAO's governing document, the Annexes of the Chicago Convention.

Sixteen out of 18 annexes to the Convention are of a technical nature, and therefore fall within the responsibilities of ANB and its subordinate offices.

16.1.3   ICAO Rulemaking and Standards[i]

As indicated on the ICAO website, rulemaking uses a language and process with an international focus and lexicon all its own. ICAO standards and other provisions are primarily developed in the following categories:

- Standards and Recommended Practices (SARPs)

- Procedures for Air Navigation Services (PANS)

- Regional Supplementary Procedures (SUPPS)

Some definitions are helpful at this point to distinguish between these three categories. The ICAO website has an excellent discussion in this area:

- A *Standard* is defined as any specification for physical characteristics, configuration, material, performance, personnel, or procedure; the uniform application of which is recognized as necessary for the safety or regularity of international air navigation and to which Contracting States must conform in accordance with the Convention. In the event of impossibility of compliance, notification to the Council is compulsory under Article 38 of the ICAO Convention.

- A *Recommended Practice* is any specification for physical characteristics, configuration, material, performance, personnel, or procedure; the uniform application of which is recognized as desirable in the interest of safety, regularity, or efficiency of international air navigation, and to which Contracting States will endeavor to conform in accordance with the Convention. Please note that states are merely invited to inform the Council of noncompliance.

- *Procedures for Air Navigation Services* (or PANS) comprise operating practices and material too detailed for Standards or Recommended Practices—they often amplify the basic principles in the corresponding SARPs. To qualify for PANS status, the material should be suitable for application on a worldwide basis.

- *Regional Supplementary Procedures* (or SUPPs) have application in the respective ICAO regions. Although the material in Regional Supplementary Procedures is similar to that in the Procedures for Air Navigation Services, SUPPs do not have the worldwide applicability of PANS, and are adapted to suit the particular region of the world, through the ICAO regional offices.

Additional guidance material is also produced to supplement the SARPs and PANS and to facilitate their implementation. Guidance material is issued as attachments to annexes or in separate documents such as manuals, circulars, and lists of designators/addresses. Usually, such guidance is approved at the same time as the related SARPs are adopted.

Rulemaking Process: The formulation of new or revised SARPs begins with a proposal for action from ICAO itself or from its Contracting member states. Proposals also may

---

[i] See **https://www.icao.int/about-icao/AirNavigationCommission/Pages/how-icao-develops-standards.aspx**

be submitted by international organizations. The SARP process has been very effective in recent years for ensuring the safe, efficient, and orderly growth of international civil aviation. The reason for success in this area lies in the four "C's" of ICAO international aviation: Cooperation, Consensus, Compliance, and Commitment; i.e., cooperation by member states in the formulation of SARPs, consensus in their approval, compliance in their application, and commitment of adherence to the ongoing process.

Proposals for highly technical SARPs are analyzed first by the Air Navigation Commission (ANC). Depending on the nature of the proposal, the commission may assign its review to specialized, expert working groups or panels. These experts act in their individual capacity and not as representatives of their member states. ICAO Council technical committees may also be established to deal with problems involving technical, economic, social, and legal aspects. Less complex issues may be assigned to the Secretariat for further examination, perhaps with the assistance of an air navigation study group.

These various groups report back to the ANC in the form of a technical proposal either for revisions to SARPs or for new SARPs, for preliminary review. This review is normally limited to consideration of controversial issues which, in the opinion of the Secretariat or the Commission, require examination before the recommendations are circulated to member states for comment. States are normally given 3 months to comment on the proposals.

Standards developed by other recognized international organizations can also be considered, provided they have been subject to adequate verification and validation. The comments of states and international organizations are analyzed by the Secretariat and a working paper detailing the comments and the Secretariat proposals for action is prepared. The Commission undertakes the final review of the recommendations and establishes the final texts of the proposed amendments to SARPs, PANS, and associated attachments. The amendments to annexes recommended by the Commission are presented to the Council for adoption under cover of a "Report to Council by the President of the Air Navigation Commission."

16.1.4 ICAO Safety Management

In 2010 ICAO determined that it was necessary to introduce Annex 19, *Safety Management*, in addition to the SARPs they had for each sector of the aviation system. The annex initially came into effect in November 2013 and established the management of safety at the state level through a state safety program (SPS), a safety management program (SMS), and protection of safety data and information. Later, Annex 19 was strengthened by amendments including the issuance of the Doc 9859, the current *ICAO Safety Management Manual* (4th edition). The individual national aviation state authorities have broad discretion in this arena. The United Kingdom had been receiving mandatory occurrence reports of safety-related events for the aviation industry for over 40 years. However, there is inconsistency in the level of other state programs. Larger states may be able to invest significant resources into their safety programs, while this may not be possible for smaller states that are just now beginning to establish safety management programs at a national level. A small state does not necessarily have to have a large-scale SMS program already in place to start a safety management program, as required by ICAO. Instead, they can:

- Use published data on worldwide accidents and causal factors and consider how those may differ in their state.

- Hold meetings to identify risks based on published data and according to their local professionals.
- Use free safety information to implement regulatory programs.

As one can see, this is not a task that can only be undertaken on a large scale; any state can start small to find suitable activities that wouldn't cause a major burden on any current safety efforts.

16.1.5   ICAO Current Events

ICAO has achieved an incredible level of regulatory conformance across the world of commercial aviation and has undoubtedly been a significant contributor to the levels of global safety performance we see today. It has demonstrated a consistent level of commitment to technical cooperation across the globe to maintain and strengthen the safety of air travel for the benefit of the world's traveling population. The traditional focus of ICAO has been shaped by scheduled airliners sharing the skies with cargo aircraft, business flights, and many varieties of privately piloted aircraft. As the organization looks forward, it faces increasingly crowded skies with all these varieties and now a whole plethora of new species of flying machines. Drones, rockets, spacecraft, and a wide variety of new and novel aircraft now need to be considered in the global regulatory process led by ICAO.

Part of ICAO's ability to cope with extensive change is embedded within the procedures it already has in place. From the original 12 annexes adopted in 1944, ICAO now has adopted 19 annexes and the organization has produced over 12,000 SARPs to accommodate the increasing scope of its activities. Since it was established in the political, regulatory, and social landscape of the post–World War II period, one of the fundamental challenges it now faces is the speed of change and evolution of the current commercial aviation system. While governance reforms have allowed ICAO to maintain its high level of effectiveness it is increasingly challenged by limited resources. To maintain its high level of effectiveness it must focus on its ability to maintain timely outputs from its existing governance structure; it has to become nimbler. The limited resources of ICAO has been a perennial issue but this problem will become increasingly acute given the speed of change it is facing. Having accumulated a wealth of expertise, it must now look on how to maximize its resources by commercializing some of its knowledge assets. The partnerships it has developed within the commercial sector is another initiative which will allow ICAO to keep up with cutting-edge technology and control its research and development budgets. Potential charging mechanisms directed at system users need further consideration to allow ICAO to pursue its strategic goal of "no country left behind."

The following sites give an up-to-date summary of ICAO challenges and initiatives:

- https://www.icao.int/Newsroom/Pages/default.aspx
- https://unitingaviation.com/
- https://www.icao.int/about-icao/NCLB/Pages/default.aspx

Useful ICAO websites:

- https://www.icao.int/Pages/default.aspx

- https://www.icao.int/safety/Pages/default.aspx
- https://www.skybrary.aero/articles/international-civil-aviation-organisation -icao
- https://en.wikipedia.org/wiki/International_Civil_Aviation_Organization
- https://www.icao.int/about-icao/Pages/default.aspx
- https://www.icao.tv/

16.2   The European Union Aviation Safety Agency (EASA)

16.2.1   EASA Background

Although the idea of creating a European Aviation institution goes back to the 1990s, the European Aviation Safety Agency (EASA) was formed in 2002 and became operational on September 28, 2003. As an independent body of the European Community, it was renamed the European Union Aviation Safety Agency in 2018. Based in Cologne, Germany, it has a remit to ensure a uniformly high level of safety in civil aviation throughout Europe. Initially taking over the responsibilities of the former Joint Aviation Authority (JAA), EASA's role has steadily grown within civil aviation regulation. It is the Regulatory Authority that uses NAAs (National Aviation Authorities) of its 31[ii] members to implement its regulations across the region. Through the implementation of EASA Regulation (EC) No 216/2008, EASA's role was extended to cover flight operations and flight crew licensing. In 2009 under Regulation (EC) No 1108/2009, EASA's remit was extended to include aerodromes' air traffic management and air navigation services.

Similar in size to ICAO, EASA employs around 800 staff and in 2022 the budget was €205 million. Most of EASA's funding (circa 60%) is paid by the aviation industry, with 20% being subsidized by the European Union (EU). It has four international representations which are in Montreal, Washington, Beijing, and Singapore.

EASA describes its mission in five principles:

- Ensure the highest common level of safety protection for EU citizens.
- Ensure the highest common level of environmental protection.
- Implement single regulatory and certification process among member states.
- Facilitate the internal aviation single market and create a level playing field.
- Work with other international aviation organizations and regulators.

EASA's responsibilities include:

- Provide expert advice to the EU on drafting new legislation.
- Develop, implement, and monitor safety rules, including inspections in the member states.

[ii] The 31 member states include all 27 European Union members plus Switzerland, Norway, Iceland, and Liechtenstein.

- Provide type-certification of aircraft and components, as well as the approval of organizations involved in the design, manufacture, and maintenance of aeronautical products.

- Certify personnel and organizations involved in the operation of aircraft.

- Certify organizations providing pan-European ATM/ANS services.

- Certify organizations located outside the territory subject to the EC law and responsible for providing ATM/ANS services or ATCO training in the member states where EC law applies.

- Authorize third-country (non-EU) operators.

- Provide safety analysis and research, including publication of an EASA Annual Safety Review.

EASA's tasks include:

- Help the Community legislature draw up common standards to ensure the highest possible levels of safety and environmental protection.

- Ensure that they are applied uniformly in Europe and that any necessary safeguard measures are implemented.

- Promote the spread of standards worldwide.

16.2.2   EASA Organization

At the head of EASA is its Executive Director who is appointed by the Agency's Management Board. The Board is composed of representatives from each member state. It is responsible for the definition of the Agency's priorities, its budget, and for monitoring its operation. Since 2013 the position of Executive Director has been held by Patrick Ky. As Executive Director, Ky holds the power to make decisions and adopt acts concerning safety and environmental protection. He initiates inspections and investigations as well as implements the budget and work program for the Agency. This powerful position is checked by an Independent Board of Appeal whose role is to ascertain that the correct European legislation has been applied. As illustrated in Figure 16-2, under the Chief Executive are a number of specialist directorates, including the Certification Directorate, the Flight Standards Directorate, the Resources & Support Directorate, and the Strategy & Safety Management Directorate.

In order to maintain competence in EASA's broad remit, the Agency needs to use relevant expertise and consultation from all stakeholders. There are special procedures in place where immediate action is required to address safety concerns and make individual decisions.

16.2.3   EASA Rulemaking

Rulemaking is the process of development and issuing of rules for the implementation of EASA's basic regulations. But to make these rules work, they must also establish "Opinions" on the scope and content of the basic regulation, on how it is to be implemented. Draft regulation also needs to be created along with a guide to how the regulation is to be used, known as an explanatory memorandum. EASA tends to create lots of abbreviations for many of the documentation and process. For example, the regulation

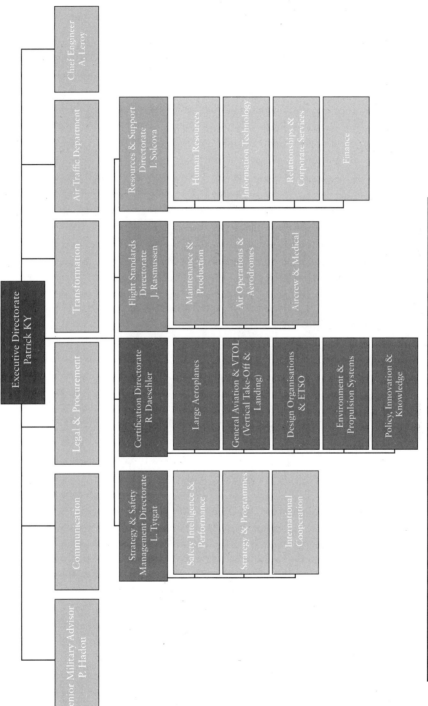

Figure 16-2 The EASA governance structure. (Source: EASA)

will need Certification Specifications (CSs),[iii] Acceptable Means of Compliance (AMC), and some Guidance Material (GM). In order to implement a basic regulation or "rule," EASA has to follow a transparent and consistent process through a number of steps.

The EASA rulemaking process:

1. *Terms of reference:* To initiate the process EASA will issue a Terms of Reference for a rulemaking task (RMT) and will then work toward the creation of a Notice of Proposed Amendment (NPA). The NPA is the output of the rulemaking group who may include subject matter experts from academia, industry, and member nation's regulatory authorities. Depending on the level of anticipated impact, a proportionate depth of analysis is undertaken which is known as a Regulatory Impact Assessment (RIA).

2. *Notice of proposed amendment:* Following this initial stage EASA issues the NPA and initiates a public consultation period where responses and comments from stakeholders are invited via a Comment Response Tool. A synoptic of the consultation output is then produced in a Comment Response Document (CRD).

3. *Opinion:* At this stage, EASA issues its position and recommendations in the form of an "Opinion." The Opinion is then evaluated by the European Commission[iv] through a process of implementation, modification an adjustment known as Comitology. Comitology refers to the process by which the European Commission exercises the implementing powers conferred on it by the EU legislator. The Commission is advised throughout the process by a series of committees made up of subject matter experts from the various member states. The whole process effectively changes "Implementing Rules" (IR) which are effectively guidance to the National Authority to "Hard Law"; that is, legally binding regulation within EU countries.

4. *Publication:* EASA produces and distributes a CRD which would typically include a "Decision." The Decision relates to the format and structure of the complied supporting documentation, which is issued as a draft of the AMC and GM. These features mean that once the IR is binding law, the NAAs maintain the ability to adapt and adjust the new regulation. NAAs are therefore able to comply by Alternative Means of Compliance (ALTMoCs) while the GM allows a purposive interpretation of the EU Regulation rather than having a one-size-fits-all regulation imposed onto NAAs. They are guiding material as the responsibility for compliance sits with the National Competent Authority, in effect EASA will tell the NAA *what* to do but not necessarily *how* to do it. The AMC and GM are normally published after the IR has been issued.

[iii] Certification Specifications (CS) are nonbinding technical standards adopted by EASA to meet the essential requirements of the Basic Regulation. Used as the basis for certification they are considered equivalent to AMC in respect of Part 121 (Design & Manufacture). Invariably CSs are aligned with FAA Federal Airworthiness Requirements (FAR).

[iv] There are three primary institutions involved in overall EU decision-making. They are the European Parliament, representing EU citizens through elected regional representatives, the Council of the European Union, representing EU governments in the form of senior ministers of each member state, and finally the European Commission, representing the EU's overall interests by producing EU legislation.

16.2.4   EASA Current Events—Response to the War in Ukraine

Although the departure of the United Kingdom from the EU and EASA (discussed in Section 16.4.1) has dominated much of the recent aviation media concerning EASA, the most significant issue currently facing EASA is the ongoing Russian military attack on Ukraine. The EU established a broad and comprehensive array of sanctions against Russia and Russian interests as a response to military aggression against Ukraine in February 2022. The most high-profile measure was the immediate banning of all Russian air carriers and aircraft from flying into, over, or out of EU territory. These measures include aviation-related services with particular focus on the banning of export of goods and technology. There is a prohibition on the provision of any technical assistance or related services to individuals or institutions that are Russian owned or for use within Russia. Measures on this scale have obviously far-reaching safety implications for EASA members.

The safety implications prompted a review of aviation-related safety issues falling under (EU) 2018/1139, which was published in April 2022.[2] The list of such a wide range of identified risks gives a good insight of the level of system complexity within modern commercial aviation.

EASA aviation safety issues arising from the war in Ukraine:

- Separation with unidentified aircraft
- Errors of civil aircraft identification by ground military forces and airborne assets outside the conflict zone
- GPS signal manipulation leading to navigation or surveillance degradation
- Continuing airworthiness-related issues due to sanctions
- Transition of a civilian airport to mixed civil-military operations
- Leased aircraft captured by the Russian Federation
- Spare parts shortages (other than aircraft)
- Increased risk of airspace infringements by military drones or aircraft spilling over from conflict zones
- Nonstandard operational air traffic routings, reservation of military areas outside the conflict zone
- Unplanned/unexpected military flights, more due regard flights
- Nonstandard military activities like drones patrolling or surveillance conducted bordering the conflict zone
- Civilian traffic unknowingly entering prohibited/restricted airspace
- Flight route congestion (hotspots)
- Skills and knowledge degradation due to lack of recent practice
- Intermediate destination stops increasing exposure to risk
- Less commonly known diversions
- A long-lasting crisis may lead to further financial strain on organizations after the COVID-19 pandemic situation
- Cosmic radiation threat associated with new (polar) routes

Useful websites about EASA:

- https://www.easa.europa.eu/en/the-agency/the-agency
- https://www.easa.europa.eu/en/home
- https://www.skybrary.aero/articles/european-union-aviation-safety-agency-ea

16.3   The Federal Aviation Administration (FAA)

The FAA is a US government agency with primary responsibility for the safety of civil aviation. The FAA's major roles and responsibilities include:

- Regulating civil aviation to promote safety
- Encouraging and developing civil aeronautics, including new aviation technology
- Developing and operating a system of air traffic control and navigation for both civil and military aircraft (Figure 16-3)
- Researching and developing the National Airspace System and civil aeronautics
- Developing and carrying out programs to control aircraft noise and other environmental effects of civil aviation
- Regulating US commercial space transportation

FIGURE 16-3   An FAA air traffic controller. (Source: FAA)

16.3.1   FAA Background

The FAA is responsible for performing several activities that are in support of the previously mentioned functions. These activities as reproduced from the FAA website include:

- *Safety regulation.* The FAA issues and enforces regulations and minimum standards covering the manufacture, operation, and maintenance of aircraft. The agency is responsible for the rating and certification of airmen and for certification of airports serving air carriers.

- *Airspace and air traffic management.* The safe and efficient utilization of the navigable airspace is a primary objective of the FAA. The agency operates a network of airport towers, air route traffic control centers, and flight service stations. It develops air traffic rules, assigns the use of airspace, and controls.

- *Air navigation facilities.* The FAA is responsible for the construction and installation of visual and electronic aids to air navigation, and for the maintenance, operation, and quality assurance of these facilities. Other systems maintained in support of air navigation and air traffic control include voice/data communications equipment, radar facilities, computer systems, and visual display equipment at flight service stations.

- *Civil aviation abroad.* The FAA promotes aviation safety and encourages civil aviation abroad. Activities include exchanging aeronautical information with foreign authorities; certifying foreign repair shops, airmen, and mechanics; providing technical aid and training; negotiating bilateral airworthiness agreements; and taking part in international conferences.

- *Commercial space transportation.* The agency regulates and encourages the US commercial space transportation industry. It licenses commercial space launch facilities and private sector launching of space payloads on expendable launch vehicles.

- *Research, engineering, and development.* The FAA does research on and develops the systems and procedures needed for a safe and efficient system of air navigation and air traffic control. The agency supports development of improved aircraft, engines, and equipment. It also conducts aeromedical research and evaluations of items such as aviation systems, devices, materials, and procedures.

- *Other programs.* The FAA provides a system for registering aircraft and recording documents affecting title or interest in aircraft and their components. Among other activities the agency administers an aviation insurance program, develops specifications for aeronautical charts, and publishes information on airways airport services, and other technical subjects in aeronautics.

16.3.2   FAA Organization

With over 35,000 air traffic controllers, technicians, engineers, and support personnel, the FAA is the largest agency of the US Department of Transportation (DOT). The FAA has authority to regulate and oversee all aspects of civil aviation, with an estimated annual budget request for 2023 of $18.6 billion. The current organization of the FAA is provided in https://www.faa.gov/about/office_org

Additional information on each of its offices together with details is readily available on the FAA website. Some of the office functions that are important to the scope and this text will be discussed in greater detail throughout the book. For the purposes of this study please see the following links to the FAA Air Traffic Organization (ATO), FAA Airports (ARP), and FAA Aviation Safety (AVS).

- ATO https://www.faa.gov/about/office_org/headquarters_offices/ato
- ARP https://www.faa.gov/about/office_org/headquarters_offices/arp
- AVS https://www.faa.gov/about/office_org/headquarters_offices/avs

16.3.3   The FAA and the Boeing 737-MAX Accidents

The two Boeing 737 MAX-8 fatal accidents of 2018 and 2019 created a significant challenge for worldwide aviation safety. A discussion of these accidents has been provided throughout this textbook, and for review, please refer to the NTSB report provided in Skybrary.aero: https://skybrary.aero/articles/ntsb-safety-recommendations-report-arising-boeing-737-max-8-fatal-accidents-2018-and-2019

In accordance with the US Federal Aviation Regulations (Airworthiness Standards—FAR Part 25), Boeing was required to certify the airworthiness of the design and production of its 737 MAX series aircraft. Subsequent aircraft crash investigations, recommendations, and redesign of the system ultimately solved the technical issues with the Maneuvering Characteristics Augmentation System; but, of course, the reputation of both Boeing and the FAA had been significantly damaged.

In like manner, the Federal Aviation Act (FA Act) specifies, in part, that when prescribing standards and regulations and when issuing certificates, the FAA shall give full consideration "to the duty resting upon air carriers (Airlines) to perform their services with the highest possible degree of safety in the public interest." The FA Act charges the FAA with the responsibility for promulgating and enforcing adequate standards and regulations. At the same time, the FA Act recognizes that the holders of air carrier certificates have a direct responsibility for providing air transportation with the highest possible degree of safety. The meaning of this statute should be clearly understood. It means that this responsibility rests directly with the air carrier, irrespective of any action taken or not taken by an FAA inspector or the FAA.

Most of the day-to-day safety inspections, reviews, and signoffs are performed by the cognizant manufacturers, airlines, and airports. In fact, the system depends on self-inspections, and it is simply not possible for the FAA to make every inspection on every airplane in every location around the world. This self-inspection, or "designee," concept is startling to many of the general public, but it has worked effectively for many decades. The airlines and the manufacturers have a great concern for the safety of their airplanes and operations; it is in their business interests to place a high priority on safety.

Before certification, the FAA's objective is to make a factual and legal determination that a prospective certificate holder is willing and able to fulfill its duties as set forth by the FA Act and complies with the minimum standards and regulations prescribed by the FAA. This objective continues after certification. If a certificate holder fails to comply with the minimum standards and regulations, Section 609 of the FA Act specifies that the FAA may reexamine any certificate holder or appliance. As a result of an inspection, a certificate may be amended, modified, suspended, or revoked, in whole or in part. Additionally, Section 605(b) generally provides that whenever an inspector finds that any aircraft, aircraft engine, propeller, or appliance used or intended to be used by any air carrier in air transportation is not in a condition for safe operation, the inspector shall notify the air carrier, and the product shall not be used in air transportation until the FAA finds the product has been returned to a safe condition.

The following conditions or situations could indicate that an air carrier's management is unable or unwilling to carry out its duties as set forth by the FA Act:

- Repetitive noncompliance with minimum standards and regulations
- Lack of knowledge and/or understanding of minimum standards and safe practices
- Lack of motivation to exercise operational and/or quality airworthiness control
- Inaccurate record-keeping procedures

The FA Act and the FARs contain the principle that air carriers offering services to the public must be held to higher standards than the general-aviation community. Inspectors must also be aware of the private rights of citizens and air carriers. Since public safety and national security are among the FAA's highest priorities, FAA inspectors must be prepared to take action when any air carrier does not, or cannot, fulfill its duty to perform services with the highest possible degree of safety.

FAA Safety Inspection Program: Aviation safety depends in part on the quality and thoroughness of the airlines' maintenance programs and on oversight and surveillance by safety inspectors of the FAA. Airline deregulation has been accompanied by increasing concern that maintenance standards might have been lowered at some carriers and that pressures of the economic marketplace might lead to unsafe operating practices. At the same time, deregulation has increased the stress on FAA inspection programs. The existing regulatory inspection program, with its local and regional structure, does not have sufficient flexibility to adapt to a dynamic industry environment.

The FAA Flight Standards District Offices (FSDOs) nationwide handle the dual functions of safety inspection and advice for airlines. In addition to scheduled airline surveillance, the local offices are responsible for safety inspections of nonscheduled air taxis and other operations, such as flight schools, engine overhaul shops, and private pilots. An air carrier's operating certificate is held at a specific flight standards office, typically the one nearest the carrier's headquarters or primary operations or maintenance base. For each carrier, a principal inspector is assigned to operations (flights, training, and dispatch), airworthiness (maintenance), and avionics (navigation and communications equipment). For large airlines, each of the principal inspectors can have one or two assistants.

Base inspections are generally pre-announced. There is also a tendency to focus on records rather than to probe deeply into the data underlying carrier records concerning maintenance, training, and flight crew logs. Inspectors at the local level try to work with air carriers to achieve compliance when they find discrepancies. Violations and fines are viewed as a last resort.

Every major airline has a reliability program that monitors maintenance activities and looks for emerging problems. For example, most airlines monitor engine temperatures, oil consumption, and the metal content of oil. They then use these tests to determine when an individual engine needs to be overhauled or repaired. Some airlines also use statistical measures such as the number of engines requiring premature overhaul, engines that are shut down in flight, the number of mechanical discrepancies that are left outstanding at flight time, and the rate at which these discrepancies are cleared. In some companies, analysts search for adverse trends that might indicate, for example, a shop procedure that needs to be revised, both to ensure safety and to reduce maintenance expenses. Some FSDOs have taken advantage of these statistical data to monitor the effectiveness of airline maintenance. In some cases, flight standards inspectors have also encouraged airlines to set up or expand their statistical reliability programs.

16.3.4　FAA Rulemaking Process[v]

Over the last 50 years, numerous reports have documented the delays in the FAA rulemaking process. There is no question that rulemaking is extremely time-consuming and complex, requiring careful consideration of the impact of proposed rules on the public, aviation industry, the economy, and the environment. The US General Accounting Office (now Government Accountability Office) has reported that after formally initiating a rule, the FAA took an average of 30 months to complete the rulemaking, and 20 percent of the rules took over 10 years or longer to complete.

The process for creating federal regulations generally has three main phases:

1. Initiating rulemaking actions
2. Developing proposed rules
3. Developing final rules

In practice, however, this process is often complex, requiring regulatory analysis, internal and interagency reviews, and opportunities for public comments. There are a number of ways to improve the transparency and effectiveness of this process.

In recent years, difficult rulemaking issues such as changes to the Helicopter Emergency Management System FAR rules pertaining to hospital MedEvac helicopter flights have taken several years in development in spite of numerous crashes, dozens of fatalities, and NTSB strong public recommendations in this area. Often it takes public outcry and political pressure to move the process forward such as seen in the wake of the Colgan Air Commuter flight crash in Buffalo, New York, in February of 2009.

16.3.4.1　FAA Office of Rulemaking (FAA ARM)

Steps in the formal process:

The formal FAA rulemaking procedures are provided in 14 CFR Part II. A diagram from the FAA's Rulemaking Process is provided in Figure 16-4.

The Secretary of Transportation reviews FAA programs for regulatory compliance. The DOT secretary notifies the FAA of schedules for certain projects, expresses general departmental policy issues, and identifies specific areas of concern. The FAA, through the DOT secretary, submits an annual draft regulatory program to the Office of Management and Budget (OMB). The program covers regulatory policies, goals, and objectives and also provides information concerning significant regulatory actions or actions that might lead to rulemaking. The OMB reviews the program to ensure that all proposed regulatory actions are consistent with administration principles.

---

[v] https://www.faa.gov/regulations_policies/rulemaking

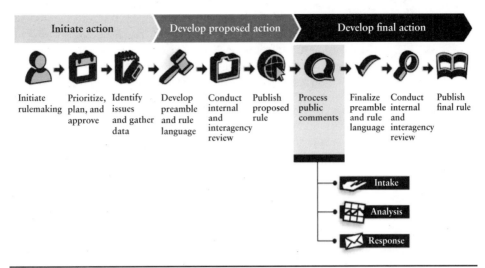

FIGURE 16-4   FAA rulemaking for significant rules. (Source: US General Accounting Office, GAO-20-383R)

After a rule is cleared to proceed, the appropriate FAA program office organizes a team to manage each rule. An Advance Notice of Proposed Rulemaking may be prepared and published at this point to obtain additional information, followed by a draft Notice of Proposed Rulemaking (NPRM) which is reviewed for regulatory impact analysis when appropriate. An impact analysis includes potential costs and benefits to society, lower-cost approaches that were not chosen and why, and an explanation of any legal preclusions from cost/benefit criteria.

When the team receives the regulatory impact analysis, it develops a new draft and briefs the principals (associate administrators, etc.) of interest. Following the principals' briefing, the team coordinates the package at the branch and division levels. After the branches and divisions concur, the package is coordinated with the principals. Next, the Associate Administrator for Aviation Safety reviews the package and forwards it to the chief counsel, who prepares the package for the Office of the Secretary of Transportation and OMB review.

After the FAA review process is complete, the FAA chief counsel submits the NPRM package to the Department of Transportation (DOT). After they sign off, the package is finally forwarded to the OMB for its review. If anyone at any stage in the process objects, the package is reworked until the problem is resolved. At that point, the FAA coordinates the package with the appropriate people. This process continues until everyone concurs or at least until all comments are considered.

After the Notice of Proposed Rulemaking is published in the Federal Register, the public may comment on the rule for a set period of time. At the close of the comment period, the FAA compiles the comments and considers appropriate changes in the rule. It develops a draft of the final rule and the whole process starts again. The final rule is reviewed by the FAA, the Office of the Secretary of Transportation (OST), and the OMB. After the rule clears the process again, the OMB releases it to the FAA. The administrator issues the final rule and has it published in the Federal Register. Once a regulation is completed and has been printed in the Federal Register as a final rule, it is "codified" by being published in the Code of Federal Regulations (CFR). The CFR is the official record of all regulations created by the federal government. These regulations are organized into 50 topics called titles or volumes. FAA regulations are found in Title 49 of the CFR.

Most analysts of the FAA rulemaking process agree that the area most in need of improvement is that of timeliness in identifying and responding to safety issues. Critics complain that excessive delays in the rulemaking process complicate and delay new or amended certification programs and make them more costly. They complain of excessive rewriting and divergent rules for different aircraft categories when such rules should be identical. Some critics blame the OST and the OMB for much of the delay.

Some critics complain that the OST and the OMB lack the necessary technical background to understand and review FAA rules. Others suggest that rulemaking quality and timeliness would improve if one person were accountable for the whole process. On the other hand, some argue that the OST reduces the delay over the long term by helping the FAA deal with the OMB. Most disagreements between the FAA and either the OST or the OMB concern economic analysis. OST officials often question the quality of the cost/benefit analysis coming from the FAA, whereas FAA officials maintain that lack of understanding hinders the quality of those evaluations.

The problem of rulemaking delay seems to lie with the whole process, not just the DOT, the OMB, or the FAA. In any case, most analysts agree that the rulemaking process needs to be streamlined before it hampers progress in the necessary changes that are needed for next-generation air traffic systems (Figure 16-5). The essential steps in developing sound public safety policy must be identified, and the rest eliminated.

FIGURE 16-5   Modern FAA next-generation air traffic control. (Source: FAA)

Mandate for commercial space operations (FAA AST)[vi]:

> When the FAA was established, its jurisdiction for rulemaking included promoting both the growth of industry and safety. As the industry grew, though, the FAA eventually moved toward focusing solely on safety. Since 1984, the FAA has been responsible for promoting spaceflight operations and overseeing its safety. Under the Office of Commercial Space Transportation, regulations are being created to ensure protection of the public, property, and national security and foreign policy interests of the United States during commercial launch or reentry activities, and to encourage, facilitate, and promote US commercial space transportation.

16.3.5 FAA Airworthiness Directives

When the need arises, the FAA releases airworthiness directives. These notices are legally enforceable rules that aim to correct an unsafe condition that arises in a particular model of aircraft, engine, avionics, or any other system in the airplane that needs to be corrected. For example, in April 2016 the FAA issued an airworthiness directive requiring Boeing 787 operators to update flight manuals due to the prevalence of bad airspeed data. The FAA made this move after there had been three safety reports of airspeed anomalies.

In each reported case, the displayed airspeed rapidly dropped significantly below the actual airplane airspeed. The directive warned the pilots not to use abrupt control inputs in case of erroneous airspeed indications, since such inputs at high speeds could structurally damage the aircraft, and also to disconnect the autopilot before making any manual flight control inputs.

16.3.6 The Colgan Air Crash and Its Aftermath

In early 2009, a regional commuter flight from Newark Liberty International Airport to Buffalo crashed while on final approach, causing 50 fatalities (Colgan Air Flight 3407). This single accident continues to have a profound effect on commercial aviation safety thinking and through the many new commercial aviation safety laws that have followed. The final NTSB report on the accident can be found in https://www.ntsb.gov /investigations/accidentreports/reports/aar1001.pdf.

The Colgan Air crash was highly influential in the thinking and shaping of aviation safety regulation in the United States and across the world, but most notably in the Airline Safety and Federal Aviation Administration Act of 2010. The Act mandates broad changes in Airline Safety and Pilot Training Improvement. Title II of the Act requires the FAA to issue new regulations in a variety of areas:

- Establishment of a comprehensive FAA Pilot Records database to ensure that airlines have access to all relevant pilot training information.
- Establishment of a special FAA Task Force on Air Carrier Safety and pilot training to evaluate best practices in the industry and provide recommendations.
- High-level review of FAR Part 121 air carriers by the Department of Transportation Inspector General.
- Establishment of a flight crewmember mentoring, professional development, and leadership program to emphasize the highest standards of professional

---

[vi] https://www.faa.gov/about/office_org/headquarters_offices/ast

performance with special focus on strict compliance with the "sterile cockpit rule" to eliminate nonessential voice communications between crewmembers during the critical phases of flight.

- A study of aviation industry best practices with regard to flight crewmember pairing, crew resource management, and pilot commuting to and from airline hubs from their place of residence.

- Implementation of specific Colgan crash NTSB remedial training recommendations regarding aircraft stall and recovery training, stick pusher, and weather event training including icing conditions, microburst, and wind shear weather events.

- Establishment of a multidisciplinary panel to recommend best practices in FAR Parts 121 and 135 training methods including initial and recurrent testing requirements for pilots; classroom instruction requirements; the best methods to allow specific academic training courses to be credited toward an Airline Transport Pilot Certification, among other priorities.

- Random, onsite safety inspections of Regional Air Carriers by the FAA at least once each year.

- Special emphasis on combating pilot fatigue through specific limitations on the hours of flight and duty time allowed for pilots. Part 121 air carriers must now submit a "Fatigue Risk Management Plan" and update it every 2 years.

- Bolstering of four industry voluntary safety programs in the following areas:
  - Aviation Safety Action Plan (ASAP).
  - Flight Operational Quality Assurance Program (FOQA).
  - Line Operations Safety Audit (LOSA).
  - Advanced Qualification Program (AQP).

- Specific development and implementation of ASAP and FOQA programs by FAR Part 121 air carriers.

- Safety Management Systems (SMS) are now legislatively required for all FAR Part 121 air carriers. This is indeed a revolutionary change which will have a profound effect on the commercial aviation industry. Under this section of the Act, each carrier's SMS program shall consider ASAP, FOQA, LOSA, and AQP concepts.

- Enhanced Air Transport Pilot (ATP) Certificate requirements. Under the act, all flight crewmembers must hold an ATP certification which is the highest pilot rating available under FAA regulations. This requirement is also a major change for the regional (commuter) airlines which previously allowed a pilot with a commercial rating to serve as first officer on their aircraft. The Act provides a default clause which requires an ATP certificate for all FAR Part 121 pilots by 3 years from enactment of this Act (July 31, 2013).

The minimum requirements for a pilot to be qualified to receive an ATP are proposed as at least 1500 flight hours and flight training, academic training, or operations experience to:

- Function effectively in a multipilot environment.
- Function effectively in adverse weather conditions, including icing.

- Function effectively during high altitude operations.
- Adhere to the highest professional standards.
- Function effectively in an air carrier operational environment.

Useful FAA websites:

- https://www.faa.gov/
- https://skybrary.aero/articles/federal-aviation-administration-faa
- https://www.skybrary.aero/search/google?keys=787
- https://www.faa.gov/about/mission/activities#:~:text=Regulating%20 civil%20aviation%20to%20promote,Airspace%20System%20and%20civil%20 aeronautics

16.4   National Aviation Authorities (NAAs)

Most countries around the world have nowhere near the level of resources deployed to regulate commercial aviation like those of the FAA and EASA. Although the FAA itself is a NAA, its sheer size and influence place it with the rankings of international regulatory bodies. Many NAAs of much more modest size rely on simply copy and paste of the regulatory output from these internationally influential institutions. There are a generic set of regulatory roles undertaken by these NAAs; they would typically include the following:

- The design of aircraft, engines, and any equipment affecting the safe operation of aircraft
- The condition and standards of manufacture and maintenance of this equipment
- Standards of operation for aircraft, equipment, and air traffic management
- Licensing of pilots, engineers, air traffic controllers, dispatchers
- Licensing of airports and navigation systems

Utilizing the legal framework established by ICAO, the NAA would typically derive its authority from a legislative instrument such as an act of parliament which empowers the authority to produce its own specialist and technically focused acts within this established framework. There has been a global trend toward harmonization, both through ICAO and bilateral agreements between individual countries allowing recognition of each other's aviation safety regimes. Some NAAs may operate their own airfields or air traffic control systems; others may prefer to delegate this function to private or quasiprivate organizations. Some NAAs are multimodal, that is they regulate all modes of transport. In some countries the NAA is responsible for the investigation of aviation accidents while others have separate specialist investigation bodies such as the NTSB in the United States or the Bureau d'Enquêtes et d'Analyses pour la Sécurité de l'Aviation Civile in France.

16.4.1   Civil Aviation Authority—CAA (UK)

The UK's Civil Aviation Authority was established in 1972 under the terms of the Civil Aviation Act 1972. Its 1200 employees are responsible for overseeing and regulating all civil aviation activities in the United Kingdom. It is a UK government corporation which is overseen by the Department for Transport (DfT). Air accidents are investigated by a separate subdivision of the DfT, that is the Air Accident Investigation Branch (AAIB).

The original plan for funding the CAA was for the organization to be self-funding, receiving income from those organizations it regulates rather than the UK taxpayer. This arrangement has occasionally prompted some criticism about the CAA's ability to maintain an appropriate level of impartiality in its role as a regulator. In recent years the CAA has been unable to maintain its funding levels and has had secured funding from the DfT. This change in funding policy has officially been declared as a result of the COVID-19 outbreak and the significant reduction in commercial aviation activity.[3] Another reason the CAA may need further funding is the administrative burden created by Brexit, although this is unlikely to be officially acknowledged due to political sensitivities.

After the United Kingdom completely left the European Union, at the end of the transition period on December 31, 2020, the law that applied to commercial aviation in the United Kingdom was all UK law. From EASA's perspective, the United Kingdom is a "third country" and therefore has to establish its safety rules, procedures, and standards that are acceptable to operate within the jurisdiction of the EU. As this change to the UK's aviation regulatory process is so substantial, to maintain continuity, the United Kingdom opted to "retain" a substantive element of EU Regulation which would gradually be replaced with UK law over a period of time.

The enacting legislation which underpins the new UK law has yet to be published. As such, the UK CAA has in the majority a regulatory structure that looks very much like the EASA regulatory structure it left behind (Figure 16-6). As an example, referring to the CAA website and searching for, as an example, the regulation covering Airworthiness and Environmental Certification, the reference is depicted as "UK Reg (EU) 748/2012 (the UK Initial Airworthiness Regulation)."[4] Following the link for further detailed guidance leads to a series of CS or Certification Specifications (AMCs) or

FIGURE 16-6   At the time of writing the CAA has retained a substantive element of EASA regulation. (Source: Dreamstime)

Acceptable Means of Compliance and (GM) Guiding Materials, are all EASA terminology that was described earlier in this chapter under the EASA rulemaking process. In summary, UK commercial aviation regulation is in the foreseeable future controlled by the UK government but looks very similar to that of EASA.

After the United Kingdom officially left the EU, under the terms of the exit agreement, aviation was granted a 2-year extension of the mutual recognition of some approvals and licenses for a period of 2 years. The hope was that a Trade and Cooperation Agreement (TCA) would be reached within that period between the UK and the EU but at the time of writing there are no signs that an agreement will be reached. The CAA has recorded some 8000 commercial aviation personnel leaving the UK system since it became apparent there would be a separation from EASA.[5] From December 31, 2022, all UK operators will be deemed third-country operators (TCOs) and have to apply to EASA for a TCO authorization certificate. Personnel that had been operating under the privileges of an EASA license such as pilots, engineers, Maintenance and Repair Organizations (MROs), and air carriers holding TCO certificates will have to apply to the UK CAA for an issue of a license to continue their activities within the United Kingdom.

Useful website for the CAA:

- CAA website: https://www.caa.co.uk

16.4.2   Transport Canada (TC)

Transport Canada (TC) is the Canadian government's ministry responsible for the country's transport policies. It is a multimodal regulator, covering aviation, rail, and shipping and sees its mission as the promotion of safe, secure, efficient, and environmentally responsible transportation. Formed in 1935 it had previously been responsible for the Canadian Coastguard and the St Lawrence Seaway, airports, and seaports. In 1994 it had a significant reduction in its portfolio and an increased focus on policy and regulation rather than operational coordination for Canada's transport infrastructure. Civilian air traffic services were transferred to a nonprofit company Nav Canada. After a period of nationalization, TC now retains control of most of the country's airfields. TC is responsible for the licensing of the country's pilots, engineers, and dispatchers.

The TC's Civil Aviation (TCCA) is a Crown corporation, responsible for the regulatory element of commercial aviation through the Canadian Aviation Regulations (CARs). It reports to parliament through the Minister of Transport. It governs the protection of specific elements of the air transport system such as passenger and airport worker screening. Based in Ottawa, it has six regional headquarters across the country.

Useful websites for TC:

- https://www.skybrary.aero/articles/transport-canada
- https://tc.canada.ca/en/corporate-services/transparency/briefing-documents -transport-canada/2021/minister/overview-regulatory-process

16.4.3   Civil Aviation Safety Authority—CASA (Australia)

CASA was established in 1995 as the independent statutory authority and governing body that regulates Australian aviation safety. The organization licenses pilots, registers aircraft, oversees aviation safety processes as well as promoting good practice in safety management. With over 800 employees and an approximate annual budget of A$200 million, it also holds responsibility for administering and classifying Australian airspace under the authority of the 2007, Airspace Act. CASA's corporate vision is "Safe skies for all" and in its Corporate Plan for 2021–2022, it defines its mission, "To promote a positive and collaborative safety culture through a fair, effective and efficient aviation safety regulatory system, supporting our aviation community."

The regulatory framework and function of CASA are defined under the Civil Aviation Act 1988. The Act states that CASA's role is the promotion of safety with particular emphasis on preventing accidents and incidents. The organization is managed by a CEO (Chief Executive Officer) who answers to the CASA Board. The Board is appointed by a government minister, that is the Minister for Infrastructure, Transport, Regional Development, and Local Government. The regulatory philosophy could be described as progressive insofar as it is pragmatic. CASA states that while it is mindful of the primacy of air safety, CASA takes account of all relevant considerations including cost. It states that safety is CASA's most important consideration, although it is not the only consideration when performing its regulatory function.

Useful website for CASA:

- https://www.skybrary.aero/articles/civil-aviation-safety-authority-casa

16.4.4   Agência Nacional de Aviação Civil—ANAC (Brasil)

Established under Federal Law number 11,182 of September 27, 2005, ANAC regulates one of the world's largest national fleets with over 10,000 Brazilian-operated civil aircraft. The agency's headquarters is based in the city of Brasilia and has four regional units in Rio de Janeiro, São Paulo, Porto Alegre, and Recife. It forms part of the Brazilian Secretariat of Civil Aviation and is a federal agency of the Brazilian government. ANAC is run through a collegiate board of directors and a CEO who will be appointed for a term of 5 years. Each of the four directors has a Superintendency which governs Operational Standards, Airport Infrastructure, Airworthiness, and Economic Regulation.

ANAC faced considerable criticism for the period between 2006 and 2008. In a period referred to as the "Apagão Aéreo" or aerial blackout, the country's civil aviation sector saw a widespread level of disruption and lack of confidence in its ability to maintain safety standards after a Gol Airlines Boeing 737 collided with an Embraer 135 Legacy business jet. The Gol 737 broke up in midair before crashing and killing all 154 passengers and crew. The accident was investigated by the Centro de Investgacão e Prevencão de Acidentes Aeronáuticos, or CENIPA. Initially the American Legacy pilots were blamed for the accident, but public anger soon turned back toward the Brazilian government and regulator. The crisis was then further accentuated by a series of ATC strikes about the country's safety standards and infrastructure which resulted in significant delays and flight cancellations.

Useful website for ANAC:

- https://skybrary.aero/territories/brazil

16.4.5   The Civil Aviation Administration of China—CAAC

Based in Beijing, the CAAC is a government-controlled agency responsible for the regulation and governance of all civil aviation in China. The CAAC is led by one administrator and supported by four deputy administrators. China is the second-largest aviation market in the world and is set to become the largest sometime in the mid-2020s.[6] CAAC is assigned tasks from the State Council and from the Ministry of Transport. as Along with the overarching responsibility for civil aviation safety, the CAAC like many other national aviation authorities holds responsibility for not only the technical regulation of civil aviation but also the economic regulation. The CAAC also conducts air traffic management on behalf of the Chinese state and oversees all elements of aviation security including the planning and construction of new airports across the country. The CAAC also investigates civil aviation accidents and incidents.

After a poor record for aviation safety before the start of the 21st century, China has now developed one of the highest performing safety records in the world, significantly lower than the average of 0.57 accidents per million flying hours. In 2021 there were no reported accidents; however, in 2022 two fatal accidents have prompted a significant review and subsequent reform of the whole of the Chinese civil aviation system. On March 21, a China Airlines Flight MU57835 crashed into the mountains of Guangxi, killing all 132 on board. Initial reports suggested the flight may have been deliberately crashed, but the effect of the highly televised crash shocked a nation that had not seen a fatal accident in over a decade and a 100 million flying hours without a major civil aviation accident. The second incident involved a Tibet Airlines Flight 9833, which overran the runway before catching fire and injuring 36 of the 122 passengers. The CAAC has now announced a far-reaching reform of the whole of the Chinese aviation system to address this lapse in safety performance.

Useful websites for CAAC:

- http://www.caac.gov.cn/en/GYMH/ZYZN/
- https://www.skybrary.aero/territories/china

16.5　The Criminalization of Aviation Accidents

One of the most important roles of government is the establishment and adherence to a system of law. The credibility of such legal systems is predicated on the understanding that all legal entities, individuals, institutions, and corporations understand their responsibilities and obligations under the law. Over the past few decades, there has been a global increase in the use of the criminal justice system as a solution to broader social and political issues. An increasing use of the criminal justice system in the aftermath of commercial aviation accidents has raised widespread concerns about the impact on overall system safety.

Throughout the international agreements of the Chicago Convention described in ICAO documentation and the various national procedures, described earlier in this chapter, a single theme prevails in the investigation process and intent of aircraft accident investigation. This single theme is that the sole purpose of an investigation into an aircraft accident is to prevent re-occurrence and preserve life. This principle stands in contrast to air accident investigation carried out by a state prosecuting authority. If there is reasonable suspicion that a criminal act has taken place, then prosecutors are duty bound to preserve evidence and actively search for any potential breach of the nation-state's criminal standards of conduct. These criminal standards are generally determined by society's collective intolerance of certain acts or omissions; these are judged by society to be unacceptable, and some form of punishment is often imposed as a deterrent, sometimes to invoke justice or occasionally as simple retribution.

These standards of conduct are sometimes but not always represented by regulatory compliance. More often, the aftermath of a multifatality accident is not an environment of calm objectivity, but one of high sociopolitical tension and the collective need of society to find some form of closure drives the agenda. The high level of political currency that can be swayed around in the aftermath of a commercial aviation accident accentuates the need for society to find closure. Over the course of the past three decades, this closure has increasingly translated into the criminal prosecution of those deemed to be responsible.

16.5.1   The Growth of Criminalization in Commercial Aviation

The criminalization of aircraft accidents describes the use of the criminal justice system against frontline operatives in aviation such as pilots, air traffic controllers, maintenance personnel, and increasingly their employers. According to a 2016 book, *Flying in the Face of Criminalization* written by Michaelides-Mateou and Mateou,[7] criminalization has grown to an annual global average of slightly over three cases per year between 2000 and 2010. With such low sample data, there can be considerable variation in annual figures, and as the criminal process takes place under sovereign authority, there is no official record of the number of incidents at international level. The difficulty in measuring the precise current growth trends in the number of events is further compounded by the invariably long periods of gestation that case investigation and compilation can take. For example, following the crash of the Air France Concorde near Charles de Gaulle Airport in Paris on July 25, 2000, the various prosecutions of individual and corporate manslaughter took over a decade to process through the French legal system. Following the crash of Air France 447, which is discussed in Chapter 1, various manslaughter charges have been brought against both companies—Air France and Airbus. The latest of these being in October 2022, over 13 years after the accident off the north coast of Brazil; at the time of writing the case is still ongoing in the French courts. What can be determined from the recorded observations of criminalization made by Michaelides-Mateou and Mateou compared with ICAO accident statistics is that while the number of aviation operators involved in a fatal accident in the first decade of the 21st century has marginally decreased since the 1990s, the number of operators, or their employees, then subsequently subjected to criminal prosecution, during that period, has at least doubled.[8]

16.5.2   Why Does Criminalization Happen?

As we discussed in Part I of this book, we humans need to understand why things happen. When we are unable to link cause and effect, we just make up a link. Criminal law is often deployed as a means of pacifying social unease. It gives an impression of regaining control after adverse events in society. Since the 1990s researchers such as Anthony Giddens[9] and Ulrich Beck[10] have noted an increasing sense of social anxiety around new technologies. We have developed an increased dependency on emerging technology so when it fails, we not only want to know why, but we want to know how we can regain control. Criminal law can sometimes give us the sense that when commercial aircraft accidents occur, they are the result of a few "bad apples" in the system. Identifying them, removing them from the system, and punishing them means we can resume normal service and get on with our lives. In this book we have explored some of the various concepts of causation but do not assume the legal systems of the world are aligned with the various philosophies of how and why things happen. The vagaries of jurisprudence mean any legal concept of causation must deal with a complex combination of social, political, and practical considerations so the philosophical purity of causation is set aside in favor of legal pragmatism. As society has learned to expect such very high standards from the commercial aviation system, when it fails, many people consider it must be someone's fault. Professor Sidney Dekker succinctly describes this change in social expectation, "Accidents are no longer accidents at all. They are failures of risk management."[11] Apart from the assumption of fault, there have been numerous explanations that have been proposed for the increase in the criminalization of aviation accidents, including the following:

- The increase in the criminal justice system to solve societal anxiety of technical failures

- An increasing awareness of the fallibility of human operators in safety critical roles

- An unevidenced faith in the reliability of complex sociotechnical systems

- The exploitation by the media of emotive imagery of major accidents

- A misuse of accident investigation material by prosecutors

In all likelihood, the reason for the rise in incidents of criminalization in aviation is part of a broader social phenomenon and will invariably be a combination of some or all of these influences. Its prominence in the sphere of commercial aviation is undoubtedly linked to the fear that many people have of aviation accidents. As discussed in Chapter 1, the emotional impact of multifatality aviation accidents has a deeper and wider ranging effect on society than any other comparable human activity. It is perhaps not unsurprising that this stimulates the law, as the "whip hand" of society to react.

16.5.3   The Impact on Safety of Criminalization

When we consider the impact on safety of criminalization there are two main areas that are affected. The first is the physical and legal restrictions that are placed around an accident investigation. Earlier in this chapter we discussed Annex 13 of the ICAO charter which emphasizes the prime goal of accident investigation is to improve safety standards rather than find blame. As such accident investigators need direct, timely, and physical access to an accident site in order to collate as much evidence as soon as they can after an accident. If a prosecuting authority such as the police or judicial services have *prima facie* evidence that a crime may have been committed, then they will see the accident scene as a crime scene and also want to secure as much evidence from the site. Now the accident scene and any other associated places that the authorities consider a source of evidence are secured. The technical investigation is now impeded if not suspended until the criminal investigation is complete.

The second main influence of criminalization on commercial aviation safety is the fear of prosecution and its effect on safety reporting. Initially this affects the investigation: the control of evidence including interviews of those connected with the accident inhibits access to crucial safety learning. Even the visible involvement of prosecuting authorities impedes the technical investigation, as those being investigated don't always appreciate the vastly different motivations between the two investigating parties.

When we looked at the influences on reporting cultures in Chapter 12 we referred to three major areas that affect reporting behavior and quality: fear, futility, and functionality. It is fear of reprisal from prosecuting authorities that is accentuated by criminalization. In his award-winning book *The Emotional Brain*, Joseph Ledoux noted that it is the emotion of fear that is our greatest motivation. In terms of the criminalization of aviation accidents it is the fear of prosecution that inhibits people from volunteering potentially incriminating information. This influence on reporting behavior does not just affect the immediate line of enquiry but as aviation accidents are so widely publicized, the process of criminalization can spread the fear of reporting right across the commercial aviation community.

16.5.4   Mitigating the Effects of Criminalization

The significant difference in approaches between investigators and the judiciary is highlighted by the Australian Transport Safety Bureau (ATSB): "Various parties may become more focused on interpreting and responding to an ATSB investigation report in the context of such legal proceedings about the occurrence rather than interpreting it as a basis for enhancing safety and providing learning opportunities for the future."[12] ATSB emphasizes the learning opportunity of aircraft accident investigation and reporting and how public attention should be exploited for the furtherance of safety best practice. However, the legal profession's perspective on investigations is not easily swayed by arguments based on the primacy of safety. Lawyers feel that their ability to conduct legitimate investigations should remain unhampered. The professional obligation of a prosecuting or plaintiff lawyer was clearly stated by Robert Lawson QC in an address at the Royal Aeronautical Society in London in 2015, which is to establish liability or blame in order to pursue criminal charges or to recover civil damages. Their professional obligation is to pursue that goal and adapt an interpretation of events in a way that best suits their client's interests. However, Lawson and many other legal professionals have conceded that the impact on safety by criminalization can have a broader safety influence and therefore should be brought into consideration when considering what evidence can be made available to the courts and what then can subsequently be released to the media.

The most effective way to deal with the dilemma between the primacy of safety and best evidence seems to be in education. Various initiatives have tried to highlight to the legal establishment the importance of maintaining psychological safety in and around reporting cultures. At the international level, ICAO has now provided guidance to accident investigators on best practice in developing an appropriate level of dialogue with investigating authorities. Annex 13 to the Chicago Convention provides guidance on recommended practice in air accident investigation, particularly on how coordination is maintained between accident investigators and judicial authorities. The simple fact is that neither type of investigation can legitimately suppress the other. The way forward would seem to be far better coordination between the technical and prosecution institutions from the outset, backed by a clear mutual understanding of each party's obligations.

16.5.5   Corporate Criminalization

Just as society's attitudes toward technology have altered in recent years, so has its relationship with the corporate world. Rather than just being perceived as an amorphous nexus of contracts, corporations have assumed or been ascribed identities of their own. Since the 17th century they have been recognized as independent legal entities, but numerous attempts to align the prosecutorial process of individual criminal liability with that of a corporate entity have been flawed and therefore unsuccessful. Some jurisdictions have abandoned the concept of corporate criminal liability altogether and attempted to prosecute senior managers within a company. The problem with this approach is that the sheer size and complexity of many modern corporations provide such a tangled web of decision-making process and intricate corporate hierarchies that it is practically impossible to link the causal chain from senior manager to the scene of the accident.[vii] Some jurisdictions, under the principle of vicarious liability have interpreted the unlawful will (*mens rea*) and the unlawful actions (*actus reus*) of a company's employees as those of the corporation.[viii] Other jurisdictions[ix] have developed a newer species of corporate criminal liability which assess the aggregate performance of the company and treats the whole organization as a single entity.[13] In this newer regime of corporate liability, the unlawful intent (or *mens rea*) of the company is replaced with a post-accident assessment of an organization's safety culture. We have discussed safety culture in Chapter 12 and while agreeing upon a regulatory definition has proved to be problematic, the recognition of a flawed safety culture in the aftermath of a multifatality accident has proved to be a more fruitful legal strategy.

16.5.6 Examples of Criminalization in Commercial Aviation

One result of this process described above has been the nature of the criminalization of air accidents—initially focused on the operating individuals but now increasingly directed toward the conduct of the organizations which have trained and molded operating standards. The rise in prominence and change in nature of the corporate body present significant challenges to the world's legal systems but an even greater challenge to the commercial aviation bodies that do not maintain a functioning safety culture. The following select examples illustrate how the focus of legal investigations of corporate conduct, in particular safety culture, is very much in the focus of prosecution bodies.

1. Air Inter Flight 148, Strasbourg, 1992: 87 Fatalities

Inquiry: BEA Report—Rapport de la commission d'enquête sur l'accident survenu le 20 janvier 1992 près du Mont Sainte-Odile (Bas Rhin) à l'Airbus A 320 immatriculé F-GGED exploité par la compagnie Air Inter.

Individual Criminal Prosecution: Six individual defendants from Airbus, air traffic control, the airline, and the French Aviation Authority, DGAC, were prosecuted then acquitted of manslaughter charges.

Corporate Criminal Prosecution: Airbus and Air France (parent company of Air Inter) were held liable for damages and were ordered to pay compensation to the relatives of the deceased (Tribunal de Grande Instance de Colmar).

Influence of Safety Culture: "The Commission therefore considers that the company culture is an important component of understanding the negative position taken by the company in respect of the GPWS."[14]

[vii] Example jurisdictions include: Brazil, Bulgaria, Germany, Greece, Luxembourg, Hungary, Japan, Mexico, Slovak Republic, Sweden.
[viii] Example jurisdictions include: Bulgaria, Denmark, France, Norway, Russia, Federal law of the United States of America.
[ix] Australia, Romania, The Netherlands, the United Kingdom.

2. ValuJet Flight 562, Florida 1996: 110 Fatalities

Inquiry: NTSB Report, 1997—"Aircraft Accident Report NTSB/AAR-97/06(PB97-910406), In Flight Fire and impact with Terrain, ValuJet Airlines Flight 592, DC-9-32, N904VJ Everglades, near Miami, Florida, May 11, 1996."

Individual Criminal Prosecution: Sabre Tech vice president of maintenance and two mechanics were charged with recklessness and numerous regulatory breaches. In 1999 all individuals were acquitted (US v Sabre Tech).[15]

Corporate Criminal Prosecution: Sabre Tech was found guilty of several regulatory and safety breaches. The convictions were eventually overturned by the Federal Appeals Court. The company was also charged by Florida state prosecutors with 110 counts of third-degree murder and 110 counts of manslaughter. All but one of the charges were dropped under a plea agreement.

Influence of Safety Culture: In a letter from ValuJet's Federal Aviation Authority inspectors: "It appears that ValuJet does not have a structure in place to handle your rapid growth, and that you may have an organizational culture that is in conflict with operating to the highest possible degree of safety."[16]

3. Air France, AFR 4590, Concorde F-BTSC, Paris, 2000: 113 Fatalities

Inquiry: BEA Report[17]—"Accident on 25 July 2000 at La Patte d'Oie in Gonesse (95) to the Concorde registered F-BTSC operated by Air France," Bureau d'Enquêtes et d'Analyses Pour la sécurité de l'aviation civile.

Individual Criminal Prosecution: Two former Concorde engineers, a former French civil aviation official, and two Continental Airlines employees were charged with various offences including manslaughter. Initially only one employee of Continental was found guilty but in 2012 his sentence was eventually overturned.

Corporate Criminal Prosecution: Continental Airlines were initially charged and convicted of corporate manslaughter. The company was fined €200,000. Although this verdict was overturned in 2012, absolving Continental of any criminal responsibility.

Influence of Safety Culture: The 2002 BEA report mentions 57 incidents of tire damage to Concorde aircraft. After implementing Airworthiness Directives during 1981–1982, no further modifications were made, suggesting the aircraft type was operated for over 17 years with a known risk to safety. The fleet incurred further incidents but the major design requirement, the strengthening of the fuel tanks was only undertaken after the fatal accident.

4. Collision of Scandinavian Airlines System Flight 686 with a Cessna Citation at the Linate Airport in Milan, 2001: 118 Fatalities

Inquiry: Agenzia Nationale Per La Sicurrezza Del Volo (ANSV): Accident involved aircraft Boeing MD-87 registration SE-DMA & Cessna 525-A, registration D-IEVX, Milano Linate airport, October 8, 2001.

Individual Criminal Prosecution: In 2004, a court convicted four defendants, including one air traffic controller and the former director of the Italian air traffic control agency (ENAV) of manslaughter and negligence and sentenced them to prison terms ranging from 6½ to 8 years.

Corporate Criminal Prosecution: Corporate criminal liability was not recognized in Italian law until 2001, namely Legislative Decrees 231/2001. This statutory liability does not extend to public bodies.

Influence of Safety Culture: Commenting on the lack of compliance of aerodrome management to Annex 14, the ICAO guide to aerodrome safety management: "The reason for this may be many; complex aerodrome management organization, lack of safety management system, unclear responsibility structure, weak safety culture, high traffic flow/intensity and physical expansion of aerodromes, etc."[18]

5. Helios Airways Flight HCY 522, Boeing 737-31S, near Athens, 2005: 112 Fatalities

Inquiry: Air Accident Investigation & Aviation Safety Board (AAIASB) Report, 2006—Aircraft Accident Report, Helios Airways Flight HCY522, Boeing 737-31S, at Grammatiko, Hellas on August 14, 2005.

Individual Criminal Prosecution: Four Helios Airways senior executives faced manslaughter charges in Cyprus and Greece. The Cypriot Court found the defendants not guilty, but the subsequent Athens High Court case dismissed the defendants appeal and upheld their sentence of 10 years imprisonment with an option to buy out their sentence for €80,000.

Corporate Criminal Prosecution: Helios Airways (later renamed "ajet") was charged by Cypriot prosecutors with manslaughter along with its senior executives; however, the company was dissolved in 2006.

Influence of Safety Culture: "The inexplicable inconsistences in the actions that were or were not performed, the actions recorded, and the actions described as having been performed by Ground Engineer No. 1 on the morning of August 14, 2005 were considered by the Board to confirm the idea that the Operator was not effectively promoting and maintaining basic elements of safety in its culture."[19]

6. Midair collision of a EMB-135 Legacy business jet and a GOL Airlines B737 8EH in Brazil, 2006: 154 Fatalities

Inquiry: Final Report A-OOX/CENIPA/2008, Occurrence: Aeronautical Accident, Aircraft Registration: PR-GTD & N600XL, July 17, 2007.

Centro de Investgacåo e Prevencåo de Acidentes Aeronåuticos (CENIPA), 2008.

Individual Criminal Prosecution: The Legacy pilots and three Brazilian air traffic controllers were charged with negligence and involuntary manslaughter. The pilots were acquitted in 2008, but that was overturned in 2010. In 2011 the pilots were sentenced to 4 years and 4 months but commuted the sentences to community service to be served in the United States.

Corporate Criminal Prosecution: Brazilian Law currently restricts corporate criminal liability to environmental and economic crimes.

Influence of Safety Culture: "The performance of the N600XL crew had a direct relationship with the decisions and organizational processes adopted by the operator, on account of culture and attitudes of informality."[20]

7. Spanair Flight 5022 accident at Madrid Barajas Airport: 2008, 154 Fatalities

Inquiry: CIAIAC (2010) Report A-032/2008 Accident involving a McDonnell Douglas DC-9-82 (MD-82) aircraft, registration EC-HFP, operated by Spanair, at Madrid-Barajas Airport, on August 20, 2008.

Individual Criminal Prosecution: The head of Spanair's maintenance department and a Spanair mechanic were charged with 154 crimes of manslaughter and 18 crimes of negligent injuries. The charges against the mechanics were subsequently dropped and blame for the accident was placed on the two deceased pilots. See Air Crash at Madrid, No. 2:10-ml-02135, Affidavit of Salvador-Coderch, D.E. No. 197, at 4.

Corporate Criminal Prosecution: Spain excludes corporate bodies from bearing direct responsibility for criminal responsibility. Criminal liability can only be imputed where the corporation acts as an accessory to an individual's criminal act.

Influence of Safety Culture: "The fact that audits conducted by the company's Quality Department were ineffective in detecting these deficiencies indicates that either the audits were not properly conceived or that the way in which technical flight records were kept was not a concern, which proves that such a culture was accepted and shared within the organization."[21]

8. Loss of Air France 447, F-GZCP, South Atlantic, 2009: 298 Fatalities

Inquiry: BEA Report, 2012—On the accident on June 1, 2009, to the Airbus A330-203, registered F-GZCP, operated by Air France flight AF 447 Rio de Janeiro.

Individual Criminal Prosecution: None.

Corporate Criminal Prosecution: Air France and Airbus had been indicted and faced possible charges of Corporate Manslaughter having been referred, in July 2014, to the French criminal court. This case was dropped but returned in October 2022 when both Air France and Airbus were charged with involuntary manslaughter. Both companies pleaded not guilty. The case is due to conclude in December 2022.

Influence of Safety Culture: "With regard to human factors, the behaviour observed at the time of an event is often consistent with, or an extension of, a specific culture, and work organisation."[22] Prior to any BEA report, Air France commissioned an independent safety review team tasked with an analysis of company culture. Although not made public, Air France had committed to implementing all the report's recommendations.

9. Loss of Lion Air Flight 610, Boeing 737 MAX 2018: 189 Fatalities

Inquiry: KNKT Report (2018),[23] Lion Mentari Airlines Boeing 737-8 (MAX); PK-LQP Tanjung Karawang, West Java, Republic of Indonesia, October 29, 2018.

10. Loss of Ethiopian Air Flight 302, Boeing 737 MAX 2019: 157 Fatalities

Inquiry: EAIB Report (2022),[24] Accident to the B737-8 (MAX) Registered ET-AVJ operated by Ethiopian Airlines, March 10, 2019.

Individual Criminal Prosecution: In October 2021 a former Boeing chief technical pilot Mark Forkner was indicted by a Texan jury for providing false, inaccurate, and incomplete information regarding the aircraft's MCAS software system. After a 4-day trial in March 2022, he was acquitted. Dennis Muilenburg, Boeing's former CEO was fined $1 million by the US Securities and Exchange Commission for making misleading public statements following the crashes of Boeing MAX aircraft.

Corporate Criminal Prosecution: In January 2021, Boeing agreed to pay $2.5 billion in fines and compensation to resolve a US Justice Department criminal investigation into the 737 MAX crashes. The settlement, which allowed Boeing to avoid prosecution, included a fine of $243.6 million, compensation to airlines of $1.77 billion, and a $500 million crash-victim fund over fraud conspiracy charges related to the plane's flawed design. In September 2022, Boeing was fined a further $200 million to settle civil charges by the US Securities and Exchange Commission for misleading its investors.

Influence of Safety Culture: Committee on Transport and Infrastructure: Extracts from the Final Committee Report: The Design, Development & Certification of the Boeing 737 MAX, September 2020.[25]

"My first concern is that our workforce is exhausted. Fatigued employees make mistakes. My second concern is schedule pressure (combined with fatigue) is creating a culture where employees are either deliberately or unconsciously circumventing established processes," Ed Pierson, Boeing Senior Manager, 737 MAX Final Assembly Plant, from an email sent to Scott Campbell, 737 General Manager, June 9, 2018, page 175.

"The effectiveness of these organizational and procedural changes that have been recommended following its internal review will be dependent on Boeing's willingness to change. However, Boeing does not appear to have fully accepted the lessons from the MAX accidents or taken responsibility for design errors. Without that recognition it is hard to believe that Boeing will make the change to improve its safety culture," page 230.

16.5.7   Resolving Criminalization

The issue of criminalization has attracted considerable interest across the commercial aviation community. Reading through much of the aviation safety literature one might assume it is a simple issue of suppressing the ambition of criminal justice systems. However, as some of the case studies suggest, there needs to be a mechanism of legal redress for individual and corporate activity which does manifestly reduce safety margins or even cost lives. It is a complex position and not easy to resolve as it sits at the juxtaposition of two fundamental principles of social justice: safety and law. Sweeping statements about the sanctity of either principle do not resolve the issue.

Useful websites on the criminalization of aviation accidents:

- https://www.skybrary.aero/search/google?keys=criminalisation
- https://skybrary.aero/bookshelf/overzealous-criminalization-adverse-effects -safety
- https://skybrary.aero/search/google?keys=Manslaughter

Key Terms

Airline Deregulation Act of 1978

Airline Safety and Federal Aviation Administration Act of 2010

Air Traffic Organization (ATO)

Aviation Safety (AVS)

Chicago Convention of 1944

Criminalization

FAA Air Traffic Organization (ATO)

FAR Part 121

FAR Part 135

Federal Aviation Administration (FAA)

Flight Standards District Offices (FSDO)

Global Positioning System (GPS)

International Civil Aviation Organization (ICAO)

International Commission for Air Navigation (ICAN)

Manslaughter

Next Generation Air Transportation System (NextGen)

Notice of Proposed Rulemaking (NPRM)

Procedures for Air Navigation Services (PANS)

Regional Supplementary Procedures (SUPPs)

Standards and Recommended Practices (SARPs)

Topics for Discussion

1. List three strategic objectives of ICAO and discuss their importance in international aviation.
2. State the purpose and name one result of the Chicago Convention of 1944.
3. Discuss the ICAO Rulemaking Process including the role of SARPs, PANS, and SUPPs.

4. Why does it take years for new guidance to be formulated through the ICAO Preliminary Review by the ANC?
5. Describe the rulemaking process of EASA.
6. Does EASA "impose" strict regulatory conformance on member states?
7. List three major functions of the FAA and discuss some of the activities that support these functions.
8. What does Section 601(b) of the FAA Act say about an air carrier's responsibility for safety?
9. Explain the tendency for inspectors to focus on records during maintenance base inspections and discuss how public complaints are handled.
10. How has inspector workload been affected since airline deregulation? Why has it been difficult attracting inspectors to major metropolitan areas?
11. Does the involvement of the criminal justice system ever improve safety standards?
12. Is being regulatory compliant the same as being legally compliant?

References

1. ICAO, Convention on International Civil Aviation (Doc 7300/9). Montreal, Canada.
2. https://www.easa.europa.eu/en/downloads/136453/en. Accessed October 2022.
3. CAA Statutory Charges Consultation Response, CAP 2329. https://publicapps.caa.co.uk/docs/33/FY22-23%20CAA%20Statutory%20Charges%20Consultation%20Response%20(CAP%202329)%20v2.pdf. Accessed October 2022.
4. CAA Basic Regulations. https://www.caa.co.uk/uk-regulations/aviation-safety/basic-regulation-the-implementing-rules-and-uk-caa-amc-gm-cs/. Accessed October 2022.
5. Flight International, Brexit EASA licence issue puts crew on countdown. https://www.flightglobal.com/flight-international/brexit-easa-licence-issue-puts-crews-on-countdown/150637.article. Accessed October 2022.
6. Dong, X., & Ryerson, M. S. (2019). Increasing civil aviation capacity in China requires harmonizing the physical and human components of capacity: A review and investigation. *Transportation Research Interdisciplinary Perspectives*, 1, 100005.
7. Michaelides-Mateou, S., & Mateou, A. (2016). *Flying in the Face of Criminalization: The Safety Implications of Prosecuting Aviation Professionals for Accidents*. Routledge.
8. Lawrenson, A. J. (2017). Safety culture: a legal standard for commercial aviation. Doctoral dissertation.
9. Giddens, A. (1999). *Runaway World: How Globalisation Is Reshaping our Lives*. St Edmundsbury Press.
10. Beck, U. (1992). *Risk Society: Towards a New Modernity*. London: Sage.
11. Dekker, S. (2007), *Just Culture, Balancing Safety and Accountability*. Farnham, UK: Ashgate Publishing.
12. ATSB (Australian Transport Safety Bureau). (2008). Analysis, causality and proof in safety investigations. https://www.atsb.gov.au/media/27767/ar2007053.pdf, p. 87. Accessed November 2022.
13. Lawrenson, A. J., & Braithwaite, G. R. (2018). Regulation or criminalisation: What determines legal standards of safety culture in commercial aviation? *Safety Science*, 102, 251–262.
14. BEA [Bureau d'Enquêtes et d'Analyses Pour la sécurité de l'aviation civile]. (1993). Rapport de la commission d'enquête sur l'accident survenu le 20 janvier 1992 près

du Mont Sainte-Odile (Bas Rhin) à l'Airbus A 320 immatriculé F-GGED exploité par la compagnie Air Inter (p. 23). http://www.bea.aero/docspa/1992/f-ed920120 /htm/f-ed920120.html

15. United States v. SabreTech, Inc., 271 F3d 1018 (11th Cir. 2001).
16. NTSB [National Transport Safety Board]. (1997). Aircraft Accident Report NTSB/ AAR-97/06(PB97-910406), In Flight Fire and impact with Terrain, ValuJet Airlines Flight 592, DC-9-32, N904VJ Everglades, near Miami, Florida, May 11, 1996 (p. 77).
17. BEA [Bureau d'Enquêtes et d'Analyses Pour la sécurité de l'aviation civile]. (2002). Accident on 25 July 2000 at La Patte d'Oie in Gonesse (95) to the Concorde registered F-BTSC operated by Air France. https://www.bea.aero/uploads /tx_elydbrapports/f-sc000725a.
18. ANSV [Agenzia Nationale Per La Sicurrezza Del Volo]. (2004). Accident involved aircraft Boeing MD-87, registration SE-DMA & Cessna 525-A, registration D-IEVX, Milano Linate airport, October 8, 2001 (p. 152).
19. AAIASB [Air Accident Investigation & Aviation Safety Board]. (2006). Aircraft Accident Report, Helios Airways Flight HCY522, Boeing 737-31S, at Grammatiko, Hellas on 14th August 2005 (p. 199).
20. CENIPA [Centro de Investgacão e Prevencão de Acidentes Aeronâuticos]. (2008). Final Report A-OOX/CENIPA/2008, Occurrence: Aeronautical Accident, Aircraft Registration: PR-GTD & N600XL, 29th September 2006 (p. 265).
21. CIAIAC. (2010). Report A-032/2008 Accident involving a McDonnell Douglas DC-9-82 (MD-82) aircraft, registration EC-HFP, operated by Spanair, at Madrid-Barajas Airport, on 20 August 2008 (p. 221).
22. BEA [Bureau d'Enquêtes et d'Analyses Pour la sécurité de l'aviation civile]. (2012). On the accident on 1st June 2009 to the Airbus A330-203 registered F-GZCP operated by Air France flight AF 447 Rio de Janeiro – Paris (p. 101).
23. KNKT [Komite Nasional Keselamatan Transpotasi Repulic of Indonesia]. (2018). 18.10.35.04, PT. Lion Mentari Airlines Boeing 737-8 (MAX); PK-LQP Tanjung Karawang, West Java Republic of Indonesia, October 29, 2018.
24. EAIB [Ethiopian Aircraft Accident Investigation Bureau]. (2022). Report No. AI-01/19, Investigation Report on Accident to the B737-MAX8, Reg. ET-AVJ Operated by Ethiopian Airlines, March 10, 2019.
25. Hearing Before the Subcommittee on Aviation of the Committee on Transportation and Infrastructure, House of Representatives:116th Congress, First Session May 15, 2019. https://www.govinfo.gov/committee/house-transportation?path= /browsecommittee/chamber/house/committee/transportation.

Further References

- Aviation Rulemaking, US General Accounting Office, http://www.gpo.gov
- Federal Aviation Administration, http://www.faa.gov
- Federal Aviation Administration rulemaking page:
- https://www.faa.gov/regulations_policies/rulemaking/
- International Civil Aviation Organization, http://www.icao.int

CHAPTER **17**

The Future of Commercial Aviation Safety

Introduction

In 2016, Klaus Schwab, the founder and executive chairman of the World Economic Forum, published an article[1] in which he coined the phrase "the Forth Industrial Revolution." Schwab differentiated this current social transformation from simply improvements in efficiency, by emphasizing the fundamental changes brought about by the introduction of new technologies such as machine-to-machine communications, gene editing, and artificial intelligence. These technologies will result in significant reduction in the need for human intervention and an increased reliance on automation. These changes that Schwab identified have already started to affect society and will invariably have a big impact on the way commercial aviation is conducted. These developments will of course create new and exciting opportunities for the sector, but it is crucial that we begin to understand that it is in the way these new technologies interrelate as this is where new types of system level threats to safety will originate.

Another highly significant way in which society is undergoing change is in its values and attitudes to resource management and initiatives in sustainability. Climate change is an existential risk to our society that not only directly influences weather patterns but also impacts business behaviors, trading patterns, and consumer attitudes toward commercial aviation. As we have discussed throughout Part I, how commercial aviation manages safety is inextricably linked to the contemporary technologies and societal values of the period. As we are now facing such rapid and momentous changes, we need to consider what potential risks and benefits commercial aviation will face to its safety in the coming years.

17.1   Future Airspace

17.1.1   The Future by Numbers

When we are talking about the near future, for the purposes of this chapter, unless otherwise stated, we are generally referring to the next two decades up to 2040. After that period, forecasts become increasingly a case of guesswork as to what the future holds for commercial aviation. One area in which there is overwhelming consensus is in the massive growth of the sector. The dip in aviation growth due to COVID-19 seems to have been largely overcome and the number of flights in 2022 resumed to over 85% of the numbers seen in 2019[2]; in 2023 it looks as though we are seeing parity to pre-COVID activity levels.

Commercial aviation is closely watched by economists as a bell-weather for the broader global economy. According to ICAO[3] commercial aviation represents 3.5% of the gross domestic product (GDP) of the world's economy, equating to some US$2.7 trillion. Over the last few decades, the sector has seen fast growth, and most economic outlooks predict further expansion. Recent estimates from IATA[4] suggest that demand for air transport will increase by an average of 3.3% over the next 20 years. If this growth level is maintained, then by 2040 the commercial aviation sector will be flying 7.8 billion passengers per year, or an average of over 20 million passengers per day. Today, the world's commercial airline fleet stands at approximately 26,000 aircraft in service, and this number is expected to double to over 50,000 by 2040. In North America alone the number of commercial aircraft is expected to grow by 46% through this period, while Europe's commercial aviation fleet is expected to grow by 76%.[5]

These numbers are impressive, but they are by no means the whole story. Airspace is already congested particularly around major conurbations around the world. This situation looks to be accentuated by the emergence of megacities, typically defined as cities with over 10 million inhabitants. The United Nations predicts that by 2050, over two-thirds of the world will live in urban areas.[6] Given the impracticality of providing adequate ground transport infrastructure, urban planners have increasingly looked toward aviation to provide transport solutions to such high-density regions. Advanced Air Mobility (AAM)[i] is one proposed solution to this problem. The use of relatively small, electrically powered, and ultimately fully autonomous aircraft are increasingly seen as the foundation of urban and regional connectivity.

If this type of technological advance sounds a little like science fiction, it's worth noting that as of summer 2022, there were 700 concept aircraft being developed by 350 companies around the world with over $6.9 billion invested in development in 2021 alone.[7] The first wave of these new types of aircraft are scheduled to receive certification in 2023. At the time of writing, American Airlines have recently placed orders on 250 Vertical Aircraft VX4 and options on a further 150 aircraft. United Airlines have ordered 200 Archer EVTOL (Electric Vertical Take-Off and Landing—an example of which is seen in Figure 17-1) aircraft; these contracts are worth approximately $1 billion each. In early 2022, Boeing invested almost $460 million in the AAM manufacturer, Wisk. This huge expansion and transformation in aviation technology is well underway.

17.1.2 Future Complexity

Unlike conventional commercial aviation, it's the ability of these new types of aircraft to rapidly scale in numbers which has attracted such interest from investors. The well-respected consultancy, Frost and Sullivan, estimated the AAM market could produce 430,000 AAM aircraft globally by 2040.[8] That rate of growth would mean that the global AAM fleet alone would be approximately 14 times the size of the current global fleet of commercial airliners. Combined with the anticipated numbers of commercial and general aviation aircraft coming onto the market within the same period challenges many

---

[i] According to NASA, the term AAM encompasses UAM (Urban Air Mobility—intracity transportation), RAM (regional air mobility—intercity transport and transportation to/from a city from a district outside a city or rural area), cargo UAS (Unmanned Aerial Systems—but not small UAS for package delivery), and other similar concepts, whether vertical or conventional takeoff and landing.

Figure 17-1   An Archer Maker EVTOL during early test-flights, in 2022. (Source: Archer Aviation Media Pack)

of the current approaches to air traffic management. It's not just a case of getting bigger; the FAA has already recognized that expansion is only the beginning of the challenge. The administration has already sought to expand the ranks of its existing 14,000 professionals working in radar facilities and ATC control towers with an ambitious recruiting drive to bring in almost 5000 new recruits by 2027.

However, the scale of expansion is just part of the issue facing the industry. This new wave of aircraft will have significantly more advanced technology than is currently in use. This new technology will involve increasing levels of automation and ultimately full autonomy both onboard the aircraft and within the air traffic management systems that will coordinate them. The challenge facing planners of the future of aviation is accentuated by the rapidly accelerating levels of these technological capabilities and particularly how these complex systems will interrelate. When introducing new technology into an already complex system it becomes increasingly difficult to anticipate what net effect on the system will be produced as these new technologies begin to combine within the whole system.

One of the biggest challenges facing AAM manufacturers is to identify who will actually fly their aircraft. According to Boeing, the existing scheduled airline sector requires 600,000 new pilots by 2040 and salaries are already increasing to attract new pilots. AAM manufacturers are promoting a pilot training scheme that requires only 500 hours to qualify to fly their new aircraft to utilize inexperienced operators as a means of expanding the AAM network and scaling their product. This level of experience would align the sector with GA (General Aviation) rather than aspiring to the safety performance standards of scheduled airlines. The difference in safety performance should not be understated with many safety experts suggesting that GA carries an accident rate 60 times that of scheduled airlines in commercial aviation.[9] If the AAM fleet is going to succeed, it needs to be seen as a trusted form of commercial aviation. To do that it must understand and adopt the high standards of design and operational principles that have resulted in such incredible levels of safety performance within the sector. While the innovative approaches taken by this new wave of aircraft manufacturers are to be applauded, this new technology should not be used as an excuse to cut corners or shortcut the certification process. The certification and regulatory approval process for new commercial aircraft are difficult—but they are difficult for a very good reason.

17.1.3   Advanced and Integrated Air Traffic Management (ATM)

To try to facilitate a smooth transition during this technology implementation, there are major government-backed projects which aim to coordinate new technological capabilities with the existing structures and protocols of ATM. EASA, the European Union Aviation Safety Agency, has embarked on a massive air traffic control infrastructure project called SESAR (Single European Sky ATM Research). The program has been running since 2004 and is now in the deployment phase, seeing widespread implementation of some of the requisite new technologies into existing ATM systems. In the United States, the FAA has embarked on NextGen, a multibillion-dollar program aimed at modernizing the national airspace system (NAS) and providing a safe and resilient air transport network. The NextGen program is recognized as one of the most ambitious infrastructure projects in United States history.

Neither NextGen nor SESAR are introducing one single technological solution to cater for future ATM demand, but both are rather a series of interlinked programs, policies, and procedures that encompass the many different technologies that recent research and development has produced. Both programs provide a skeletal structure for an anticipated future demand. These demands include a modernized air traffic infrastructure, changes in aviation communication, navigation, and surveillance. The programs both aim to provide safety and security enhancements and to cater for emerging considerations regarding the environment.

The programs will also have to cater for entirely new species of air transport such as commercial space operations (discussed later in this chapter) and wide use of AAM. These are modes of transport that were the stuff of dreams and science fiction are now about to become a part of our daily lives. Maintaining the public's confidence in emergent technology is crucial; these programs are publicly funded so must be cognizant of political and social considerations. Studies of public perception relating to the introduction of these new technologies generally indicate positive support, however, the most consistent and highest priority concern for people is that of safety.

17.1.4   Challenges on the Near Horizon

While the architects of these new and advanced systems repeatedly refer to their safety and reliability, the simple fact is that in terms of safety management of commercial aviation, this is uncharted territory. Over its history, commercial aviation has not only embraced change, but has pushed the boundaries of technological capabilities. For the most part, the sector has done this successfully; not only maintaining its enviable safety record but demonstrating improvements over the years. However, the astonishing pace of change brought about by this fourth industrial revolution is unprecedented, the sheer scale and pace of change, in itself, provides a significant challenge to the maintenance of flight safety standards.

The high demand on existing resources looks set to increase but has already placed stresses on the existing system of commercial aviation as demand for more intense flying schedules has increased. In August 2022, *Aviation Herald*[10] reported that the pilots of an Ethiopian Airlines Boeing 737-800, flight number ET343, enroute from Khartoum to Addis Ababa, fell asleep and missed the commencement of their descent. In May, a similar event occurred on an ITA[ii] flight from New York to Rome's Fiumicino Airport. The captain of Flight AZ609 was subsequently dismissed by the airline for misconduct. Ever-increasing demands on productivity mean flight schedules are resulting in cases of extreme fatigue.

Once seen as an inevitable and acceptable feature of commercial jet operation, fatigue has been increasingly recognized as a serious threat to safe operations. In a survey involving 800 commercial pilots, carried out by the British Airline Pilots Association (BALPA), 55% of respondents cited fatigue as the biggest threat to flight safety, exceeding that of terrorism. In another survey involving Delta Airlines pilots,[11] less than a third of the 1108 respondents said they considered their employer receptive and responsive when they reported to the company that they were too fatigued to operate. Although there were variations between different fleets, all the Delta pilot respondents referred to examples where operational pressure contributed to their fatigue. Many national and regional regulators have now engaged Fatigue Risk Management Systems (FRMS) which provide science-based and contextual mechanisms to balance operational efficiency with the overwhelming need to avoid the intolerable risk associated with pilot fatigue.

With growing pressure on commercial aviation to meet resurgent post-COVID air transport demands, other threats to the system are developing. In line with Boeing's projected expansion of the commercial aviation market, there is a growing need for operational and support staff; the sector simply does not have enough qualified personnel. In their 2022 statistical review,[12] Boeing predict a demand for 602,000 newly qualified airline pilots by 2041 (prior to COVID-19, this number had been 800,000). The large number of new aircraft coming on to the market set a requirement for 610,000 new maintenance technicians and 899,000 newly qualified cabin crew. To give some sense of perspective, in 2019 there were 330,000 commercial pilots operating globally, recognized as an aging workforce with reduced training facilities and far fewer military pilots available to replenish numbers. The surge in demand has impacts right across the supply chain to produce well-trained and highly qualified personnel. The cost of training is growing with increasingly sophisticated training aircraft required to prepare candidates for today's highly automated commercial aircraft. As things stand today there are simply insufficient training resources to meet this level of demand.

The economic demand to fill this number of vacancies raises questions about maintaining the quality of training that can be realistically provided. This is accentuated by the need to bring in new safety-critical personnel with lower levels of experience. The industry has mitigated some of the risks of higher cockpit workload by the extensive use of automation. However, this higher level of sophistication has also raised some concerns in the aftermath of accidents such as Air France 447 and the Air Asiana crash at San Francisco discussed earlier in this chapter. The complex interaction between human pilot and autonomy has not always improved safety standards. There needs to be a continued focus on pilot's ability to fly with or without automation during their training, particularly during technical malfunction or other adverse circumstances. A deeper knowledge of these advanced technical systems would help to avoid a decrease in situation awareness during critical flight phases, but this quality and depth of preperation and conditioning require even more of these stretched training resources.

[ii] ITA is the Italian National Airline which took over from its predecessor Alitalia in October 2021.

Other recent technical phenomena affecting the integrity of modern aircraft navigation are frequency cluttering and cross-interference issues. As more aircraft enter the airspace system, being able to electronically track and identify them (conspicuity) becomes increasingly important if ATM is to maintain its effectiveness. Both NextGen and SESAR have placed considerable dependence on ADS-B as an obvious and low-cost solution to aircraft conspicuity with such rapidly expanding numbers. This issue will become increasingly poignant as massive numbers of uncrewed aircraft begin to enter the airspace system, coordinated by Uncrewed Traffic Management or UTM systems. The ADS-B system is significantly cheaper than having ground-based radar systems that can interrogate aircraft returns to check their integrity; with ADS-B this is based on trust. This leaves the system open to simple error and a broad range of nefarious activity. There are some technological developments that are trying to mitigate this system weakness. One solution is to triangulate positional data with primary radar. The limiting factor here is the huge amount of energy required for primary radar to emit sufficient energy and to "bounce" enough energy back to make objects detectable by the radar antennae. One solution is to use primary radar simply as the transmitter and to use a network of receiver antennae to detect the reflected radar energy, thus extending the system range. However, another other significant weakness is simply the numeric capacity of ADS-B systems. The simple fact is that the frequencies that ADS-B operate on have limited bandwidth. When maximum capacity is reached in areas of high density, when conspicuity is crucial, signals can become erroneous or simply lost.

Another and more recent example of frequency cluttering is on the impact of some 5G transmitters operating in the vicinity of major airports. The 5G band operates on a similar frequency bandwidth to aircraft radio altimeters. Verizon and AT&T have been rolling out 5G broadband services since early 2022, but interference in the 3.7–3.98 GHz band (5G C-Band) can affect the radio altimeters when transmission takes place in the vicinity of airports. Many modern aircraft use radio altimeters: a type of radar that gives the aircraft very precise height readings above the ground during the landing phase. These reading activate various systems during ILS approaches, required in poor visibility, such as speedbrake deployment, autothrottle, and go-around modes for the aircraft's flight director. The FAA issued a series of Airworthiness Directives[13] throughout 2022 to highlight the issue and is now collaborating with broadband service providers to restrict activity near to major airports. The longer-term solution is likely to be the gradual introduction of broadband transmission filters retrofitted to thousands of aircraft across the world.

17.1.5 Challenges on the Far Horizon

A common feature of new-generation ATMs is an increasing use of and dependency on satellite-based navigation aids. ICAO has proposed this approach, termed Performance Based Navigation (PBN),[iii] as a solution to the increasing demands placed on ATM. PBN offers the ability to provide more flexible and efficient routing solutions, ultimately

---

[iii] ICAO PBN Manual (Doc 9613) defines PBN as: Area navigation based on performance requirements for aircraft operating along an ATS route, on an instrument approach procedure or in a designated airspace. Where: Airborne performance requirements are expressed in navigation specifications in terms of accuracy, integrity, continuity, and functionality needed for the proposed operation in the context of a particular airspace concept. Within the airspace concept, the availability of GNSS Signal-In-Space (SIS) or that of some other applicable navigation infrastructure has to be considered in order to enable the navigation application.

removing the need for rigid airspace structures such as airways and permanent blocks of controlled airspace. As part of the SESAR program, Functional Airspace Blocks (FABs) were introduced in 2014 which negate the need for discrete crossing points and therefore reducing the probability of air-to-air collision. These FABs are potentially the basis of all future airspace systems and represent the most significant conceptual change in airspace design.

The use of satellite-based navigation using systems such as GPS (Global Positioning System) has many other advantages in addition to the efficient and flexible use of airspace. The FAA has been gradually reducing the number of navaids through a program called MON (Minimum Operational Network) increasing system dependence on GPS navigation. Convention airport approach systems such as ILS (Instrument Landing Systems) have enabled operations in low and even no-visibility conditions with suitably equipped airports and aircraft. However, these systems require expensive installation and maintenance processes. PBN, using GPS allows satellite-based approaches[iv] to almost the same accuracy as ILS systems at a fraction of the cost. The issue arises when we consider the reliability of GPS-based systems. Any satellite-based system is susceptible to solar weather changes which can dramatically reduce accuracy performance. They have also been the deliberate target of jamming systems, either accidentally or through deliberate acts; this has been an almost day-to-day occurrence over the Eastern Mediterranean and Middle East: Out of all MENA (Middle East and North African) states, 73.6% of all GPS jamming incidents occurred in Turkey and Iraq.[14]

The FBI has also reported numerous incidents of GPS jamming devices used by car thieves in order to avoid detection. In August 2017 at Nantes Airport, France, a man was fined €2000 for leaving a GPS jammer accidentally active in his car while he was away on vacation. His simple error caused days of disruption to the airport's operations. The Russian satellite system GLONASS failed in 2014 and in 2016 critical timing errors in GPS due to installation errors resulted in unreliable system output. The fundamental issue is that PBN could potentially be a massive single point of failure for the global ATM system. This has necessitated the establishment of degraded GPS procedures for ATC and non-normal checklists for flight crew; however, there is no provision for total loss or corruption of the whole system.

Another significant challenge to commercial aviation safety comes from the new operating types that are starting to enter the sector. We have already discussed the massive potential scale of the AAM, or air taxi market but there is another associated sector to consider. As of summer 2022, the FAA had registered 865,505 drones; 314,689 of these were registered as commercial drones and 538,172 as recreational.[15] Drones have already demonstrated their potential benefit and utility for the future as seen in Figure17-2, where research drones developed at the university of Bath, UK were used to help the prediction of avalanches in Andorra. However, drones have also demonstrated their ability to cause extensive disruption if inadequately managed.[15] While military drone pilots are well trained and disciplined, the same cannot be said for the majority of recreational and even some commercial operators. The quality of selection, training, and supervision involved in the training of a multimillion-dollar aircraft is not justified as an investment to operate a $100 or even a $10,000 UAV. However, these aircraft will ultimately need to share the same airspace and will require a significant level of

---

[iv] These satellite-based instrument approaches have little in the way of standardization when it comes to naming conventions. They are known as inter alia RNAV, LPV, GNSS, GPS, or PBN approaches.

FIGURE 17-2 Small drones are increasingly demonstrating utility such as helping to predict avalanches such as in this research drone in Andorra. (Source: Authors)

coordination and supervision given the current and the predicted massive increase in the numbers of aircraft.

17.2 Digital Transformation

17.2.1 Possibilities and Challenges

The envisaged scale and complexity of new ATMs will require a considerable amount of data exchange, far exceeding the current level of IT (Information Technology) dependence. The airline industry is heavily reliant on IT systems not only for booking, ticketing, and administration, but increasingly for many safety-critical and operational activities. These include up-to-date meteorological data, critical flight performance calculations, flight planning, and essential maintenance data. The current focus on digitalization is in effect an increasing awareness that commercial aviation can gain more value from the increased availability of massive data sets, sensor information, and storage capabilities.

In addition to the immediate availability of data to enhance situation awareness and therefore improve operational decision-making, digitalization opens considerable possibilities in processing the vast amount of safety-related data that commercial aviation produces. One output from enhanced digitalization of data is that it can be manipulated to detect patterns and in the case of safety management, emergent risks. This capability has significant potential when we consider the increased complexity of future air transport systems. However, there has to be some consideration of the potential to become overly dependent on this new technology. One of the fundamental principles concerning the introduction of new technology is the law of unintended consequences. While technology is introduced in order to improve capabilities or efficiencies, the complex nature of commercial aviation means it is notoriously difficult to foresee all potential outcomes. Dr. Kathy Abbott is the Chief Scientific and Technical Advisor to the FAA. She notes that the effect of the introduction of new technology is particularly difficult even for those who are subject matter experts. As we have discussed in Part IV of this book, this issue is accentuated when technical expertise is not matched with knowledge of complexity science, systems thinking and human factors.[16]

The fragility of these IT systems has been frequently witnessed across the global sector. A 2019 report from the Government Accounting Office (GAO) documented 34 IT outages across the US airline industry. The report noted that 85% of these outages resulted in some form of flight delay, and 14% caused the total halt of all operations for several hours or more. These IT failures continue to the point of being a regular feature during peak periods of demand. All new ATM systems place considerably more dependence on the free flow of encrypted information and increasingly service safety-critical operations, sometimes referred to as the "internet of things" or IoS. Although these systems promise considerable improvements in efficiency, they also present a classic example of a "tightly coupled" system described in Chapter 3. The increasing dependency on certain core technologies is one of the main reasons that safety researchers have increasingly looked to systems theory as a way of managing emergent risks in such complex environments.

17.2.2 Artificial Intelligence

The term Artificial Intelligence or AI first appeared in the late 1950s following ideas proposed by scientists such as the famous Alan Turing.[17] The need to manage increasingly large volumes of digital information within massively expanded and highly complex ATM systems has brought considerable attention to AI from the commercial aviation community. With its capability to recognize patterns within huge amounts of data, AI has brought considerable benefits to the ability of computers to understand natural human language, computer vision, synthetic data generation, and has accelerated data-rich scientific research. However, there are considerable differences in opinions between aviation professionals as to the extent AI can play a role in such a safety-critical environment as aviation. But before we progress with any discussion as to what AI can and should do, it would be of some considerable value to clarify what AI is.

AI is a broad term that encompasses a wide range of digital capabilities. The term has evolved as each technology strand has progressed and created further applications. EASA's definition is appropriately broad, as it defines AI as "any technology that appears to emulate the performance of a human."[18] It may be thought of a discipline of computer science that mimics aspects of human decision-making. Earlier versions of AI included simple, rule-based systems that captured human knowledge (such as primitive medical diagnostic systems) whereas today, significantly more complex behaviors are captured through machine learning and synthesized through deep artificial neural networks[v] which enable higher tasks such as self-driving cars.

Fundamentally, the basic ATM function is to enable massive amounts of data to flow to the right place in an understandable format. In facilitating this complex data flow between the air and ground-based decision-making elements, then articulating emergent patterns within that data, AI can, in theory, enhance the basic ATM function. This "big-data" environment is where AI and machine learning can potentially provide

---

[v] The term "deep" refers to the expanded number of layers of neural networks within the computing architecture that have been enabled with increased computational power and memory.

the most benefit by reducing human workload and increasing system capabilities in complex environments.

In February 2020, EASA published its roadmap to the introduction of AI into commercial aviation.[19] The roadmap envisages a three-stage rollout of AI technology into the sector: the first phase is in the format of pilot-assist (2022–2025); the second is a form of human–machine interface in operating the aircraft through enhanced situation awareness and onboard decision-making aides (2025–2030); the third phase is the development of fully autonomous commercial aircraft that can operate themselves (2035 and beyond). In its roadmap EASA highlights four major questions raised by the introduction of this new technology:

- How to establish public trust in AI-based systems?
- How to integrate the ethical dimension of AI (transparency, nondiscrimination, fairness, etc.) in safety certification processes?
- How to prepare the certification of AI systems?
- What standards, protocols, and methods do we need to develop to ensure that AI will further improve the current level of safety of air transport?

There are no easy answers to any of these questions and in fact they highlight how challenging and complex the introduction of AI will be for the global commercial aviation community. Each step along the EASA roadmap must reach an acceptable level of trustworthiness, not only for each individual or component that uses AI to enhance capability, but for the whole system (including all ancillary elements) to demonstrate an acceptable level of reliability.

17.2.3   New Technologies Meet Old Rules

This latter point perhaps underlines the primary concern of regulators concerning the introduction of AI systems. The well-proven safety standards that have been established over the past decades are built on a different set of assumptions than the radical approach taken by AI. The way in which AI software learns from patterns in data can result in occasional anomalies in its outputs. Given the complexity of the process, these "edge cases" are practically impossible for humans to anticipate. The capability of AI to absorb and manipulate such vast quantities of data means that unpredictability at some point in the system life cycle is inevitable. When multiple AI systems are interacting, each with different parameters, then the consequences of undesirable outcomes become, to use the language of Johari's Window[20], "unknown unknowns."

Traditional software used in safety-critical applications in aviation are certified to be deterministic, which is wholly predictable. That principle has been the bedrock of all avionic standards since computers were first introduced into modern aircraft. The most influential institution in the development of these standards is the RTCA.[vi] The RTCA produces many highly detailed guidelines such as DO-178C (avionics software) and DO-254 (avionics hardware) ensuring that the output of these systems remain predictable in all envisaged scenarios. Machine Learning AI cannot achieve the same predictability as the software is programmed to learn and may produce a different output, or

---

[vi] The RTCA, founded in 1935, was originally known as the Radio Technical Commission for Aeronautics. This rather anachronistic title has now been shortened to simply the RTCA: https://www.rtca.org

within the scope of awareness of the software, a "better" output. Whether this alternate or "better" output is actually desirable in a complex and safety-critical environment is the fundamental concern.

However, there have been some proposals to attempt to improve the alignment between AI and existing regulatory principles. These proposals include increasing redundancy, producing external monitors to assess decisions outputs from AI systems, designing in default safe modes, and restricting the ability of AI programs to evolve during live operations. EASA has also proposed three guiding principles to ensure safety in AI systems:

- Keeping a human either in command or in the loop
- Developing an ability for independent AI systems to monitor operational AI activity
- Providing constant monitoring through a traditional backup system or safety net

Despite these preemptive mitigations and despite the massive potential that AI could potentially give to commercial aviation, the operational deployment of AI into safety-critical operations is likely to face considerable skepticism if not resistance from commercial aviation regulators. Certainly, for the foreseeable future, the only form of AI that would even be considered by aerospace regulators is deterministic.

17.2.4   Automation—With Humans

A crucial feature of the current system of air transport is its ability to adapt to inherent imperfections. This is not just a case of waiting for things to go wrong where human intervention stops an incident deteriorating into an accident, but on a more day-to-day level. Remember in the theory section of this chapter we started to ask the question of why does it all go so right? We are not necessarily talking about human intervention such as Captain Chesley Sullenberger and First Officer Jeff Skiles, skillfully bringing a stricken aircraft to a safe landing in the Hudson River[21] but rather the "ordinary" stuff that keeps this complex system stable.

In any quotidian day, there are a vast number of subtle tweaks and adjustments that expert operators (pilots, cabin crew, engineers, dispatchers, air traffic controllers, and operations staff) make to influence safe outcomes. However, when the autothrottle function of a Turkish Airlines Flight TK1951, malfunctioned in February 2009, the three-man crew failed to notice the reducing airspeed until it was too late to recover. This image of the PFD (Primary Flying Display) in Figure 17-3, of the type used by flight TK1951, shows the large amount of critical data that flight crews must monitor. Although the "SINGLE CH" (Single Channel) caption is prominent, the technical implications were not obvious to the crew, particularly as it appeared during final approach when their workload was particularly high.

The aircraft's automated stall warning system activated but there was insufficient time and altitude for the crew to initiate a recovery. The aircraft stalled and crashed on short finals to Amsterdam's Schiphol airport killing all three pilots and six passengers.[vii]

[vii] A reconstruction of the accident by the Dutch Air Safety Board can be found here: https://www.youtube.com/watch?v=4r9oy_rx4hM

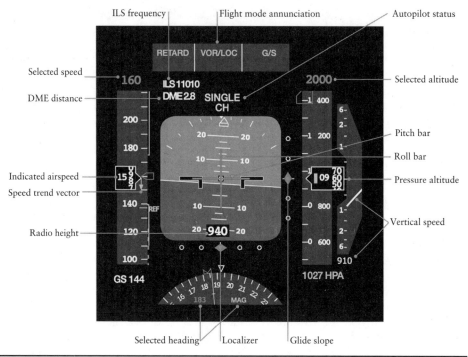

FIGURE 17-3 The instrument display prior to the TK1951 crash. Humans are not good at monitoring complex systems for long periods. (Source: Dutch Safety Board)

The incident and subsequent report provide a comprehensive illustration of the dependency modern pilots have on automation in order to deal with the complexity of modern aircraft in their day-to-day operation and, what can happen when it malfunctions.

On the whole, it is generally accepted that automation has played a crucial role in improving safety standards across commercial aviation. However, as the level of automation in cockpits has increased, the fundamental role that humans have played in the cockpit of modern aircraft has largely reduced to that of systems managing and monitoring. Humans are generally poor at monitoring complex systems effectively, particularly for long periods of time, such as on long-haul commercial flights. But the reality of the current state of commercial aviation is that today's volumes of traffic could not operate without high levels of automation. Automation has enabled crew numbers to reduce on the flight deck from complements of typically five for long-haul flights to today's typical crew of two pilots.[viii] The human element has been the consistent feature of the commercial aviation system, but could it be replicated: if we take human intelligence out of this system, what do we replace it with?

17.2.5 Autonomy—Without Humans

There is no clear line of distinction between what constitutes autonomy from automation. In many ways, one might consider the difference as a position along a spectrum of complexity rather than as a binary distinction. For our purposes, we can define

---

[viii] The concept of single pilot operation for commercial flights is discussed in Part III.

automatic as machine behavior which is highly predictable and driven by relatively simple rules such as a traffic light or a domestic oven temperature control. Although increasingly sophisticated, an autopilot or Traffic Collison Avoidance System (TCAS) will still fall into this category; it uses the aircraft's basic sensor data to provide rudimentary guidance.

In contrast to an automatic guidance system that would offer enhanced situation awareness and a suggested resolution, an autonomous guidance system would decide for itself. Such a system would need various forms of complex information such as *inter alia* the aircraft's own current (geodetic[ix]) position, the predicted future aircraft position, distance to obstacles and other air vehicles, topography, weather, and risk to our own aircraft and to other air and ground users. The system would then need to calculate resolution through an algorithm or series of algorithms to keep the aircraft and other air system users acceptably safe. It would also need to keep track of its own onboard system health as this directly affects the aircraft's capabilities. The aircraft would also need the capability to deal with non-normal situations such as loss of propulsion or degraded navigational capability.

Although this sounds an incredibly complex system and one which presents a big challenge to aircraft and avionic manufacturers, the technology existing today for those problems can be anticipated. As described in Chapter 13, the most effective methodology we are aware of, which can identify these hazardous scenarios, is STPA. The problem appears deceptively simple when using reliability theory. Even if these challenges could be met, there is still the challenge of finding ways to integrate this technology into existing airspace systems, regulate this technology and achieve public trust as to its reliability. Some in the commercial aviation industry have been dismissive that these hurdles can be overcome, but history would suggest that the realization of such a technological revolution is not a question of if, but of when.

Let's consider the capability of a human pilot as if it were such a new technology. The onboard management tasks such as fuel and position monitoring, contingency planning, communication, and navigation can all be automated, in fact many of these tasks are already automated on modern crewed aircraft. However, to keep the aircraft safe, it must also be able to detect and avoid other aircraft in its vicinity and take appropriate avoidance action. This task may be delegated to ATC radar and procedure in controlled airspace but most of the world's airspace is uncontrolled and relies on aircraft achieving their own safe separation. The primary means of avoidance in uncontrolled airspace (and sometimes within controlled airspace) is for the pilot to see and avoid other aircraft.

Various studies have assessed the effectiveness of the see-and-avoid principle. Possibly the most cited is the work carried out by J. W. Andrews,[22] which assessed pilots' ability to see and avoid other aircraft that they considered a potential hazard. Despite the pilots being briefed in advance that these encounters would take place during a short flight, the pilots in the study only sighted 36 out of 64 encounters. Of the encounters where the aircraft was sighted, the average range of acquisition was just under a mile. With even light aircraft, where closing speeds could regularly exceed four miles per minute, this leaves a pilot with only a few seconds to react. With larger

---

[ix] Guidance requires very accurate positional reference information based on the aircraft's position to Earth. Geodetic is derived from the science of geodesy: the Earth science of accurately measuring and understanding Earth's figure, orientation in space, and gravity.

faster aircraft this issue presents a significant risk to safe operation when solely relying on the see-and-avoid principle in busy airspace. In a typical aircraft a pilot can only see forward through the forward windshield with no view above, below or behind the aircraft. Even within the sector of airspace covered by the forward cockpit screen, it takes a human approximately one minute to fully scan through 180° far too much time when reaction requirements are in seconds. The inadequacy of the human-dependent see-and-avoid principle for today's air space system is well established.[23]

However, a significantly greater threat lies in future air space systems. Considering the numeric expansion of small uncrewed aircraft coming into our skies, one study considered the see-and-avoid principle regarding small UAVs.[24] The study suggests that the probability of detecting a small UAV in time to avoid a collision in all the cases modeled was far less than 50% and for the smaller hobbyist UAVs, the probability was well under 10%. In summary, the report recommended that newer and more capable systems be mandated in order to prevent an increase in potential midair collisions.

One solution could be through some form of AI-based systems, but as we have discussed there are significant technological and regulatory issues with the use of AI in safety-critical operations. If adequate collision avoidance capability is to be mandated in commercial aviation, a minimum and mandated capability that can detect potential threats in every direction above, below, and around an aircraft is inevitable. Given the limited ability of humans to monitor large amounts of information consistently for long periods of time, the use of autonomous systems to monitor information may well be a way in which safety standards in commercial aviation could not only be maintained but potentially improved. For now, the use of autonomy will be of limited application. There is considerable work to be done in understanding not only how autonomy works onboard aircraft but how it works within the wider system of systems in commercial aviation.

17.3   Sustainable Aviation

17.3.1   Changing Climate

The Intergovernmental Panel on Climate Change (IPCC) released the third part of its sixth evaluation of the world's climate in 2022[25]; it reports that today the world's climate has reached 1.1°C above pre-industrial levels and looks set to reach 1.5°C by 2040. There is strong academic consensus,[26] represented by successive metastudies of peer-reviewed academic journals, that the global climate is warming and that this warming is caused by human activity.[27] According to climate expert, Professor David Lee of Manchester Metropolitan University, England, aviation produces approximately 3.5% of all global warming caused by humanity. Commercial aviation is facing what may amount to its greatest ever challenge in adapting to its impact on the environment.

In a 2019 survey ICAO asked all 193 members to report on the impact to operations of climate change.[28] Of the total replies, 65% of respondents considered these effects had already impacted air transport in their respective state. They report detailed eight areas of direct impact: sea level rise, increased intensity of storms, temperature changes, changing precipitation, icing conditions, wind directions, desertification, and changes to biodiversity. The overall picture is one of an accelerating rate of climate change, but that rate seems to vary considerably from region to region. This accelerating change makes critical weather forecasting and modeling increasingly difficult, operations less resilient, and ultimately compromises commercial aviation to deliver a consistently safe level of operation.

17.3.2 The Operational Impact of Climate Change

The effects of climate change have a direct impact on the operation of commercial aviation safety. According to expert meteorologists, these effects have already started. In July 2019, at least 35 passengers were injured on board an Air Canada Flight after the aircraft encountered severe air turbulence. Air Canada Flight 33, enroute between Vancouver and Sydney was forced to divert to Honolulu after some passengers were seriously injured after being thrown around the aircraft's cabin, some hitting the ceiling. Climate change means that the level of sheer, where air masses of different speeds intermingle and cause turbulence, is not only increasing numerically but is becoming harder to accurately predict. In numerous scientific studies meteorologists, are suggesting we will have to strap ourselves in—more turbulence is on its way.

We discussed the various levels of turbulence in Chapter 1; turbulence is a regular phenomenon in aviation but severe turbulence, the sort that can occasionally injure or kill aircraft occupants, is currently rare. Turbulence in general is currently responsible for around 65% of weather-related aviation accidents but according to meteorologists, this figure is set to dramatically increase over the coming decades. Clear Air Turbulence (CAT) is a particular threat as it is particularly difficult to predict and therefore avoid. As it generally occurs at high altitudes, passengers are often unbuckled from their seatbelts so are particularly vulnerable to injury. Successive scientific studies[29] into the changing nature of upper altitude turbulence suggest that between 2050–2080, instances of severe CAT will double. In areas such as the North Atlantic, severe CAT will become as common as moderate CAT is today, becoming a significant risk to almost every transatlantic flight.

There are however many more types of safety implications from our rapidly changing climate. Extreme thunderstorms produced from the expanding tropical region of the planet mean that not only will the violence and intensity of these storms increase but the areas in which they occur will considerably expand.[30] Normally associated with the ITCZ,[x] recent monitoring of these powerful storms in the Gulf of Mexico has shown a gradual progression north into the mid-United States. Associated weather includes not only extreme turbulence and downdraughts, but lightening, severe icing, ice crystal icing, intense rainfall, and hailstone.

Increasing temperatures at airports have a very significant effect on aircraft performance. While the mean temperature increases projected by climatologists are only in the 1–1.5°C range, local increases are likely to be far more severe, in some areas these changes could be in the region of 20°C; in warmer climates, these temperatures would make many airports partially or totally inoperable. Another emerging threat to airport operations is rising sea levels. At coastal airports, particularly during storm surges, flooding can force delays and occasionally total closure (Figure 17-4). 2016 estimates suggested sea levels have risen approximately 0.2 meters since 1900 but in 2020, it was noted that the rate of rising waters had doubled over the previous seven years. According to the US National Oceanic and Atmospheric Administration, sea levels are projected to rise 10–12 inches (0.25–0.3 meters) over the next 30 years, with damaging flooding occurring more than 10 times more often than it does today.[31] This scenario

---

[x] The Intertropical Convergence Zone, or ITCZ, produced the storm which played a significant role in the loss of AF447 (see Chapter 1).

Figure 17-4 Coastal airports will face increased threats from frequent flooding and rising sea levels. (Source: iStock Images)

clearly threatens low-lying runways such as those located at Bangkok Suvarnabhumi (1.52 meters above mean sea level—AMSL), Miami International (2.44 meters), and Amsterdam Schiphol (–3.35 meters).

Most of the changes that are being predicted by the world's scientific community can be measured in terms of decades. From this perspective, commercial aviation would have time to adapt and evolve into a new climatic environment. However, one significant characteristic of climate change is that it is neither consistent nor evenly spread across geographic regions. Although scientists have witnessed the impact of climate change in terms of degrees of temperature change or millimeters of sea level change, it is likely that aviation will experience climate change through a series of extreme weather events. The aircraft we fly and the airports and facilities we operate in and around are only built and designed to tolerate within a certain bandwidth of weather activity. This bandwidth looks set to expand. As climate change accelerates, we have seen extreme weather events, not all of which can be accurately predicted. If these "edge cases" become the new norm, we can no longer rely on existing systems to protect us and maintain current commercial aviation safety standards.

17.3.3   Changing Attitudes and Rules

The level of public interest in the influence of commercial aviation on the environment is high and growing. For example, if you try to book a flight using the Google search engine, you may well have used Google Flight. This tool is designed to calculate the environmental impact of the specific flight you intend to take and highlights the specific impact that individuals can have on climate change. It is one example of a growing social awareness of the direct impact that flying can make on climate change and how the behavior of society is likely to continue to change. Flygskam or "flight shame" originated as a social movement in Sweden in 2017 and, according to BBC,[32] resulted in a very rare reduction in air passengers of 4% for that year. Changing attitudes and ultimately behaviors are likely to accelerate as society begins to experience more and more of the immediate and direct effects of these global phenomena.

Some of the world's largest airlines are having to commit to sustainable strategies in order to mitigate the impact they are having on the environment. In September 2021, Delta Airlines pledged $ 1 billion to become carbon neutral by 2030. Jet Blue has targeted 2040 and United Airlines by 2050.[33] These changes are not simply in reaction to social media or a potential increase in flying shame but as a necessary business strategy.

Recent social attitudes have tended toward direct action rather than simple campaigning and this action has resulted in new regulatory and reporting requirements, initially for the financial and insurance sectors.[34] These global institutions which underpin the viability of commercial aviation have not only to report where they are trying to reduce or mitigate the impact that they have on the environment but what negative impact they have directly had within the reporting period; they are compelled to report the environmental damage they are doing. By the end of 2022 progressive European environmental legislation will require an even wider group of companies to report their impact on climate change.[35] The anticipated change to commercial aviation is unprecedented. The industry has been almost exclusively dependent on fossil fuels since its inception and now must consider viable alternatives if it is to prosper and provide a safe and sustainable transport network for the global economy.

17.3.4 Technological Solutions

Airlines have traditionally looked to top technological innovation to improve their capabilities and their response to the growing pressure to become more sustainable is no different. By far the highest profile issue is the sheer amount of carbon dioxide that aircraft emissions put into the atmosphere. In 2019, before the airline industry was dramatically reduced by the COVID-19 pandemic, the global airline fleet consumed 360 billion liters fossil fuels and pumped over 885 million tons (approximately 900 metric tons) of carbon emissions into the atmosphere.[36]

But given the inherent energy efficiency of hydrocarbon fuels, this not an easy problem to solve. Short-range aircraft such as air taxis are not as energy hungry as larger long-haul aircraft and therefore electricity is a workable solution. For anything other than the shortest-range commercial aircraft, the only current and viable alternative is switching to alternative sustainable aviation fuels or SAFs. SAFs aim to reduce the carbon footprint of the fuel through the lifecycle although production, refinement to combustion. These fuels must be mixed with convention aviation fuel to make them sufficiently combustible which, as with all aviation fuel, is a tightly regulated and controlled process. While SAFs are a means to reduce aviation's carbon they are only an interim solution. If the global scheduled airline fleet is set to double by 2050, then halving emissions doesn't solve the problem. Targets such as the Green Deal of the European Commission sets a goal of 2050 for the continent to reach carbon neutrality by 2050 so further technological progress needs to be made requiring massive changes to the infrastructure that supports commercial aviation.

Battery-electric powered aircraft charged with renewable energy offer the advantage of causing no in-flight climatic impact. For larger aircraft, battery-electric powered propulsion is not a feasible option. A fully fueled Boeing 787-10 can fly for roughly 8000 miles with 300 passengers and their luggage but if you wanted to produce the same energy from a battery it would weigh 6.6 million pounds (approximately 3000 metric tons). However, batteries as a means of propulsion are feasible on smaller scale and short-range aircraft. Although these air taxis or AAM are an exciting area of growth for commercial aviation, the business plans to scale to hundreds of thousands of aircraft proposed by their manufacturers are inhibited by the need for a human pilot. In lager airliners, the marginal cost of a pilot is relatively small but in the air taxi world, having one of maybe six seats occupied by a pilot is the difference between profit and loss. This means that to make this area of aviation a success, fully autonomous guidance systems need to be implemented before these aircraft can thrive. This presents a far greater challenge than getting these aircraft certified by national regulators. Although autonomy might be the next logical step in the evolution of commercial aviation, overcoming the public's aversion to flying a pilotless aircraft is a significantly greater challenge than making them legally safe. Making them socially trusted takes even more time and resources.

Hydrogen has long been seen as an alternative to kerosene-based fuel. The US Airforce was experimenting in the 1950s with hydrogen as a means to propel their high-altitude spy aircraft. There are two ways the energy of hydrogen can be translated into propulsion. The first is through electrochemical conversion in fuel cells, which can support shorter-range aircraft and the second is through thermal conversion in combustion engines which could support longer range aircraft. Hydrogen has lots of advantages, not least the fact that it is clean and produces water as a by-product of combustion. Under their ZEROe concept program, Airbus claim they can produce the world's first zero-emission aircraft in production by 2035. Universal Hydrogen is a start-up company producing conversion kits for regional aircraft and has programs in place in Denmark and Jersey in the Channel Islands that could have hydrogen-powered commercial aircraft in service by 2025.

The greatest technical challenge is probably in how and where to store liquid hydrogen both on the ground and onboard the aircraft. Hydrogen has lower energy density when compared with conventional kerosene-based aviation fuel and requires a heavy cryogenic tank which incurs significant performance penalties. To overcome these penalties, commercial aviation will have to progress from its traditionally conservative approach to aircraft design and consider concepts such as blended-wing body aircraft. These changes in design philosophy bring new risks and challenges to the management of safety in commercial aviation.

17.4   Managing Future Risks

As aviation expands and new technologies are introduced, the complexity of the global aerospace system increases. The increase in the use of digital (and analog) computing will change processes, workflows, and capabilities. The future looks set to be a combination of evolving technological risks, which are emerging just as environmental risks are becoming more widespread. Although climate change and longer-term global aggregate temperature changes can be predicted, climate change will also create local changes with no historical trends from which to make estimates of what might happen in the future. Some areas might see more storms, others drought and others rapid changes. System failures tend to arise when these short-term and unforeseen adjustments combine with the introduction of new technologies. A systems vision of future risks suggests new methods of risk management will be required. In the past, the system of commercial aviation has predominantly relied on a combination of regulation and insurance as its primary risk management tools; however, these facilities may not be appropriate in the future.

17.4.1   Compliance Plus

The role of government to ensure that the aviation industry is regulated such that it be safe and promote aviation will likely continue. Regulators should ensure that the system remains safe without curtailing its growth. A reduction in regulation that allows for more innovation may sound appealing, but if it results in more accidents, the public could lose its hard-won trust in commercial aviation. Increasing or tightening regulations run the risk of inhibiting growth and investment in the sector. It is almost inevitable that as aviation grows in size and complexity it will become more difficult to manage. There are two main challenges: complexity and resource.

The increasing level of technical complexity associated with future commercial aviation systems means that it is rapidly becoming impossible for regulators to maintain an adequate level of specialism in order to monitor the safety performance of emerging technologies. To compensate for this challenge, many regulatory authorities have embarked on a philosophical transformation toward performance-based regulation. Rather than demand a manufacturer or operator demonstrate how a system works, the challenge is to provide sufficient evidence that it does work. This process can no longer be an entirely human function. It is inevitable that AI systems will increasingly help regulators to identify emerging trends and patterns that could indicate system weakness.

One solution to limited resources might be found in the FAA's Organization Designation Authorization (ODA) program (FAA, 2022). Due to limited resources, the FAA has the authority to designate individuals to act on its behalf for pilot certification, medical exams, and certification of aircraft. This has occurred over time, for example, initially FAA only delegated some experienced pilots to conduct private pilot checkrides, but over time that has been expanded so designated pilots can conduct checkrides up through the Air Line Transport level.

This has been a similar expansion for the certification of aircraft, from initially allowing just basic inspections to smaller modifications and then to major design aspects. FAA oversight is then limited to the audits and checks on the inspections designated individuals themselves. However, the FAA's widespread use of the ODA program came under considerable scrutiny in the aftermath of the Boeing MCAS crashes. Maintaining objectivity is a crucial characteristic of an effective regulator. When you are effectively regulating your own employer's product it is easy to imagine how a conflict of interest might emerge.

One solution to mitigate the reduction of regulatory resource that has been used in other industries is a "fee for intervention" regime. Under this process, the operator or manufacturer is periodically audited for safety compliance and performance. If the regulator identifies areas that require further regulatory intervention, then the organization is charged a fee. This effectively maintains the regulator's budget and so allows resources to be distributed elsewhere.

17.4.2   Insuring the Future

One of the very earliest and effective ways to manage risk is through the use of insurance products. The insurance industry can influence the behavior of companies through their efforts to minimize their own losses. People and companies buy liability insurance to protect themselves from the cost of having to pay damages after being found at-fault after an accident. The liability is shaped by tort law and, if a company is found guilty of negligence or malicious conduct, a civil court can award large financial penalties against them. These amounts can be high enough to drive a weak company out of business. Insurance companies can exercise leverage of how a company conducts business through a combination of increasing insurance rates or even refusing to provide coverage absent certain assurances or changes. The insurance industry uses probability to measure risk. Using formulas, the insurance industry can convert the risk into an expected cost over time and identify the extent the risk propagates across an industry. A bit analogous to how a casino wins makes money, the risk of a loss is smaller than the total amount of insurance premiums that need to be paid. As explained previously, this can result in missing risk in a complex system.

The principle of the insurance industry tends to sway toward risk sharing rather than risk management. A company will often carry some of the burden of the risk based on its own beliefs of exposure, but when the cost of the insurance is low enough that a less risky approach is to just pay for more insurance coverage. This has led to the assertion that the current pricing mechanisms and risk-sharing protocols provide too much protection from real liability and not enough real-world exposure to risk. If you consider that over the last 10 years the total premium revenue from the world's scheduled airline fleet has averaged at around $1.5 billion *per annum,* compared with the potentially trillions of dollars it is underwriting, we begin to see how distorted an insurance risk-sharing mechanism can become. One single multifatality hull loss under the current insurance regime could wipe out all profit from that market for two or perhaps three years.

17.5   New Looks

If we consider someone visiting a major airport periodically over the past few decades, as casual observer, they may not have hardly noticed the change in the look and shape of the aircraft they saw. This is not to say that there have not been significant improvements in the technological capability of the aircraft, or that the power and efficiency of these aircraft have not advanced. However, there has been little change in the outward appearance of the majority of commercial aircraft. In the very near future, this is all about to change.

17.5.1   Supersonic Flight

Since the Concorde era, there have been many discussions of whether or not a viable civilian supersonic transport might reappear, such as the concept aircraft featured in Figure 17-5. Although popular in the public imagination, numerous challenges exist, these involve a combination of noise issues, environmental concerns, economic concerns, and design aspects. Flying at faster than the speed of sound results in a phenomenon known as a sonic boom. As aircraft exceed the speed of sound a pressure wave develops that sounds like thunder. Consequently, many countries around the world have banned supersonic overflight. There has been considerable work since the time of Concorde to identify possible ways that the sonic boom can be reduced.

The basic concept is to disperse the shock wave such that the "shock" is not concentrated in one place. This is coupled with flying at very high altitudes, so the sound attenuates before reaching the ground.

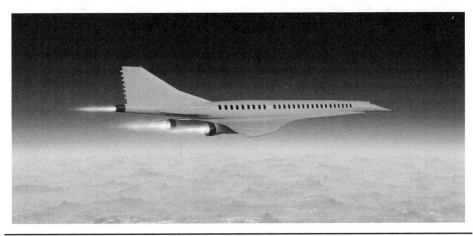

Figure 17-5 Concept of a future supersonic aircraft. (Source: Dreamstime)

The addition of jet exhaust into the upper atmosphere has been an ongoing environmental concern. Supersonic flight requires substantially more thrust than subsonic flight, on average. This results in an increase in fuel burn of about five to seven times per passenger. This higher fuel burn results in significantly reduced range, a difficult challenge. Military aircraft work around this via aerial refueling—not something likely to be viable for a civilian transport due to the risks involved.

Aircraft design will include new constraints. The aircraft skin will need to be able to withstand much higher temperatures due to the temperature increase from friction at high speeds. This, in turn, results in structural constraints, plus the demands of supersonic flight require narrow wings and fuselages, another constraint. To be able to operate out of current airports the design needs to accommodate a wide variety of speeds. The issue of pilot visibility also needs to be addressed.

Highly swept wings at low speeds require a high angle of attack, and hence aircraft body angle, to be able to sustain flight. Supersonic aircraft need to have a long and pointy nose section to reduce drag. The pilots need to be able to see to land. Concorde met these challenges through a nose section that dropped for takeoff and landing (Figure 17-6), but recent concepts have not adopted this solution. Several of the proposals require the pilots to be able to see forward through synthetic vision—problematic in itself due to potential visual latencies creating time lags between what a pilot is seeing and what is really occurring. The variety of potential risks with such a scenario are beyond the scope of this book, but the reader is encouraged to research this factor further.

A final design issue is the powerplant. Although some designs might be able to accommodate an existing engine design that has been used for supersonic military aircraft, this also creates limitations on the design. To move beyond this, a new design would require a new engine. Engine manufacturers have been loath to invest in the development of a powerplant for an uncertain and quite possibly a niche market.

This leads us to the economic aspects. If supersonic transport is to be viable, customers must be willing to pay for the reduction in flight time supersonic travel can afford. This must not be a one-time splurge for the novelty, but a regular affair. One of the chief advantages of Concorde was that at the time the aircraft operated there was no in-flight internet access. People would be completely cut off often unable to work or

FIGURE 17-6   An Air France Concorde with the nose lowered for landing. (Source: Dreamstime)

communicate during flight. This mattered and cutting off three hours on a flight from Europe to North America could be justified. Today many people use flight time effectively and are able to maintain contact and work during the flight. The incentive for high-speed flight, romantic as it might be, is not as high. Despite this, several airlines have expressed an interest should a viable aircraft become available.

17.5.2   Commercial Spaceflight

The concept of space flight has been captured by writers for centuries. Although the majority of the writings involved fantasy, some authors, notably in 1945, Arthur C. Clarke, speculated and wrote about the possibility of using geosynchronous orbit to allow communications satellites to be able to remain stationary over ground stations and allow for global communications. Still, the viability of launching vehicles into space did not come about until the refinement of rocket technology for weapons systems during World War II. The Soviet Union (Russia) launched the first small object, Sputnik, into space in 1957, which spurred competition with the United States, which continued through landing humans on the Moon in 1969.

By the late 1950s satellites were being used for military reconnaissance, and the first military navigation satellites were launched in the early 1960s. For civilian applications, the first weather satellite was launched in 1960, and the first communications satellite in 1962. Communication was an early commercial use of satellites, and through the 1960s, Clarke's concepts came to fruition. Although these latter satellites were launched using government-funded rockets, commercial enterprises were paying for their equipment to be launched.

Although initially funded by governments, the European Ariane rocket (Figure 17-7) was the first viable commercial space flight operator, launching satellites into space, with launches starting in the early 1980s. During this time companies were also paying to launch satellites on both Russian and American launch vehicles. The end of the US NASA Space Shuttle program opened an opportunity for nongovernment organizations to enter the space launch business.

FIGURE 17-7   Ariane 5 lift-off, on July 1, 2009, in Kourou, French Guyana. (Source: Dreamstime)

Seeing an opportunity, several private firms have jumped into the launch business. The most prominent is SpaceX, although Boeing, United Space Alliance, Lockheed, Blue Horizon, and others have entered the fray. To be clear, although these are private firms, they are bidding on government contracts and a large percentage of their revenue is currently from government. This leaves us with a current situation that was not all that different than before. The Space Shuttle was built by private firms, and multiple vendors were involved. Is this really a private venture or is it a government one? What defines that difference?

One of the concerns with commercial launches is space debris. Colliding with a small object at 50,000 km/h is a serious concern. As more entrants launch vehicles, the more bits end up in Earth orbit creating more hazards. Each additional piece of debris adds to the risk for a future spacecraft or satellite. Coupled with this can be flawed analysis. For example, in February 2022, Starlink launched 49 satellites into a planned orbit, but they did not consider the impact of solar activity. The solar activity expanded the atmosphere—a well-known phenomenon. Starlink lost 40 satellites that were in orbits after they were impacted.

In addition to the launches into orbit, companies like Branson's Virgin Galactic are capitalizing on space tourism. Here the goal is to launch passengers high enough that they can experience space, although they are not actually in orbit. Although perhaps not viable for more than tourism, this certainly provides interesting material for media events!

One of the unintended consequences of space launches has been the impact on commercial airline traffic. The launches out of the Kennedy Space complex in Florida have been particularly problematic. With multiple companies, as well as NASA and the US military, there is now a launch about every 7 days. Each time the airspace has to be shut down air traffic control needs to reroute aircraft around broad swaths of airspace. This is in a part of the United States where air traffic is already constrained between military-restricted airspace and frequent convective weather. The consequence can be significant flight delays, additional fuel, and diverts. This can increase the risk to an operation as well as safety.

17.6   Future Challenges

Commercial aviation is embarking on an exciting age of innovation. We are at the forefront of a technological revolution that will see the consolidation of commercial space flight, supersonic flight, autonomous aircraft, artificial intelligence, robotics, and automated air traffic control. But we are also facing an exponential increase in system complexity and resource demand. There is a growing recognition that whatever we can achieve in air transport should not supersede the need to protect our environment. Many of the traditional approaches to technology are increasingly restricted because of their harmful effects on our environment. The strategic decisions of future air transport are not now simply a case of "can we do this"? Is the activity safe? We now must consider what the long-term effects are on our quality of life. When we consider the fundamental purpose of commercial aviation safety, to preserve human life, it is quite logical that we need to consider the broader implications and the future effects of commercial flight.

Air travelers have come to expect incredible levels of safety assurance within our industry, but when the system very occasionally fails, and that trust has been broken, it has increasingly translated into a societal backlash and occasionally through legal process. Some of the assumptions we have come to rely on may need to be reconsidered if as an industry we are to maintain or even exceed the exceptionally high levels of safety performance which have underpinned our industry. One of the most important ways we can keep these standards high is through education. Many of the innovations and initiatives that underpin safety standards are invisible, even to those that travel regularly by air and even to many of those that work within the industry. It's not just a case of being safe but being seen to be safe. Hopefully this book will contribute to that process. It lays down what we have done before, but it also calls for ideas to meet the challenges for the future of commercial aviation safety. We hope that some of you will join us and take up that challenge.

Key Terms

AAM (Advanced Air Mobility)

Determinism

EVTOL

FRMS (Fatigue Risk Management Systems)

IPCC (Intergovernmental Panel on Climate Change)

Next Gen

ODA (Organization Designation Authorization)

SAF (Sustainable Aviation Fuel)

SESAR (Single European Sky ATM Research)

Topics for Discussion

1. How can the emergent EVTOL operators train and retain sufficient pilots?
2. What are the direct threats to aviation from climate change?
3. What are the indirect threats to aviation from climate change?
4. What new threats to safety come from flying uncrewed aircraft within the same airspace as crewed aircraft?
5. Can spaceflight become as safe as commercial aviation is today?
6. What risks are associated with supersonic flight?

References

1. World Economic Forum. https://www.weforum.org/agenda/2016/01/the-fourth-industrial-revolution-what-it-means-and-how-to-respond/. Accessed August 2022.
2. IMF. (2022). *World Economic Outlook Update.* July.
3. ICAO. https://www.icao.int/Meetings/FutureOfAviation/Pages/default.aspx. Accessed August 2022.
4. IATA. https://www.iata.org/en/policy/future-of-airlines-2035/. Accessed August 2022.
5. Statistica. https://www.statista.com/statistics/262971/aircraft-fleets-by-region-worldwide/. Accessed August 2022.
6. United Nations. https://www.un.org/development/desa/publications/2018-revision-of-world-urbanization-prospects.html.Accessed August 2022.
7. McKinsey & Company, disclosed funding as of June 2022. https://www.mckinsey.com/industries/aerospace-and-defense/our-insights/future-air-mobility-blog/future-air-mobility-funding-continues-to-flow-after-outlier-year-2021. Accessed August 2022
8. Frost & Sullivan. https://www.frost.com/news/press-releases/frost-sullivan-presents-the-evolving-urban-air-mobility-landscape-up-to-2040/. Accessed August 2022.
9. Boyd, D. D., Scharf, M., & Cross, D. (2021). A comparison of general aviation accidents involving airline pilots and instrument-rated private pilots. *Journal of Safety Research, 76,* 127–134.
10. CNN Report. https://edition.cnn.com/travel/article/pilots-reported-to-fall-asleep-ethiopian-airlines/index.html. Accessed August 2022.
11. Gander, P., Mangie, J., Phillips, A., Santos-Fernandez, E. Wu, L. J. (2018). Monitoring the effectiveness of fatigue risk management: a survey of pilots' concerns. *Aerospace Medicine and Human Performance, 89,* 889–995.
12. Boeing Pilot & Technician Outlook. https://www.boeing.com/commercial/market/pilot-technician-outlook/. Accessed August 2022.
13. FAA Statements on 5G. https://www.faa.gov/newsroom/faa-statements-5g. Accessed August 2022.
14. ICAO GNSS/GPS Interference Report. https://www.icao.int/MID/Documents/2020/CNS%20SG10/CNS%20SG%20PPT8-%20GPS%20Interference.pdf. Accessed August 2022.

15. FAA. https://www.faa.gov/uas/resources/by_the_numbers/. Accessed August 2022.
16. Abbott, K. FAA. https://skybrary.aero/sites/default/files/bookshelf/32615.pdf. Accessed August 2022.
17. Turing, A. M. (1950). Computing machinery and intelligence. *Mind.*
18. EASA Concept Paper: First usable guidance for Level 1 machine learning applications. https://www.easa.europa.eu/downloads/134357/en. Accessed August 2022.
19. EASA AI Roadmap. https://www.easa.europa.eu/downloads/134357/en. Accessed August 2022.
20. Luft, J., & Ingham, H. (1955). The Johari window: a graphic model of interpersonal awareness. *Proceedings of the Western Training Laboratory in Group Development.* University of California, Los Angeles.
21. NTSB Report US Airways 1549. https://www.ntsb.gov/investigations/accidentreports/reports/aar1003.pdf
22. Andrews, J. W. (1991). *Unalerted Air-to-Air Visual Acquisition.* MIT Lexington Lincoln Lab.
23. Hobbs, A. (1991). Limitations of the see-and-avoid principle. *Australia Bureau of Air Safety Investigation.*
24. Woo, G. S., Truong, D., & Choi, W. (2020). Visual detection of small unmanned aircraft system: modelling the limits of human pilots. *Journal of Intelligent & Robotic Systems*, 99(3), 933–947.
25. IPCC Report 2022. https://www.ipcc.ch/report/ar6/wg2/. Accessed August 2022.
26. National Aeronautics and Space Administration. (n.d.). Climate change evidence: how do we know? Climate Change: Vital Signs of the Planet. https://climate.nasa.gov/evidence. Accessed August 2022.
27. Myers, K. F., Doran, P. T., Cook, J., Kotcher, J. E., Myers, T. A. (2021). Consensus revisited: quantifying scientific agreement on climate change and climate expertise among Earth scientists 10 years later. *Environmental Research Letters.* October 20.
28. ICAO Secretariat, Burbidge, R., Freeburg, A., Scavuzzi, J., Lacoin, S., & Ozeren, U. (2019). 2019 Environmental Report, Aviation and Environment, Destination Green. The Next Chapter, Ch7 Climate Adaptation Synthesis. Doc 10126. Accessed August 2022.
29. Williams & Storer. (2022). Climate change and clear air turbulence. https://rmets.onlinelibrary.wiley.com/doi/full/10.1002/qj.4270. Accessed August 2022.
30. US Government Report. (USGCRP). (2017). Climate Science Special Report: Fourth National Climate Assessment, Volume I. In: Wuebbles, D. J., Fahey, D. W., Hibbard, K. A., Dokken, D. J., Stewart, B. C., & Maycock, T. K. (eds.), *U.S. Global Change Research Program*, Washington, DC, USA, 470 pp, doi: 10.7930/J0J964J6. Accessed August 2022.
31. National Oceanic and Atmospheric Administration. https://oceanservice.noaa.gov/hazards/sealevelrise/sealevelrise-tech-report.html. Accessed August 2022.
32. BBC. Sweden sees rare fall in passenger numbers as flight shaming takes off. https://www.bbc.com/news/world-europe-51067440. Accessed August 2022.

33. https://cen.acs.org/environment/sustainability/Airlines-want-make-flight-sustainable/99/i32. Accessed August 2022.
34. Chatham House. (2021). Climate change risk assessment 2021. London: Chatham House.
35. EU Sustainability Directive. https://www.ey.com/en_kz/assurance/how-the-eu-s-new-sustainability-directive-will-be-a-game-changer
36. IATA: https://www.iata.org/en/programs/environment/sustainable-flying-blog/why-saf-is-the-future-of-aviation/: accessed Sept 2022.

Index

Note: Figures, notes, and tables are denoted by *f*, n, and *t*, respectively